Complex Carbohydrates of Nervous Tissue

Complex Carbohydrates of Nervous Tissue

Edited by

Richard U. Margolis

New York University Medical Center
New York, New York

and

Renée K. Margolis

State University of New York
Downstate Medical Center
Brooklyn, New York

Plenum Press · New York and London

Library of Congress Cataloging in Publication Data

Main entry under title:

Complex carbohydrates of nervous tissue.

Includes bibliographical references and index.
1. Carbohydrates. 2. Complex compounds. 3. Nerve tissue. I. Margolis, Richard U. II. Margolis, Renee K. [DNLM: 1. Neurochemistry. 2. Gangliosides. 3. Glycoproteins. 4. Glycosaminoglycans. 5. Anions. WL104 C737]
QP701.C58 599'.01'88 78-26881
ISBN 0-306-40135-5

A Division of Plenum Publishing Corporation
227 West 17th Street, New York, N.Y. 10011

Printed in the United States of America

Contributors

Richard T. Ambron • Department of Anatomy, Division of Neurobiology and Behavior, College of Physicians and Surgeons, Columbia University, New York, New York

Glyn Dawson • Departments of Pediatrics and Biochemistry, Joseph P. Kennedy, Jr., Mental Retardation Research Center, University of Chicago, Chicago, Illinois

J. S. Elam • Department of Biological Science, Florida State University, Tallahassee, Florida

Giuliana Giannattasio • Department of Pharmacology and CNR Center of Cytopharmacology, University of Milan, Milan, Italy

George W. Jourdian • Rackham Arthritis Research Unit and Department of Biological Chemistry, University of Michigan, Ann Arbor, Michigan

O. K. Langley • Institute of Neurology, London, England (Present address: Institut de Chimie Biologique de la Faculté de Medecine, Strasbourg, France)

Robert W. Ledeen • Saul R. Korey Department of Neurology and Department of Biochemistry, Albert Einstein College of Medicine, Bronx, New York

Su-Chen Li • Department of Biochemistry and Delta Primate Research Center, Tulane Medical Center, New Orleans, Louisiana

Yu-Teh Li • Department of Biochemistry and Delta Primate Research Center, Tulane Medical Center, New Orleans, Louisiana

Henry R. Mahler • Molecular and Cellular Biology Program and Department of Chemistry, Indiana University, Bloomington, Indiana

Renée K. Margolis • Department of Pharmacology, State University of New York, Downstate Medical Center, Brooklyn, New York

Richard U. Margolis • Department of Pharmacology, New York University School of Medicine, New York, New York

Marie-France Maylié-Pfenninger • Department of Developmental Genetics, Sloan-Kettering Institute for Cancer Research, New York, New York

Barbara J. McLaughlin • Department of Anatomy, University of Tennessee Center for the Health Sciences, Memphis, Tennessee

Jacopo Meldolesi • Department of Pharmacology and CNR Center of Cytopharmacology, University of Milan, Milan, Italy

Karl H. Pfenninger • Department of Anatomy, College of Physicians and Surgeons, Columbia University, New York, New York

Richard H. Quarles • Section on Myelin and Brain Development, Developmental and Metabolic Neurology Branch, NINCDS, National Institutes of Health, Bethesda, Maryland

Abraham Rosenberg • Department of Biological Chemistry, M. S. Hershey Medical Center, Hershey, Pennsylvania

M. G. Scher • Department of Biochemistry, University of Maryland School of Medicine, Baltimore, Maryland

James H. Schwartz • Departments of Physiology and Neurology, Division of Neurobiology and Behavior, College of Physicians and Surgeons, Columbia University, New York, New York

Robert R. Townsend • Department of Biochemistry and Delta Primate Research Center, Tulane Medical Center, New Orleans, Louisiana

C. J. Waechter • Department of Biochemistry, University of Maryland School of Medicine, Baltimore, Maryland

John G. Wood • Department of Anatomy, University of Tennessee Center for the Health Sciences, Memphis, Tennessee

Antonia Zanini • Department of Pharmacology and CNR Center of Cytopharmacology, University of Milan, Milan, Italy

Preface

It is only relatively recently that neurochemists and neurobiologists have shown appreciable interest in the class of macromolecules now generally referred to as complex carbohydrates, although gangliosides were, of course, first identified and studied in brain. The glycosaminoglycans fell chiefly within the province of connective-tissue biochemists, and earlier information concerning the structure and metabolism of glycoproteins was largely limited to the more accessible glycoproteins and oligosaccharides (such as those found in plasma, milk, and urine), or ones which are relatively simple to prepare in a soluble and manageable form. Techniques were later devised for the isolation and purification of tightly bound membrane glycoproteins, where initial studies concentrated mainly on the erythrocyte, for which large amounts of a single cell population are available.

Because of the structural complexity of nervous tissue and the large numbers, low concentrations, and membrane-bound form of many of its complex carbohydrates, progress has occurred more slowly in this area. Although methods developed for the study of these compounds in other tissues have often been of considerable value in the investigation of nervous tissue complex carbohydrates, one cannot expect the properties and functional roles of glycoconjugates to be similar throughout the body. For this reason, attempts to directly carry over to nervous tissue certain supposedly "general" principles gained from the study of cartilage or immunoglobulins, for example, may be somewhat analogous to exploring Italy with a map of France. The neurochemist studying complex carbohydrates must therefore keep a very open mind, but in compensation for the delays involved in covering largely uncharted territory, there are also numerous instances in which structural features, localizations, or other properties of these compounds appear to be unique to nervous tissue or were first identified there.

Up to now much of the research on nervous tissue glycoconjugates has necessarily been of a somewhat preliminary nature, including studies aimed at obtaining initial information on their structure, distribution, and metabolism. Only by building on a firm foundation of this type will it be possible to formulate and test relevant hypotheses concerning their functional roles. However, the great strides made in the past 5–10 years suggest that this is an opportune time to take stock of where we stand now and which lines of investigation appear most profitable to pursue further. It is therefore hoped that the present volume will be of value not only to neurochemists and neurobiologists, but also to the larger group of investigators interested in complex carbohydrates from other standpoints, and for whom a summary of current knowledge in this area may provide pertinent insights into the biological roles of these highly versatile compounds.

Richard U. Margolis and Renée K. Margolis

New York

Contents

Chapter 3

Structure and Distribution of Glycoproteins and Glycosaminoglycans

Renée K. Margolis and Richard U. Margolis

Chapter 4

Biosynthesis of Glycoproteins

C. J. Waechter and M. G. Scher

Chapter 5

Biosynthesis of Glycosaminoglycans

George W. Jourdian

Chapter 6

Brain Glycosidases

Robert R. Townsend, Yu-Teh Li, and Su-Chen Li

Chapter 7

Histochemistry and Cytochemistry of Glycoproteins and Glycosaminoglycans

John G. Wood and Barbara J. McLaughlin

Chapter 8

Glycoproteins of the Synapse

Henry R. Mahler

Chapter 9

Surface Glycoconjugates in the Differentiating Neuron

Karl H. Pfenninger and Marie-France Maylié-Pfenninger

Chapter 10

Histochemistry of Polyanions in Peripheral Nerve

O. K. Langley

Chapter 11

Glycoproteins in Myelin and Myelin-Related Membranes

Richard H. Quarles

Chapter 12

Axonal Transport of Complex Carbohydrates

J. S. Elam

Chapter 13

Regional Aspects of Neuronal Glycoprotein and Glycolipid Synthesis

Richard T. Ambron and James H. Schwartz

Chapter 14

Complex Carbohydrates of Cultured Neuronal and Glial Cell Lines

Glyn Dawson

Chapter 17

Perspectives and Functional Implications

Richard U. Margolis and Renée K. Margolis

1

Structure and Distribution of Gangliosides

Robert W. Ledeen

1. Introduction

The carbohydrate-rich glycocalyx that surrounds mammalian cells and determines surface properties includes a variety of sialic-acid-containing oligosaccharides linked to either ceramide or a peptide chain. The CNS appears to be unique in containing over half its sialic acid in lipid-bound form, with the result that gangliosides reach their highest concentration in this tissue. Organs such as lung, intestine, mammary gland, and thyroid are reported to have total sialic acid concentrations equal to or greater than that of brain (Warren, 1959; Puro *et al.*, 1969), but in these and other extraneural tissues, gangliosides account for less than 10% of the total (Puro *et al.*, 1969). Although sialoglycocongugates are often viewed as membrane-specific compounds, it is noteworthy that brain, despite its abundance of membranous processes, does not greatly exceed other tissues in total sialic acid content.

The high concentration of gangliosides in neuronal elements has fostered speculation of a functional role for gangliosides in nerve conduction and/or synaptic transmission, and this has appeared consistent with the hypothesis that their primary localization is at the nerve ending. However, reexamination of the latter question has led to the suggestion that gangliosides may not be localized in this manner but rather may be distributed over a large portion of the neuronal surface (Ledeen, 1978). If verified, this would require reassessment of the view that their neural function is linked specifically to events at the synapse. A somewhat different outlook has developed in recent years with the discovery that gangliosides are receptors for a number of bacterial toxins (reviewed by Fishman and Brady, 1976; Ledeen and Mellanby, 1977). This has led to consideration of gangliosides

Robert W. Ledeen • Saul R. Korey Department of Neurology and Department of Biochemistry, Albert Einstein College of Medicine, Bronx, New York.

serving as receptors for various peptides with structures bearing some resemblance to those of the toxins (Mullin *et al.*, 1976). In this manner, they might function by providing high-affinity transport mechanisms for taking polypeptide signals into the cell. It is possible that a variety of functions will eventually be found for gangliosides, consistent with their occurrence in diverse components of the nervous system and the impressive diversity of structure that is now becoming apparent. Some aspects of structure and distribution are reviewed in this chapter.

2. Structures

2.1. Classification

Close to 40 different ganglioside structures have been identified in vertebrate tissues to date. A majority of these have been isolated from brain, but many occur in extraneural tissues as well. Still others have been obtained from the latter source only. One factor contributing to structural diversity is variation in type of sialic acid, the identifying carbohydrate of gangliosides. Approximately 20 different sialic acids are known to exist in nature (reviewed by Ledeen and Yu, 1976a; see also Haverkamp *et al.*, 1976), of which only a few have been detected thus far in gangliosides. The major types are *N*-acetylneuraminic acid (NeuAc) and *N*-glycolylneuraminic acid (NeuGc). The actual number in gangliosides may be greater, however, considering that *O*-acylated forms are often converted to simpler (deacylated) sialic acid during isolation. The fact that polysialogangliosides sometimes contain two different types of sialic acid within the same molecule further contributes to structural complexity.

Glycosphingolipid structures may be classified on the basis of the carbohydrate immediately linked to ceramide. The group derived from galactosylceramide (Fig. 1) is a relatively small family containing a single ganglioside (G_{M4} = G_7).* Virtually all other vertebrate gangliosides are derived from the family originating with glucosylceramide (Fig. 2). This family divides into four main branches from lactosylceramide, based on the nature of the third attached sugar. Of these four, the globo series is the only one not yet found to include gangliosides. The definition of ganglioside adopted here includes all forms of lipid-bound sialic acid, so that "hematoside" is considered a subgroup that lacks hexosamine.

*The symbol G_{M4} is an extrapolation of the Svennerholm (1963) nomenclature system, used throughout most of the chapter. G_7, the symbol originally assigned to this ganglioside (Ledeen *et al.*, 1973), is based on the system of Korey and Gonatas (1963). Systematic nomenclature recommended by the IUPAC-IUB Commission on Biochemical Nomenclature [*J. Lipid Res.* **19**:114 (1978)] assigns the following symbols to the four major gangliosides of mammalian brain: (G_{M1}) II^3NeuAc-GgOse$_4$Cer; (G_{D1a}) IV^3NeuAc,II^3NeuAc-GgOse$_4$Cer; (G_{D1b}) II^3(NeuAc)$_2$-GgOse$_4$Cer; (G_{T1b}) IV^3NeuAc, II^3(NeuAc)$_2$-GgOse$_4$Cer.

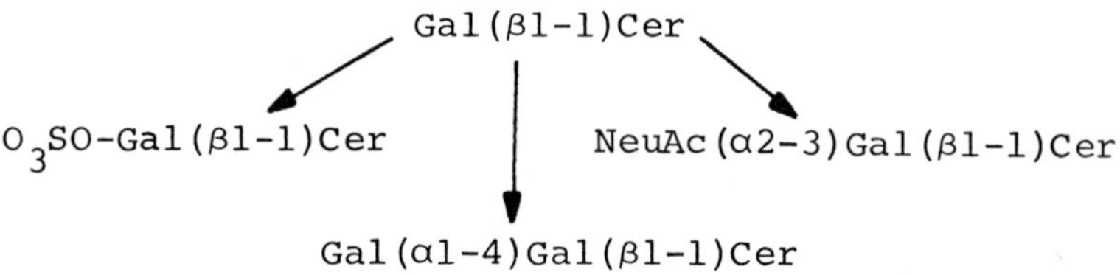

Fig. 1. Family of glycosphingolipids derived from galactosylceramide.

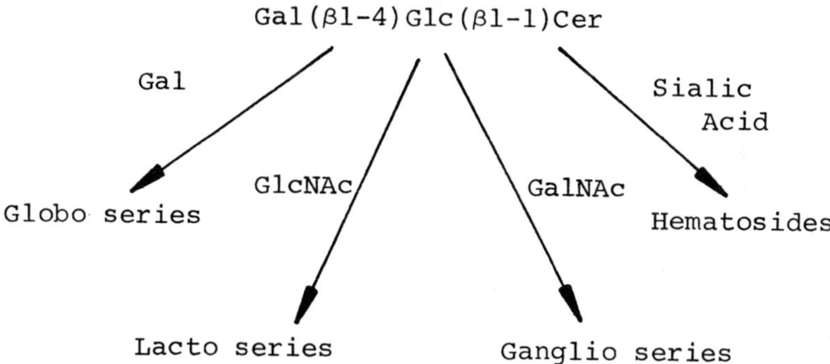

Fig. 2. Family of glycosphingolipids derived from glucosylceramide.

2.2. Ganglio Series

The majority of known gangliosides, including most of those in brain, belong to the ganglio series (Table 1). Since sialic acid and *N*-acetylgalactosamine are both substituted on the same galactose, these all contain branched structures. Biosynthetic evidence (Roseman, 1970) indicates that sialic acid is attached first, followed by the hexosamine. Most mammals contain four major brain gangliosides based on G_{M1} as the basic structural unit (Fig. 3). The three-dimensional projection shown was arrived at by manipulation of molecular models, which indicated reduced steric crowding with sialic acid perpendicular to the oligosaccharide backbone. The additional sialic acids attached to this structure in G_{D1a}, G_{D1b}, and G_{T1b} are removed by neuraminidase, yielding G_{M1}, which is resistant to further hydrolysis except in the presence of detergent (Wenger and Wardell, 1973). The disialosyl grouping NeuAc(2-8)NeuAc- present in G_{D1b}, G_{T1b}, and other species reacts very sluggishly with this enzyme in comparison to the NeuAc(2-3)Gal structure present in G_{D1a}, G_{M3}, etc. It is not certain whether the NeuAc(2-8)-NeuAc- grouping would show similarly slow reactivity when attached to terminal galactose, as in G_{T1a} and G_{Q1b}. This disialosyl unit, long known to be present in gangliosides, was only recently detected in glycoproteins (Finne *et al.*, 1977).

Table I. Structures of Vertebrate Gangliosides: Ganglio Series

Structure	Symbol[a]	Reference
Monosialo		
1. GalNAc(β1-4)Gal(β1-4)Glc(β1-1)Cer 3 ↑ α 2 NeuAc	G_{M2}	Kuhn and Wiegandt (1964) Ledeen and Salsman (1965)
2. GalNAc(β1-4)Gal(β1-4)Glc(β1-1)Cer 3 ↑ α 2 NeuGc	G_{M2}(NeuGc)	Wiegandt (1973)
3. Gal(β1-3)GalNAc(β1-4)Gal(β1-4)Glc(β1-1)Cer 3 ↑ α 2 NeuAc	G_{M1}	Kuhn and Wiegandt (1963*a*) Wiegandt (1973)
4. Gal(β1-3)GalNAc(β1-4)Gal(β1-4)Glc(β1-1)Cer 3 ↑ α 2 NeuGc	G_{M1}(NeuGc)	Wiegandt (1973)
5. Gal(β1-3)GalNAc(β1-4)Gal(β1-4)Glc(β1-1)Cer 2 3 ↑ α ↑ α 1 2 Fuc NeuAc		Ghidoni *et al.* (1976) A. Suzuki *et al.* (1975)

Structure	Designation	Reference
6. Gal(β1-3)GalNAc(β1-4)Gal(β1-4)Glc(β1-1)Cer 2 ↑α 1 Fuc (on Gal); 3 ↑α 2 NeuGc (on Gal)		Wiegandt (1973)
7. Gal(β1-3)GalNAc(β1-4)Gal(β1-4)Glc(β1-1)Cer 3 ↑α 1 Fuc (on Gal); 3 ↑α 2 NeuAc (on Gal)		Ohashi and Yamakawa (1977)
Disialo		
8. GalNAc(β1-4)Gal(β1-4)Glc(β1-1)Cer 3 ↑α 2 NeuAc(8⟵2α)NeuAc (on Gal)	G_{D2}	Kuhn and Wiegandt (1964) Klenk and Naoi (1968)
9. Gal(β1-3)GalNAc(β1-4)Gal(β1-4)Glc(β1-1)Cer 3 ↑α 2 NeuAc (on Gal); 3 ↑α 2 NeuAc (on Gal)	G_{D1a}	Kuhn and Wiegandt (1963*b*) Klenk and Gielen (1963) Price *et al.* (1975)
10. Gal(β1-3)GalNAc(β1-4)Gal(β1-4)Glc(β1-1)Cer 3 ↑α 2 NeuAc (on Gal); 3 ↑α 2 NeuGc (on Gal)	G_{D1a}(NeuAc/NeuGc)	Ghidoni *et al.* (1976) Price *et al.* (1975)
11. Gal(β1-3)GalNAc(β1-4)Gal(β1-4)Glc(β1-1)Cer 3 ↑α 2 NeuGc (on Gal); 3 ↑α 2 NeuAc (on Gal)	G_{D1a}(NeuGc/NeuAc)	Ghidoni *et al.* (1976)

(continued)

Table I. (*continued*)

Structure	Symbol[a]	Reference
12. Gal(β1-3)GalNAc(β1-4)Gal(β1-4)Glc(β1-1)Cer 3 ↑α 2 NeuGc (on Gal(β1-3)); 3 ↑α 2 NeuGc (on Gal(β1-4))	$G_{D1a}(NeuGc)_2$	Wiegandt (1973)
13. Gal(β1-3)GalNAc(β1-4)Gal(β1-4)Glc(β1-1)Cer 3 ↑α 2 (on Gal(β1-4)) NeuAc(8←2α)NeuAc	G_{D1b}	Kuhn and Wiegandt (1963*b*) Klenk *et al.* (1967)
14. Gal(β1-3)GalNAc(β1-4)Gal(β1-4)Glc(β1-1)Cer 2 ↑α 1 Fuc (on Gal(β1-3)); 3 ↑α 2 NeuAc(8←2α)NeuAc (on Gal(β1-4))		Ando and Yu (1977*c*)
15. GalNAc(β1-4)Gal(β1-3)GalNAc(β1-4)Gal(β1-4)Glc(β1-1)Cer 3 ↑α 2 NeuAc (on Gal(β1-3)); 3 ↑α 2 NeuAc (on Gal(β1-4))		Svennerholm *et al.* (1973)
Trisialo		
16. GalNAc(β1-4)Gal(β1-4)Glc(β1-1)Cer 3 ↑α 2 (on Gal) NeuAc(8←2α)NeuAc(8←2α)NeuAc	G_{T2}	Yu and Ando (1978)
17. Gal(β1-3)GalNAc(β1-4)Gal(β1-4)Glc(β1-1)Cer 3 ↑α 2 NeuAc(8←2α)NeuAc (on Gal(β1-3)); 3 ↑α 2 NeuAc (on Gal(β1-4))	G_{T1a}	Ando and Yu (1977*b*)

```
18. Gal(β1-3)GalNAc(β1-4)Gal(β1-4)Glc(β1-1)Cer
     3                    3
     ↑ α                  ↑ α
     2                    2
    NeuAc                NeuAc(8←2α)NeuAc
```

G_{T1b} — Kuhn and Wiegandt (1963*b*); Klenk *et al.* (1967)

```
19. Gal(β1-3)GalNAc(β1-4)Gal(β1-4)Glc(β1-1)Cer
                          3
                          ↑ α
                          2
                         NeuAc(8←2α)NeuAc(8←2α)NeuAc
```

G_{T1c} — Ishizuka and Wiegandt (1972)

Tetrasialo

```
20. Gal(β1-3)GalNAc(β1-4)Gal(β1-4)Glc(β1-1)Cer
     3                    3
     ↑ α                  ↑ α
     2                    2
    NeuAc(8←2α)NeuAc     NeuAc(8←2α)NeuAc
```

G_{Q1b} — Ando and Yu (1977*a*)

```
21. Gal(β1-3)GalNAc(β1-4)Gal(β1-4)Glc(β1-1)Cer
     3                    3
     ↑ α                  ↑ α
     2                    2
    NeuAc                NeuAc(8←2α)NeuAc(8←2α)NeuAc
```

G_{Q1c} — Ishizuka and Wiegandt (1972)

Pentasialo

```
22. Gal(β1-3)GalNAc(β1-4)Gal(β1-4)Glc(β1-1)Cer
     3                    3
     ↑ α                  ↑ α
     2                    2
    NeuAc(8←2α)NeuAc     NeuAc(8←2α)NeuAc(8←2α)NeuAc
```

G_{P1} — Ishizuka and Wiegandt (1972)

---

[a] The symbols are those of the Svennerholm (1963) system. Additional symbols beyond those originally proposed have been added in a manner thought to be consistent with the system as a whole. Where more than one type of sialic acid is present in the same molecule, the first designated in parentheses is that most distal from ceramide.

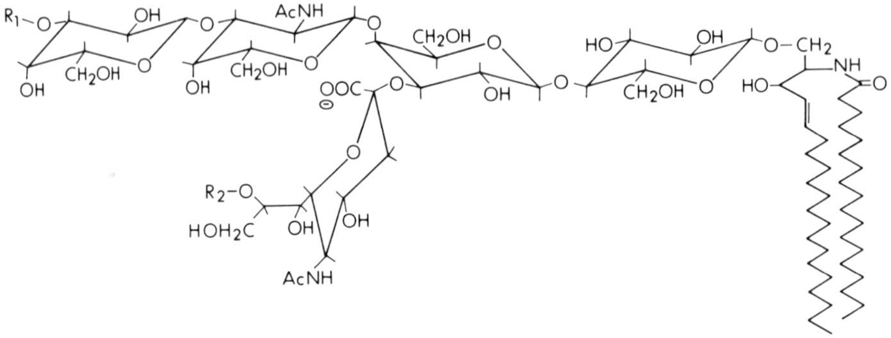

Fig. 3. Structures of the four major gangliosides of mammalian brain. (G_{M1}) $R_1 = R_2 = H$; (G_{D1a}) R_1 = NeuAc, R_2 = H; (G_{D1b}) R_1 = H, R_2 = NeuAc; (G_{T1b}) $R_1 = R_2$ = NeuAc.

Comparative studies have revealed significant departures from the typical mammalian brain pattern among lower vertebrates such as fish. These have relatively little monosialoganglioside, while tetra- and penta-sialogangliosides comprise the major types (Ishizuka *et al.*, 1970; Avrova, 1971). Frog was suggested to have a similar pattern, but the finding that ray and lamprey contain relatively more monosialogangliosides than mammals (Avrova, 1971) indicated that such variations may not conform to a simple phylogenetic progression. Interesting structural differences have recently come to light with the discovery (Ando and Yu, 1977*a*) that G_{Q1b} (Table I, entry 20), the tetrasialoganglioside of human brain, has a disialosyl NeuAc-(2-8)NeuAc unit attached to each galactose of the $GgOse_4Cer$ backbone, while G_{Q1c} (Table I, entry 21) from fish brain has a trisialosyl unit on the internal galactose (Ishizuka and Wiegandt, 1972). The fact that the newly discovered gangliosides G_{T2} (Table I, entry 16) and G_{T3} (Table III, entry 9) were found in fish but not mammalian brain has led to a proposed new biosynthetic pathway present in fish and possibly other lower vertebrates (Yu and Ando, 1978).

The major brain gangliosides of vertebrates thus derive in most instances from the tetraglycosyl structure $GgOse_4Cer$, while minor gangliosides are based on this or a smaller core (e.g., $GgOse_3Cer$, LacCer). An exceptional case is the disialoganglioside discovered by Svennerholm *et al.* (1973) to have a gangliopentaose structure equivalent to G_{D1a} with an additional GalNAc attached (β1-4) to terminal galactose (Table I, entry 15; systematic name: IV^3NeuAc, II^3NeuAc-$GgOse_5Cer$). A minor ganglioside of normal human brain, this was recently found to be elevated in Tay–Sachs disease (Yu, 1978), and is thus apparently degraded by the same hexosaminidase as G_{M2}. Two fucose-containing gangliosides have been found in brain (Table I, entries 5 and 14) in both of which fucose is attached (α1-2) to terminal galactose. These may be considered another form of pentaose [gangliofucopentaose (see Wiegandt, 1973)]. Two fucose-containing gangliosides have been found in extraneural tissues (Table I, entry 7; Table II, entry 4).

Table II. Structures of Vertebrate Gangliosides: Lacto Series

Structure	Symbol[a]	Reference
1. Gal(β1-4)GlcNAc(β1-3)Gal(β1-4)Glc(β1-1)Cer 3 ↑ α 2 NeuAc	IV^{3}NeuAc-nLcOse$_4$Cer	Wherrett (1973) Li *et al.* (1973) Ando *et al.* (1973)
2. Gal(β1-4)GlcNAc(β1-3)Gal(β1-4)Glc(β1-1)Cer 6 ↑ α 2 NeuAc	IV^{6}NeuAc-nLcOse$_4$Cer	Wiegandt (1973)
3. Gal(β1-4)GlcNAc(β1-3)Gal(β1-4)Glc(β1-1)Cer 3 ↑ α 2 NeuGc	IV^{3}NeuGc-nLcOse$_4$Cer	Wiegandt (1973)
4. Gal(β1-4)GlcNAc(β1-3)Gal(β1-4)Glc(β1-1)Cer 3 3 ↑ α ↑ α 2 1 NeuAc Fuc	III^{3}Fucα, IV^{3}NeuAc-nLcOse$_4$Cer	Rauvala (1976*a*)
5. Gal(β1-4)GlcNAc(β1-3)Gal(β1-4)Glc(β1-1)Cer 3 ↑ β 1 GlcNAc(4 ←1β)Gal(3 ←2α)NeuAc	IV^{3}NeuAc-nLcOse$_6$Cer	Wiegandt (1974)

[a] Recommendations of the IUPAC-IUB Commission on Biochemical Nomenclature (1976): *Hoppe-Seyler's Z. Physiol. Chem.* **358:**617 (1977); *J. Lipid Res.* **19:**114 (1978).

Table III. Structures of Vertebrate Gangliosides: Hematoside Series

Structure	Symbol[a]	Reference
1. NeuAc(α2-3)Gal(β1-1)Cer	G_{M4}	Kuhn and Wiegandt (1964) Ledeen *et al.* (1973)
2. NeuAc(α2-3)Gal(β1-4)Glc(β1-1)Cer	G_{M3}	Kuhn and Wiegandt (1964) Wiegandt (1973) Seyfried *et al.* (1978)
3. NeuGc(α2-3)Gal(β1-4)Glc(β1-1)Cer	G_{M3}(NeuGc)	Yamakawa and Suzuki (1951) Klenk and Lauenstein (1953) Wiegandt (1973)
4. OAc-NeuGc(α2-3)Gal(β1-4)Glc(β1-1)Cer		Hakomori and Saito (1969)
5. NeuAc(α2-3)NeuAc(α2-3)Gal(β1-4)Glc(β1-1)Cer	G_{D3}	Kuhn and Wiegandt (1964) Wiegandt (1973) Handa and Burton (1969)
6. NeuGc(α2-8)NeuAc(α2-3)Gal(β1-4)Glc(β1-1)Cer	G_{D3}(NeuGc/NeuAc)	Wiegandt (1973)
7. NeuAc(α2-8)NeuGc(α2-3)Gal(β1-4)Glc(β1-1)Cer	G_{D3}(NeuAc/NeuGc)	Wiegandt (1973)
8. NeuGc(α2-8)NeuGc(α2-3)Gal(β1-4)Glc(β1-1)Cer	G_{D3}(NeuGc)$_2$	Handa and Yamakawa (1964)
9. NeuAc(α2-8)NeuAc(α2-8)NeuAc(α2-3)Gal(β1-4)Glc(β1-1)Cer	G_{T3}	Yu and Ando (1978)

[a]The symbols are those of the Svennerholm (1963) system. Additional symbols beyond those originally proposed have been added in a manner thought to be consistent with the system as a whole. Where more than one type of sialic acid is present in the same molecule, the first designated in parentheses is that most distal from ceramide.

In addition to the unusual pattern of fish, frog, etc. mentioned above, certain other vertebrates depart from the typical mammalian brain pattern in having considerable amounts of sialosylgalactosyl ceramide (G_{M4} = G_7). In human white matter, this was found to be the third most abundant ganglioside on a molar basis (Ledeen *et al.*, 1973), while recent work has revealed appreciable levels in other primates and an unusually high concentration in chicken and pigeon brains (Cochran *et al.*, 1977). In all cases, its primary locus was found to be CNS myelin, none being detected in the peripheral nervous system (PNS) (Fong *et al.*, 1976).

2.3. Lacto Series

A growing number of gangliosides have been detected in the lacto series (Table II), most of which exist in extraneural tissues (Wiegandt, 1973; Wherrett, 1976). Two basic tetraglycosyl structures occur in this series:

Gal(β1-3)GlcNAc(β1-3)Gal(β1-4)Glc(β1-1)Cer
Lacto-*N*-tetraosylceramide
Type I

Gal(β1-4)GlcNAc(β1-3)Gal(β1-4)Glc(β1-1)Cer
Lacto-*N*-neotetraosylceramide
Type II

Virtually all the well-characterized gangliosides in this group are based on lacto-*N*-neotetraosylceramide (Type II). The structure IV^3NeuAc-nLcOse$_4$ Cer* (Table II, entry 1) was shown to be the major ganglioside in human peripheral nerve (Li *et al.*, 1973) and human erythrocytes (Ando *et al.*, 1973; Wherrett, 1973). It is a minor ganglioside of human brain and thus far the only one in that organ found to contain glucosamine.

2.4. Hematosides

Hematosides (Table III) of the G_{M3} and G_{D3} types normally occur as minor gangliosides in brain, but become greatly elevated in certain disease states. The discovery of G_{T3} in fish brain (Yu and Ando, 1978) marks the first demonstration of a trisialo type. The term "hematosides" arose from their initial detection in erythrocytes (Yamakawa and Suzuki, 1951), but they have since been shown to occur to one degree or another in virtually every vertebrate tissue. Because of their relative abundance in organs other than brain, they have come to be regarded as the characteristic gangliosides of extraneural tissues. Hematosides are indeed abundant in certain of these tissues such as equine erythrocytes (Yamakawa and Suzuki, 1951; Klenk and

*The Svennerholm system has not yet been adapted to glucosamine-containing gangliosides. These are designated in the text and Table II according to the recommendation of the IUPAC-IUB Commission on Biochemical Nomenclature [*J. Lipid Res.* **19**:114 (1978)].

Lauenstein, 1953), human spleen (Wiegandt, 1973), bovine adrenal medulla (Ledeen *et al.*, 1968), human liver (Seyfried *et al.*, 1978), human skeletal muscle (Svennerholm *et al.*, 1972), human plasma (Yu and Ledeen, 1972), and human kidney (Rauvala, 1976*b*). On the other hand, hexosamine-containing gangliosides were also found to be prominent in other extraneural tissues such as human erythrocytes (Ando and Yamakawa, 1973), rabbit plasma (Yu and Ledeen, 1972), pig adipose tissue (Ohashi and Yamakawa, 1977), and pig intestine (Puro *et al.*, 1969). These and other examples invalidate the presumption that hematosides invariably occur as the predominant ganglioside types outside the CNS.

2.5. Lipophilic Components

The lipophilic components of brain gangliosides are unique in containing a mixture of C_{18} and C_{20} long-chain bases. Other brain sphingolipids contain the former only, and the same has been generally true for all sphingolipids in extraneural tissues. Stearate is the predominant fatty acid for most of the brain gangliosides, while a mixture ranging from C_{16} to C_{24} is found outside the CNS. Gangliosides from certain peripheral nerve sources present an interesting hybrid of CNS and extraneural properties (Fong *et al.*, 1976). Ganglioside G_{M4} (G_7) from human myelin has unique lipophilic components closely matching those of myelin cerebrosides (Ledeen *et al.*, 1973), with an absence of d20 sphingosines.*

A developmental study in the rat revealed that brain gangliosides have only d18 sphingosine at birth, but subsequently accumulate d20 sphingosine until the two are approximately equivalent by 6–8 weeks (Rosenberg and Stern, 1966). In human brain, the proportion of d20 sphingosine was found to increase rapidly the first 10 years of life and more slowly thereafter, eventually reaching a proportion of 60% (Naoi and Klenk, 1972). A more detailed study in the human revealed similar changes during development for individual gangliosides (Månsson *et al.*, 1978). For G_{D1a} and G_{T1}, the proportion of d20 sphingosines increased from 10% at birth to a final value of 70% by 23 years of age. G_{M1} and G_{D1a} reached approximately the same final value but required longer times. These findings were corroborated in one report that analyzed individual gangliosides from a 23-year-old person (Kawamura and Taketomi, 1975), but differed from other studies (Schengrund and Garrigan, 1969; Yohe *et al.*, 1976) that reported larger differences in the percentages of d20 sphingosines. Individual gangliosides also had similar fatty acid patterns that gradually changed from a stearate composition of approximately 92–94% in the newborn to 86–89% in old age (Månsson *et al.*, 1978).

*Long-chain base nomenclature system of Karlsson (1970).

3. Distribution

3.1. Gross Compartmentalization in the Nervous System

Early in the development of ganglioside biochemistry, it was recognized (Klenk and Langerbeins, 1941; Svennerholm, 1957) that gray matter has a substantially higher concentration of gangliosides than white matter, and from this arose the concept of a neuron-specific substance. This idea has since been refuted by several studies demonstrating the presence of gangliosides in most cellular and subcellular fractions of the CNS, although the proposition that the bulk of brain ganglioside occurs in neurons remains valid. Their mode of distribution within these complex cells, however, is still unclear (see Section 3.2).

Various studies have revealed small but significant differences in total ganglioside concentration and pattern among different gray matter regions, and also in comparing white matter tracts (Lowden and Wolfe, 1964; Suzuki, K., 1967; Dominick and Gielen, 1968). Cerebellum, for example, differs from the cerebral cortex in having an unusually high level of trisialoganglioside (G_{T1b}) and greatly reduced monosialoganglioside (G_{M1}). The latter is elevated in most white matter regions due to its predominance in myelin (Suzuki, K., *et al.*, 1967; Ledeen *et al.*, 1973). Various components of the optic system were found to differ significantly from CNS in general (for a review, see Ledeen and Yu, 1976*b*). Retina, for example, has one eighth the ganglioside concentration of gray matter, and G_{D3} is the major type in the case of mammals (Holm *et al.*, 1972; Handa and Burton, 1969; Edel-Harth *et al.*, 1973). Developmental changes and species differences have been noted (Dreyfus *et al.*, 1975).

Peripheral nerve contains much less total ganglioside than brain, but roughly the same amount as whole spinal cord (Table IV). Values for typical extraneural tissues are shown in Table IV for comparison. On the basis of the limited amount of structural work that has been done, it seems likely that significant differences will be found between CNS and PNS. Thus, the glucosamine-containing ganglioside $IV^3NeuAc\text{-}nLcOse_4Cer$ (Table II, entry 1), shown to be the major ganglioside in human femoral nerve, is a minor ganglioside of brain (Li *et al.*, 1973). Bovine intradural root myelin was found to have both glucosamine- and galactosamine-containing gangliosides, and two of the latter were suggested to have structures corresponding to G_{M1} and its NeuGc analogue (Fong *et al.*, 1976). Adrenal medulla, derived embryologically from the neural crest and considered a component of the sympathetic nervous system, was shown for the cow (Ledeen *et al.*, 1968) and other species (Price and Yu, 1976) to have high proportions of hematosides. In addition, three structures belonging to the ganglio series were characterized from the bovine tissue (Price *et al.*, 1975).

Table IV. Comparison of Ganglioside Concentrations in CNS, PNS, and Extraneural Tissue

Tissue	Lipid-bound sialic acid (μg/g fresh tissue)	Reference
Central nervous system		
Human gray matter	900	Yu and Ledeen (1970)
Human white matter	267	Yu and Ledeen (1970)
Human spinal cord	87	Ueno *et al.* (1978)
Peripheral nervous system		
Human sciatic nerve	80[a]	MacMillan and Wherrett (1969)
Human femoral nerve	34	Svennerholm *et al.* (1972)
Bovine intradural root	93	Fong *et al.* (1976)
Rabbit sciatic nerve	83	Yates and Wherrett (1974)
Rat sciatic nerve	11[a]	Klein and Mandel (1975)
Rat superior cervical ganglion	93	Harris and Klingman (1972)
Extraneural		
Human liver	66	Seyfried *et al.* (1978)
Human kidney	10	Rauvala (1976*b*)

[a]Recalculated from original reference.

CSF has a very low ganglioside concentration, reported at 1 μg ganglioside/ml (Bernheimer, 1969). This is equivalent to the value of approximately 25 μg lipid-bound sialic acid/100 ml reported for human lumbar CSF (Ledeen and Yu, 1972). The thin-layer chromatography (TLC) pattern was similar to that of brain, but with relatively more tri- and tetrasialogangliosides. Human ventricular CSF, on the other hand, had substantially less total ganglioside than the lumbar fluid and a TLC pattern resembling that of plasma (e.g., enriched G_{M3}). Fatty acid and long-chain base compositions indicated CSF gangliosides to be derived from both brain and plasma (Ledeen and Yu, 1972).

3.2. Brain Cells

In recent years, attention has focused on distribution among cellular elements of brain, and several laboratories have now analyzed the gangliosides of isolated neuronal and glial cells (Table V). The results vary considerably, but in all studies the ganglioside concentration of isolated cells was found to be well below that of cerebral cortex of the same animal. Species differences may explain part of the variability, but it is likely that differences in cell isolation and assay procedures are more important determinants. Among neuronal preparations, the relatively high value for neurons obtained from ox Dieter's vestibular nucleus (Derry and Wolfe, 1967) is possibly due to the presence of ganglioside-rich processes contaminating the hand-dissected preparation. Neurons isolated by bulk centrifugation give generally lower

Table V. Gangliosides of Isolated Brain Cells

Tissue	Sialic acid (μg/mg protein)[a]	Distribution of sialic acid[b]									
		G_{Q1}	G_{T1b}	G_{D1b}	G_{D2}	G_{D1a}	G_{D3}	G_{M1}	G_{M2}	G_{M3}	Ref.[c–l]
Neurons											
Ox Deiter's nucleus	5.8										*c*
Rat cortex	1.2										*d*
Rat cortex		10	15	11	6	34		14	6	5	*e*
Rabbit cortex	2.9	2	17	14	–	42		25			*f*
Pig brainstem	2.0										*g*
Hamster cortex		5	26	9		41	6	12			*h*
Human brain	1.0										*i*
Rat brain	0.8										*j*
Astroglia											
Rat cortex	3.2										*d*
Rat cortex		9	13	11	6	36		15	5	5	*e*
Rat cortex	2.1										*j*
Rat astroblasts	0.6						4			96	*h*
Rabbit cortex	5.6	1	12	15	–	48		23			*f*
Hamster cortex		6	18	18		39	4	15			*h*
Hamster astroglial nodule	1.5										*k*
Oligodendroglia											
Calf white matter	1.3										*l*
Human white matter	0.4										*i*

[a] Some values have been recalculated from the original data, and some represent the average of a range.
[b] Distribution expressed as percentage of lipid-bound sialic acid. The symbol – denotes less than 1% detected. Some distributions have been recalculated.
[c–l] References: [c] Derry and Wolfe (1967); [d] Norton and Poduslo (1971); [e] Abe and Norton (1974); [f] Hamberger and Svennerholm (1971); [g] Tamai *et al.* (1971); [h] Robert *et al.* (1975); [i] Yu and Iqbal (1977); [j] Skrivanek *et al.* (1978); [k] Shein *et al.* (1970); [l] Poduslo and Norton (1972).

values; the improved methods of recent years have yielded neuronal cell body preparations of high purity with ganglioside levels in the vicinity of 1 μg sialic acid/mg protein. The fact that cerebral cortex of most mammals contains approximately 8–10 μg/mg protein indicates that the bulk of gray matter ganglioside occurs in the processes (see Section 3.3).

Most astroglial preparations were reported to have higher values than isolated neurons. In general, when neuron and astrocyte fractions are prepared and analyzed by the same laboratory, the latter fraction is found to contain approximately twice the ganglioside concentration of the former, even though the absolute values vary. TLC patterns of the two fractions are very similar to each other and to those of whole brain (Table V). However, since the astrocytes suffer greater contamination, it remains uncertain whether this might account for the high concentrations and patterns. It is equally possible that the gangliosides seen are true constituents of those cells and the relatively high ratio of surface area to volume in astrocytes provides more plasma membrane per cell (Norton and Poduslo, 1971). Support for the former hypothesis appeared to come from the observation (Robert *et al.*, 1975) that primary cultures of rat astroblasts contain much lower levels and a pattern with 96% G_{M3}. As the authors point out, however, these are embryonic cells that may differ significantly from differentiated astrocytes in ganglioside composition. The question thus remains unresolved.

Oligodendroglia were originally reported (Poduslo and Norton, 1972) to have a slightly higher level than neurons from the same laboratory (Norton and Poduslo, 1971), but a more recent study (Yu and Iqbal, 1977) indicates the ganglioside concentration in these cells to be less than half that of neurons. TLC patterns were similar for the two preparations, with the difference that G_{M4} (G_7) was clearly observable in the oligodendrocytes but appeared only as a trace in neurons (Yu and Iqbal, 1977).

3.3. Brain Subcellular Fractions

Subcellular fractions of both neurons and glia have been intensively studied in regard to ganglioside content and pattern. The nerve-ending complex has aroused particular interest as the putative site for localization of neuronal gangliosides. This belief arose from the results of one or two studies that claimed an unusually high concentration in the synaptic plasma membrane (SPM), in the range 40–45 μg NeuAc/mg protein. However, a survey of the literature (Table VI) reveals that this has not been a universal finding, and the majority of reported values are considerably lower. Because of the sixfold variation between the highest and lowest assays, the true value remains uncertain.

The majority of ganglioside in the nerve-ending complex exists in the SPM, and hence the concentration in isolated synaptosomes is bound to be considerably less. The variation in reported values for these organelles,

Table VI. Gangliosides of Neuronal Subfractions

Fraction	Sialic acid (μg/mg protein)[a]	Reference
Synaptosomes		
Ox	16.3	Wiegandt (1967)
Rabbit	13.8	Hamberger and Svennerholm (1971)
Rabbit	11.9	Tettamanti *et al.* (1973)
Rat	10.3	Caputto *et al.* (1974)
Rat	9.7	Dekirmenjian and Brunngraber (1969)
Rat	9.2	Avrova *et al.* (1973)
Rat	9.2	Seminario *et al.* (1964)
Rat	7.0	Yohe *et al.* (1977)
Guinea pig	8.5	Eichberg *et al.* (1964)
Human	7.0	Kornguth *et al.* (1974)
Synaptic plasma membranes		
Rat ("cholinergic")	45.2	Avrova *et al.* (1973)
Rat	44.6	Breckenridge *et al.* (1972)
Rat	19.3	Brunngraber (*et al.* (1967)
Rat ("noncholinergic")	18.5	Avrova *et al.* (1973)
Rat	18.0	Rapport and Mahadik (1978)
Guinea pig	17.0	Whittaker (1969)
Rat ("cholinergic")	16.7	Lapetina *et al.* (1968)
Rat	15.5	Skrivanek *et al.* (1978)
Rat ("noncholinergic")	7.3	Lapetina *et al.* (1968)
Synaptic vesicles		
Rat	5.1	Lapetina *et al.* (1968)
Rat	2.9	Breckenridge *et al.* (1973)
Synaptosomal soluble		
Rat	0.56	Lapetina *et al.* (1968)
Rat	0.34	Ledeen *et al.* (1976)
Axons		
Ox[b]	0.38	DeVries and Norton (1974)

[a]Some values have been recalculated from the original data and some represent the average of a range.
[b]This preparation contained primarily axoplasm with little or no axolemma.

particularly for the rat, is not nearly as great as for SPM (Table VI), and this has permitted calculation of the fraction of total cortical ganglioside that is present in the combined nerve endings (Ledeen, 1978). In making the calculation for rat cortex, a value of 9 μg NeuAc/mg protein was used for the concentration in nerve endings and 21×10^{11} for the number of nerve endings per gram of cortex (Cragg, 1972). From these values and the estimated protein content of nerve endings, it was calculated that these structures contain approximately 12% of total ganglioside in the rat cortex. This was considered an upper-limit estimate. Similar calculations for neuronal and glial cell bodies showed that these together account for less than 10% of the total.

The conclusion reached from these and other considerations is that gangliosides are most abundant in neuronal processes but not localized specifically at nerve endings. Rather, they seem likely to be distributed over a considerable portion of the neuronal surface that would encompass cell bodies, dendrites, nerve endings, and possibly axons. Despite the low concentration in whole perikarya, calculations show that the plasma membrane would have a concentration at least equal to that of SPM if it contained 50–80% of the total present in the cell body. Histochemical study of rat cerebellum (DeBaecque *et al.*, 1976) revealed intense staining with anti-G_{M1} antibody in the granular layer, which is composed primarily of neuronal perikarya. Further support comes from the fact that microsomes, which originate from a diversity of plasma and reticular membranes, have a relatively high ganglioside content (Wolfe, 1961; Wherrett and McIlwain, 1962; Seminario *et al.*, 1964; Eichberg *et al.*, 1964), approaching in some cases the value for SPM. Aside from concentration, the ganglioside patterns of microsomes, SPM, and neuronal cell bodies are virtually indistinguishable (Skrivanek *et al.*, 1978).

Synaptic vesicles contain relatively little ganglioside, and it is often assumed that the small quantities detected represent contaminants. However, even the purest preparations were found to contain small but possibly significant amounts (Breckenridge *et al.*, 1973). Chromaffin granules, considered to be morphological and functional analogues of synaptic vesicles, were recently shown to have appreciable ganglioside in the membrane (Geissler *et al.*, 1977). These results suggest that it might be profitable to examine adrenergic and cholinergic vesicles separately. The soluble compartment of the synaptosome contains a very low concentration, comparable to that in axoplasm (Table VI). Its metabolic properties were quite distinctive, leading to the suggestion that part of the oligosaccharide chain is synthesized locally at the nerve ending, in contrast to the much larger pool of SPM gangliosides that derives entirely from axonal transport (Ledeen *et al.*, 1976). Thus, while gangliosides appear to exist in every subfraction that has been carefully examined, their primary localization within the nerve-ending complex is undoubtedly the SPM. How they are distributed within this membrane is not yet clear, since one study (Lapetina and DeRobertis, 1968) found an even distribution between junctional complex and adjoining membrane, while another (Ledeen *et al.*, 1976) reported a higher concentration in the latter.

The subfraction of glial cells that has received major attention is myelin, and while gangliosides always appear as minor components in this membrane, the level in CNS is considerably greater than that in PNS (reviewed by Ledeen and Yu, 1976*b*). For rat brain myelin, a developmental study showed G_{M1} to be the predominant species, increasing eventually to 90% of the total (Suzuki, K., *et al.*, 1967). Similar results were obtained in mouse brain, where additionally it was shown that the total quantity of ganglioside and the level of G_{M4} (G_7) both increase slowly over several months (Yu and Yen, 1975). While the latter ganglioside is relatively minor in rodents and most other mammals, it is considerably more abundant in primate and avian myelin (Cochran *et al.*, 1977). A recent study (Fong *et al.*, 1976) pointed out that the concentration of G_{M1}, while small compared to that of other lipids, is approximately equimolar with respect to myelin basic protein. Other subfractions of the oligodendrocyte are now being isolated (Poduslo, 1975), but these have not yet been analyzed for ganglioside content.

ACKNOWLEDGMENTS. This work was supported by grants NS 04834, NS 03356, and NS 10931 from the National Institutes of Health, United States Public Health Service.

References

Abe, T., and Norton, W. T., 1974, The characterization of sphingolipids from neurons and astroglia of immature rat brain, *J. Neurochem.* **23**:1025.

Ando, S., and Yamakawa, T., 1973, Separation of polar glycolipids from human red blood cells with special reference to blood group-A activity, *J. Biochem.* **73**:387.

Ando, S., and Yu, R. K., 1977*a*, Isolation and characterization of human and chicken brain tetrasialoganglioside, *Proc. Int. Soc. Neurochem.* **6**:535 (abstract).

Ando, S., and Yu, R. K., 1977*b*, Isolation and characterization of a novel trisialoganglioside, G_{T1a}, from human brain, *J. Biol. Chem.* **252**:6247.

Ando, S., and Yu, R. K., 1977*c*, Isolation and structural study of a novel fucose-containing disialoganglioside from human brain, *Fourth International Symposium on Glycoconjugates* (abstract), p. 1.

Ando, S., Kon, K., Isobe, M., and Yamakawa, T., 1973, Structural study on tetraglycosyl ceramide and gangliosides isolated from red blood cells, *J. Biochem.* (*Tokyo*) **73**:893.

Avrova, N. F., 1971, Brain ganglioside patterns of vertebrates, *J. Neurochem.* **18**:667.

Avrova, N. F., Chenykaeva, E. Y., and Obukhova, E. L., 1973, Ganglioside composition and content of rat-brain subcellular fractions, *J. Neurochem.* **20**:997.

Bernheimer, H., 1969, Zur Kenntnis der Ganglioside im Liquor cerebrospinalis des Menschen, *Klin. Wochenschr.* **47**:227.

Breckenridge, W. C., Gombos, G., and Morgan, I. G., 1972, The lipid composition of adult rat brain synaptosomal plasma membranes, *Biochim. Biophys. Acta* **266**:695.

Breckenridge, W. C., Morgan, I. G., Zanetta, J. P., and Vincendon, G., 1973, Adult rat brain synaptic vesicles. II. Lipid composition, *Biochim. Biophys. Acta* **320**:681.

Brunngraber, E. G., Dekirmenjian, H., and Brown, B. D., 1967, The distribution of protein-bound *N*-actylneuraminic acid in subcelluar fractions of rat brain, *Biochem. J.* **103**:73.

Caputto, R., Maccioni, H. J., and Arce, A., 1974, Biosynthesis of brain gangliosides, *Mol. Cell. Biochem.* **4**:97.

Cochran, F., Yu, R. K., and Ledeen, R. W., 1977, Comparison of CNS myelin gangliosides in several vertebrate species, *Proc. Int. Soc. Neurochem.* **6**:540 (abstract).

Cragg, B. G., 1972, The development of cortical synapses during starvation in the rat, *Brain* **95**:143.

DeBaecque, C., Johnson, A. B., Naiki, M., Schwarting, G., and Marcus, D. M., 1976, Ganglioside localization in cerebellar cortex: An immunoperoxidase study with antibody to G_{M1} ganglioside, *Brain Res.* **114**:117.

Dekirmenjian, H., and Brunngraber, E. G., 1969, Distribution of protein-bound *N*-acetylneuraminic acid in subcellular particulate fractions prepared from rat whole brain, *Biochim. Biophys. Acta* **177**:1.

Derry, D. M., and Wolfe, L. S., 1967, Gangliosides in isolated neurons and glial cells, *Science* **158**:1450.

DeVries, G. H., and Norton, W. T., 1974, The lipid composition of axons from bovine brain, *J. Neurochem.* **22**:259.

Dominick, V., and Gielen, W, 1968, Über unterschiedliche Gehalte einzelner Gehirnregionen an Gangliosiden, *Hoppe-Seyler's Z. Physiol. Chem.* **349**:731.

Dreyfus, H., Urban, P. F., Edel-Harth, S., and Mandel, P., 1975, Developmental patterns of gangliosides and of phospholipids in chick retina and brain, *J. Neurochem.* **25**:245.

Edel-Harth, S., Dreyfus, H., Bosch, P., Rebel, G., Urban, P. F., and Mandel, P., 1973, Gangliosides of whole retina and rod outer segments, *FEBS Lett.* **35**:284.

Eichberg, J., Whittaker, V. P., and Dawson, R. M. C., 1964, Distribution of lipids in subcellular particles of guinea-pig brain, *Biochem. J.* **92**:91.

Finne, J., Krusius, T., Rauvala, H., and Hemminki, K., 1977, The disialosyl group of glycoproteins, *Eur. J. Biochem.* **77**:319.

Fishman, P. H., and Brady, R. O., 1976, Biosynthesis and function of gangliosides, *Science* **194**:906.

Fong, J. W., Ledeen, R. W., Kundu, S. K., and Brostoff, S. W., 1976, Gangliosides of peripheral nerve myelin, *J. Neurochem.* **26**:157.

Geissler, D., Martinek, A., Margolis, R. U., Margolis, R. K., Skrivanek, J. A., Ledeen, R., König, P., and Winkler, H., 1977, Composition and biogenesis of complex carbohydrates of ox adrenal chromaffin granules, *Neuroscience* **2**:685.

Ghidoni, R., Sonnino, S., Tettamanti, G., Wiegandt, H., and Zambotti, V., 1976, On the structure of two new gangliosides from beef brain, *J. Neurochem.* **27**:511.

Hakomori, S., and Saito, T., 1969, Isolation and characterization of a glycosphingolipid having a new sialic acid, *Biochemistry* **8**:5082.

Hamberger, A., and Svennerholm, L., 1971, Composition of gangliosides and phospholipids of neuronal and glial cell enriched fractions, *J. Neurochem.* **18**:1821.

Handa, S., and Burton, R. M., 1969, Lipids of retina. 1. Analysis of gangliosides in beef retina by thin layer chromatography, *Lipids* **4**:205.

Handa, S., and Yamakawa, T., 1964, Chemistry of lipids of posthemolytic residue or stroma of erythrocytes. XII. Chemical structure and chromatographic behavior of hematosides obtained from equine and dog erythrocytes, *Jpn. J. Exp. Med.* **34**:293.

Harris, J. U., and Klingman, J. D., 1972, Detection, determination, and metabolism *in vitro* of gangliosides in mammalian sympathetic ganglia, *J. Neurochem.* **19**:1267.

Haverkamp, J., Schauer, R., and Wember, M., 1976, Neuraminic acid derivatives newly discovered in humans: *N*-acetyl-9-*O*-L-lactoylneuraminic acid and *N*-acetyl-2,3-dehydro-2-deoxyneuraminic acid, *Hoppe-Seyler's Z. Physiol. Chem.* **357**:1699.

Holm, M., Månsson, J.-E., Vanier, M.-T., and Svennerholm, L., 1972, Gangliosides of human, bovine and rabbit retina, *Biochim. Biophys. Acta* **280**:356.

Ishizuka, I., and Wiegandt, H., 1972, An isomer of trisialoganglioside and the structure of tetra- and pentasialogangliosides from fish brain, *Biochim. Biophys. Acta* **260**:279.

Ishizuka, I., Kloppenburg, M., and Wiegandt, H., 1970, Characterization of gangliosides from fish brain, *Biochim. Biophys. Acta* **210**:299.

Karlsson, K., 1970, Sphingolipid long chain bases, *Lipids* **5**:878.

Kawamura, N., and Taketomi, T., 1975, Further study on cerebral sphingolipids including gangliosides in two cases of juvenile amaurotic family idiocy (Spielmeyer–Vogt Type) using a new analytical procedure of sphingolipids, *Jpn. J. Exp. Med.* **45**:489.

Klein, F., and Mandel, P., 1975, Gangliosides of the peripheral nervous system of the rat, *Life Sci.* **16**:751.

Klenk, E., and Gielen, W., 1963, Über ein zweites hexosaminhaltiges Gangliosid aus Menschengehirn, *Hoppe-Seyler's Z. Physiol. Chem.* **330**:218.

Klenk, E., and Langerbeins, H., 1941, Über die Verteilung der Neuraminsäure im Gehirn (mit einer Mikromethode zur quantitativen Bestimmung der Substanz im Nervengewebe), *Hoppe-Seyler's Z. Physiol. Chem.* **270:**185.

Klenk, E., and Lauenstein, I., 1953, Über die Glykolipoide und Sphingomyeline des Stromas der Pferdeerythrocyten, *Hoppe-Seyler's Z. Physiol. Chem.* **295:**164.

Klenk, E., and Naoi, M., 1968, Übër eine Komponente des Gemisches der Gehirnganglioside, die durch Neuraminidaseein-Wirkung in das Tay–Sachs-Gangliosid übergeht, *Hoppe-Seyler's Z. Physiol. Chem.* **349:**288.

Klenk, E., Hof, L., and Georgias, L., 1967, Zur Kenntnis der Gehirnganglioside, *Hoppe-Seyler's Z. Physiol. Chem.* **348:**149.

Korey, S. R., and Gonatas, J., 1963, Separation of human brain gangliosides, *Life Sci.* **2:**296.

Kornguth, S., Wannamaker, B., Kolodny, E., Geison, R., Scott, G., and O'Brien, J. F., 1974, Subcellular fractions from Tay–Sachs brains: Ganglioside, lipid, and protein composition and hexosaminidase activities, *J. Neurol. Sci.* **22:**383.

Kuhn, R., and Wiegandt, H., 1963*a*, Die Konstitution der Ganglio-*N*-tetraose und des Gangliosides G_I. *Chem. Ber.* **96:**866.

Kuhn, R., and Wiegandt, H., 1963*b*, Die Konstitution der Ganglioside G_{II}, G_{III}, and G_{IV}, *Z. Naturforsch.* **18b:**541.

Kuhn, R., and Wiegandt, H., 1964, Weitere Ganglioside aus Menschenhirn, *Z. Naturforsch.* **19b:**256.

Lapetina, E.G., and DeRobertis, E., 1968, Action of Triton X-100 on lipids and proteolipids of nerve-ending membranes, *Life Sci.* **7:**203.

Lapetina, E. G., Soto, E. F., and DeRobertis, E., 1968, Lipids and proteolipids in isolated subcellular membranes of rat brain cortex, *J. Neurochem.* **15:**437.

Ledeen, R. W., 1978, Ganglioside structures and distribution: Are they localized at the nerve ending?, *J. Supramol. Struct.* **8:**1.

Ledeen, R. W., and Mellanby, J., 1977, Gangliosides as receptors for bacterial toxins, in: *Current Topics in Toxinology* (A. Bernheimer, ed.), pp. 16–42, John Wiley and Sons, New York.

Ledeen, R., and Salsman, K., 1965, Structure of the Tay–Sachs' ganglioside, *Biochemistry* **4:**2225.

Ledeen, R. W., and Yu, R. K., 1972, Gangliosides of CSF and plasma: Their relation to the nervous system, in: *Sphingolipids, Sphingolipidoses and Allied Disorders* (B. W. Volk and S. M. Aronson, eds.), pp. 77–93, Plenum Press, New York.

Ledeen, R. W., and Yu, R. K., 1976*a*, Chemistry and analysis of sialic acid, in: *Biological Roles of Sialic Acid* (A. Rosenberg and C.-L. Schengrund, eds.), pp. 1–57, Plenum Press, New York.

Ledeen, R. W., and Yu, R. K., 1976*b*, Gangliosides of the nervous system, in: *Glycolipid Methodology* (L. A. Witting, ed.), pp. 187–214, Supelco, Bellefonte, Pennsylvania.

Ledeen, R., Salsman, K., and Cabrera, M., 1968, Gangliosides of bovine adrenal medulla, *Biochemistry* **7:**2287.

Ledeen, R. W., Yu, R. K., and Eng, L. F., 1973, Gangliosides of human myelin: Sialosylgalactosylceramide (G_7) as a major component, *J. Neurochem.* **21:**829.

Ledeen, R. W., Skrivanek, J. A., Tirri, L. J., Margolis, R. K., and Margolis, R. U., 1976, Gangliosides of the neuron: Localization and origin, *Adv. Exp. Med. Biol.* **71:**83.

Li, Y.-T., Månsson, J. E., Vanier, M.-T., and Svennerholm, L., 1973, Structure of the major glucosamine-containing ganglioside of human tissues, *J. Biol. Chem.* **248:**2634.

Lowden, J. A., and Wolfe, L. S., 1964, Studies on brain gangliosides. III. Evidence for the location of gangliosides specifically in neurones, *Can. J. Biochem.* **42:**1587.

MacMillan, V. H., and Wherrett, J. R., 1969, A modified procedure for the analysis of mixtures of tissue gangliosides, *J. Neurochem.* **16:**1621.

Månsson, J.-E., Vanier, M.-T., and Svennerholm, L., 1978, Changes in the fatty acid and sphingosine composition of the major gangliosides of human brain with age, *J. Neurochem.* **30:**273.

Mullin, B. R., Fishman, P. H., Lee, G., Aloj, S. M., Ledley, F. D., Winand, R. J., Kohn, L. D., and Brady, R. O., 1976, Thyrotropin–ganglioside interactions and their relationship to the structure and function of thyrotropin receptors, *Proc. Natl. Acad. Sci. U.S.A.* **73:**842.

Naoi, M., and Klenk, E., 1972, The sphingosine bases of the gangliosides from developing human brain and from brains of amaurotic idiots, *Hoppe-Seyler's Z. Physiol. Chem.* **353:**1677.

Norton, W. T., and Poduslo, S. E., 1971, Neuronal perikarya and astroglia of rat brain: Chemical composition during myelination, *J. Lipid Res.* **12:**84.

Ohashi, M., and Yamakawa, T., 1977, Isolation and characterization of glycosphingolipids in pig adipose tissue, *J. Biochem. (Tokyo)* **81:**1675.

Poduslo, S. E., 1975, The isolation and characterization of a plasma membrane and a myelin fraction derived from oligodendroglia of calf brain, *J. Neurochem.* **24:**647.

Poduslo, S. E., and Norton, W. T., 1972, Isolation and some chemical properties of oligodendroglia from calf brain, *J. Neurochem.* **19:**727.

Price, H. C., and Yu, R. K., 1976, Adrenal medulla gangliosides: A comparative study of some mammals, *Comp. Biochem. Physiol.* **54B:**451.

Price, H., Kundu, S., and Ledeen, R., 1975, Structures of gangliosides from bovine adrenal medulla, *Biochemistry* **14:**1512.

Puro, K., Maury, P., and Huttunen, J. K., 1969, Qualitative and quantitative patterns of gangliosides in extraneural tissues, *Biochim. Biophys. Acta* **187:**230.

Rapport, M. M., and Mahadik, S. P., 1978, Personal communication.

Rauvala, H., 1976*a*, The fucoganglioside of human kidney, *FEBS Lett.* **62:**161.

Rauvala, H., 1976*b*, Isolation and partial characterization of human kidney gangliosides, *Biochim. Biophys. Acta* **424:**284.

Robert, J., Freysz, L., Sensenbrenner, M., Mandel, P., and Rebel, G., 1975, Gangliosides of glial cells: A comparative study of normal astroblasts in tissue culture and glial cells isolated on sucrose–Ficoll gradients, *FEBS Lett.* **50:**144.

Roseman, S., 1970, The synthesis of complex carbohydrates by multiglycosyltransferase systems and their potential function in intercellular adhesion, *Chem. Phys. Lipids* **5:**71.

Rosenberg, A., and Stern, N., 1966, Changes in sphingosine and fatty acid components of the gangliosides in developing rat and human brain, *J. Lipid Res.* **7:**122.

Schengrund, C.-L., and Garrigan, O. W., 1969, A comparative study of gangliosides from the brains of various species, *Lipids* **4:**488.

Seminario, L. M., Hren, N., and Gomez, C. J., 1964, Lipid distribution in subcellular fractions of the rat brain, *J. Neurochem.* **11:**197.

Seyfried, T. N., Ando, S., and Yu, R. K., 1978, Isolation and characterization of human liver hematoside, *J. Lipid Res.* **19:**538.

Shein, H., Britva, A., Hess, H. H., and Selkoe, D. J., 1970, Isolation of hamster brain astroglia by *in vitro* cultivation and subcutaneous growth, and content of cerebroside, ganglioside, RNA and DNA, *Brain Res.* **19:**497.

Skrivanek, J., Ledeen, R., Norton, W., and Farooq, M., 1978, Ganglioside distribution in rat cortex, *Trans. Am. Soc. Neurochem.* **9:**133.

Suzuki, A., Ishizuka, I., and Yamakawa, T., 1975, Isolation and characterization of a ganglioside containing fucose from boar testis, *J. Biochem. (Tokyo)* **78:**947.

Suzuki, K., 1967, Ganglioside patterns of normal and pathological brains, in: *Inborn Disorders of Sphingolipid Metabolism* (S. M. Aronson and B. W. Volk, eds.), pp. 215–230, Pergamon Press, New York.

Suzuki, K., Poduslo, S. E., and Norton, W. T., 1967, Gangliosides in the myelin fraction of developing rats, *Biochim. Biophys. Acta* **144:**375.

Svennerholm, L., 1957, Quantitative estimation of sialic acids. II. A colorimetric resorcinol–hydrochloric acid method, *Biochim. Biophys. Acta* **24:**604.

Svennerholm, L., 1963, Chromatographic separation of human brain gangliosides, *J. Neurochem.* **10:**613.

Svennerholm, L., Bruce, A., Månsson, J.-E., Rynmark, B.-M., and Vanier, M.-T., 1972, Sphingolipids of human skeletal muscle, *Biochim. Biophys. Acta* **280:**626.

Svennerholm, L., Månsson, J.-E., and Li, Y.-T., 1973, Isolation and structural determination of a novel ganglioside, a disialosylpentahexosylceramide from human brain, *J. Biol. Chem.* **248:**740.

Tamai, Y., Matsukawa, S., and Satake, M., 1971, Gangliosides in neuron, *J. Biochem.* (*Tokyo*) **69:**235.

Tettamanti, G., Bonali, F., Marchesini, S., and Zambotto, V., 1973, A new procedure for the extraction, purification and fractionation of brain gangliosides, *Biochim. Biophys. Acta* **296:**160.

Ueno, K., Ando, S., and Yu, R. K., 1978, Gangliosides of human, cat, and rabbit spinal cords and cord myelin, *J. Lipid Res.* **19:**863.

Warren, L., 1959, The thiobarbituric acid assay of sialic acids, *J. Biol. Chem.* **234:**1971.

Wenger, D. A., and Wardell, S., 1973, Action of neuraminidase (EC 3.2.1.18) from *Clostridium perfringens* on brain gangliosides in the presence of bile salts, *J. Neurochem.* **20:**607.

Wherrett, J. R., 1973, Characterization of the major ganglioside in human red cells and of a related tetrahexosyl ceramide in white cells, *Biochim. Biophys. Acta* **326:**63.

Wherrett, J. R., 1976, Gangliosides of the lacto-*n*-glycaose series (glucosamine containing gangliosides), in: *Glycolipid Methodology* (L. A. Witting, ed.), pp. 215–232, Supelco, Bellefonte, Pennsylvania.

Wherrett, J. R., and McIlwain, H., 1962, Gangliosides, phospholipids, protein and ribonucleic acid in subfractions of cerebral microsomal material, *Biochem. J.* **84:**232.

Whittaker, V. P., 1969, The synaptosome, in: *Handbook of Neurochemistry*, Vol. 2 (A. Lajtha, ed.), pp. 327–364, Plenum Press, New York.

Wiegandt, H., 1967, The subcellular localization of gangliosides in the brain, *J. Neurochem.* **14:**671.

Wiegandt, H., 1973, Gangliosides of extraneural organs, *Hoppe–Seyler's Z. Physiol. Chem.* **354:**1049.

Wiegandt, H., 1974, Monosialo-lactoisohexaosylceramide: A ganglioside from human spleen, *Eur. J. Biochem.* **45:**367.

Wolfe, L. S., 1961, The distribution of gangliosides in subcellular fractions of guinea-pig cerebral cortex, *Biochem. J.* **79:**348.

Yamakawa, T., and Suzuki, S., 1951, The chemistry of the lipids of posthemolytic residue or stroma of erythrocytes. I. Concerning the ether-insoluble lipids of lyophilized horse blood stroma, *J. Biochem.* (*Tokyo*) **38:**199.

Yates, A. J., and Wherrett, J. R., 1974, Changes in the sciatic nerve of the rabbit and its tissue constituents during development, *J. Neurochem.* **23:**993.

Yohe, H. C., Roark, D. E., and Rosenberg, A., 1976, C_{20}-sphingosine as a determining factor in aggregation of gangliosides, *J. Biol. Chem.* **251:**7083.

Yohe, H. C., Chang, N. C., Glaser, G. H., and Yu, R. K., 1977, Ganglioside changes associated with drug induced seizures, *Trans. Am. Soc. Neurochem.* **8:**185.

Yu, R. K., 1978, Personal communication.

Yu, R. K., and Ando, S., 1978, Novel gangliosides of fish brain, *Trans. Am. Soc. Neurochem.* **9:**135.

Yu, R. K., and Iqbal, K., 1977, Gangliosides of human myelin, oligodendroglia and neurons, *Proc. Int. Soc. Neurochem.* **6:**547.

Yu, R. K., and Ledeen, R. W., 1970, Gas–liquid chromatographic assay of lipid-bound sialic acids: Measurement of gangliosides in brain of several species, *J. Lipid Res.* **11:**506.

Yu, R. K., and Ledeen, R. W., 1972, Gangliosides of human, bovine, and rabbit plasma, *J. Lipid Res.* **13:**680.

Yu, R. K., and Yen, S. I., 1975, Gangliosides in developing mouse brain myelin, *J. Neurochem.* **25:**229.

2

Biosynthesis and Metabolism of Gangliosides

Abraham Rosenberg

1. Introduction: Location and Types of Gangliosides

It is pertinent to an understanding of the metabolism of brain gangliosides to consider their concentration and their more or less specific location within neural cells. The CNS of vertebrates characteristically has a relatively high level of gangliosides (Wiegandt, 1968; Brunngraber *et al.*, 1972). Those gangliosides containing the more highly sialylated (tri- and tetrasialosyl) gangliotetraoses, e.g., $II^3(NeuAc)_3$-$GgOse_4Cer$, or G_{T1c}; $IV^3(NeuAc)II^3(NeuAc)_2$-$GgOse_4Cer$, or G_{T1b}; $IV^3(NeuAc)_2II^3(NeuAc)_2$-$GgOse_4$ Cer, or G_{Q1b}, appear to occur specifically in the brain of animals, and additionally, those containing $IV^3(NeuAc)_2$ $II^3(NeuAc)_3$-$GgOse_4Cer$, or G_{P1c} , are found in fish brain (Avrova, 1971). Sialosyl gangliotetraosyl ceramides also occur to a varying degree in cellular membranes in organs other than brain. Sialosyl lactosyl and gangliotriaosyl ceramides (G_{M3} and G_{M2}) occur largely in nonneural as well as neural cells. For a discussion of the structures and distribution of the ("brain") gangliosides, see Chapter 1.

In best attempts to prepare purified subcellular fractions of whole brain, the highest amount of gangliosides repeatedly has been reported to occur in the "microsomal," or small-membrane-fragment, fraction and in the "synaptosomal," or resealed torn-off nerve-ending, fraction (Derry and Wolfe, 1967; Wiegandt, 1967; Breckenridge *et al.*, 1972; Schengrund and Rosenberg, 1970; Yohe and Rosenberg, 1977). In the latter, the gangliosides are almost exclusively synaptic plasma membrane components. Neuronal gangliosides are not restricted entirely to the perisynaptic region; neuronal soma and nonneuronal neuroglial cells appear rarely if ever to be devoid of gangliosides of a composition similar to that of synaptic membranes (Norton and Poduslo,

Abraham Rosenberg • Department of Biological Chemistry, M. S. Hershey Medical Center, Hershey, Pennsylvania.

1970; Schengrund and Nelson, 1976). Synaptic concentration of gangliosides, nevertheless, must figure in the extrapolation of *in vitro* findings, so as to arrive at a generalized concept of ganglioside metabolism in brain.

While discussion in any detail is beyond the purview of this chapter, consideration should be given to the ample occurrence in brain gangliosides of the 20-carbon sphingoid component, icosasphingosine, in addition to the 18-carbon component, sphingosine. The latter is generally the sphingoid component of all other brain sphingolipids, and of the sphingolipids, except for peripheral nerve myelin (Fong *et al.*, 1976), of organs other than brain. It is also remarkable that most brain ganglioside molecules are stearoyl compounds and thus bear throughout organismic life the fatty acid composition that often typifies fetal brain sphingolipids generally (Rosenberg and Stern, 1966; Vanier *et al.*, 1973). It may be reasonably presumed, by analogy with the established precursor role for palmitoyl-CoA in the biosynthesis of sphingosine (Stoffel, 1970), that the obligate precursor for the hydrocarbon chain of icosasphingosine will be stearoyl-CoA. The preference for stearic acid in the biosynthesis of both lipophilic chains of the brain gangliosides is noteworthy, but remains a little-understood metabolic phenomenon (Yohe *et al.*, 1976).

A distinguishing characteristic of the nonneuronal ganglioside components of brain, when this organ is considered as a whole, is the enrichment of monosialogangliosides in white matter myelin (Yu and Yen, 1975). Peripheral nervous system (PNS) has a glucosamine-containing analogue (Fong *et al.*, 1976). Especially interesting is the unique occurrence in CNS but not PNS (Fong *et al.*, 1976) myelin of the compound sialosyl galactosyl ceramide (Kuhn and Wiegandt, 1964; Ledeen *et al.*, 1973). Unlike the neuronal synaptic membrane gangliosides, the monosialo compounds of brain white matter myelin appear to have as their exclusive sphingoid component 18-carbon sphingosine, the fatty acyl component of the sialosyl galactosyl ceramide being frequently of the 2-hydroxy and longer-chain fatty acid species found also in myelin galactosyl and sulfogalactosyl ceramides, but not in synaptic brain gangliosides.

In view of the foregoing, it is within the framework of heterogeneity of cell types in the nervous system and with awareness of a degree of specificity of structure, cellular occurrence, and subcellular concentration of different brain gangliosides that detailed information concerning the metabolism of nervous system gangliosides should properly be arranged.

2. Biosynthesis of Brain Gangliosides

2.1. Synthesis by Glycosyltransferases

It is generally accepted that nervous system gangliosides are synthesized by the action of membrane-bound multiglycosyltransferase systems that catalyze the sequential transfer of monosaccharide units from their activated forms as nucleoside diphosphoryl sugars (or, in the case of sialic acid, a nucleoside monophosphoryl sugar) to a ceramide-containing acceptor molecule (Roseman, 1970). Sialosyl galactosyl ceramide, or G_{M4}, of myelin, the unique and structurally the simplest nonneuronal ganglioside, may be derived from the sialosylation of galactosyl ceramide, the major glycosphingolipid component of myelin, and a precursor relationship for galactosyl ceramide has been described (Ledeen *et al.*, 1973). For construction of the gangliotetraose backbone of the brain gangliosides, the initial sugar transfer is that of glucose from UDP-glucose to the lipid, ceramide, to form glucosyl ceramide (glucocerebroside), the structural base of the synaptic membrane group of gangliosides.

Cellular location of the membrane-bound glycosyltransferases has been, for some time, the subject of controversy. Occasional reports that nerve-ending preparations and some of the subfractions prepared from them possess glycosyltransferase activities capable of building at least in part the gangliotetraose moiety of the brain gangliosides (Den *et al.*, 1975; Bosmann, 1972; DiCesare and Dain, 1972) are difficult to accept fully because of the complexity of the incompletely separated subcellular fractions studied. The ganglioside glycosyltransferases in liver are found to be enriched in Golgi membranes (Keenan *et al.*, 1974). There is, as yet, no compelling evidence that would substantiate a different arrangement for neuronal cells. Indeed, it was reported (Fishman, 1974) that ganglioside glycosyltransferases may be demonstrated to accompany the myelin subfraction of rat brain even before myelination has proceeded to any considerable extent in such animals. Several reports have described the presence of glycosyltransferases in practically all major subcellular fractions prepared from brain, subject to the difficulties with purification of fractions inherent in currently available technology (Arce *et al.*, 1971; Hildebrand *et al.*, 1970). Recently, evidence was presented (Ng and Dain, 1977) that strongly supports the extant presumption that in brain, the ganglioside glycosyltransferases may reside, as in liver, in Golgi and perhaps in smooth endoplasmic reticulum but not, in general, in the nerve-ending membranes.

Extensive study concerning the nature of membrane-bound glycosyltransferases that are responsible for the biosynthesis of brain gangliosides has been made over the past two decades. Generally, these enzymes show a requirement, *in vitro*, for activation by detergents such as Triton-CF54, Triton X-100, Cutscum, or Tween 80. Apparent K_m values for exogenously supplied substrates reportedly range between 10^{-4} and 10^{-5} M. Divalent manganese ion emerges quite frequently as a metal ion requirement, although a generalization in this regard for all the glycosyltransferases in the gangliotetraose-synthesizing membrane-bound systems in brain is risky: the first enzymatic transfer of a sugar, to ceramide from UDP-Glc, to produce glucosyl ceramide (glucocerebroside), reportedly has no divalent cation requirement; also, sialosylation of the growing oligosaccharide chain by the enzymatic transfer of sialic acid from CMP-NeuAc frequently displays no apparent divalent cation requirement, or else such ions provide only weak stimulation (Van den Eijnden and Van Dijk, 1974). A tabulation of the K_m values, metal ion requirements, and pH optima for the brain glycosyltransferases involved in ganglioside biosynthesis was made several years ago, and is available in a review article (Burton, 1974).

2.2. Sequential Glycosylations

It is generally acknowledged that the detailed steps in the glycosylation of ceramide to produce the various brain ganglioside molecules have emerged mainly from the initial contributions from the laboratory of Roseman (Kaufman *et al.*, 1967). Chick embryo brain preparations were utilized in these studies. It should be noted in passing that the danger of contamination by retinal cells in whole embryonic chicken brain preparations is considerable. Chick retinal cells, and to a great degree chick brain cells, unlike those of some other animal species, are now known to retain a relatively extraordinary capacity for the synthesis of the disialoganglioside $II^3(NeuAc)_2$-LacCer- (Dreyfus *et al.*, 1975). In the pathway proposed by Roseman, sialosylation of II^3(NeuAc)-LacCer to produce the latter disialo compound conceptually represents the point of divergence in the biosynthesis of those gangliosides in which $II^3(NeuAc)_2$-$GgOse_4$ is a structural component, i.e., those gangliosides with two sialosyl groups appended at the inner galactosyl residue of the gangliotetraose backbone, as distinct from the II^3NeuAc-$GgOse_4$-containing compounds in which only one sialosyl residue is appended. It may be suggested that the special features of embryonic chicken brain be kept in view in the interpretation that a particular glycosylation step represents a key selective event for ganglioside biosynthesis in neural cells. Also, it should be noted that to this day, detailed evidence for the occurrence of strict multiglycosyltransferase systems or complexes is not extensive. However, studies from Caputto's laboratory (Maccioni *et al.*, 1971) have been instrumental in demonstrating that expected straightforward precursor relationships do not hold true for exogenous glycosyl sphingolipid substrates when they are supplied for further glycosylation in order to yield the full sialosyl gangliotetraosyl ceramide structures.

Step 1. The fundamental glycosyl lipid of the lactosyl and gangliosyl ceramide family of compounds, glucosyl ceramide, may be synthesized by the glucosylation of sphingosine:

$$\text{Sphingosine} + \text{UDP-Glc} \xrightarrow{\text{Glucosyltransferase}} \text{glucosylsphingosine} + \text{UDP}$$

followed by fatty acylation:

$$\text{Glucosyl sphingosine} + \text{stearoyl-CoA} \longrightarrow \text{glucosyl ceramide} + \text{CoA}$$

These reactions were convincingly demonstrated to occur in brain microsomes (Curtino and Caputto, 1974).

Alternatively, ceramide may be glucosylated directly:

$$\text{Cer} + \text{UDP-Glc} \xrightarrow{\text{Glucosyltransferase}} \text{glucosylceramide} + \text{UDP}$$

This latter is the usually accepted pathway, which was demonstrated in rat brain and in embryonic chicken brain particles (Basu *et al.*, 1968; Shah, 1971).

Step 2. The next stage is the further glycosylation of glucosyl ceramide. Rat brain particles (Basu *et al.*, 1968) and chicken brain particles (Hildebrand *et al.*, 1970) were first shown to carry out this reaction:

$$\text{GlcCer} + \text{UDP-Gal} \xrightarrow{\text{Galactosyltransferase}} \text{LacCer} + \text{UDP}$$

Step 3. Conceptually, the next glycosylation step in the reaction sequence evokes the special characteristics of sialosylation, as compared with the transfer of neutral sugars from the appropriate UDP-sugars. The sialosylation of lactosyl ceramide

$$\text{LacCer} + \text{CMP-NeuAc} \xrightarrow{\text{Sialosyltransferase}} \text{II}^3\text{NeuAcLacCer} + \text{CMP}$$

was demonstrated, for example, in particles prepared from rat brain (Arce *et al.*, 1966; Kaufman *et al.*, 1967). Certain characteristics of the sialosyltransferases of rat brain were elucidated recently (Ng and Dain, 1977). It would appear that the foregoing reaction is the favored one with respect to endogenous glycolipid substrate; i.e., incorporation of sialic acid from CMP-sialic acid into an endogenous acceptor, undoubtedly lactosyl ceramide, produces II^3NeuAcLacCer (G_{M3} ganglioside) in sufficient quantity to account for more than 60% of the synthesized sialosyl glycosyl ceramides.

A divergent reaction was shown to be catalyzed by the brain of young rats (Handa and Burton, 1969; Yip and Dain, 1969):

$$\text{LacCer} + \text{UDP-GalNAc} \xrightarrow{N\text{-acetylgalactosaminyltransferase}} \text{GgOse}_3\text{Cer}$$

i.e., GalNAc(β1–4)Gal(β1–4)GlcCer + UDP. The gangliotriaose chain may be subsequently elongated and sialosylated.

The source of CMP-sialic acid as sialosyl donor for the foregoing reaction and for subsequent sialosylation reactions in the synthesis of brain gangliosides represents a puzzle. The enzyme that synthesizes this compound by the unique reaction

$$\text{NeuAc} + \text{CTP} \xrightleftharpoons{\text{Pyrophosphorylase}} \text{CMP-NeuAc} + \text{PP}_i$$

has been shown, for neural cells, to be localized in the nuclei (Van den Eijnden, 1973). If the glycosyltransferases reside in the Golgi apparatus, as developed above, then the mode of translocation of CMP-NeuAc remains to be elucidated. The inefficiency of its possible transport by axoplasmic flow to the nerve ending, for ganglioside synthesis there, represents a further conceptual barrier to the possibility of ganglioside synthesis in this region of the neuron.

On the basis of heat stabilities (Ng and Dain, 1977) and other criteria, (Basu *et al.*, 1968), the sialosyltransferase that converts LacCer to II^3NeuAcLacCer (G_{M3} ganglioside of "hematoside") differs from those sialosyltransferases that subsequently attach additional NeuAc residues to the oligosaccharide moiety of the ganglioside molecule. At any rate, the transfer of sialic acid to lactosyl ceramide appears generally to be the predominant cellular synthetic reaction over the transfer of *N*-acetylgalactosamine, and indeed, hematoside itself appears to be the preferred acceptor for this latter transfer:

$$II^3\text{NeuAcLacCer} + \text{UDP-GalNac} \longrightarrow II^3\text{NeuAc-GgOse}_3\text{Cer} + \text{UDP}$$

This product is the so-called "Tay–Sachs ganglioside," or G_{M2} which accumulates in the catabolic sphingolipidystrophy so named, and bears a terminal (β1–4) *N*-acetylgalactosaminyl residue in its GgOse oligosaccharide moiety. It has been suggested that in fetal brain, the sialosylation of LacCer takes predominance, while in late neonatal rat brain, *N*-acetylgalactosaminylation has precedence (compare Handa and Burton, 1969, and Arce *et al.*, 1966).

A putative anabolic ganglioside storage disease, representing a unique metabolic derangement, in which oversynthesis and storage of II^3NeuAc-LacCer occurs, has been uncovered and characterized as due to a deficiency in UDP-GalNac: II^3NeuAcLacCer *N*-acetylgalactosaminyltransferase activity. In this case, II^3NeuAcLacCer is not normally glycosylated, but accumulates and in substantial amount is further sialosylated by the action of a second sialosyltransferase to produce $II^3(\text{NeuAc})_2$LacCer, or G_{D3} ganglioside. This disialosyl compound also accumulates (Max *et al.*, 1974; Okada and O'Brien, 1968). The latter reaction leads back to a controversial conceptual juncture in the established sequence of reactions for the normal biosynthesis of the brain gangliosides.

Step 4. As mentioned earlier, the branch point proposed by Roseman (1970) for the divergence of biosynthesis of Lac($\leftarrow$2αNeuAc)-containing and Lac($\leftarrow$2αNeuAc8)$_2$-containing gangliosides lies in the alternative *N*-acetylgalactosaminylation or else sialosylation of hematoside:

$$II^3\text{NeuAcLacCer} \begin{array}{l} \text{CMP-NeuAc} \\ (1)\longrightarrow \\ (2)\longrightarrow \\ \text{UDP-GalNac} \end{array}$$

(1) NeuAc-(α2$\rightarrow$8) NeuAc-(α2$\rightarrow$3) Gal (β1$\rightarrow$4) Glc (β1$\rightarrow$1) Cer
(2) GalNAc(β1$\rightarrow$4) Gal (3$\leftarrow$2α-NeuAc)(β1$\rightarrow$4) Glc(β1$\rightarrow$1) Cer

Specificity for synthesis of the anomeric linkages shown, as well as for the different glycose acceptors, residues in the individual glycosyltransferases involved. Activity of the enzyme that catalyzes reaction (2), i.e., UDP-GalNAc:II^3NeuAcLacCer *N*-acetylgalactosaminyltransferase, is salient in mammalian brain, especially during the early development of this organ (Cumar *et al.*, 1971, DiCesare and Dain, 1971).

While any extended consideration is beyond the purview of this chapter, it is of interest that cells from extraneural tissues, in which II^3NeuAcLacCer and $II^3(NeuAc)_2$LacCer generally are the major components of the cellular gangliosides, have low and often barely detectable levels of this GalNac transferase (Keenan, 1974; Keenan *et al.*, 1974; DiCesare and Dain, 1971; Den *et al.*, 1971). Such nonneural cells resemble, in this respect, the metabolic feature of the hematoside sphingolipidystrophy touched on above.

With low GalNac transferase activity in a given cell, it is conceivable that an unimpaired synthesis of II^3NeuAcLacCer provides a more or less abundant store of this compound as a precursor for the biosynthesis of the disialo compound, as outlined in reaction (1) above. That this latter disialo compound, $II^3(NeuAc)_2$LacCer, or G_{D3} ganglioside, along with the monosialo compound, II^3NeuAcLacCer, or G_{M3} ganglioside, is a preferred substrate for GalNac transferase activity was confirmed by Cumar *et al.* (1971) and Keenan *et al.* (1974).

The transfer of *N*-acetylgalactosamine, alternatively, either to monosialosyl or to disialosyl lactosyl ceramide had been considered the branch point for the biosynthesis of those gangliosides of the $GgOse_4$ variety bearing either one or two sialosyl residues on the galactosyl residue proximal to the lipophilic Cer moiety of the molecules. Further glycosylations will produce the completed mono-, di-, and polysialo ganglioside molecules:

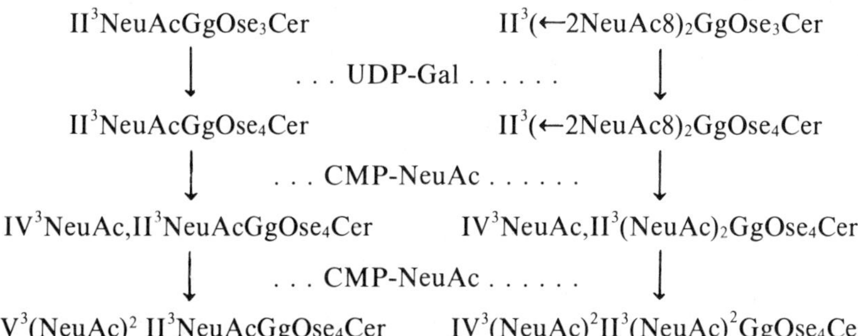

Additional elongation of the $GgOse_4$ chain, e.g., to $GgOse_5$, is not unknown (see Chapter 1). There is good evidence that the individual glycosyltransferases, particularly the sialosyltransferases, represent different enzymatic activities; notably, one may put forward the generalization that the structure of the carbohydrate chain of the gangliosides is determined by the precise specificities of these transferases (Roseman, 1970). It is still open to question and to study whether within a given multiglycosyltransferase system different individual proteins catalyze the transfer of the positionally different sialosyl and galactosyl residues or whether there are only single catalytically active polypeptides for each sugar transfer, specificity of sugar transfer being conferred by environmental, allosteric, or other, e.g., aggregative, effects.

2.3. Glycosylation *in Vivo*

There appears to be a real difference between the effective sequence of multiglycosyltransferase reactions when exogenous lipids are supplied as substrates, as in the biosynthetic scheme outlined above, and when the activities of endogenous lipid substrates in the native membrane are utilized as the natural acceptors of sugar transfer from the appropriate exogenously supplied nucleotide-sugars. Studies have indicated that during the biosynthesis of gangliosides from endogenous lipid acceptors, the intermediates do not leave the membrane system and they are not freely exchangeable with exogenously supplied lipid substrates (Arce *et al.*, 1971). These studies (Caputto *et al.*, 1974) demonstrated convincingly that $II^3NeuAcGgOse_4Cer$ is, by sialosylation, converted to $II^3(\leftarrow 2NeuAc8)_2GgOse_4Cer$. Consequently, the scheme outlined above, as derived by the use of exogenously supplied substrates in detergent-laden media, and in which $II^3(NeuAc)_2LacCer$ appears to be an obligate precursor for $II^3(NeuAc)_2GgOse_4Cer$, may possibly be contrived because of the experimental conditions and does not represent the favored cellular biosynthetic pathway. It has been further shown in Caputto's laboratory that the introduction of the IV^3NeuAc residue into the ganglioside molecule in the endogenous membrane-bound lipid precursor is in all probability catalyzed by a single sialosyl transferase; i.e., the difference between the II^3NeuAc- and $II^3(NeuAc)_2$- residues is not recognized by the native enzyme. With the enzyme in chick embryo brain, which transfers sialic acid from CMP-NeuAc to *exogenously supplied* $II^3(NeuAc)_2GgOse_4Cer$ to form IV^3NeuAc, $(II^3NeuAc)_2GgOse_4Cer$, but will not sialosylate IV^3NeuAc, $II^3NeuAcGgOse_4Cer$ to yield this latter compound, there is a clear distinction between the transfer of NeuAc to an existing NeuAc residue in the molecule or to the terminal Gal residue (Mestrallet *et al.*, 1974).

Further evidence that ganglioside biosynthesis, at least for neurons, is not a synaptic membrane function, but probably occurs in the endoplasmic reticulum and Golgi membranes, was recently adduced (Maccioni *et al.*, 1977). Noting that the incubation of brain membranes with a particular nucleotide sugar does not substantially decrease the amount of the presumed endogenous oligosaccharidyl ceramide precursor and that after treatment with labeled sugar, the labeling of repeating, e.g., galactosyl and sialosyl, units in the ganglioside molecule is relatively the same for all components regardless of whether they are more proximal or distal from the ceramide residue, these investigators concluded that the ganglioside molecule, once built, passes into an end-product pool, presumably the perisynaptic membranes or other membrane structures, and is no longer accessible to the glycosylating system. It may serve to establish a frame of reference for the importance of the distinctions made above concerning certain biosynthetic steps as determined by the use of exogenously supplied substrates that an extremely small fraction of a percentage of the total gangliosides in brain preparations is actually labeled in these experiments with exogenous substrates.

2.4. From Synthetic Machinery to Membrane Structure

The mode of passage of biosynthesized ganglioside molecules from the putative multiglycosyltransferase apparatus in the neuron to the nerve ending is still unclear. It was reported (Forman and Ledeen, 1972) and disclaimed (Rosner *et al.*, 1973; Holm, 1972) that gangliosides are supplied from the neuronal perikaryon to the nerve ending by fast exoplasmic flow. Speculatively, movement from Golgi to plasma membrane and lateral phase mobility in this membrane may provide a simple mechanism. Neuronal cell bodies were shown to be capable of sialosyl gangliotetraosyl ceramide biosynthesis (Jones *et al.*, 1972; Radin *et al.*, 1972), but surprisingly, convincing evidence is not yet forthcoming for a biosynthetic capability in glial cells, although these latter cells in brain may be rich in such gangliosides (see Chapter 1). Metabolic details have not yet been elaborated for the rarer, fucose-containing gangliosides and for those with longer than the $GgOse_4$ oligosaccharide backbone or with more than 4 sialosyl residues. For structures of these compounds, see Chapter 1.

3. Turnover of Brain Gangliosides

3.1. Turnover *in Toto*

Early studies on the turnover of brain gangliosides using labeled hexoses or hexosamines as precursors *in vivo* found that the turnover rates of the various gangliotetraose-containing gangliosides are rather similar (Suzuki, 1967). The half-life reportedly ranges around 3 weeks, with a slow initial turnover rate immediately postpartum, an increase during the period of rapid synaptogenesis, then a decline. It was demonstrated (Maccioni *et al.*, 1971; Holm and Svennerholm, 1972) that the different sugar residues in the ganglioside molecule all have similar half-lives. Furthermore, partial desialylation of gangliosides in nerve-ending preparations with *Clostridium perfringens* sialidase and subsequent attempts at resialosylation showed that endogenous sialosyltransferases are incapable of rebuilding the partially degraded ganglioside molecules in the membrane. These findings are strong evidence that the ganglioside molecules probably turn over *in toto* in the living cell (Maccioni *et al.*, 1974).

3.2. Developmental Correlations

Given the heterogeneity of cell types in whole brain, it is difficult to make correlations between developmental stages of this organ and the accumulation of one or another species of ganglioside. It is a generally accepted hypothesis that the rates of synthesis of individual types of ganglioside molecules are controlled by the activities of the specific glycosyltransferases involved. Salient among correlations that may be made currently are the increases in the monosialogangliosides, NeuAcGalCer, or G_{M4}, and II^3NeuAcGgOse$_4$, or Cer G_{M1}, in normally developing myelin (Yu and Yen, 1975; Geel and Gonzales, 1977) and increases in the disialoganglioside, IV^3NeuAc, II^3NeuAc-GgOse$_4$Cer, or G_{D1a}, as the predominant ganglioside in perisynaptic membranes (Yohe and Rosenberg, 1977) concurrently with the period of active synaptogenesis (Vanier *et al.*, 1971). A perinatal spike in the trisialoganglioside, IV^3NeuAc, II^3(NeuAc)$_2$GgOse$_4$Cer, or G_{T1b}, was reported (Yusuf *et al.*, 1977).

4. Catabolism of Nervous System Gangliosides

4.1. Lysosomal Glycohydrolases and Plasma Membrane Sialidase (Neuraminidase)

It has been presumed generally that the glycohydrolases that sequentially remove the glycose units from the hydrophilic component of the gangliosides during their degradation in the cell are lysosomal enzymes. The sialidase (neuraminidase) that removes sialic acid from the higher gangliotetraose-containing gangliosides in synaptic membranes, leaving only that enzyme-resistant sialosyl group linked to the inner galactose residue proximal to the ceramide moiety of the ganglioside molecule, and thus reducing these compounds to the sialidase-resistant monosialoganglioside, II^3NeuAcGgOse$_4$-Cer, or G_{M1}, is an intrinsic protein component of synaptic membranes and is closely associated in these membranes with the gangliosides that act as substrates (Schengrund and Rosenberg, 1970; Yohe and Rosenberg, 1977). This finding raises the intriguing possibility that gangliosialidase complexes operate under conditions to: (1) release sialic acid that may be taken up by other cells; (2) expose β-D-galactosyl end groups in the surface of the synaptic membrane on removal of sialic acid; and (3) modulate negative charge density by removing bound sialic acid.

Further degradation of the gangliosides may take place by the action of lysosomal acid hydrolases. It is not unlikely that lactosyl ceramide and the partially desialylated higher gangliosides are hydrolyzed by a sequential removal of sugar units from these compounds as they occur in former surface-membrane fragments incorporated into secondary lysosomes. Rarely is the inner sialosyl residue (i.e., that residue bound in $\alpha2\rightarrow3$ linkage to an equatorial hydroxyl group of that inner β-galactosyl component that constitutes the second glycose unit away from the lipophilic ceramide residue in the higher gangliosides) removable by the action of sialidase. Only when the adjacent *N*-acetylhexosamine at position 4 has been removed yielding sialosyl lactosyl ceramide, does the sialosyl residue become susceptible to easy removal by all known sialidases. This phenomenon may have important biological consequences. In the synaptic membrane, when IV^3NeuAc, $II^3NeuAcGgOse_4Cer$ is reduced by the action of sialidase to II^3NeuAc-$GgOse_4Cer$, this unique compound maintains a negative charge due to the remaining sialosyl component that is located close to the lipophilic ceramide residue of the molecule. Thus, this negative charge should be anchored near the hydrocarbon core of the membrane. There is no compelling evidence that further removal of sugar units from G_{M1} normally takes place at the cell surface, their removal being effected internally and presumably lysosomally by the sequential action of (1) an acid ganglioside β-galactosidase; (2) a hexosaminidase, designated A, to distinguish it from an isoenzyme, termed B, that appears to act preferentially on asialo *N*-acetylhexosaminides; (3) a sialidase; (4) a β-galactosidase; and finally (5) a glucocerebroside β-glucosidase that yields the lipophilic ceramide residue, this compound being cleaved by ceramidase to fatty acid and sphingosine. Details are obtainable in a recent review (Suzuki, 1976).

4.2. Deficiencies in Catabolic Glycohydrolases

Inherited enzyme deficiencies, leading to storage of undegraded glycosphingolipid and causing severe neuropathology, are known for many of the lysosomal glycosidases that are responsible for the degradation of the brain gangliosides. Excluded are $II^3NeuAcLacCer$ sialidase (neuraminidase) and LacCer galactosidase. The best known are $II^3NeuAcGgOse_4Cer$, or "G_{M1}," gangliosidosis (in which a β-galactosidase is deficient in activity); II^3NeuAc-$GgOse_3Cer$, or "G_{M2}," gangliosidosis, variant B (in which hexosaminidase A is deficient) and variant O (in which both hexosaminidases A and B are deficient); and Gaucher's disease (in which glucocerebrosidase, i.e., glucosyl ceramide β-glucosidase, is deficient). This topic was reviewed recently (Tallman and Brady, 1976).

4.3. Speculations on Sialidase Resistance of G_{M1}

The potential biological importance of the phenomenon in which II^3Neu-$AcGgOse_4Cer$ (G_{M1} ganglioside) is protected from enzymatic removal of its sialosyl group has prompted certain thoughts about the mechanism by which this enzymatic resistance is achieved. While steric hindrance by the adjacent *N*-acetyl galactosaminyl residue has from time to time been proposed by others, examination of either steric, Dreiding, or space-filling, Godfrey, models of the molecule do not show occlusion of the sialosyl group of the G_{M1}

residue by the adjacent *N*-acetylgalactosaminyl group. The pK_a of sialic acid is near 2.7. Observation of a pK_a value for G_{M1} of roughly 5.5 (unpublished results), coupled with structural considerations, prompt the following conceptual speculations: (1) Hydrogen bonding may form between the protonated carboxyl group of the inner sialic acid residue and the polarizable carbonyl group of the *N*-acetyl substituent (at carbon-2 of the pyranose ring) of the β-D-galactosaminyl residue. This residue is linked to galactose at position 4, adjacent to the sialosyl group at position 3. (2) Such bonding would impart a rigidity to the inner portion of the oligosaccharide backbone, locking the sialosyl group in a position such that the α-ketosidic bond is not readily available for enzymatic hydrolysis. (3) The axial hydroxyl group at carbon-4 on the pyranose ring of the *N*-acetylgalactosaminyl residue apparently may present a further impairment of rotation of the sialosyl residue. Continuing with this speculation, it may be suggested that (4) under the proper conditions, lipophilic association of acidic lipid molecules with the ganglioside may occur in such a way as to bring an anionic group into a position to compete with the hydrogen bonding of the sialic acid carboxyl group and the carbonyl group of *N*-acetylgalactosamine, freeing the sialosyl residue from its rigid position and allowing rotation around the α-ketosidic bond; it should then be possible to enzymatically cleave the sialic acid. It has been reported that bile salts can render this ganglioside susceptible to sialidase (neuraminidase). The sialic acid is readily cleaved both from II^3NeuAc-$GgOse_3Cer$ and from $II^3NeuAcGgOse_4Cer$ (G_{M1} and G_{M2} gangliosides) by *Clostridium* sialidase when combined with bile salts (Wenger and Wardell, 1973). This purely speculative proposal is open to experiment. It would be of more than passing interest to investigate the effects of naturally occurring acidic lipid components of certain membranes, e.g., sulfogalactosyl ceramide of myelin, on the susceptibility of G_{M1} ganglioside to sialidase. See Chapter 1 for depiction of ganglioside structures.

5. Influence of Physicochemical Properties on Metabolism

5.1. Possibility of Self-Aggregation

The sphingosine residues of bovine synaptic membrane gangliosides have an increased content of icosasphingosine corresponding to an increased number of sialic acid residues in the hydrophilic headgroup (Yohe *et al.*, 1976). Micellar size (Yohe and Rosenberg, 1972) is found to be decreased considerably by increasing sialic acid content. The presence of icosasphingosine has an opposing effect on the free energy of aggregation and increases micellar size. More than 60% of the trisialoganglioside molecules of mature bovine synaptic membranes contain icosasphingosine, as do some 55% of disialogangliosides and less than 40% of monosialogangliosides. The free energies of association per monomer, RT ln (critical micelle concentration), that have been deduced for the naturally occurring gangliosides in bovine synaptic membranes are similar: (−9.4 RT) for trisialoganglioside, (−9.3 RT) for disialoganglioside, and (−9.2 RT) for monosialoganglioside. In mammalian brain, icosasphingosine is synthesized and, apparently, is built exclusively into nonmyelin gangliosides, and to no appreciable extent into other sphingolipids. Biosynthetic control mechanisms appear to operate so as to provide a balanced self-aggregating arrangement for synaptic membrane gangliosides. It has been known for some time that in early fetal mammalian brain, icosasphingosine is practically nonexistent, but appears perinatally and reaches essentially full content concurrently with the completion of synaptogenesis (Rosenberg and Stern, 1966).

5.2. Receptor Functions

An exciting metabolic feature of the gangliosides is their putative function as specific receptor molecules in cellular plasma membranes. Substances bound by ganglioside molecules will probably be restricted relatively closely to the hydrocarbon core of the membrane. The studies of Haywood (1974*a*, *b*) with artificial lipid vesicles (liposomes) demonstrated the necessity for gangliosides in such structures in order for Sendai virus to bind to their surface. With regard to the action of this myxovirus in effecting the fusion of mammalian cells, it has been proposed that binding of the virus to gangliosides can produce a destabilization of the membrane, leading to cell fusion. Such destabilization was, in fact, observed for ganglioside-containing liposomes (Hill and Lester, 1972). In preliminary studies aimed at unraveling the mechanism of myoblast fusion, an important normal phenomenon in muscle development, we recently discerned the presence of substantial amounts of $II^3NeuAcGgOse_4Cer$ and a burst of synthesis of $II^3(NeuAc)LacCer(G_{M3})$ in prefusion myoblasts, followed by the rapid disappearance of those gangliosides coincident with the completion of spontaneous cell fusion in culture. Artificial enrichment of the outer cell membrane in the former ganglioside greatly augments the rate of cell fusion. Removal of sialic acid from the latter prevents cellular aggregation (Kemp, 1970).

High specificity of G_{M1} ganglioside as a receptor for cholera toxin, which displays adenylate-cyclase-activating (choleragenic) activity, is suggested by careful recent elaboration of this phenomenon in several laboratories (Sattler *et al.*, 1975; Gill and King, 1975; Holmgren *et al.*, 1975). Certain mechanistic details of the role of G_{M1} ganglioside in the action of cholera toxin have been elucidated. It would appear that of the two subunits of the 86×10^3-dalton choleragen molecule, the smaller 32×10^3-dalton "A" protomer (which is itself comprised of two disulfide-linked polypeptides) possesses choleragenic activity but cannot alone penetrate the plasma membrane of the cell. The larger, "B" protomer, an oligomer of noncovalently linked, roughly 10^4-dalton polypeptides, binds to G_{M1} gangliosides and apparently perturbs the plasma membrane of the cell, allowing penetration by the choleragenic subunit. Activation of adenylate cyclase by the A protomer requires NAD and other substances and may involve the prior splitting of NAD in the cell membrane (Gill, 1975). In a manner similar to that described previously for diphtheria toxin (Kandel *et al.*, 1974), the A protomer of cholera toxin was shown to possess NADase activity, cleaving this compound to nicotinamide and ADP-ribose (Moss *et al.*, 1977). The terminal β-D-galactosyl residue and the carboxyl group of the α-sialosyl residue bound to the inner galactose of G_{M1} ganglioside were found to be necessary for high-affinity binding of the B promoter of cholera toxin (Sattler *et al.*, 1977). The effectiveness of B-protomer-binding of G_{M1} ganglioside as an agent in membrane perturbation has been shown by the release of trapped glucose by the B protomer, or else the entire choleragen molecule, from liposomes containing G_{M1}. Cholera toxin was found to inhibit the establishment of antiviral activity by interferon in mouse L-cells (Friedman and Kohn, 1976), but in the case of interferon from mouse fibroblasts and also human leukocytes and fibroblasts, the trisialoganglioside, $IV^3NeuAcII^3(NeuAc)_2GgOse_4Cer$, or G_{T1}, and the monosialoganglioside G_{M2} (which lacks the terminal β-D-galactosyl residue of G_{M1}) were the most effective binding substances: G_{M1}, as well as the disialoganglioside IV^3NeuAc, $II^3NeuAcGgOse_4Cer$, or G_{D1a}, were without effect (Vengris *et al.*, 1976). Artificial enrichment of cell membranes with G_{M1} enhances cholera-toxin-binding, while extracellular G_{M1} competes with cellular G_{M1} for binding and offsets cellular intoxication. Naturally, these findings lead to speculation as to whether G_{M1} ganglioside may act as a determinant receptor for similar nonpathogenic adenylate-cyclase-activating proteins that normally may be elaborated by cells, as exogenous regulators of cell growth. Indeed, certain analogies have been suggested and indirect evidence has been forthcoming to support the contention that thyrotropin (TSH), for example, interacts with thyroid plasma membrane in a manner analogous to the binding of choleragen. Sequence homologies have been detected between the β subunit of TSH and the B protomer of choleragen. In the case of TSH, the most effective binding gangliosides reportedly are the trisialoganglioside G_{T1} and, better, the disialoganglioside $II^3(2\leftarrow NeuAc8)_2GgOse_4Cer$, or G_{D1b}, which latter bears a sialosyl($\alpha 2\rightarrow 8$)sialosyl($\alpha 2\rightarrow 3$) residue on the inner β-D-galactosyl unit of the tetraglycosyl backbone (Mullin *et al.*, 1976). Interestingly, the steroidogenic effect of both cholera and *Escherichia coli* enterotoxins on adrenal tumor cells is inhibited by extracellular G_{M1} ganglioside, suggesting that this ganglioside may function as a receptor for *E. coli* toxin as well (Donta and Viner, 1975). Tetanus toxin, which may be split by thiol reduction into a 105×10^3-dalton, β, polypeptide and a 55×10^3-dalton, α, polypeptide, also appears to bind effectively to G_{D1b} and to G_{T1} gangliosides (Van

Heyningen, 1976). Good evidence was presented that a particular region on the heavy chain of tetanus toxin contains the site that binds to these gangliosides (Helting *et al.*, 1977). For comparison, the α-toxin of staphylococcus was found specifically to bind to a rarer, glucosamine-containing ganglioside, NeuAc(β2→3)Gal(β1→4)GlcNac(β1→3)Gal(β1→4)Glc(β1→1)Cer, in the surface of rabbit erythrocytes. It remains for future work to determine whether the proposed analogies between the binding of bacterial toxins to gangliosides in membranes and that of certain glycoprotein hormones e.g., TSH and human chorionic gonadotropin (hCG) are forced, and whether some important normal function for the binding of adenylate-cyclase-activating hormones may be ascribed to the high concentration of gangliosides in nervous system cells. It may not be simply fortuitous that extracellular G_{D1b} inhibits TSH-binding to cells. The monovalent lectin-like activity displayed by TSH, and possibly hCG, and their binding to glycoproteins (Bellesino and Bahl, 1975; Tate *et al.*, 1975) in the plasma membrane suggests that if G_{D1b} in the plasma membrane acts as a receptor, it may do so after capture of the hormone molecule by glycoprotein more superficially placed in the membrane. It is not inconceivable that subsequent interaction with the deeply located gangliosides determines the penetrability of the membrane by such adenylate-cyclase-activating messenger molecules. In most mammalian cells, sialoglycosphingolipids comprise less than one third of the sialo compounds, the greater proportion being sialoglycoproteins. In synaptic membranes of the rat, however, this situation is reversed; roughly two thirds of the sialo compounds are comprised by sialoglycosphingolipids (Yohe and Rosenberg, 1977). Synaptic membranes thus may present a high potential for perturbability and penetration on interaction of the highly concentrated higher gangliosides with molecules that have an affinity for their sialosyloligosaccharidyl components.

References

Arce, A., Maccioni, H., and Caputto, R., 1966, Enzymic binding of sialyl groups to ganglioside derivatives by preparations from the brain of young rats, *Arch. Biochem. Biophys.* **116**:52.

Arce, A., Maccione, H., and Caputto, R., 1971, The biosynthesis of gangliosides, *Biochem. J.* **121**:483.

Avrova, N., 1971, Brain ganglioside patterns of vertebrates, *J. Neurochem.* **18**:667.

Basu, S., Kaufman, B., and Roseman, S., 1968, Enzymatic synthesis of ceramide lactose by glycosyltransferases from embryonic chicken brain, *J. Biol. Chem.* **243**:5802.

Bellisino, R., and Bahl, O., 1975, Human chorionic gonadotropin. V. Tissue specificity of binding and partial characterization of soluble human chorionic gonadotropin–receptor complexes, *J. Biol. Chem.* **250**:3837.

Bosmann, H., 1972, Synthesis of glycoproteins in brain: Identification, purification and properties of glyosyl transferases from purified synaptosomes of guinea pig cerebral cortex, *J. Neurochem.* **19**:763.

Breckenridge, W., Gombos, G., and Morgan, I., 1972, The lipid composition of adult rat brain synaptosomal plasma membranes, *Biochem. Biophys. Res. Commun.* **266:**695.

Brunngraber, E., Whitting, L., Haberlund, C., and Brown, D., 1972, Glycoproteins in Tay–Sachs disease: Isolation and carbohydrate composition of glycopeptides, *Brain Res.* **38:**151.

Burton, R., 1974, Glycolipid metabolism, in : *Fundamentals of Lipid Chemistry* (R. M. Burton and F.C. Guerra, eds.), p. 392, BI-Science Publications Division, Webster Groves, Maryland.

Caputto, R., Maccioni, H., and Arce, A., 1974, Biosynthesis of brain gangliosides, *Mol. Cell. Biochem.* **4:**97.

Cumar, F., Fishman, P., and Brady, R., 1971, Analogous reactions for the biosynthesis of monosialo- and disialo-gangliosides in brain, *J. Biol. Chem.* **246:**5075.

Curtino, J., and Caputto, R., 1974, Enzymic synthesis of cerebroside from glycosylsphingosine and stearoyl-CoA by an embryonic chicken brain preparation, *Biochem. Biophys. Res. Commun.* **56:**142.

Den, H., Schultz, A., Basu, S., and Roseman, S., 1971, Glycosyltransferase activities in normal and polyma-transformed BHK cells, *J. Biol. Chem.* **246:**2721.

Den, H., Kaufman, B., and Roseman, S., 1975, The sialic acids. XVIII. Subcellular distribution of several glycosyltransferases in embryonic chicken brain, *J. Biol. Chem.* **250:**739.

Derry, D., and Wolfe, L., 1967, Gangliosides in isolated neurons and glial cells, *Science* **158:**1450.

DiCesare, J., and Dain, J., 1971, The enzymatic synthesis of gangliosides. IV. UDP-*N*-acetylgalactosamine: (*N*-acetyl-neuraminyl)-galactosylglucosyl ceramide-*N*-acetylgalactosaminyl-transferase in rat brain, *Biochim. Biophys. Acta* **231:**385.

DiCesare, J., and Dain, J., 1972, Localization, solubilization and properties of *N*-acetyl galactosaminyl and galactosyl ganglioside transferases in rat brain, *J. Neurochem.* **19:**403.

Donta, S., and Viner, J., 1975, Inhibition of the steroidogenic effects of cholera and heat-labile *Escherichia coli* enterotoxin by G_{M1} ganglioside: Evidence for a similar receptor site for the two toxins, *Infect. Immun.* **11:**982.

Dreyfus, H., Urban, P., Edel-Harth, S., and Mandel, P., 1975, The gamma-aminobutyric acid (GABA) system in brain during acute and chronic ethanol intoxication, *J. Neurochem.* **25:**45.

Fishman, P., 1974, Normal and abnormal biosynthesis of gangliosides, *Chem. Phys. Lipids* **13:**305.

Fong, J., Ledeen, R., Kundu, S., and Brostoff, S., 1976, Gangliosides of peripheral nerve myelin, *J. Neurochem.* **26:**157.

Forman, D., and Ledeen, R., 1972, Axonal transport of gangliosides in the goldfish optic nerve, *Science* **177:**1630.

Friedman, R., and Kohn, L., 1976, Cholera toxin inhibits interferon action, *Biochem. Biophys. Res. Commun.* **70:**1078.

Geel, S., and Gonzales, L., 1977, Cerebral cortical ganglioside and glycoprotein metabolism in immature hypothyroidism, *Brain Res.* **128:**515.

Gill, D., 1975, Involvement of nicotinamide adenine dinucleotide in the action of cholera toxin *in vitro, Proc. Natl. Acad. Sci. U.S.A.* **72:**2064.

Gill, D., and King, C., 1975, The mechanism of action of cholera toxin in pigeon erythrocyte lysates, *J. Biol. Chem.* **250:**6424.

Handa, S., and Burton, R., 1969, Biosynthesis of glycolipids: Incorporation of *N*-acetyl galactosamine by a rat brain particulate preparation, *Lipids* **4:**589.

Haywood, A., 1974a, Characteristics of Sendai virus receptors in a model membrane, *J. Mol. Biol.* **83:**427.

Haywood, A., 1974b, Fusion of Sendai viruses with model membranes (letter to the editor), *J. Mol. Biol.* **87:**625.

Helting, T., Zwisler, O., and Wiegandt, H., 1977, Structure of tetanus toxin. II. Toxin binding to ganglioside, *J. Biol. Chem.* **252:**194.

Hildebrand, J., Stoffyn, P., and Hauser, G., 1970, Biosynthesis of lactosyl ceramide by rat brain preparations and comparison with formation of ganglioside G_{M1} and psychosine during development, *J. Neurochem.* **17**:403.

Hill, M., and Lester, R., 1972, Mixtures of gangliosides and phosphatidylcholine in aqueous dispersions, *Biochim. Biophys. Acta* **282**:18.

Holm, M., 1972, Gangliosides in the optic pathway: Biosynthesis and biodegradation studied *in vivo, J. Neurochem.* **19**:673.

Holm, M., and Svennerholm, R., 1972, Synthesis and biodegradation of rat brain gangliosides *in vivo, J. Neurochem.* **19**:609.

Holmgren, J., Lonnroth, I., Månsson, J.-E., and Svennerholm, L., 1975, Interaction of cholera toxin and membrane G_{M1} ganglioside of small intestine, *Proc. Natl. Acad. Sci. U.S.A.* **72**:2520.

Jones, J., Ramsey, R., Aexel, R., and Nicholas, H., 1972, Lipid biosynthesis in neuron-enriched fractions of rat brain: Ganglioside biosynthesis, *Life Sci.* **11**:309.

Kandel, J., Collier, R., and Chung, D., 1974, Interaction of fragment A from diphtheria toxin with nicotinamide adenine dinucleotide, *J. Biol. Chem.* **249**:2088.

Kaufman, B., Basu, S., and Roseman, S., 1967, Studies on the biosynthesis of gangliosides, in: *Inborn Disorders of Sphingolipid Metabolism*, Proceedings of the 3rd International Symposium on the Cerebral Sphingolipidoses (S.M. Aronson and B.W. Volk, eds.), p. 193, Pergamon Press, New York.

Keenan, T., 1974, Composition and synthesis of gangliosides in mammary gland and milk of the bovine, *Biochin.. Biophys. Acta* **255**:70.

Keenan, T., Morré, D., and Basu, S., 1974, Ganglioside biosynthesis: Concentration of glycosphingolipid glycosyltransferases in Golgi apparatus from rat liver, *J. Biol. Chem.* **249**:310.

Kemp, R., 1970, The effect of neuraminidase (3:2:1:18) on the aggregation of cells dissociated from embryonic chick muscle tissue, *J. Cell Sci.* **6**:751.

Kuhn, R., and Wiegandt, H., 1964, Weitere Ganglioside aus Menschenhirn, *Z. Naturforsch.* **19**:256.

Ledeen, R., Yu, R., and Eng, L., 1973, Gangliosides of human myelin: Sialosylgalactosylceramide (G_7) as a major component, *J. Neurochem.* **21**:829.

Lowden, J., and Wolfe, L., 1964, Studies on brain gangliosides. 3. Evidence for the location of gangliosides specifically in neurons, *Can. J. Biochem.* **42**:1587.

Maccioni, A., Arce, A., and Caputto, R., 1971, The biosynthesis of gangliosides: Labelling of rat brain gangliosides *in vivo*, *Biochem. J.* **125**:1131.

Maccioni, H., Arce, A., Landa, C., and Caputto, R., 1974, Rat brain microsomal gangliosides: Accessibility to a neuraminidase preparation and the possible existence of different pools in relation to their biosynthesis, *Biochem. J.* **138**:291.

Maccioni, H., Landa, C., Arce, A., and Caputto, R., 1977, The biosynthesis of brain gangliosides–Evidence for a "transient pool" and an "end product pool" of gangliosides, *Adv. Exp. Biol. Med.* **83**:267.

Max, S.R., Maclaren, N., Brady, R., Bradley, R., Rennels, M., Tanaka, J., Garcia, J., and Cornblath, M., 1974, GM_3 (hematoside) sphingolipodystrophy, *N. Engl. J. Med.* **291**:929.

Mestrallet, M., Cumar, F., and Caputto, R., 1974, On the pathway of biosynthesis of trisialogangliosides, *Biochem. Biophys. Res. Commun.* **59**:1.

Moss, J., Osborne, J., Fishman, P., Brewer, H., Jr., Vaughan, M., and Grady, R., 1977, Effect of gangliosides and substrate analogues on the hydrolysis of nicotinamide adenine dinucleotide by choleragen, *Proc. Natl. Sci. U.S.A.* **74**:74.

Mullin, B., Fishman, P., Lee, G., Aloj, S., Ledley, F., Winand, R., Kohn, L., and Brady, R., 1976, Thyrotropin–ganglioside interactions and their relationship to the structure and function of thyrotropin receptors, *Proc. Natl. Acad. Sci. U.S.A.* **72**:842.

Ng, S., and Dain, J., 1977, Sialyltransferases in rat brain: Ultracellular localization and some membrane properties, *J. Neurochem.* **29**:1085.

Norton, W., and Poduslo, S., 1970, Neuronal soma and whole neuroglia of rat brain: A new isolation technique, *Science* **167:**1144.

Okada, S., and O'Brien, J., 1968, Generalized gangliosidosis: Beta-galactosidase deficiency, *Science* **160:**1002.

Radin, N., Brenkert, A., Arora, R., Sellinger, O., and Flangas, A., 1972, Glial and neuronal localization of cerebroside-metabolizing enzymes, *Brain Res.* **39:**163.

Roseman, S., 1970, The synthesis of complex carbohydrates by multiglycosyltransferase systems and their potential function in intercellular adhesion, *Chem. Phys. Lipids* **5:**270.

Rosenberg, A., and Stern, N., 1966, Changes in sphingosine and fatty acid components of the gangliosides in developing rat and human brain, *J. Lipid Res.* **7:**122.

Rosner, H., Wiegandt, H., and Ratmann, H., 1973, Sialic acid incorporation into gangliosides and glycoproteins of the fish brain, *J. Neurochem.* **21:**655.

Sattler, J., Wiegandt, H., Stark, J., Kranz, T., Ronneberger, H.-J., Schmidtberger, R., and Zilg, H., 1975, Studies of the subunit structure of choleragen, *Eur. J. Biochem.* **57:**309.

Sattler, J., Schwarzman, G., Staerk, J., Ziegler, W., and Wiegandt, H., 1977, Studies of the ligand binding to cholera toxin. II. The hydrophilic moiety of sialoglycolipids, *Z. Physiol. Chem.* **358:**159.

Schengrund, C.-L., and Nelson, J., 1976, Ganglioside sialidase activity in bovine neuronal perikarya, *Neurochem. Res.* **1:**171.

Schengrund, C.-L., and Rosenberg, A., 1970, Intracellular location and properties of bovine brain sialidase, *J. Biol. Chem.* **245:**6196.

Schengrund, C.-L., Jensen, D., and Rosenberg, A., 1971, Localization of sialidase in the plasma membrane of rat liver cells, *J. Biol. Chem.* **247:**2742.

Shah, S., 1971, Glycosyl transferases of microsomal fractions from brain: Synthesis of glucosyl ceramide and galactosyl ceramide during development and the distribution of glucose and galactose transferase in white and gray matter, *J. Neurochem.* **18:**395.

Stoffel, W., 1970, Studies on the biosynthesis and biodegradation of sphingosine bases, *Chem. Phys. Lipids* **5:**139.

Suzuki, K., 1967, Formation and turnover of the major brain gangliosides during development, *J. Neurochem.* **14:**917.

Suzuki, K., 1976, Catabolism of sialyl compounds in nature, in: *Biological Roles of Sialic Acid* (A. Rosenberg and C.-L. Schengrund, eds.), p. 159, Plenum Press, New York.

Tallman, J., and Brady, R., 1976, Disorders of ganglioside catabolism, in: *Biological Roles of Sialic Acid* (A. Rosenberg and C.-L. Schengrund, eds.), p. 183, Plenum Press, New York.

Tate, R., Holmes, J., Kohn, L., and Winand, R., 1975, Characteristics of a solubilized thyrotropin receptor from bovine thyroid plasma membranes, *J. Biol. Chem.* **250:**6527.

Van den Eijnden, D., 1973, The subcellular location of cytidine 5′-mono-phospho-*N*-acetyl neuraminic acid synthetase in calf brain, *J. Neurochem.* **21:**949.

Van den Eijnden, D., and Van Dijk, W., 1974, Properties and regional distribution of cerebral CMP-*N*-acetylneuraminic acid: Glycoprotein sialyltransferase, *Biochim. Biophys. Acta* **362:**136.

Van Heyningen, 1976, Binding of ganglioside by the chains of tetanus toxin, *FEBS Lett.* **68:**5.

Vanier, M., Holm, M., Öhman, R., and Svennerholm, L., 1971, Developmental profiles of gangliosides in rat and human brain, *J. Neurochem.* **18:**581.

Vanier, M., Holm, M., Månsson, J., and Svennerholm, L., 1973, The distribution of lipids in the human nervous system. V. Gangliosides and allied neutral glycolipids of infant brain, *J. Neurochem.* **21:**1375.

Vengris, V., Reynold, F., Jr., Hollenberg, M., and Pitha, P., 1976, Interferon action: Role of membrane gangliosides, *Virology* **72:**486.

Wenger, D., and Wardell, S., 1973, Action of neuraminidase (E.C.3.2.1.18) from *Clostridium perfringens* on brain gangliosides in the presence of bile salts, *J. Neurochem.* **20:**607.

Wiegandt, H., 1967, The subcellular localization of gangliosides in the brain, *J. Neurochem.* **14:**671.

Wiegandt, H., 1968, Struktur und Funktion der Ganglioside, *Angew. Chem.* **80**:89.

Yip, M., and Dain, J., 1969, The enzymic synthesis of ganglioside. 1. Brain uridine diphosphate D-galactose: *N*-acetyl-galactosaminyl-galactosyl-glucosyl-ceramide galactosyl transferase, *Lipids* **4**:270.

Yohe, H., and Rosenberg, A., 1972, Interaction of triiodide anion with gangliosides in aqueous iodine, *Chem. Phys. Lipids* **9**:279.

Yohe, H., and Rosenberg, A., 1977, Action of intrinsic sialidase of rat brain synaptic membranes on membrane sialolipid and sialoprotein components, *in situ*, *J. Biol. Chem.* **252**:2412.

Yohe, H., Roark, D., and Rosenberg, A., 1976, C_{20}-sphingosine as a determining factor in the aggregation of gangliosides, *J. Biol. Chem.* **251**:7083.

Yu, R., and Yen, S., 1975, Gangliosides in developing mouse brain, *J. Neurochem.* **25**:229.

Yusuf, H., Merat, A., and Dickerson, J., 1977, Effect of development on the gangliosides of human brain, *J. Neurochem.* **28**:1299.

3

Structure and Distribution of Glycoproteins and Glycosaminoglycans

Renée K. Margolis and Richard U. Margolis

1. Introduction

The glycosaminoglycans (called mucopolysaccharides in the older nomenclature) are high-molecular-weight linear carbohydrate polymers that are generally composed of disaccharide repeating units of a uronic acid (D-glucuronic acid or L-iduronic acid) and a hexosamine (*N*-acetylglucosamine or *N*-acetylgalactosamine). In the case of the sulfated glycosaminoglycans, the polysaccharide chains are *O*-glycosidically linked at their reducing ends to a protein moiety through the characteristic -glucuronosyl-galactosyl-galactosyl-xylosyl-serine sequence.

The glycoproteins differ from the glycosaminoglycans in the following major respects: (1) they do not contain uronic acid; (2) they lack a serially repeating unit; (3) they contain a relatively low number of sugar residues in the heterosaccharide, which is often branched; and (4) they contain several sugars that are not characteristic components of the glycosaminoglycans (e.g., fucose, sialic acid, galactose, mannose). However, the glycosaminoglycans and glycoproteins share many features, and in certain cases such as keratan sulfate (which has a repeating unit consisting of galactose and *N*-acetylglucosamine, but no uronic acid), the distinction between these two classes of compounds is not well defined.

The early literature on the glycosaminoglycans and glycoproteins of nervous tissue was reviewed by R. U. Margolis (1969) and Brunngraber (1972). However, much information in this area has been obtained quite

Reneé K. Margolis • Department of Pharmacology, State University of New York, Downstate Medical Center, Brooklyn, New York. **Richard U. Margolis** • Department of Pharmacology, New York University School of Medicine, New York, New York.

recently, and this chapter will therefore emphasize new findings concerning the structure and distribution of nervous tissue glycoproteins and glycosaminoglycans. The reader is also referred to the comprehensive treatise on the glycoproteins edited by Gottschalk (1972), while other aspects of this subject are covered by Balazs (1970), Ginsburg (1972), R. U. Margolis and R. K. Margolis (1972*b*), Montreuil (1974), Hughes (1976), Morgan *et al.* (1977), Comper and Laurent (1978), and Gombos and Zanetta (1978).

2. Structure

2.1. Composition and Structure of Nervous Tissue Glycoproteins

2.1.1. Sugar Composition

Although there were several earlier reports concerning the nature of the carbohydrate–protein linkages, the *O*-glycosidically linked oligosaccharides, and the presence of sulfated galactose and glucosamine residues in brain glycoproteins (Margolis, R. K., and Margolis, R. U., 1970, 1973; Margolis, R. U., *et al.*, 1972), much of our information concerning their oligosaccharide structure has been obtained only within the last five or so years. Brunngraber and co-workers originally demonstrated the presence of the usual glycoprotein sugars (i.e., glucosamine, galactosamine, mannose, galactose, fucose, and sialic acid) in glycopeptides prepared by protease digestion of a lipid-free protein residue from brain, and later reported the average sugar composition of glycopeptide fractions obtained by gel filtration and column electrophoresis (Di Benedetta *et al.*, 1969; Brunngraber *et al.*, 1973). These procedures yield a number of glycopeptide fractions differing in molecular size and the molar ratios of their constituent sugars, but do not by themselves provide homogeneous glycopeptides or oligosaccharides suitable for structural studies.

The sialic acid in rat and rabbit brain glycoproteins was shown by gas chromatography to consist exclusively of *N*-acetylneuraminic acid (R. K. Yu, unpublished results; Margolis, R. U., *et al.*, 1972). There have also been occasional reports of the presence in brain glycoproteins of significant amounts of other sugars such as rhamnose (Brunngraber and Brown, 1964), xylose (Bogoch, 1968), and mannosamine (Brunngraber *et al.*, 1970; Balsamo and Lilien, 1975), which are not established constituents of animal glycoproteins. These sugars were usually identified only on the basis of paper or thin-layer chromatography in a single solvent system, and xylose could have originated from contaminating glycosaminoglycans (see Section 2.2.1). In a search for mannosamine using a highly sensitive and specific ion-exchange chromatographic system that is capable of separating glucosamine, galactosamine, and mannosamine, we could not detect this amino sugar in hydrolysates of rat brain glycopeptides, although traces (1.5% of the total hexosamine) were found in glycopeptides prepared from rabbit brain glycoproteins (Margolis, R. U., *et al.*, 1972). In view of the sparse data supporting the reports cited above, we are inclined to discount the possibility that these "unusual" sugars are present in brain glycoproteins.

The situation concerning glucose is somewhat more complicated. There have been numerous reports of the presence of small amounts of glucose in brain glycopeptide preparations, although this sugar is not an established constituent of any animal glycoprotein oligosaccharide except for the glucosyl-galactosyl-hydroxylysine sequence that occurs in collagen, basement membranes, and the C1q protein of complement. However, it was recently found by two laboratories that in early postnatal rat brain, there are relatively large amounts of a high-molecular-weight glucose polymer, which decreases in concentration to approximately 3% of the total glycoprotein carbohydrate in adult brain, and can be separated from the glycopeptides by gel filtration on Sephadex G-50 (Krusius *et al.*, 1974; Margolis, R. K., *et al.*, 1975, 1976*b*). This material, which is almost completely depolymerized to monomeric glucose by digestion with amyloglucosidase from *Aspergillus niger*, appears to be a metabolically stable form of glycogen or limit dextrin (Margolis, R. K., *et al.*, 1976*b*), and is a likely source for the small amounts of glucose found in brain glycopeptides and glycoproteins.

Biosynthetic studies have demonstrated the incorporation of glucose into both a lipid-linked oligosaccharide (presumably a dolichol phosphate derivative) and a protein-bound form in brain (see Chapter 4). The synthesis of similar dolichol-linked glucose-containing oligosaccharides had previously been reported in other tissues such as thyroid. The relevance of these data to questions of glycoprotein composition and structure is difficult to evaluate in the absence of conclusive evidence that significant amounts of glucose occur in animal glycoproteins (with the specific exceptions mentioned above). However, at present, it appears most likely that while a small number of highly labeled glucose residues may become incorporated into a protein-bound form, these glucose (and probably also some mannose) units serve mainly as a regulatory signal for further "processing" of the still-incomplete oligosaccharide chains, and that they are subsequently removed by specific endo- or exoglycosidases.

Glucose is reported to be the major sugar in alkaline phosphatase from sheep brain, and also to be present in several other brain glycosidases and sulfatases (for references, see Dorai and Bachhawat, 1977). However, since the purification of these enzymes involved several steps of ion-exchange chromatography and gel filtration on cellulose and dextran-containing media, it appears quite possible that the glucose might be derived from these sources. The large amounts of glucose reported in the brain glycoprotein named GP-350 are probably also due to the presence of one or more non-glycoprotein components, since the procedure for preparing GP-350 was recently shown to result in a very heterogenous product (see Section 3.1.2).

Finally, with respect to glucose, very high concentrations of this sugar are usually found in glycopeptides prepared from subcellular fractions of brain and other tissues when the fractionation procedure involves the use of sucrose density gradients. We believe that this glucose (the concentration of which is often proportional to the molarity of sucrose from which the fraction was obtained) is probably derived from a dextrin-type impurity present in sucrose.

2.1.2. Carbohydrate–Protein Linkages and O-Glycosidically Linked Oligosaccharides

Approximately 85–90% of the carbohydrate in brain glycoproteins is linked through *N*-acetylglucosamine at the reducing ends of the oligosaccharide chains to the amide nitrogen of asparagine residues in the protein moiety (Margolis, R. U., *et al.*, 1972). This type of carbohydrate-protein linkage is characteristic of many soluble and membrane glycoproteins present in a wide variety of animal and plant tissues (Kornfeld and Kornfeld, 1976). The glycosylasparagine linkage fragment [2-acetamido-1-(L-β-aspartamido)-1,2-dideoxy-β-D-glucose] derived from these *N*-glycosidically linked oligosaccharides was also identified in partial acid hydrolysates of glycopeptides prepared from rat and rabbit brain glycoproteins (Margolis, R. U., *et al.*, 1972).

In addition to the predominant type of *N*-glycosidic carbohydrate–protein linkages, *N*-acetylgalactosamine at the reducing end of other oligosaccharide chains is linked *O*-glycosidically to the hydroxyl groups of serine and threonine residues. The presence of this type of linkage, which in other tissues is most commonly found in mucins and in certain erythrocyte glycoproteins, was demonstrated by the finding that alkaline-borohydride treatment of glycopeptides prepared from brain glycoproteins leads to the destruction of a portion of the serine, threonine, and galactosamine present, and the appearance in acid hydrolysates of alanine, α-aminobutyric acid, and galactosaminitol* (Margolis, R. U., *et al.*, 1972). Smith degradation studies of di- and trisaccharides isolated after alkali and alkaline-borohydride treatment of brain glycopeptides demonstrated that most of the *O*-glycosidically linked oligosaccharides contained the sequence galactosyl-(1→3)[*N*-acetylneuraminyl-(2→6)]-*N*-acetylgalactosamine (in which the galactosamine was originally linked to serine and threonine residues in the protein moiety), and

*Alkali treatment of glycoproteins containing *O*-glycosidic linkages to 3-hydroxy amino acids results in a base-catalyzed β-elimination of the glycosyl residue and the conversion of serine and threonine to α-aminoacrylic and α-aminocrotonic acid, respectively. When alkali treatment is carried out in the presence of a reducing agent such as sodium borohydride, varying amounts (depending on the specific conditions) of the unsaturated amino acids are reduced to alanine and α-aminobutyric acid, and the liberated *O*-glycosyl group is converted to the corresponding alcohol.

Table I. *O*-Glycosidically Linked Oligosaccharides of Rat Brain Glycoproteins

Oligosaccharide	Percentage
α-Galactosyl-(1→3)-*N*-acetylgalactosamine	13
β-Galactosyl-(1→3)-*N*-acetylgalactosamine	16
β-Galactosyl-(1→3)-[*N*-acetylneuraminyl-(2→6)]-*N*-acetylgalactosamine	9
N-Acetylneuraminyl-(2→3)-β-galactosyl-(1→3)-*N*-acetylgalactosamine	12
N-Acetylneuraminyl-(2→3)-β-galactosyl-(1→3)-[*N*-acetylneuraminyl-(2→6)]-*N*-acetylgalactosamine	50
	100

that the galactose appeared to be di-substituted with sialic acid (Margolis, R. K., and Margolis, R. U., 1973). Later, more detailed studies employing methylation analysis, chromium trioxide oxidation, and gas–liquid chromatography (GLC)–mass spectrometry demonstrated the presence of five different *O*-glycosidically linked oligosaccharides (Finne, 1975; Finne and Rauvala, 1977), the structures and relative proportions of which are given in Table I. These *O*-glycosidically linked oligosaccharides comprise approximately 10–15% of the carbohydrate in brain glycoproteins (Margolis, R. K., and Margolis, R. U., 1973; Finne and Krusius, 1976), and it is of interest that the previously unreported α-galactosyl-(1→3)-*N*-acetylgalactosamine sequence (which appears to exist only in a nonsialylated form) may be a brain-specific disaccharide, since it was detected in rat, rabbit, and chicken brain, but not in five other rat tissues examined (Finne and Krusius, 1976).

The concentration of *O*-glycosidically linked glycoprotein oligosaccharides in the purified chondroitin sulfate proteoglycan of brain (see Section 2.2.2) is much greater than in the lipid-free residue of whole rat brain. In addition to α- and β-galactosyl(1→3)*N*-acetylgalactosaminitol and their mono- and disialyl derivatives (see Table I), novel di- and tetrasaccharides containing mannitol at their proximal ends were also obtained by mild alkaline borohydride treatment of the proteoglycan glycopeptides. These were identified as *N*-acetylglucosaminyl(β1→3)mannitol and Gal(β1→4)[Fuc(α1→3)]GlcNAc(β1→3)Manol (Finne *et al.*, 1979). Sialylated and possibly sulfated derivatives of the GlcNAc(β1→3)Manol "core" disaccharide sequence, as well as significant amounts of free mannitol, were also detected. We consider it likely that these oligosaccharides reflect the presence in brain glycoproteins of a novel type of carbohydrate–protein linkage involving *O*-glycosidically linked mannose residues.

2.1.3. *Structural Features of Asparagine-Linked Oligosaccharides*

A general method for the fractionation of glycopeptides from brain and other tissues by gradient elution from a concanavalin A affinity column has recently been described (Krusius and Finne, 1977). Although previous studies of brain glycoproteins and glycopeptides had employed affinity chromatography using this lectin, the resulting fractions were not as well defined or as suitable for subsequent structural studies. By the use of their method, Krusius and Finne were able to isolate four structurally distinct types of glycopeptides or oligosaccharides, consisting of *O*-glycosidically linked oligosaccharides, two fractions (A and B) of acidic-type *N*-glycosidic glycopepides, and *N*-glycosidic glycopeptides of neutral type (Fraction C). (For reviews of glycoprotein structure and terminology, see Montreuil, 1975; Kornfeld and Kornfeld, 1976.)

The *O*-glycosidically linked oligosaccharides of brain have been discussed above. Although the complete structures of the *N*-glycosidically linked oligosaccharides are not known, there is considerable information concerning their sugar sequences and substitution patterns. Thus, the acidic type of carbohydrate unit contains a core composed of mannose and *N*-acetylglucosamine to which varying numbers of sialosylgalactosyl-*N*-acetylglucosamine branches are attached. Fraction A glycopeptides obtained by affinity chromatography probably contain an average of 3 to 4, and Fraction B glycopeptides 2, peripheral branches, while the neutral-type glycopeptides present in Fraction C contain only mannose and *N*-acetylglucusamine (Krusius and Finne, 1977). It should, however, be emphasized in this connection that the acidic-*type* glycopeptides of Fractions A and B may not contain sialic acid residues or other acidic groups (e.g., sulfate), and are therefore not necessarily acidic in the sense that they are retained by anion-exchange columns.

2.1.3a. Disialosyl Groups. Rauvala and Kärkkäinen (1977) developed a mass fragmentographic method for the analysis of the permethylated methyl glycoside methyl esters of *N*-acetyl- and *N*-glycolylneuraminic acids and of *N*-acetylneuraminic acid 8-acetate. Fragmentation of the neuraminic acid derivatives in electron-impact mass spectrometry was studied by deuterium labeling. Using this method, Finne *et al.* (1977) were able to demonstrate that the previously held view that disialosyl [α-*N*-acetylneuraminyl-(2→8)-*N*-acetylneuraminyl] groups are found only in glycolipids is incorrect, since this sequence is also present in glycoproteins from brain and other tissues. Although most of the sialic acid in brain glycoproteins is present in the usual terminal nonreducing position on the oligosaccharide chains, 8.5% of the total glycoprotein sialic acid in adult rat brain occurs in the form of disialosyl groups (measured as 8-*O*-substituted neuraminic acid). This proportion is severalfold greater than that found in other tissues such as muscle, intestinal mucosa, and liver (2.6–3.1%), while glycoproteins from kidney, gastric mucosa, erythrocytes, and plasma have no disialosyl groups. (The proportion of 8-*O*-substituted sialic acid in brain gangliosides is approximately 17%.) Soluble glycoproteins of brain contain a considerably lower proportion of disialosyl groups (1.6%) than those present in the particulate fraction ($\approx$9.5%), while much higher percentages are found in certain microsomal subfractions (15–16%) and in plasma membranes from 8-day-old rat brain (19.8%) (Finne *et al.*, 1977; Krusius *et al.*, 1978).

2.1.3b. Substitution Patterns of Sugars in Brain Glycoprotein Oligosaccharides. By the application of methylation analysis and mass fragmentography, Krusius and Finne have elucidated a number of structural features of brain glycoproteins, which were also compared with those of liver and kidney (Krusius and Finne, 1977). Although the data obtained from this type of study give mostly a "statistical" description of the substitution patterns of the constituent sugars in glycoprotein oligosaccharides rather than actual sugar sequences, it is possible to draw several general conclusions concerning the structure of brain glycoproteins.

In the neutral-type glycopeptides isolated by affinity chromatography on Con A–Sepharose (Fraction C, as described in Section 2.1.3 above), the branch points are composed of 3,6-di-*O*-substituted mannose residues, with peripheral mannose residues linked to each other by 1→2 and 1→6 linkages. In the more complex acidic type oligosaccharides of Fractions A and B, containing two to four peripheral branches, there are also significant amounts of 2,4- and 2,6-di-*O*-substituted mannose residues, as well as a 3,4,6-tri-*O*-substituted mannose.

Most of the galactose in brain glycoproteins is either 3-*O*-substituted or terminal, together with much smaller amounts of 6-*O*-substituted and traces of 3,6-di-*O*-substituted galactose. More than 85% of the galactose substituents at C-3 and C-6 are removed by neuraminidase treatment, indicating that galactose is largely substituted by sialic acid residues. The presence of significant amounts of nonsubstituted galactose and glucosamine (see below) indicates that the peripheral branches of brain glycoproteins are often "incomplete," and that an appreciable proportion have galactose or glucosamine as terminal sugars, rather than fucose or sialic acid as is more common in the plasma glycoproteins or in those of liver and kidney (Krusius and Finne, 1977).

Chromium trioxide oxidation of brain glycopeptide fractions destroyed all the galactose and *N*-acetylglucosamine, indicating that these sugars have a β-configuration, while 50% of the mannose and 80% of the fucose were resistant to oxidation (Krusius and Finne, 1978). These latter results are consistent with the known presence of α-linked mannose residues in the "core" region of other glycoprotein oligosaccharides, and with the usual α-configuration of fucosyl linkages.

In all three tissues studied, the major methylation product of *N*-acetylglucosamine was derived from 4-*O*-substituted glucosamine. In addition, brain glycoproteins contain smaller but significant amounts of 3,4- and 4,6-di-*O*-substituted glucosamine, as well as terminal (unsubstituted) glucosamine residues. Neither of the di-*O*-substituted glucosamine residues was affected by neuraminidase treatment, whereas most of these residues had disappeared after removal of fucose by fucosidase treatment or mild acid hydrolysis. The methylation product derived from 4,6-di-*O*-substituted glucosamine is expected from the fucose attached at C-6 of the asparagine-linked glucosamine residue. Although fucose substitution in this position is known to occur in a number of glycoproteins, substitution at C-3 had not previously been reported.

In subsequent investigations by this laboratory, rat brain glycopeptides were fractionated by affinity chromatography on wheat germ agglutinin–Sepharose as well as Con A–Sepharose (Krusius and Finne, 1978). A glycopeptide fraction containing 65% of the total brain glycoprotein fucose was purified by this procedure. Structural studies involving neuraminidase treatment, partial acid hydrolysis, chromium trioxide oxidation, and uronic acid degradation established that approximately 80% of the fucose in these glycopeptides is present in an oligosaccharide with the structure NeuAc(α2→3)Gal(β1→4)[Fuc(α1→3)]GlcNAc(β1-. The structure of the desialylated trisaccharide sequence was also confirmed by GLC-mass spectrometry after nitrous acid deamination. Only 29% of the total fucose is linked to the C-6 of *N*-acetylglucosamine, and less than 1–2% to galactose (at C-2).

Glycoprotein oligosaccharides containing both fucose and sialic acid on the same peripheral branch have not previously been reported, although the sugar sequence described above was recently characterized in a ganglioside purified from human kidney (Rauvala, 1976). It also represents a sialosylated form of the so-called X antigen structure present in glycolipids found in high concentration in adenocarcinoma and epithelial glandular tissue, but below the level of detection (by immunological methods) in parenchymatous tissues such as liver, spleen, and kidney (Yang and Hakomori, 1971). In brain, the oligosaccharide occurs mainly in membrane glycoproteins, although small amounts are also present in the soluble fraction.

Tsay *et al.* (1976) reported on the isolation and characterization of a decasaccharide (Fig. 1) and a disaccharide from the brain and spleen of a patient with fucosidosis, a genetically determined inborn error of glycoprotein catabolism due to a deficiency of α-L-fucosidase (see Chapter 16). The disaccharide, fucosyl-α-D-(1→6)-*N*-acetylglucosamine, evidently originates from the *N*-acetylglucosaminylasparagine carbohydrate–protein linkages of certain fucosyl glycoproteins, of which the larger decasaccharide was probably also a part.

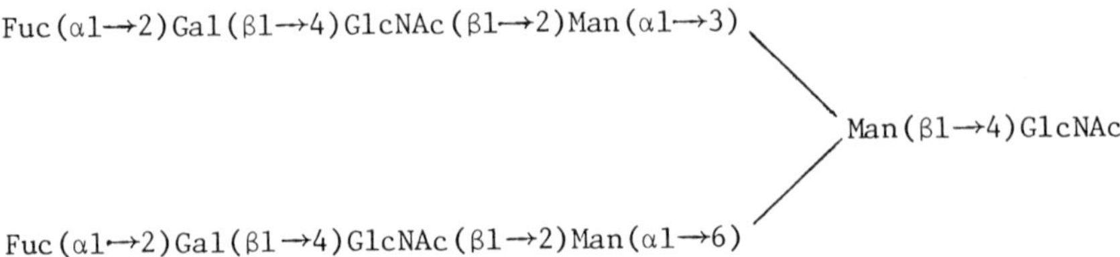

Fig. 1. Structure of an accumulating oligosaccharide from fucosidosis brain (Tsay *et al.*, 1976).

With the exception of the terminal fucose residues, the decasaccharide is very similar in structure to those present in the immunoglobulins, which usually have galactose or sialic acid at the nonreducing ends, raising the possibility that the isolated decasaccharide may not originate in brain but rather by transglycosylation of accumulated plasma glycoproteins. Such an explanation might appear especially likely since in rat brain glycoproteins almost all the fucose is linked to glucosamine residues as described above, and less than 1–2% to C-2 of galactose. However, while there may be major species differences in the oligosaccharide structure of rat and human brain glycoproteins, it appears more probable that the decasaccharide isolated by Tsay *et al.* (1976) originates from a very minor fraction of brain fucosyl glycoproteins. Such an explanation is supported by the finding that the deficiency in fucosidosis is chiefly of enzymes cleaving Fuc(α1→2)Gal and Fuc(α1→6) GlcNAc linkages, whereas there is a considerably higher activity of fucosidases acting on α1→3 and α1→4 linkages (Dawson and Tsay, 1977). This selective deficiency of fucosidases with limited specificities would account for the preservation (and release by a brain endoglucosaminidase) of an oligosaccharide that is present in only very small proportion in brain glycoproteins, but the further catabolism of which is prevented by the inability to remove terminal α1→2 fucose residues in fucosidosis.

The only other structural data on the (presumably asparagine-linked) oligosaccharides of brain glycoproteins are from the earlier report of Katzman (1972), who described the isolation of *N*-acetyllactosamine and galactosyl-β-D-(1→4)-*N*-acetyllactosamine from a partial acid hydrolysate of bovine brain glycopeptides. *N*-acetyllactosamine is a common structural component of glycoproteins from a number of sources, such as blood group substances, gastric mucin, and certain plasma glycoproteins, but the galactosyl-(1→4)-galactosyl linkage present in the trisaccharide had not previously been reported. It should also be noted that no evidence was found for the presence of 4-substituted galactose residues in rat brain glycoproteins (Krusius and Finne, 1977).

2.1.3c. Sulfate and Phosphate Groups in Brain Glycoproteins. The presence of sulfated sugars in the glycosaminoglycans and their presumed absence in the glycoproteins was earlier considered to be one of the major features distinguishing these two classes of complex carbohydrates. Although sulfated glycoproteins were later described in gastrointestinal mucosa and certain mucous secretions, most parenchymal glycoproteins including those of brain were not thought to contain sulfate. Ion-exchange chromatography of the desialylated glycopeptides from rat brain glycoproteins demonstrated that approximately 40% of the hexosamine and neutral sugar is in oligosaccharide chains containing sulfate (but *no* uronic acid), and that the sulfated and nonsulfated glycopeptides have similar molar ratios of hexosamine and neutral sugars (Margolis, R. K., and Margolis, R. U., 1970). The sulfated sugars in the glycoproteins were tentatively identified as galactose 6-sulfate and *N*-acetylglucosamine 6-sulfate, based on analysis of the sulfated monosaccharides released by partial acid hydrolysis of the glycopeptides. These conclusions were recently confirmed in studies on bovine brain glycoproteins (Allen *et al.*, 1976).

The sulfated glycoproteins from brain appear to constitute a distinct class of compounds that differ from any of the known glycosaminoglycans. They are similar in several respects to keratan sulfate; however, the sulfated glycoproteins from brain have a higher content of fucose and mannose and a lower content of sulfate than has been reported for keratan sulfate. Moreover, while in keratan sulfate the disaccharide repeating unit contains galactose and *N*-acetylglucosamine, with sulfate in varying amounts on both sugars, approximately one half the sulfate in the sulfated glycopeptides from brain could be isolated in the form of a *trisaccharide* containing galactose, glucosamine, and mannose. Since keratan sulfate contains only small amounts of mannose (presumably in the carbohydrate–protein linkage region), the sulfated glycopeptides of brain must be considered as being structurally distinct from keratan sulfate in the usual meaning of this term.

Sulfated glycoproteins are rather widely distributed in brain and do not appear to be uniquely associated with any particular cellular or subcellular fraction. In rat brain myelin, they are largely confined to the predominant high-molecular-weight glycoprotein, and in sciatic nerve, the major myelin protein is both glycosylated and sulfated (Matthieu *et al.*, 1975*a*,*b*). Sulfated glycoproteins are also present in synaptic plasma membranes (Simpson *et al.*, 1976).

It was recently reported that phosphoglycoproteins occur in brain, and nuclear magnetic resonance studies indicate that the phosphorus is present in a monoester linkage to a hexose (Davis *et al.*, 1976*a*,*b*). However, no information was provided concerning the concentration of phosphorus in the glycopeptides or its molar ratio to the oligosaccharide sugars, and although it was stated that there was no ribose or absorbance above 230 nm, the sample size and concentration used for these measurements were not given. This information would be very useful in evaluating these reports, since it was previously demonstrated that the content of phosphorus and ribose in bovine brain glycopeptide fractions both decrease to very low levels during successive purification steps (Katzman, 1972). Definitive evidence for the presence of phosphorus in brain glycoproteins will therefore have to await the isolation and characterization of phosphorylated oligosaccharides.

2.1.3d. Developmental Changes in Brain Glycoproteins. In distinction to the developmental changes of the glycosaminoglycans (Section 3.2.2), the level of glycoprotein carbohydrate increases continuously after birth, before reaching approximately adult values at the same time as the glycosaminoglycans [i.e., 30 days (see Appendix I)]. However, the increase in glucosamine and mannose, which usually occupy inner positions in the oligosaccharide chains of glycoproteins, is much greater than that of fucose and sialic acid, which are located mostly at terminal sites on the oligosaccharides. Double-label studies involving simultaneous analysis of the glycoproteins from young and adult brain, together with sugar analyses of the fractionated glycopeptides, indicated that the relatively greater increase in glucosamine and mannose (as compared to fucose and sialic acid) in brain glycoproteins during development is due to the preferential synthesis in older animals of a population of glycoproteins with oligosaccharide chains consisting largely of "core" sugars, and containing little fucose and sialic acid. The smaller "core"-type oligosaccharides were also found to have a somewhat more rapid turnover (Margolis, R. K., *et al.*, 1976*b*).

Other studies have demonstrated differences in the polyacrylamide gel electrophoretic behavior of proteins and glycoproteins prepared from brain at various stages of development (Dutton and Barondes, 1970; Quarles and Brady, 1971). However, these data reflect mainly the molecular size of entire glycoprotein molecules, rather than the structure or composition of their carbohydrate moieties.

2.2. Composition and Structure of Brain Glycosaminoglycans

2.2.1. General Structural Characteristics and Carbohydrate–Protein Linkages

The structures of the glycosaminoglycans (mucopolysaccharides) are summarized in Table II. The concentration of total glycosaminoglycans in adult rat brain (in terms of hexosamine) is 0.4 μmol/100 mg lipid-free dry weight, comprising 63% chondroitin sulfate, 25% hyaluronic acid, and 12% heparan sulfate (see Appendix I). Data on the concentration and composition of brain glycosaminoglycans in the cow, human, and other species can be found in reports by R. U. Margolis (1967), Singh and Bachhawat (1968), and Singh *et al.* (1969).

Approximately 90% of the chondroitin sulfate in whole adult rat and bovine brain is chondroitin 4-sulfate and the remainder mostly chondroitin 6-sulfate, together with small amounts of under- and oversulfated polysaccharide chains as revealed by chondroitinase treatment (Margolis, R. U., 1967; Saigo and Egami, 1970; Margolis, R. U., and Margolis, R. K., 1972a). However, white matter contains only 60% chondroitin 4-sulfate and 40% chondroitin 6-sulfate (Margolis, R. U., 1967).

Several laboratories have reported that only very small amounts of dermatan sulfate are present in brain (Saigo and Egami, 1970; Margolis, R. U., and Margolis, R. K., 1972*a*., Dietrich *et al.*, 1976; Toledo and Dietrich, 1977). Moreover, iduronic acid, which is present in both dermatan sulfate and heparan sulfate, comprises 5% or less of the total uronic acid in brain glycosaminoglycans (Margolis, R. U., and Margolis, R. K., 1972*b*). However, in recent studies using more sensitive radiochemical methods, we have obtained

Table II. Structures of the Glycosaminoglycans

Glycosaminoglycan[a]	Sugars present in disaccharide repeating unit, and linkage	Minor or atypical monosaccharide components[b]
Hyaluronic acid	D-Glucuronic acid ($\beta 1 \rightarrow 3$) *N*-acetyl-D-glucosamine ($\beta 1 \rightarrow 4$)	
Chondroitin 4-sulfate (chondroitin sulfate A)	D-Glucuronic acid ($\beta 1 \rightarrow 3$) *N*-acetyl-D-galactosamine 4-*O*-sulfate ($\beta 1 \rightarrow 4$)	
Chondroitin 6-sulfate (chondroitin sulfate C)	D-Glucuronic acid ($\beta 1 \rightarrow 3$) *N*-acetyl-D-galactosamine 6-*O*-sulfate ($\beta 1 \rightarrow 4$)	
Dermatan sulfate (chondroitin sulfate B)	L-Iduronic acid ($\alpha 1 \rightarrow 3$)[c] *N*-acetyl-D-galactosamine 4-*O*-sulfate ($\beta 1 \rightarrow 4$)	D-Glucuronic acid ($\beta 1 \rightarrow 3$) *N*-acetyl-D-glucosamine 6-*O*-sulfate
Keratan sulfate (keratosulfate)	D-Galactose and D-galactose 6-*O*-sulfate ($\beta 1 \rightarrow 3$) *N*-acetyl-D-glucosamine 6-*O*-sulfate ($\beta 1 \rightarrow 4$)	L-Fucose D-Mannose Sialic acid
Heparan sulfate (heparitin sulfate)	D-Glucuronic acid ($\alpha 1 \rightarrow 4$)[d] *N*-acetyl-D-glucosamine ($\alpha 1 \rightarrow 4$) *N*-sulfo-D-glucosamine and *N*-sulfo-D-glucosamine 6-*O*-sulfate ($\alpha 1 \rightarrow 4$)	L-Iduronic acid *N*-acetyl-D-glucosamine 6-*O*-sulfate
Heparin	L-Iduronic acid ($\alpha 1 \rightarrow 4$) *N*-sulfo-D-glucosamine and *N*-sulfo-D-glucosamine 6-*O*-sulfate ($\alpha 1 \rightarrow 4$)	D-Glucuronic acid D-Glucuronic acid 2-*O*-sulfate L-Iduronic acid 2-*O*-sulfate *N*-acetyl-D-glucosamine 6-*O*-sulfate

[a] Older nomenclature in parentheses.
[b] Chondroitin 4-sulfate, chondroitin 6-sulfate, dermatan sulfate, heparin, and heparan sulfate also contain D-galactose and D-xylose in the carbohydrate–peptide linkage (-glucuronosyl-galactosyl-galactosyl-xylosyl-*O*-serine).
[c] L-Iduronic acid is the 5-epimer of D-glucuronic acid, and therefore the iduronosyl group has α-configuration according to the convention of carbohydrate chemistry. Linkage is equatorial, carboxyl group axial.
[d] Linkage details of glucuronidic and glucosaminidic bonds of heparin and heparan sulfate are still incomplete, but all linkages are thought to be $\alpha 1 \rightarrow 4$.

evidence that as much as 13% of the chondroitin sulfate *radioactivity* in brain 18 hours after labeling the glycosaminoglycans with [^{35}S]sulfate may reside in dermatan sulfate (unpublished results). If the specific activity of dermatan sulfate after 18 hr labeling is the same as that of chondroitin-4-sulfate (the major component), it can be calculated that dermatan sulfate accounts for 8% of the total glycosaminoglycans of rat brain. Although Glimelius *et al.* (1978) recently reported a much larger proportion of labeled dermatan sulfate in the glycosaminoglycans of cultured normal glial cells, this finding may be attributable to the effects of the culture conditions or to the presence of connective tissue elements such as fibroblasts and endothelial cells.

Heparan sulfate has a more complex composition than the other glycosaminoglycans insofar as it contains *N*-acetylglucosamine, glucosamine-*N*-sulfate, glucosamine-*N*,*O*-disulfate, D-glucuronic acid, L-iduronic acid, and probably *N*-acetylglucosamine-*O*-sulfate residues. Moreover, there is known to be a considerable variation in the structures of heparan sulfates isolated from different sources (Linker and Hovingh, 1973). Heparan sulfate is distinguished from the closely related glycosaminoglycan heparin (a product of mast cells) mainly in having a lower degree of sulfation, a larger proportion of *N*-acetylated as compared to *N*-sulfated glucosamine residues, and in containing mainly glucuronic acid as its uronic acid rather than the 70–80% iduronic acid found in most preparations of heparin.

We have found that the heparan sulfate from brain can be separated into two fractions of similar molecular size and proportions of *N*- and *O*-sulfate, but which differ mainly in total sulfate content (Margolis and Atherton, 1972). Nitrous acid degradation studies indicated that in contrast to the molecular structure proposed for heparan sulfate from the few other tissues studied, such as lung, the heparan sulfate from brain contains a relatively smaller proportion of uronosyl-*N*-sulfohexosamine repeating units, and that much of the heparan sulfate of brain consists of large, completely *N*-acetylated segments containing a relatively high content of ester sulfate.

Traces of heparin probably also occur in brain as a component of mast cells which are known to be associated with blood vessels and meninges, but contrary to one or two early reports, there is no evidence that heparin is present in brain parenchyma.

Alkaline-borohydride treatment of chondroitin sulfate and heparan sulfate isolated from brain following pronase digestion of the lipid-free protein residue demonstrated that the polysaccharide chains in both of these sulfated glycosaminoglycans are covalently linked to protein by the usual type of *O*-glycosidic xylosyl-serine linkages (Margolis, R. U., *et al.*, 1972), and are therefore of the same type in brain as previously established for proteoglycans from other tissues (Lindahl and Rodén, 1972).

Small amounts of protein are usually found associated with hyaluronic acid purified by a variety of techniques, but it has not yet been possible to demonstrate a covalent linkage of hyaluronic acid to protein in brain or any other tissue. This point has therefore remained somewhat controversial. Part of the difficulty arises from the relatively high molecular weight of hyaluronic acid [140,000 in brain (Margolis, R. U., 1967), and up to 1,000,000 or more when isolated from other sources] and the correspondingly minor portion of the molecule which would constitute a possible carbohydrate–protein linkage region. Although there have been reports of the presence of small amounts (0.2% or less by weight) of a sugar resembling arabinose in brain hyaluronic acid, later and more detailed investigations using combined gas chromatography/mass spectrometry yielded conflicting results (Katzman, 1974; Varma *et al.*, 1977). Moreover, it has recently been reported that papain preparations are contaminated by arabinose-containing polysaccharides which copurify with hyaluronic acid in the usual procedures employed for the isolation of this glycosaminoglycan (Chandrasekaran *et al.*, 1978). Since papain was used by most of the investigators who reported the presence of arabinose, it will be important to exclude this possible source of contamination. The reported concentration of arabinose would represent approximately one residue per hyaluronic acid chain, so that if this sugar is actually present it could be involved in a postulated carbohydrate–protein linkage in a role similar to that of xylose in the sulfated glycosaminoglycans. It should be emphasized, however, that despite the extensive literature on this subject there is no generally accepted evidence either for the presence of arabinose in hyaluronic acid or for any type of covalent hyaluronate–protein linkage.

Vitello *et al.* (1978) have recently described age-related changes in the levels of a "keratan sulfate-like glycosaminoglycan" in rat brain. However, this material has a sulfate/glucosamine molar ratio of 6.6, a galactose/glucosamine ratio of 2.3, and a fucose/glucosamine ratio of 1.5, whereas well-characterized preparations of keratan sulfate (from either cartilage or cornea) have approximately equimolar amounts of glucosamine, galactose, and sulfate, and only traces of fucose. Therefore, the authors' conclusion that the fraction studied was "keratan sulfate-like" rested almost entirely on the fact that it was soluble in excess cetylpyridinuum chloride, which precipitates other glycosaminoglycans but not glycopeptides derived from the sulfated glycoproteins present in brain (Section 2.1.3c). Although the chemical composition reported by Vitello *et al.* (1978) for their "keratan sulfate-like" material does not resemble that of any known glycosaminoglycan or glycoprotein, it is likely that sulfated glycopeptides comprise a significant part of this fraction, since there are no reliable data supporting the occurrence of keratan sulfate in brain if one defines this glycosaminoglycan by the usually accepted criteria. Moreover, in mucopolysaccharidoses characterized by a genetically determined inability to degrade keratan sulfate and an accumulation of this glycosaminoglycan in visceral organs and tissues, it is significant that similar substances are not detected in the CNS, nor is there any other type of neurological involvement (see Chapter 16).

2.2.2. Brain Proteoglycans

As mentioned above, there is good evidence to indicate that the polysaccharide chains of the sulfated glycosaminoglycans of brain are covalently linked to a protein core through the conventional -glucuronosyl-galactosyl-galactosyl-xylosyl-serine sequence, and that they therefore occur as proteoglycans. Approximately 40% of the glycosaminoglycans in brain are soluble, and much of that present in the particulate portion in association with very-low-density microsomal membrane subfractions can be solubilized by various types of mild washing procedures (Kiang *et al.*, 1978). We have isolated and partially purified the soluble chondroitin sulfate proteoglycan from rat brain by ion-exchange chromatography and gel filtration, and described some of its properties (Margolis, R. U., *et al.*, 1976, 1978).

The brain proteoglycan is homogeneous in the analytical ultracentrifuge, where it shows a sharp peak with a sedimentation coefficient of 6.5 S, and elutes as a single peak of uniform composition on Sepharose (CL-4B in the presence of 4 M guanidine. It contains 56% protein, 24% glycosaminoglycans, and 20% glycoprotein carbohydrate. The last two components appear to be covalently linked to a common protein core since they cannot be separated on the basis of size, charge, or density. The glycosaminoglycan portion is composed of 93% chondroitin sulfate (of which 90% is chondroitin 4-sulfate and the remainder 6-sulfate), together with 4.5% hyaluronic acid and 2.4% heparan sulfate.

The chondroitin sulfate proteoglycan of brain, which is largely present as a cytoplasmic component of neurons and glia (Margolis, R. K., *et al.*, 1979), therefore differs considerably from the extracellular protein–polysaccharide complexes of cartilage. These exist in the form of large (90–95 S) aggregates containing hyaluronic acid and specific "link proteins." After reduction and alkylation, or in solvents such as 4 M guanidine, the cartilage proteoglycan aggregates dissociate into smaller subunits of approximately 22 S, which usually contain less than 10% protein. In contrast, reduction and alkylation of the brain proteoglycan promotes aggregation with other proteins to yield a larger-size complex containing approximately 70% protein but with no appreciable change in its glycosaminoglycan or glycoprotein composition.

Branford White and Hudson (1977) also described a procedure for the isolation of a chrondroitin sulfate proteoglycan by dissociative extraction (4 M guanidine) of a lipid-free protein residue of brain. Since yields and other relevant data were not included in their report, we tested this procedure and found that the fraction claimed to contain only chondroitin sulfate is in fact largely composed of hyaluronic acid, and that the purified product accounts for only 2.7% of the chondroitin sulfate present in brain (Margolis, R. K., *et al.*, 1978). It therefore does not appear from our experience that dissociative extraction of a brain protein residue under these conditions yields either a significant proportion or a representative sample of proteoglycans.

3. Distribution

3.1. Glycoproteins

3.1.1. Cellular and Subcellular Distribution

It can be stated in general terms that the glycoproteins are largely membrane constituents in brain, with only 10-15% present in the soluble fraction, whereas much of the total glycosaminoglycan content of brain is either soluble or easily extractable. The concentrations of glycoproteins and glycosaminoglycans in neuronal cell bodies, axons, astrocytes, and oligodendroglia isolated in bulk from bovine brain are summarized in Table III. Comparable data for their distribution and metabolism in rat brain have also been reported (Margolis, R. U., and Margolis, R. K., 1974; Margolis, R. K. *et al.*, 1975).

Studies on the localization and composition of glycoproteins in synaptic and myelin membranes are considered in detail in Chapters 8 and 11, and will therefore not be covered here. Similarly, the localization of glycoproteins and glycosaminoglycans by histo- and cytochemical techniques (Chapters 7, 9, and 10), and their presence in the axon (Chapters 12 and 13) and in secretory organelles (Chapter 15), are the subjects of individual reviews.

One direction of future studies on brain glycoproteins will undoubtedly be the attempt to identify specific structural features of these molecules with particular morphological entities in nervous tissue. Since glycoproteins are distinguished from other proteins by the presence of one or more oligosaccharide units, it is usually their carbohydrate moiety that has attracted the greatest interest. Some useful data concerning this aspect can be obtained from studies employing plant lectins with different carbohydrate specificities. However, while such investigations may reveal differences in lectin affinity

Table III. Distribution of Glycoproteins and Glycosaminoglycans in Bovine Neurons and Glia

	Neuronal perikarya	Axons	Astrocytes	Oligodendroglia
Glycoprotein carbohydrate (% of lipid-free dry wt.)	1.1	1.1	1.9	2.4
Total glycosaminoglycans (μmol hexosamine/100 mg lipid-free dry wt.)	0.830	0.508	0.565	0.144
Percentage of total glycosaminoglycans				
Hyaluronic acid	19	43	34	48
Chondroitin sulfate	64	41	48	24
Heparan sulfate	17	16	18	27

between glycoproteins localized at morphologically identifiable sites or at different stages of development, they are usually very limited in their ability to provide structural information concerning the oligosaccharides to which the lectins bind.

One of the ultimate aims of neurochemistry is of course to obtain knowledge of the structure of nervous tissue constituents at the molecular level. In the case of glycoproteins, this implies an understanding of the structures and roles of their oligosaccharides in the various cellular and subcellular components of nervous tissue. It will obviously be many years before such a goal is attained, and a full understanding of the functional roles of nervous tissue complex carbohydrates will require that this type of strictly structural data be integrated with parallel studies on their metabolism, developmental changes, and other characteristics.

Without a revolution in methodology that greatly reduces the amount of material required for fractionation and structural analysis, information concerning glycoprotein structure in particular cellular and subcellular entities necessitates our first obtaining more complete data on the properties of glycoproteins in nervous tissue as a whole. Only with this type of foundation will it be possible to identify structural features present at particular locations, such as at the cell surface, in subsynaptic specializations, or as components of various intracellular organelles. We recently applied this approach to the study of soluble glycoproteins derived from the cell bodies as compared to those in nerve endings, and to glycoproteins present in mitochondria, synaptic membranes, and five microsomal subfractions of brain (Krusius *et al.*, 1978). These investigations revealed significant differences in the distribution of a number of structural characteristics of glycoproteins, including the relative proportions of several oligosaccharides whose structures had previously been defined in studies of whole brain.

The recent availability of a number of highly specific endoglycosidases capable of removing oligosaccharides from glycoproteins has greatly expanded the possibilities for studies on glycoprotein (and glycosaminoglycan) structure and localization. Using these enzymes for the analysis of cell-surface components present on neurons isolated in bulk from brain (after the administration of radioactive sugar precursors), we have found it possible to obtain considerable structural data from relatively small amounts of cellular material, and anticipate that this approach can be extended to the analysis of changes in the structure and distribution of complex carbohydrates that occur in processes such as brain development.

3.1.2. Identified Glycoproteins of Nervous Tissue

Because of the difficulties associated with the isolation, purification, and characterization of insoluble membrane proteins, much of the work on glycoprotein structure has employed glycopeptides resulting from proteolytic digestion of a lipid-free residue of whole brain or subcellular fractions. Such glycopeptide preparations are usually adequate when one is interested primarily in the oligosaccharide moiety, and are also often more amenable to further fractionation and structural analysis. Although similar oligosaccharides may be present on a number of different glycoproteins, the great heterogeneity and complexity of brain in terms of its cell types and histological organization has hindered attempts to isolate and characterize specific intact glycoproteins, especially in view of their large numbers and the small amounts of any individual species. There are many enzymes present in brain and other tissues that are known to be glycoproteins. These include acetylcholinesterase, dopamine β-hydroxylase, ribonuclease, alkaline phosphatase, most of the glycosidases, and [Na^+–K^+]ATPase (for references and further discussion, see Chapters 6 and 7). However, the contribution of these enzymes to the total glycoprotein carbohydrate of nervous tissue is probably quite small.

Brain microtubule protein (tubulin) has been shown to contain small amounts of sugar (1.3% by weight), and certain features of its carbohydrate composition and structure have been described (Margolis, R. K., *et al.*, 1972). Although it is possible to prepare tubulin essentially free of carbohydrate (Eipper, 1972), and it is recognized that the asymmetric, dimeric tubulin molecules are associated with a number of other proteins, certain of which may be essential for its polymerization into microtubules, the carbohydrate detected would appear to be a component of tubulin itself, since it cannot be separated on the basis of either molecular size or net charge (Feit and Shelanski, 1975). Moreover, cytochemical studies using Con A demonstrated that this lectin binds to brain microtubules purified by repeated cycles of polymerization–depolymerization (Behnke, 1975; Hüttich *et al.*, 1977). Con A binds preferentially to the outer surface of the microtubule walls but is also partly present in the lumen, and it was suggested that the apparent surface coating of carbohydrate may be responsible for the peritubular "clear zone" observed morphologically (Behnke, 1975). Published data on the presence or absence of carbohydrate in purified tubulin may be only apparently contradictory if tubulin exists in both glycosylated and nonglycosylated forms, the latter of which was selected by the purification procedure employed by Eipper (1972). It is also possible that the nonglycosylated form may be concentrated at such sites as synaptic membranes or postsynaptic densities rather than in the soluble fraction of brain, and that glycosylation or deglycosylation is important for the insertion of tubulin into neural membranes.

Van Nieuw Amerongen and co-workers have published a number of reports on a brain glycoprotein designated GP-350 [because of its elution from a DEAE–cellulose column with 350 mM NaCl (Van Nieuw Amerongen *et al.*, 1972)]. It was isolated from bovine, rat, and human brain, but was reported to be absent from liver and kidney. As originally described, GP-350 accounts for approximately 3% of the total soluble protein and 20% of the soluble glycoprotein sialic acid in bovine cortical grey matter. It is highly acidic, with an isoelectric point of 2, and has a molecular weight of approximately 11,600 as estimated by sodium dodecyl sulfate–polyacrylamide gel electrophoresis (SDS–PAGE)(Van Nieuw Amerongen and Roukema, 1973). Purified GP-350 is reported to contain 17% of carbohydrate, including 4% (by weight) of glucose, which is the major sugar component but probably represents contamination with nonglycoprotein glucose-containing materials (see Section 2.1.1 and below).

A particulate form of GP-350 was also purified from the crude mitochondrial fraction of brain, and claimed to be identical in composition and properties to the soluble GP-350. In most brain regions, approximately two thirds of the GP-350 is present as a soluble glycoprotein, and subcellular distribution studies revealed that the particulate portion was detectable only in the synaptosomal fraction (Van Nieuw Amerongen and Roukema, 1974). On the basis of immunofluorescence studies (Van Nieuw Amerongen *et al.*, 1974), it was concluded that GP-350 is localized mainly, if not exclusively, in neuronal structures (i.e., in axons, and to a lesser extent in neuronal perikarya). No fluorescence was visible in glial cells or in gliomas. Fluorescence was also present in long parallel nerve fibers of human sciatic nerve, and in densely packed ring-shaped and oval structures of the pituitary and pineal gland.

However, recent studies showed that GP-350 is not a homogeneous glycoprotein but rather is a mixture of low-molecular-weight proteins and lipids whose relative proportions are determined by the particular isolation procedure employed (Simonian *et al.*, 1978*a*). Moreover, it does not appear to be specifically localized in nervous tissue, since immunologically and electrophoretically identical glycoproteins can be demonstrated in a number of other tissues such as liver, kidney, and thymus.

After reduction with mercaptoethanol and electrophoresis in 19% SDS–polyacrylamide gels, the GP-350 from brain could be resolved into nine protein bands with apparent molecular weights of 4000–15,000. Similar electrophoretic patterns were obtained from unreduced samples of brain GP-350, and, with the exception of a few bands, after electrophoresis of GP-350 prepared from liver. GP-350 isolated from brain and other tissues can also be fractionated into several components by gel filtration on Sephadex G-50.

Both brain and liver GP-350 were found to be immunologically heterogeneous, containing inert as well as five immunologically active components, demonstrated by the precipitin reaction after double immunodiffusion of the multiple GP-350 components separated by either gel filtration or disc-gel electrophoresis without SDS. Reactivity with GP-350 antiserum prepared by Van Nieuw Amerongen and co-workers was present primarily in the major peak isolated by gel filtration on Sephadex G-50. This fraction gave three precipitin lines and eluted at a position corresponding to a molecular size of 9500–11,500.

Lipid analysis of brain GP-350 demonstrated the presence of approximately 17% by weight of mostly neutral lipids together with traces of glycolipids, and the immunological reactivity of the GP-350 protein components was not altered in delipidated preparations. However, the carbohydrate content in the brain GP-350 prepared by Simonian *et al.* (1978*a*,*b*) was only 2% after removal of lipids by chloroform–methanol extraction (and 3% before delipidation), as compared to the 17% carbohydrate reported by Van Nieuw Amerongen *et al.* (1972). In view of these recent data, it will be necessary to have more complete information concerning the actual structure, distribution, and metabolism of the various components of GP-350 before it will be possible to evaluate the numerous studies on what was originally considered to be a homogeneous and well-characterized brain-specific glycoprotein.

Reports have also appeared concerning other "brain-specific" glycoproteins such as the α_2-glycoprotein isolated from EDTA-saline extracts of white matter by affinity chromatography using immunoadsorbents, and claimed to originate in glial cells (Warecka *et al.*, 1972). However, these preparations have been found to vary in composition from one time to another and to contain five or more protein components, one of which is a glycoprotein with an apparent molecular weight of approximately 50,000 as estimated by SDS–PAGE (Brunngraber *et al.*, 1974, 1975). Further purification and characterization of this glycoprotein will be required before it is possible to draw any conclusions concerning its composition, structure, and localization in brain.

3.2. Glycosaminoglycans

3.2.1. Cytoplasmic and Membrane Localization of Hyaluronic Acid, Chondroitin Sulfate, and Heparan Sulfate

In contrast to the brain glycoproteins, which are chiefly membrane constituents, 40% of the total glycosaminoglycans can be obtained in the soluble fraction after high-speed centrifugation of a brain homogenate in 0.3 M sucrose. The percentage of the individual glycosaminoglycans found in the sucrose supernatant ranges from 17% of the heparan sulfate to 51% of the chondroitin sulfate (Table IV). Although most of the glycosaminoglycans are originally found in the microsomal fraction (Table IV), additional amounts can be solubilized by washing with 0.3 M sucrose, resulting in the total solubilization of approximately one third of the heparan sulfate and 50–75% of the hyaluronic acid and chondroitin sulfate (Kiang *et al.*, 1978).

When rat brain microsomes are subfractionated on a sucrose density gradient, almost all the hyaluronic acid and chondroitin sulfate are found associated with a very-low-density subfraction (sedimenting on 0.5 M sucrose) consisting exclusively of smooth membranes derived from plasma membranes and smooth endoplasmic reticulum (Kiang *et al.*, 1978). In contrast, most of the glycoproteins and a large portion of the heparan sulfate are

Table IV. Distribution of Glycosaminoglycans in Subcellular Fractions Prepared by Differential Centrifugation of a Rat Brain Homogenate

Fraction	Percentage[a]		
	Hyaluronic acid	Heparan sulfate	Chondroitin sulfate
Nuclei and cell debris	19	21	12
Crude mitochondrial fraction	5	3	1
Microsomes[b]	54	59	36
Soluble fraction	22	17	51

[a]Represents the percentage of the total in whole brain that is found in each fraction (sum of figures in each column = 100%). Whole brain concentrations (micromoles of glucosamine or galactosamine per gram wet weight of brain) of hyaluronic acid, heparan sulfate, and chondroitin sulfate are 0.102, 0.062, and 0.302, respectively. For the *concentration* of glycosaminoglycans in each fraction, see R. K. Margolis *et al.* (1975).
[b]Twenty-five percent of the heparan sulfate and 60–65% of the total microsomal hyaluronic acid and chondroitin sulfate can be removed by washing the microsomal membranes [see Kiang *et al.* (1978) and Section 3.2.1].

present in membranes of greater density (sedimenting between 0.7 and 1.1 M sucrose).

Essentially all (97–98%) of the hyaluronic acid and chondroitin sulfate present in the very-low-density microsomal subfraction can be removed by simple washing of the membranes, and is similar metabolically to the large pool of these glycosaminoglycans initially present in the soluble fraction of brain. Although the exact nature of their association with this smooth membrane subfraction is still unclear, it is apparent that almost all the hyaluronic acid and chondroitin sulfate in brain is either soluble or occurs in an easily solubilizable form. [This conclusion assumes that they are essentially present in the soluble and microsomal fractions obtained by differential centrifugation, ignoring the small amounts found in the crude mitochondrial fraction, and ascribing most of the hyaluronic acid and chondroitin sulfate in the initial low-speed pellet to the presence there of unbroken cells (see Section 3.2.3)]. It should also be noted that glycosaminoglycans are not present in myelin (Margolis, R. U., 1967; Adams and Bayliss, 1968; Matthieu *et al.*, 1975*a*).

Most of the heparan sulfate is firmly attached to membranes in subfractions that appear to be enriched in plasma membranes. Such a localization is supported by our preliminary studies indicating that heparan sulfate can be removed from the surface of isolated neurons using a specific endoglycosidase (heparitinase), and is consistent with other reports that heparan sulfate is present on the surface of many types of cells (Kraemer, 1971; Kraemer and Smith, 1974; Buonassisi and Root, 1975; Akasaki *et al.*, 1975; Underhill and Keller, 1977; Oldberg *et al.*, 1977).

The soluble chondroitin sulfate occurs in the form of a proteoglycan (together with glycoproteins and small amounts of hyaluronic acid and heparan sulfate) as described above. Its exact localization and function are not yet clear, although histochemical studies indicate that chondroitin sulfate and other glycosaminoglycans are present in both the cytoplasm (Castejón, 1970; Castejón and Castejón, 1976) and extracellular space (Bondareff, 1967; Tani and Ametani, 1971) of brain. Their cytoplasmic localization has recently been directly demonstrated by studies using isolated neurons, where it was found that 82% of the chondroitin sulfate and 55% of the heparan sulfate can be recovered, together with lactate dehydrogenase, in the cell-free supernatant following hypotonic lysis (Margolis, R. K., *et al.*, 1979). In contrast, only 25% of the hyaluronic acid and glycoprotein was released in a soluble form. These findings are also of general biological interest since chondroitin sulfate is considered to be a component of the extracellular matrix or is possibly present in part on the cell surface, but has not previously been found as a cytoplasmic constituent of any tissue.

3.2.2. *Developmental Changes*

In studies on complex carbohydrates during rat brain development, it was found that the levels of all three glycosaminoglycans increased postnatally to reach a peak at 7 days, after which they declined steadily, attaining by 30 days concentrations within 10% of those present in adult brain (see Appendix I). The greatest change occurred in hyaluronic acid, which decreased by 50% between 7 and 10 days, and declined to adult levels (28% of the peak concentration) by 18 days of age. In 7-day-old rats almost 90% of the hyaluronic acid in brain is water-extractable, as compared to only 15% in adult animals, and this large amount of soluble hyaluronic acid in young brain is relatively inactive metabolically. On the basis of these and other data, it appears plausible that the high levels of soluble hyaluronic acid in very young brain may function in the retention of water and thereby form an easily penetrable extracellular matrix through which neuronal migration and differentiation may take place during brain development. [For a more detailed discussion of developmental aspects of glycosaminoglycan distribution and metabolism, see R. U. Margolis *et al.* (1975), Toole (1976), and R. U. Margolis and R. K. Margolis (1977).]

3.2.3. *Analyses of Nerve Endings and Nuclei*

Although earlier reports suggested the presence of considerable amounts of sulfated glycosaminoglycans in synaptosomes, we found that a once-washed crude mitochondrial fraction contains only 1–3% of the total chondroitin sulfate and heparan sulfate of rat brain, and that purified synaptosomes or subsynaptosomal fractions prepared from them have both a very low concentration of sulfated glycosaminoglycans and an insignificant incorporation of labeled precursors into these complex carbohydrates (Margolis, R. K., *et al.*, 1975). Similar results were later reported by Simpson *et al.* (1976). However, from preliminary studies it appears that hyaluronic acid may be present in synaptic vesicles of brain (cf. also Vos *et al.*, 1969). Many of the previous reports concerning complex carbohydrates of nerve endings were based on studies using synaptosomes prepared before the importance of thorough washing of the crude mitochondrial fraction to remove microsomal and other contaminants was fully recognized. [For a more detailed discussion of this question, see Barondes (1974), Morgan (1976), Morgan and Gombos (1976), and Chapter 8.] The problem of possible microsomal contamination is particularly difficult in relation to studies of glycosaminoglycans, since these are present in high concentrations in certain classes of microsomal membranes, as discussed above.

It has recently been reported that unidentified glycosaminoglycans are present in highly purified cholinergic synaptic vesicles isolated from the electric organ of *Torpedo marmorata*, and that the major vesicle "protein," earlier named vesiculin, is in fact composed of glycosaminoglycans (Stadler and Whittaker, 1978). The vesicles contain 48 μmol of glycosaminoglycan hexosamine (81% GlcNAc and 19% GalNAc) per 100 mg protein. This concentration is over ten times that present in chromaffin granules (see Chapter 15), and would account for 3–4% of the vesicle dry weight (Tashiro and Stadler, 1978).

Although 12–21% of the total brain glycosaminoglycans are found in a fraction containing nuclei, unbroken cells, and cell debris (Table IV), analysis of purified nuclei from brain demonstrated that the nuclei themselves contain only 10–12% of the hyaluronic acid and chondroitin sulfate, and 4% of the heparan sulfate, present in the crude "nuclei and cell debris" fraction (Margolis, R. K., *et al.*, 1976*a*). Control experiments demonstrated that these small amounts of glycosaminoglycans did not originate from the nonspecific binding of soluble acidic proteoglycans to basic nuclear proteins. Studies in other laboratories, involving radioactive labeling of cultured mouse melanoma cells and HeLa S_3 cells, had suggested that glycosaminoglycans and glycoproteins may be components of the eukaryotic genome, but whether the relatively small amounts of glycosaminoglycans and glycoproteins found in brain nuclei are biologically significant and represent a feature common to all animal cell nuclei remains to be determined. However, while complex carbohydrates on the cell surface are likely to be involved in determining intercellular recognition and adhesion, contact inhibition, antigenicity, and similar surface-related phenomena, it is possible that nuclear glycosaminoglycans and glycoproteins may play a role in quite different types of processes, such as control of the cationic environment at the nuclear membrane, transport selectivity into the nuclei, regulation of nuclear enzyme activity, and control of template activity in chromatin (Bhavanandan and Davidson, 1975). The predominantly nuclear localization of cytidine 5′-monophosphoro-*N*-acetylneuraminic acid synthetase (Van Den Eijnden, 1973), a key enzyme in the biosynthesis of glycoproteins and gangliosides, suggests that the nucleus may also be directly involved in regulating the levels and turnover of these complex carbohydrates in brain.

4. Conclusions

Over the last decade, there has been a marked increase in our knowledge of the structure and distribution of nervous tissue glycoproteins and glycoaminoglycans. To a certain extent, advances in the neurochemistry of complex carbohydrates have paralleled and benefited from the great strides made in the study of glycoproteins and glycosaminoglycans in other tissues. However, it is becoming increasingly apparent that there are a number of properties peculiar to these compounds in nervous tissue, and that analogies with data from other sources may be risky.

Reliable information is now available concerning the glycosaminoglycan composition of brain and of nervous tissue organelles such as chromaffin granules (see Chapter 15), and much is also known regarding their localization at the cellular and subcellular levels. These data have at least partially answered a number of earlier questions and hypotheses about their possible functional roles in myelin, at nerve endings and other sites, and have indicated that chondroitin sulfate is largely present in brain as a soluble, cytoplasmic constituent of neurons and glia, rather than as an extracellular ground substance.

Considerable progress has also been made in studies on the structure and composition of brain glycoproteins. These investigations have been hampered by the insoluble nature of most nervous tissue glycoproteins, which occur largely as components of plasma and intracellular membranes, and because the great cellular heterogeneity of brain makes it difficult to obtain sufficient amounts of any one species of glycoprotein. However, the recent development of techniques for the structural analysis of oligosaccharides employing combined gas chromatography and mass spectrometry has provided a tool of great sensitivity and specificity. Application of these newer methods to the study of brain glycoproteins has revealed much detailed information concerning their structure, including the elucidation of certain features that had not previously been reported for other tissues, and some that appear to be unique to brain. The extension of this approach to well-characterized subcellular fractions can be expected to ultimately provide an understanding of the structure and function of nervous tissue glycoproteins at the membrane and molecular levels.

References

Adams, C. W. M., and Bayliss, O. B., 1968, Histochemistry of myelin. VII. Analysis of lipid–protein relationships and absence of acid mucopolysaccharide, *J. Histochem. Cytochem.* **16:**119.

Akasaki, M., Kawasaki, T., and Yamashina, I., 1975, The isolation and characterization of glycopeptides and mucopolysaccharides from plasma membranes of normal and regenerating livers of rats, *FEBS Lett.* **59:**100.

Allen, W. S., Otterbein, C., Varma, R., Varma, R. S., and Wardi, A. H., 1976, Nondialyzable sulfated sialoglycopeptide fractions derived from bovine heifer brain glycoproteins: Isolation, characterization, and carbohydrate–peptide linkage studies, *J. Neurochem.* **26**:879.

Balazs, E. A. (ed.), 1970, *Chemistry and Biology of the Intercellular Matrix*, Vol. 2, *Glycosaminoglycans and Proteoglycans*, Academic Press, London.

Balsamo, J., and Lilien, J., 1975, The binding of tissue-specific adhesive molecules to the cell surface: A molecular basis for specificity, *Biochemistry* **14**:167.

Barondes, S. H., 1974, Synaptic macromolecules: Identification and metabolism, *Annu. Rev. Biochem.* **43**:147.

Behnke, O., 1975, Studies on isolated microtubules: Evidence for a clear space component, *Cytobiologie* **11**:366.

Bhavanandan, V. P., and Davidson, E. A., 1975, Mucopolysaccharides associated with nuclei of cultured mammalian cells, *Proc. Natl. Acad. Sci. U.S.A.* **72**:2032.

Bogoch, S., 1968, *The Biochemistry of Memory*, Oxford University Press, New York.

Bondareff, W., 1967, Demonstration of an intercellular substance in mouse cerebral cortex, *Z. Zellforsch.* **81**:366.

Branford White, C. J., and Hudson, M., 1977, Characterization of a chondroitin sulphate proteoglycan from ovine brain, *J. Neurochem.* **28**:581.

Brunngraber, E. G., 1972, Biochemistry, function and neuropathology of glycoproteins in brain tissue, in: *Functional and Structural Proteins of the Nervous System* (A. N. Davison, P. Mandel, and I. G. Morgan, eds.), pp. 109-133, Plenum Press, New York.

Brunngraber, E. G., and Brown, B. D., 1964, Fractionation of brain macromolecules. II. Isolation of protein-linked sialomucopolysaccharides from subcellular, particulate fractions from rat brain, *J. Neurochem.* **11**:449.

Brunngraber, E. G., Aro, A., and Brown, B. D., 1970, Differential determination of glucosamine, galactosamine, and mannosamine in glycopeptides derived from brain tissue glycoproteins, *Clin. Chim. Acta* **29**:333.

Brunngraber, E. G., Hof, H., Susz, J., Brown, B. D., Aro, A., and Chang, I., 1973, Glycopeptides from rat brain glycoproteins, *Biochim. Biophys. Acta* **304**:781.

Brunngraber, E. G., Susz, J. P., and Warecka, K., 1974, Electrophoretic analysis of human brain-specific proteins obtained by affinity chromatography, *J. Neurochem.* **22**:181.

Brunngraber, E. G., Susz, J. P., Javaid, J., Aro, A., and Warecka, K., 1975, Binding of concanavalin A to the brain-specific proteins obtained from human white matter by affinity chromatography, *J. Neurochem.* **24**:805.

Buonassisi, V., and Root, M., 1975, Enzymatic degradation of heparin-related mucopolysaccharides from the surface of endothelial cell cultures, *Biochim. Biophys. Acta* **385**:1.

Castejón, H. V., 1970, Histochemical demonstration of acid glycosaminoglycans in the nerve cell cytoplasm of mouse central nervous system, *Acta Histochem.* **35**:161.

Castejón, H. V., and Castejón, O. J., 1976, Electron microscopic demonstration of hyaluronidase sensible proteoglycans at the presynaptic area in mouse cerebellar cortex, *Acta Histochem.* **55**:300.

Chandrasekaran, E. V., BeMiller, J. N., and Leer, S.-C. D., 1978, Isolation, partial characterization, and biological properties of polysaccharides from papain, *Carbohydr. Res.* **60**:105.

Comper, W. D., and Laurent, T. C., 1978, Physiological function of connective tissue polysaccharides, *Physiol. Rev.* **58**:255.

Davis, L. G., Costello, A. J. R., Javaid, J. I., and Brunngraber, E. G., 1976*a*, ^{31}P nuclear magnetic resonance studies on the phosphoglycopeptides obtained from rat brain glycoprotein, *FEBS Lett.* **65**:35.

Davis, L. G., Javaid, J. I., and Brunngraber, E. G., 1976*b*, Identification of phosphoglycoproteins obtained from rat brain, *FEBS Lett.* **65**:30.

Dawson, G., and Tsay, G., 1977, Substrate specificity of human α-L-fucosidase, *Arch. Biochem. Biophys.* **184**:12.

Di Benedetta, C., Brunngraber, E. G., Whitney, G., Brown, B. D., and Aro, A., 1969, Compositional patterns of sialofucohexosaminoglycans derived from rat brain glycoproteins, *Arch. biochem. Biophys.* **131**:404.

Dietrich, C. P., Sampaio, L. O., and Toledo, O. M. S., 1976, Characteristic distribution of sulfated mucopolysaccharides in different tissues and their respective mitochondria, *Biochem. Biophys. Res. Commun.* **71**:1.

Dorai, D. T., and Bachhawat, B. K., 1977, Purification and properties of brain alkaline phosphatase, *J. Neurochem.* **29**:503.

Dutton, G. R., and Barondes, S. H., 1970, Glycoprotein metabolism in developing mouse brain, *J. Neurochem.* **17**:913.

Eipper, B. A., 1972, Rat brain microtubule protein: Purification and determination of covalently bound phosphate and carbohydrate, *Proc. Natl. Acad. Sci. U.S.A.* **69**:2283.

Feit, H., and Shelanski, M. L., 1975, Is tubulin a glycoprotein?, *Biochem. Biophys. Res. Commun.* **66**:920.

Finne, J., 1975, Structure of the *O*-glycosidically linked carbohydrate units of rat brain glycoproteins, *Biochim. Biophys. Acta* **412**:317.

Finne, J., and Krusius, T., 1976, *O*-glycosidic carbohydrate units from glycoproteins of different tissues: Demonstration of a brain-specific disaccharide, α-galactosyl-(1→3)-*N*-acetylgalactosamine, *FEBS Lett.* **66**:94.

Finne, J., and Rauvala, H., 1977, Determination (by methylation analysis) of the substitution pattern of 2-amino-2-deoxyhexitols obtained from *O*-glycosylic carbohydrate units of glycoproteins, *Carbohydr. Res.* **58**:57.

Finne, J., Krusius, T., Rauvala, H., and Hemminki, K., 1977, The disialosyl group of glycoproteins: Occurrence in different tissues and cellular membranes, *Eur. J. Biochem.* **77**:319.

Finne, J., Krusius, T., Margolis, R. K., and Margolis, R. U., 1979. Novel mannitol-containing oligosaccharides obtained by mild alkaline borohydride treatment of a chondroitin sulfate proteoglycan from brain (submitted for publication).

Glimelius, B., Norling, B., Westermark, B., and Wasteson, Å., 1978, Composition and distribution of glycosaminoglycans in cultures of human normal and malignant glial cells, *Biochem. J.* **172**:443.

Gombos, G., and Zanetta, J. P., 1978, Recent methods for the separation and analysis of central nervous system glycoproteins, in: *Research Methods in Neurochemistry*, Vol. 4 (N. Marks and R. Rodnight, eds.), pp. 307–343, Plenum Press, New York.

Gottschalk, A. (ed.), 1972, *Glycoproteins—Their Composition, Structure and Function*, 2nd ed., Elsevier, Amsterdam.

Hughes, R.C., 1976, *Membrane Glycoproteins—A Review of Structure and Function*, Butterworths, London.

Hüttich, K., Müller, H., and Unger, E., 1977, Concanavalin A-Bindung an zyklisch reassemblierte Mikrotubuli aus Schweinehirn-Tubulin, *Acta Histochem.* **58**:324.

Katzman, R. L., 1972, Isolation of *N*-acetyllactosamine and galactosyl-β-D-(1→4)-*N*-acetyllactosamine from beef brain glycopeptides, *J. Biol. Chem.* **247**:3744.

Katzman, R. L., 1974, Absence of arabinose in bovine brain hyaluronic acid as analysed by gas–liquid chromatography and mass spectrometry, *Biochim. Biophys. Acta* **372**:52.

Kiang, W.-L., Crockett, C. P., Margolis, R. K., and Margolis, R. U., 1978, Glycosaminoglycans and glycoproteins associated with microsomal subfractions of brain and liver, *Biochemistry* **17**:3841.

Kornfeld, R., and Kornfeld, S., 1976, Comparative aspects of glycoprotein structure, *Annu. Rev. Biochem.* **45**:217.

Kraemer, P. M., 1971, Heparan sulfates of cultured cells. I. Membrane-associated and cell-sap species in Chinese hamster cells, *Biochemistry* **10**:1437.

Kraemer, P. M., and Smith, D. A., 1974, High molecular weight heparan sulfate from the cell surface, *Biochem. Biophys. Res. Commun.* **56**:423.

Krusius, T., and Finne, J., 1977, Structural features of tissue glycoproteins—Fractionation and methylation analysis of glycopeptides derived from rat brain, kidney and liver, *Eur. J. Biochem.* **78**:369.

Krusius, T., and Finne, J., 1978, Characterization of a novel sugar sequence from rat brain glycoproteins containing fucose and sialic acid, *Eur. J. Biochem.* **84**:395.

Krusius, T., Finne, J., Kärkkainen, J., and Järnefelt, J., 1974, Neutral and acidic glycopeptides in adult and developing rat brain, *Biochim. Biophys. Acta* **365**:80.

Krusius, T., Finne, J., Margolis, R. U., and Margolis, R. K., 1978, Structural features of microsomal, synaptosomal, mitochondrial and soluble glycoproteins of brain, *Biochemistry* **17**:3849.

Lindahl, U., and Rodén, L., 1972, Carbohydrate-peptide linkages in proteoglycans of animal, plant and bacterial origin, in: *Glycoproteins—Their Composition, Structure and Function*, 2nd ed. (A. Gottschalk, ed.), pp. 491–517, Elsevier, Amsterdam.

Linker, A., and Hovingh, P., 1973, The heparitin sulfates (heparan sulfates), *Carbohydr. Res.* **29**:41.

Margolis, R. K., and Margolis, R. U., 1970, Sulfated glycopeptides from rat brain glycoproteins, *Biochemistry* **9**:4389.

Margolis, R. K. and Margolis, R. U., 1973, Alkali-labile oligosaccharides of brain glycoproteins, *Biochim. Biophys. Acta* **304**:421.

Margolis, R. K., Margolis, R. U., and Shelanski, M. L., 1972, The carbohydrate composition of brain microtubule protein, *Biochem. Biophys. Res. Commun.* **47**:432.

Margolis, R. K., Margolis, R. U., Preti, C., and Lai, D., 1975, Distribution and metabolism of glycoproteins and glycosaminoglycans in subcellular fractions of brain, *Biochemistry* **14**:4797.

Margolis, R. K., Crockett, C. P., Kiang, W.-L., and Margolis, R. U., 1976*a*, Glycosaminoglycans and glycoproteins associated with rat brain nuclei, *Biochim. Biophys. Acta* **451**:465.

Margolis, R. K., Preti, C., Lai, D., and Margolis, R. U., 1976*b*, Developmental changes in brain glycoproteins, *Brain Res.* **112**:363.

Margolis, R. K., Crockett, C. P., and Margolis, R. U., 1978, Dissociative extraction of brain proteoglycans, *J. Neurochem.* **30**:1177.

Margolis, R. K., Thomas, M. D., Crockett, C. P., and Margolis, R. U., 1979, Presence of chondroitin sulfate in the neuronal cytoplasm, *Proc. Natl. Acad. Sci. U.S.A.* (in press).

Margolis, R. U., 1967, Acid mucopolysaccharides and proteins of bovine whole brain, white matter and myelin, *Biochim. Biophys. Acta* **141**:91.

Margolis, R. U., 1969, Mucopolysaccharides, in: *Handbook of Neurochemistry*, Vol. 1 (A. Lajtha, ed.), pp. 245–260, Plenum Press, New York.

Margolis, R. U., and Atherton, D. M., 1972, The heparan sulfate of rat brain, *Biochim. Biophys. Acta* **273**:368.

Margolis, R. U., and Margolis, R. K., 1972*a*, Sulfate turnover in mucopolysaccharides and glycoproteins of brain, *Biochim. Biophys. Acta* **264**:426.

Margolis, R. U., and Margolis, R. K., 1972*b*, Mucopolysaccharides and glycoproteins, in: *Research Methods in Neurochemistry*, Vol. 1 (N. Marks and R. Rodnight, eds.), pp. 249–284, Plenum Press, New York.

Margolis, R. U., and Margolis, R. K., 1974, Distribution and metabolism of mucopolysaccharides and glycoproteins in neuronal perikarya, astrocytes, and oligodendroglia, *Biochemistry* **13**:2849.

Margolis, R. U., and Margolis, R. K., 1977, Metabolism and function of glycoproteins and glycosaminoglycans in nervous tissue, *Int. J. Biochem.* **8**:85.

Margolis, R. U., Margolis, R. K., and Atherton, D. M., 1972, Carbohydrate-peptide linkages in glycoproteins and mucopolysaccharides from brain, *J. Neurochem.* **19**:2317.

Margolis, R. U., Margolis, R. K., Chang, L., and Preti, C., 1975, Glycosaminoglycans of brain during development, *Biochemistry* **14**:85.

Margolis, R. U., Lalley, K., Kiang, W.-L., Crockett, C., and Margolis, R. K., 1976, Isolation and properties of a soluble chondroitin sulfate proteoglycan from brain, *Biochem. Biophys. Res. Commun.* **73**:1018.

Margolis, R. U., Margolis, R. K., Kiang, W.-L., and Crockett, C. P., 1978, Soluble proteoglycans and glycoproteins of brain, in: *Glycoconjugate Research: Proceedings of the Fourth International Symposium on Glycoconjugates* (J. D. Gregory and R. W. Jeanloz, eds.), Academic Press, New York.

Matthieu, J.-M., Quarles, R. H., Poduslo, J. F., and Brady, R. O., 1975*a*, [^{35}S]Sulfate incorporation into myelin glycoproteins. I. Central nervous system, *Biochim. Biophys. Acta* **392**:159.

Matthieu, J.-M., Everly, J. L., Brady, R. O., and Quarles, R. H., 1975*b*, [^{35}S]Sulfate incorporation into myelin glycoproteins. II. Peripheral nervous system, *Biochim. Biophys. Acta* **392**:167.

Montreuil, M. J. (ed.), 1974, *Actes Colloq. Int. No. 221* (2è Symposium International sur les Glycoconjugués, "La Méthodologie de la Structure et du Métabolisme des Glycoconjugués," organisé à Villeneuve d'Ascq, France), June 1973, Editions du CNRS, Paris.

Montreuil, J., 1975, Recent data on the structure of the carbohydrate moiety of glycoproteins—Metabolic and biological implications, *Pure Appl. Chem.* **42**:431.

Morgan, I. G., 1976, Synaptosomes and cell separation, *Neuroscience* **1**:159.

Morgan, I. G., and Gombos, G., 1976, Biochemical studies of synaptic macromolecules, in: *Neuronal Recognition* (S. Barondes, ed.), pp. 179–202, Plenum Press, New York.

Morgan, I. G., Gombos, G., and Tettamanti, G., 1977, Glycoproteins and glycolipids of the nervous system, in: *The Glycoconjugates*, Vol. 1 (M. I. Horowitz and W. Pigman, eds.), pp. 351–383, Academic Press, New York.

Oldberg, Å., Höök, M., Öbrink, B., Pertoft, H., and Rubin, K., 1977, Structure and metabolism of rat liver heparan sulphate, *Biochem. J.* **164**:75.

Quarles, R. H., and Brady, R. O., 1971, Synthesis of glycoproteins and gangliosides in developing rat brain, *J. Neurochem.* **18**:1809.

Rauvala, H., 1976, The fucoganglioside of human kidney, *FEBS Lett.* **62**:161.

Rauvala, H., and Kärkkäinen, J., 1977, Methylation analysis of neuraminic acids by gas chromatography–mass spectrometry, *Carbohydr. Res.* **56**:1.

Saigo, K., and Egami, F., 1970, Purification and some properties of acid mucopolysaccharides of bovine brain, *J. Neurochem.* **17**:633.

Simonian, S., Heijlman, J., and Hooghwinkel, G. J. M., 1978*a*, Studies on the low molecular weight glycoprotein GP-350: Molecular and immunological heterogeneity, *J. Neurochem.* **31**:103.

Simonian, S., Heijlman, J., and Hooghwinkel, G. J. M., 1978*b*, Carbohydrate composition of the soluble and membrane-bound GP-350 fractions, *Scand. J. Immunol.* (in press).

Simpson, D. L., Thorne, D. R., and Loh, H. H., 1976, Sulfated glycoproteins, glycolipids and glycosaminoglycans from synaptic plasma and myelin membranes: Isolation and characterization of sulfated glycopeptides, *Biochemistry* **15**:5449.

Singh, M., and Bachhawat, B. K., 1968, Isolation and characterization of glycosaminoglycans in human brain of different age groups, *J. Neurochem.* **15**:249.

Singh, M., Chandrasekaran, E. V., Cherian, R., and Bachhawat, B. K., 1969, Isolation and characterization of glycosaminoglycans in brain of different species, *J. Neurochem.* **16**:1157.

Stadler, H., and Whittaker, V. P., 1978, Identification of vesiculin as a glycosaminoglycan. *Brain Res.* **153**:408.

Tani, E., and Ametani, T., 1971, Extracellular distribution of ruthenium red positive substance in the cerebral cortex, *J. Ultrastruct. Res.* **34**:1.

Tashiro, T., and Stadler, H., 1978, Chemical composition of cholinergic synaptic vesicles from *Torpedo marmorata* based on improved purification, *Eur. J. Biochem.* **90**:479.

Toledo, O. M. S., and Dietrich, C. P., 1977, Tissue specific distribution of sulfated mucopolysaccharides in mammals, *Biochim. Biophys. Acta* **498**:114.

Toole, B. P., 1976, Morphogenetic role of glycosaminoglycans in brain and other tissues, in: *Neuronal Recognition* (S. Barondes, ed.), pp. 275–329, Plenum Press, New York.

Tsay, G. C., Dawson, G., and Sung, S.-S. J., 1976, Structure of the accumulating oligosaccharide in fucosidosis, *J. Biol. Chem.* **251**:5852.

Underhill, C. B., and Keller, J. M., 1977, Heparan sulfates of mouse cells: Analysis of parent and transformed 3T3 cell lines, *J. Cell. Physiol.* **90**:53.

Van den Eijnden, D. H., 1973, The subcellular localization of cytidine 5′-monophospho-*N*-acetylneuraminic acid synthetase in calf brain, *J. Neurochem.* **21**:949.

Van Nieuw Amerongen, A., and Roukema, P. A., 1973, Physico-chemical characteristics and regional distribution studies of GP-350, a soluble sialoglycoprotein from brain, *J. Neurochem.* **21**:125.

Van Nieuw Amerongen, A., and Roukema, P.A., 1974, GP-350, a sialoglycoprotein from calf brain: Its subcellular localization and occurrence in various brain areas, *J. Neurochem.* **23**:85.

Van Nieuw Amerongen, A., Van den Eijnden, D. H., Heijlman, J., and Roukema, P. A., 1972, Isolation and characterization of a soluble glucose-containing sialoglycoprotein from the cortical grey matter of calf brain, *J. Neurochem.* **19**:2195.

Van Nieuw Amerongen, A., Roukema, P. A., and Van Rossum, A. L., 1974, Immunofluorescence study on the cellular localization of GP-350, a sialoglycoprotein from brain, *Brain Res.* **81**:1.

Varma, R., Vercellotti, J. R., and Varma, R. S., 1977, On arabinose as a component of brain hyaluronate—Confirmation by chromatographic, enzymatic and chemical ionization–mass spectrometric analyses, *Biochim. Biophys. Acta* **497**:608.

Vitello, L., Breen, M., Weinstein, H. G., Sittig, R. A., and Blacik, L. J., 1978, Keratan sulfate-like glycosaminoglycan in the cerebral cortex of the brain and its variation with age, *Biochim. Biophys. Acta* **539**:305.

Vos, J., Kuriyama, K., and Roberts, E., 1969, Distribution of acid mucopolysaccharides in subcellular fractions of mouse brain, *Brain Res.* **12**:172.

Warecka, K., Möller, H. J.,Vogel, H.-M., and Tripatzis, I., 1972, Human brain-specific alpha$_2$-glycoprotein: Purification by affinity chromatography and detection of a new component; localization in nervous cells, *J. Neurochem.* **19**:719.

Yang, H.-J., and Hakomori, S.-I., 1971, A sphingolipid having a novel type of ceramide and lacto-*N*-fucopentaose III, *J. Biol. Chem.* **246**:1192.

4

Biosynthesis of Glycoproteins

C. J. Waechter and M. G. Scher

1. Introduction

Studies on the wide diversity of biological functions performed by soluble and membrane-bound glycoproteins in mammalian tissues have become a major theme in the research of a large number of biochemists and cell biologists, as well as neurochemists. One focal point in these investigations has been the possibility that cell-surface glycoproteins and glycosyltransferases play a role in cell–cell contact relationships (Roseman, 1970). An extensive list of investigations in nervous tissue has been stimulated by the postulated roles of membrane glycoproteins and membrane surface glycosyltransferases in interneuronal recognition (Barondes, 1970) and the myelination of axon plasma membranes (Brady and Quarles, 1973). A thorough knowledge of the structures, metabolism, and regional and subcellular location of the brain membrane glycoproteins will be prerequisite to understanding their precise neurological functions. Many of these aspects of glycoprotein neurochemistry are covered in detail elsewhere in this volume.

The objective of this chapter is to describe the current knowledge of the two established enzymatic mechanisms responsible for the biosynthesis of the oligosaccharide side chains linked to glycoproteins in nervous tissue. The sugar nucleotide pathway has been known for nearly 20 years in mammalian tissues, and proceeds by the transfer of single sugar residues to growing carbohydrate chains with nucleotide sugars serving as the direct glycosyl donors (Gottschalk, 1972). The reactions catalyzed by sugar nucleotide glycosyltransferases involved in proteoglycan biosynthesis are discussed in Chapter 5. For many years, it was believed that the carbohydrate chains of

Abbreviations used in this chapter: (Dol-P) dolichyl monophosphate: (GalNAc) *N*-acetylgalactosamine; (GlcNAc) *N*-acetylglucosamine; (GPD) glucosylphosphoryldolichol; (MPD) mannosylphosphoryldolichol; (NeuAc) *N*-acetylneuraminic acid; (SDS-PAGE) sodium dodecyl sulfate–polyacrylamide gel electrophoresis; (TM) tunicamycin.

C. J. Waechter and M. G. Scher • Department of Biochemistry, University of Maryland School of Medicine, Baltimore, Maryland.

all glycoproteins were elaborated by the sugar nucleotide pathway. However, during the past eight years, it has become evident that at least one type of oligosaccharide unit of glycoproteins in animal tissues is synthesized by another mechanism in which polyisoprenoid glycolipids participate as intermediary glycosyl carriers (Lennarz, 1975; Lucas and Waechter, 1976; Waechter and Lennarz, 1976; Parodi and Leloir, 1976; Hemming, 1977). The lipid intermediate pathway for glycoprotein biosynthesis in animal tissues is analogous to the biosynthesis of complex polysaccharides in bacterial cell envelopes, in which polyprenol-linked sugars are known to function as activated glycosyl donors (Lennarz and Scher, 1972). On the basis of studies conducted with membrane preparations from many mammalian tissues including brain, it now appears that dolichol-linked glycosyl units are involved in the assembly of a mannosylated *N,N′*-diacetylchitobiose inner core. This general structure is a common inner core region found *N*-glycosidically linked to asparagine residues in many glycoproteins in eukaryotic cells (Kornfeld and Kornfeld, 1976). In this chapter, special emphasis has been given to *in vitro* studies conducted with cell-free, particulate enzyme preparations from brain tissue, which have provided information about the enzymatic properties of both sugar nucleotide and lipid-mediated glycosyltransferases. While an attempt has been made to be comprehensive in our literature survey, the list of contributors to this field may be incomplete due to limitations in space and time, or an inadvertent oversight by the authors.

2. Sugar Nucleotide: Glycoprotein Glycosyltransferases

The now classic multiglycosyltransferase model proposed by Roseman (1970) consists of a complex of glycosyltransferase enzymes each of which catalyzes transfer of a single sugar from a specific sugar nucleotide donor to the appropriate acceptor molecule. The product of each reaction becomes the substrate for the subsequent reaction. Thus, if the necessary conditions for any particular enzyme are not met, the oligosaccharide will not be further elongated. The origin of the well-known microheterogeneity of oligosaccharides attached to glycoprotein is accounted for by this mechanism. Much of the available information on the enzymatic properties and specificity of the individual sugar nucleotide glycosyl transferases is discussed in this section. All the sugar nucleotide glycosyltransferases described below catalyze general reactions analogous to the three single sugar transfers illustrated in Fig. 1. (For an outline of the biosynthesis of nucleotide monosaccharides, see Chapter 5.)

2.1. *N*-Acetylgalactosaminyltransferases

In many glycoproteins containing *O*-glycosidically linked oligosaccharides, *N*-acetylgalactosamine (GalNAc) is the linkage sugar (Kornfeld

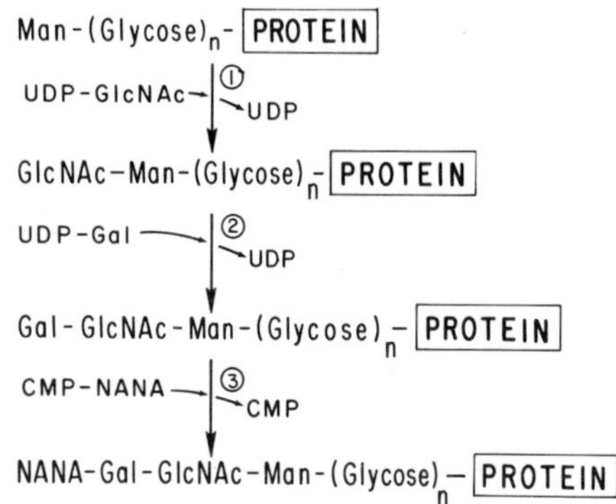

Fig. 1. Sugar nucleotide:glycoprotein glycosyltransferase reactions involved in the biosynthesis of the terminal trisaccharide sequence NANA (= NeuAc)-Gal-GlcNAc.

and Kornfeld, 1976). Ko and Raghupathy (1972) obtained a microsomal fraction from rat brain that catalyzes transfer of [^{14}C]-GalNAc from UDP-[^{14}C]-GalNAc to endogenous acceptors or to a variety of exogenous glycoproteins, including myelin basic protein from bovine brain and spinal cord and bovine submaxillary glycoprotein with sialic acid and galactosamine removed. Treatment of the [^{14}C]-GalNAc-labeled glycoprotein with alkaline-borohydride released [^{14}C]galactosaminitol, and the authors concluded that [^{14}C]-GalNAc was linked by an alkali labile *O*-glycosidic bond to serine (or threonine). The K_m for UDP-GalNAc varies from 1.52 μM in the presence of endogenous acceptor to 3.68 μM with the submaxillary protein. Use of Triton X-100 greatly stimulates activity. Stimulation of activity by the addition of detergent has been observed in many studies of brain sugar nucleotide:glycoprotein glycosyltransferases. It is quite likely that the detergents facilitate the interaction between the membrane-bound enzymes and the exogenous glycoprotein acceptors. Divalent cation was required, and Mn^{2+} was the most effective ion. The pH optimum was determined to be 6.8.

Because of the many possible functions of glycoproteins, the regulation of glycoprotein synthesis is a significant factor to consider while studying the sugar nucleotide glycosyltransferases. Ko and Raghupathy (1973) reported that GTP inhibited GalNAc transfer, although Gal transfer was stimulated. β-γ-Methylene GTP could substitute for GTP, as could other phosphorylated guanosine derivatives, but not guanosine itself. The inhibition of *N*-acetylgalactosaminyltransferase activity increased with the number of phosphate groups on guanosine. Sulfhydryl blocking agents minimized the effect of GTP on Gal transfer. The authors suggested that the increase in free nucleotide reported by others in developing mouse brain and the differing effect of GTP on two enzymes of the same multiglycosyltransferase system may constitute an allosteric regulatory phenomenon in glycoprotein biosynthesis.

Bosmann (1972) localized *N*-acetylgalactosaminyltransferase activity in the plasma membrane and mitochondrial subfractions of synaptosomes from guinea pig cerebral cortex. The enzyme, purified 45-fold in the presence of detergent, was inactive with other sugar nucleotides tested, and 99% of the radioactivity in the product was recovered as GalNAc. Bovine submaxillary mucin, with *N*-acetylneuraminic acid (NeuAc) and GalNAc removed, served as exogenous acceptor. The partially purified enzyme showed a high degree of specificity for the acceptor. Other glycoproteins including intact bovine submaxillary mucin and desialylated bovine submaxillary mucin were ineffective acceptors, as were mono- and disaccharides. The enzyme preparation was optimally active at pH 7.7 and required divalent cations, with Mn^{2+} being the most effective. The K_m for UDP-GalNAc was determined to be 25 μM.

2.2. Mannosyltransferases

Mannose constitutes a large portion of several types of carbohydrate chains in glycoproteins (Kornfeld and Kornfeld, 1976). Mannosyltransferases were discovered in guinea pig cerebral cortex in the synaptosomal fraction and in intraneural mitochondria prepared from the synaptosomes (Bosmann and Hemsworth, 1970). Transfer of mannose from GDP-mannose was dependent on Mn^{2+}, and EDTA inhibited the transfer. Enzyme assays, correlated with electron microscopy, revealed that rat brain mannosyltransferase activity was associated specifically with synaptic vesicles (Broquet and Louisot, 1971). Enzyme activity in the rat brain microsomal preparation appeared to be optimal at two different pHs, 5.9 and 7.4 (Broquet *et al.*, 1971). Activity was stimulated by Mg^{2+} and Mn^{2+}, and the K_m for GDP-mannose at either pH was 0.07 μM. As with most of the other glycosyltransferases, mannosyltransferases were stimulated by Triton X-100. Maximum stimulation occurred at a Triton/protein ratio of 0.15:1.0 (wt./wt.). A mitochondrial mannosyltransferase with somewhat different physicochemical properties was isolated by Broquet *et al.* (1975*a*). The mitochondrial enzyme preparation displayed optimal activity at pH 6.5, and the K_m value for GDP-mannose was found to be 3.3 μM. The enzymatically formed product was sensitive to trypsin, but the extent of proteolytic digestion was not reported.

Similar to the transfers catalyzed by the *N*-acetylgalactosaminyltransferases (Section 2.1), the transfer of mannose catalyzed by microsomal and mitochondrial enzymes was inhibited by GTP and β-γ-methylene GTP (Broquet *et al.*, 1971). The mitochondrial and microsomal enzymes showed different responses to sulfhydryl reagents. The mitochondrial enzyme, but not the microsomal enzyme, was inhibited by 5,5′-dithio-*bis*(2-nitrobenzoic acid). Mersalyl inhibited the mitochondrial enzyme to a greater extent than the microsomal enzyme. *N*-Ethylmaleimide had little effect on either preparation (Broquet *et al.*, 1975*b*).

The results of studies on membrane-bound mannosyltransferases in which the glycoprotein products were not thoroughly characterized must be carefully scrutinized. Since brain membranes catalyze the enzymatic transfer of mannose and glucose from their nucleotide derivatives into dolichol-linked oligosaccharides, which are not extracted with $CHCl_3$–CH_3OH (2:1), it is possible that [^{14}C]mannose- or [^{14}C]glucose-labeled oligosaccharide lipid intermediates could have been present in the glycoprotein fraction.

2.3. *N*-Acetylglucosaminyltransferases

In the "complex" type of asparagine-linked glycoproteins, the distal portions of oligosaccharide chains linked to the Man(GlcNAc)$_2$ inner core often contain the terminal trisaccharide sequence NeuAc ⟶ Gal ⟶ GlcNAc (Kornfeld and Kornfeld, 1976). The biosynthesis of this sequence common to serum-type glycoproteins is known to occur by successive monosaccharide transfers from sugar nucleotides to a glycoprotein acceptor (Roseman, 1970), as depicted in Fig. 1.

N-Acetylglucosaminyltransferase is present in the synaptosomal fraction of embryonic chicken brain and, to a very small extent, in embryonic brain fluid (Den *et al.*, 1970). Free mannose was ineffective as an acceptor, but orosomucoid pretreated with sialidase, β-galactosidase, and β-*N*-acetylglucosaminidase to expose mannosyl end groups was the most suitable acceptor. Only minimal transfer of *N*-acetylglucosamine (GlcNAc) from UDP-GlcNAc was obtained with intact orosomucoid (a serum glycoprotein), probably due to microheterogeneity of acceptor glycoprotein, or with orosomucoid pretreated with sialidase. The embryonic chicken brain *N*-acetylglucosaminyltransferase was assayed routinely with 12.5 mM Mn^{2+} and at pH 7.4. Enzymatic activity was very low in the absence of Triton X-100. Reaction 1 in Fig. 1 was also demonstrated in a synaptosomal preparation from guinea pig brain, purified 34-fold and separated from endogenous acceptor (Bosmann, 1972). The enzyme preparation was also minimally active with UDP-Glc, GDP-Man, GDP-Fuc, and UDP-GalNAc. The most effective exogenous acceptor was fetuin, with NeuAc, Gal, and GlcNAc removed to expose a terminal mannosyl residue. Of the radioactivity in the enzymatic product labeled by incubation with UDP-[^{14}C]-GlcNAc, 98% was recovered as [^{14}C]-GlcNAc. The K_m for UDP-GlcNAc in this reaction was 12 μM. On addition of cations, more GlcNAc was transferred in the presence of Mn^{2+} than in the presence of Mg^{2+}. The addition of Co^{2+}, Hg^{2+}, Fe^{2+}, Cd^{2+}, and EDTA caused inhibition of transfer. A pH optimum of 7.4–8.0 was found for this system.

Calf brain white matter membranes catalyze the transfer of GlcNAc from UDP-GlcNAc to endogenous glycoprotein acceptors (Waechter and Harford, 1977). The [^{14}C]-GlcNAc-labeled glycoprotein product was degraded to lower-molecular-weight glycopeptides by pronase or trypsin digestion. Treatment of [^{14}C]-GlcNAc-labeled glycopeptides with an exo-β-*N*-acetylglucosaminidase released all the radioactivity from the glycoprotein, indicating that the [^{14}C]-GlcNAc residues had been attached to the oligosaccharide chain by β-glycosidic linkages at nonreducing termini. Analysis of the intact glycoprotein by sodium dodecyl sulfate–polyacrylamide gel electrophoresis (SDS-PAGE) revealed that four polypeptides with apparent molecular weights of 145,000, 105,000, 54,000, and 35,000 had been labeled. With the calf brain preparation, Mn^{2+} enhanced glycoprotein labeling more effectively than Mg^{2+} or Ca^{2+}. For this system, a broad pH optimum, from 7.0 to 9.5, was observed. Enzymatic transfer of [^{14}C]-GlcNAc from UDP-[^{14}C]-GlcNAc to endogenous acceptors was also demonstrated with neuronal membranes from rat brain (Broquet and Louisot, 1971).

2.4. Galactosyltransferases

One type of galactosyltransferase reaction is illustrated in Fig. 1. In this sequence, the product from the *N*-acetylglucosaminyltransferase reaction (reaction 1) serves as substrate for the galactosyltransferase (reaction 2). Embryonic chicken brain synaptosomal preparations catalyze the transfer of galactose from UDP-galactose to four different types of acceptor glycoproteins (Den *et al.*, 1970). Transfer of sugar was detected by utilizing bovine submaxillary mucin (pretreated with sialidase to expose terminal GalNAc residues), Tay–Sachs ganglioside (also terminal GalNAc), xylosylserine (the carbohydrate–protein linkage in proteoglycans), and orosomucoid (pretreated with sialidase and β-galactosidase to expose GlcNAc residues). Thus, glycoproteins of both the mucin and serum types are galactosylated by brain membranes. Free GlcNAc also served as acceptor. The particulate brain preparation appears to consist of four separate galactosyltransferases based on substrate competition experiments. This preparation also required Mn^{2+}, was routinely assayed at pH 7.4, and required Cutscum for activity.

A soluble galactosyltransferase in embryonic chick brain fluid (Den *et al.*, 1970) is active only with serum-type glycoproteins and GlcNAc. After transfer of galactose catalyzed by the soluble enzyme preparation to desialylated, degalactosylated orosomucoid, the product was digested to [^{14}C]glycopeptides with pronase. Treatment with β-galactosidase released 78% of the radioactivity (as galactose) from glycoprotein.

Broquet and Louisot (1971) detected galactosyltransferase activity in rat brain synaptic vesicles. Ko and Raghupathy (1971) also found, in a microsomal preparation from rat brain, galactosyltransferase activity that required a divalent cation. Optimal activity was observed at pH 6.3. Triton X-100 could be used to solubilize the rat brain microsomal preparation, but the activity with endogenous acceptor was reduced by about 67%. With the use of desialylated, degalactosylated fetuin as exogenous acceptor in the presence of Triton X-100, galactosylation proceeded more efficiently than in the absence of detergent. During development, the rat brain galactosyltransferase activity, measured with endogenous acceptors, increased to an adult level that was 4–6 times greater than the level in newborns. However, in the presence of exogenous acceptors, considerable enzymatic activity was detected even in the preparations from the youngest animals, and maximal activity was obtained at 22 days. In addition, in assaying membranes from young animals, the heat-treated microsomes from older animals provided a greater stimulation of enzymatic activity than heat-treated microsomes from younger animals. The results suggest that during maturation, the observed increase in enzyme activity is due, at least partially, to an increase in the appropriate endogenous acceptors. Garfield and Ilan (1976) also observed developmental variations in galactosyltransferase levels in embryonic chick tissue. With increasing embryonic age from 7 to 15 days, utilizing exogenous acceptor, they observed decreased enzymatic activity in the embryonic chick retina, optic tectum, and telencephalon. The developmental studies of Jato-Rodriguez and Mookerjea (1974) utilizing rat brain preparations and endogenous acceptor produced results different from those of Ko and Raghupathy (1971), since the former report indicated a decrease in galactosyltransferase activity as the age of the animals increased. Mookerjea and Schimmer (1975) compared galactosyltransferase activity in rat brain homogenates and cultured glial tumor cell homogenates. Both preparations required Mn^{2+} and had a pH optimum between 5 and 7. K_m values of 0.22 and 0.13 mM were found for the whole brain homogenate and the glial tumor cell homogenate, respectively. The specific activity of the tumor cells was 4–8 times higher than that of the brain homogenate. Triton X-100 stimulated the activity in the brain homogenate 25-fold, and in the glial tumor preparation 8-fold. Rat brain homogenates and glial tumor cell homogenates showed low incorporation with exogenous fetuin, ovalbumin, and GlcNAc. Asialofetuin was more active than untreated fetuin, but the highest activity was obtained with GlcNAc serving as acceptor.

Bosmann (1972) showed that galactosyltransferase activity can be released from guinea pig cerebral cortex membranes in the presence of Triton X-100. The preparation had a pH optimum of 6.7 and a K_m for UDP-Gal of 8 μM. A requirement for Mn^{2+} was demonstrated. Mono- and disaccharides would not serve as acceptors for the purified guinea pig galactosyltransferase, but orosomucoid or fetuin with sialic acid and galactose removed were substrates. Of the radioactivity in the glycoprotein product labeled by incubation with UDP-[^{14}C]galactose, 93% was recovered as [^{14}C]galactose.

2.5. *N*-Acetylneuraminyltransferases

In the biosynthesis of the terminal trisaccharide unit characteristic of serum-type glycoprotein, transfer of NeuAc to the growing oligosaccharide chain completes the elongation process (Fig. 1, reaction 3).

N-Acetylneuraminyltransferase activity exhibiting an absolute requirement for detergent was found in embryonic chicken brain preparations. Effective acceptors containing galactosyl termini were desialylated orosomucoid, desialylated fetuin, *N*-acetyllactosamine, and lactose (Den *et al.*, 1970). Experiments in which substrates were combined indicated that one enzyme catalyzed the reaction with glycoprotein or the disaccharide acceptors. The K_m values for desialylated orosomucoid, *N*-acetyllactosamine, and lactose were 2.6, 1.0, and 1.5 mM, respectively. The embryonic chick brain *N*-acetylneuraminyltransferase was not inhibited by EDTA.

A procedure used to purify synaptosomal *N*-acetylneuraminyltransferase from guinea pig cortex (Bosmann, 1973) resulted in a preparation that was dependent on exogenous acceptors bearing a terminal Gal residue. These included fetuin (minus NeuAc) and orosomucoid (minus NeuAc). However, desialylated bovine submaxillary mucin, which contains a terminal GalNAc unit, was unacceptable. The enzyme displayed no requirement for a cation, and was not inhibited by EDTA. Optimal activity was obtained at pH 6.3. With the use of the Cleland bisubstrate model, K_m values for desialylated fetuin and CMP-NeuAc were determined to be 35 and 3 μM, respectively.

Van den Eijnden and Van Dijk (1974) also reported *N*-acetylneuraminyltransferase activity in rat and calf brain systems. Neither the rat brain nor the calf brain system displayed requirements for cation, although 10–20% stimulation in the presence of Mn^{2+}, Mg^{2+}, and Ca^{2+} was observed. EDTA was not inhibitory to these systems. Triton X-100 (1%, vol./vol.) stimulated the transfer of NeuAc to desialylated exogenous acceptors 3-fold with rat brain and calf brain preparations. In the presence of endogenous acceptors, Triton produced minimal stimulation of enzymatic activity. Other detergents were less effective in facilitating the interaction between the membrane-associated glycosyltransferase and the acceptor. With exogenous acceptors, rat and calf brain preparations had pH optima between 6.5 and 6.9; with endogenous acceptors, between 6.0 and 6.2. With the rat brain preparation in the presence of exogenous acceptors, a K_m value for CMP-NeuAc was reported to be 0.53 mM. The radioactive products formed with endogenous and desialylated exogenous acceptors were stable to alkaline-borohydride treatment. Presumably, transfer to the desialylated glycoproteins occurs at the nonreducing termini of Gal(β1$\rightarrow$4)GlcNAc units. The nonreducing terminal trisaccharide unit illustrated in Fig. 1 has been found on alkali-stable oligosaccharide chains. On the other hand, transfer to native fetuin was also observed, and at least part of the radioactivity incorporated into fetuin was attached to alkali-labile units, possibly at the nonreducing termini of Gal(β1$\rightarrow$3)GalNAc units. Van den Eijnden *et al.* (1975) also evaluated developmental changes in terms of variations in *N*-acetylneuraminyltransferase activity as well as the level of the endogenous acceptors in rat brain.

The properties of glycoprotein *N*-acetylneuraminyltransferase activity in rat cerebra were studied by Ng and Dain (1977*a*). An apparent K_m of 0.13 mM for CMP-NeuAc was obtained utilizing either endogenous glycoprotein or desialylated fetuin as the acceptor. Since desialylated fetuin did not compete for sialylation with either endogenous or exogenous glycolipid acceptors, it was concluded that the glycoprotein sialyltransferase is a separate enzyme.

2.6. Fucosyltransferases

Fucosyltransferase activity was reported in membranous preparations from mouse brain (Zatz and Barondes, 1971). Solubilization of the membrane-bound activity with Triton X-100 resulted in enhanced incorporation of fucose from GDP-fucose into exogenous substrate, but activity with endogenous acceptors was lost. Porcine plasma glycoprotein was an effective acceptor after pretreatment with mild acid or neuraminidase to remove terminal NeuAc residues. If the desialylated acceptor was also treated with β-galactosidase, the observed activity was markedly reduced. It is possible that the brain fucosyltransferase transfers fucose to galactosyl residues attached to GalNAc, but not GlcNAc. The particulate fucosyltransferase did not have an absolute metal requirement, but was stimulated with added Mg^{2+}. Triton-solubilized enzyme was inhibited by Mg^{2+} or Mn^{2+}. The particulate enzyme had a broad pH optimum from 5.0 to 7.0, whereas the Triton-solubilized enzyme had a narrow optimum around 5.0. Nucleotide inhibition of fucosyltransferase was observed with GTP, GDP, and GMP. Other nucleotides inhibited to a smaller extent.

Studies on fucosyltransferase activity in rat brain have provided evidence for several isozymes (Belon *et al.*, 1975). The reaction kinetics with GDP-fucose and fetuin were consistent with an ordered "Bi–Bi" mechanism, and GDP was observed to be a competitive inhibitor (Louisot *et al.*, 1976).

2.7. Glucosyltransferases

Guinea pig intraneural mitochondria and synaptosomal preparations were shown to contain glucosyltransferase as well as galactosyl- and mannosyltransferases by Bosmann and Hemsworth (1970). In the partially purified synaptosomal preparations of Bosmann (1972), a collagen glucosyltransferase was isolated that was specific for UDP-glucose. The enzyme preparation required a terminal galactose unit on the acceptor. Intact guinea pig skin collagen was ineffective, but collagen with glucose removed was an effective acceptor. The enzyme had a pH optimum of 7.7, and the K_m for UDP-glucose was 4 μM. The glucosyltransferase preparation required a cation, and Mn^{2+} and Mg^{2+} were more effective than Ca^{2+} or Co^{2+}.

Finally, it should be noted that in studies of membrane-bound glucosyltransferases, as well as mannosyltransferases, it is necessary to extract the glycoprotein fractions with $CHCl_3$–CH_3OH–H_2O (10:10:3) to remove the dolichol-bound oligosaccharides that would be enzymatically labeled by incubating brain membranes with GDP-[^{14}C]mannose or UDP-[^{14}C]glucose (see Section 3.2).

3. Lipid-Mediated Glycosyltransferases

This section describes what is currently understood about the biosynthesis, structure, and function of dolichol-linked glycosyl groups in the assembly of asparagine-linked oligosaccharide units in brain glycoproteins. Although definitive proof for every step has not been obtained at this writing, it appears that the postulated reaction scheme for lipid-mediated glycoprotein biosynthesis depicted in Fig. 2 is operative in central nervous tissue. Undoubtedly, the final stages in this pathway will require modifications and additions as more details are learned through further research. The hypothetical conversion of dolichyl esters to dolichyl monophosphate (Dol-P) and the presumptive regeneration of Dol-P from the pyrophosphate form are presented as potentially fertile areas for future studies.

3.1. Biosynthesis and Structure of Monosaccharide and Disaccharide Derivatives of Dolichol

The first clue that mannosylphosphoryldolichol (MPD) was synthesized by brain membranes appeared in 1969, when Zatz and Barondes (1969)

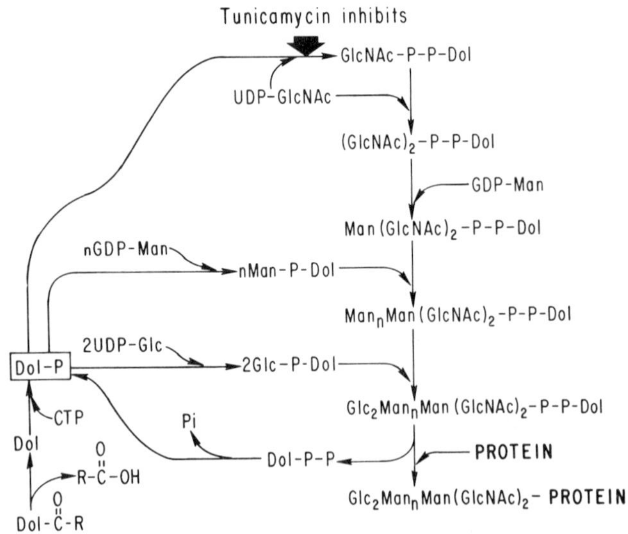

Fig. 2. Postulated reaction scheme for lipid-mediated glycosyltransferases and possible steps in dolichol metabolism.

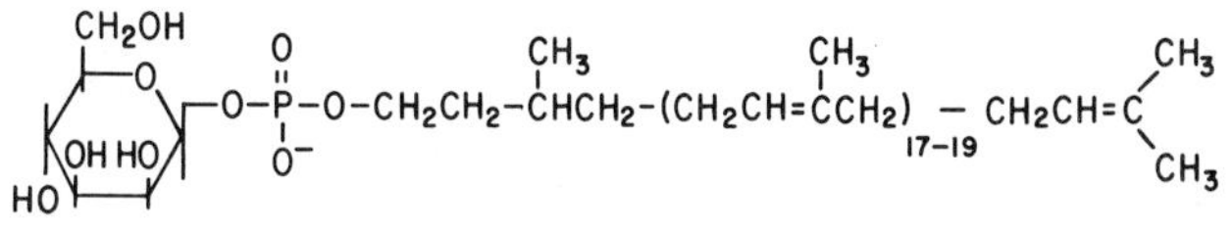

Mannosylphosphoryldolichol

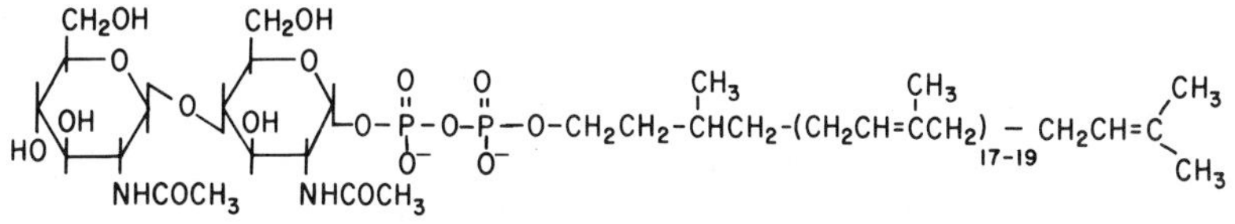

N,N'-Diacetylchitobiosylpyrophosphoryldolichol

Fig. 3. Chemical structures of mannosylphosphoryldolichol and *N,N'*-diacetylchitobiosylpyrophosphoryldolichol.

reported that mouse microsomes catalyzed the transfer of mannose from GDP-mannose into a mannolipid that was sensitive to mild acid, but stable to mild base treatment. This *in vitro* system for mannolipid synthesis required Mn^{2+} and, inexplicably, phosphoenolpyruvate (PEP). Wolfe *et al.* (1974) later showed that when homogenates from embryonic chick brain and adult rat brain were incubated with GDP-mannose, mannose was incorporated into a mannolipid with the properties of a mannosylphosphorylpolyisoprenol. The embryonic chick brain system also exhibited a requirement for a divalent metal ion, and mannolipid synthesis was stimulated by exogenous Dol-P. While maximal activity was observed in the presence of Mn^{2+}, mannolipid synthesis also occurred at a lower rate in the presence of Mg^{2+}. PEP had no effect in embryonic chick brain. More recently, calf white matter membranes were shown to catalyze the transfer of mannose from GDP-mannose into MPD in the presence of a divalent metal ion (Waechter *et al.*, 1976). The order of effectiveness of five metal ions was $Mn^{2+} > Co^{2+} > Ni^{2+} > Mg^{2+} > Hg^{2+}$. A broad concentration optimum for Mn^{2+} was observed between 5 and 50 mM. An apparent K_m of 0.15 μM was determined for GDP-mannose, and at a saturating concentration of sugar nucleotide, the synthesis of mannolipid was stimulated by exogenous Dol-P. The formation of MPD by white matter was also shown to be freely reversible on the addition of GDP, but not of GMP. This result indicates that the free energy for transfer in the sugar nucleotide is preserved in the dolichol-bound monosaccharide. Extensive chemical and chromatographic analyses including cochromatography with authentic, chemically synthesized β-MPD (Warren *et al.*, 1975) in three chromatography systems were consistent with the white matter mannolipid having the structure of a β-mannosylphosphoryldolichol. The structure of MPD is illustrated in Fig. 3. The chain length of the polyprenol is based on a structural analysis of dolichols isolated from calf brain (Breckenridge *et al.*, 1973).

The synthesis of GPD was first detected in brain by Behrens *et al.* (1971*a*). Preliminary evidence for the synthesis of glucosylphosphoryldolichol (GPD) was obtained with membrane preparations from rat brain (Jankowski and Chojnacki, 1972) and embryonic chick brain (Breckenridge and Wolfe, 1973). In these studies, it was found that brain membranes catalyzed the transfer of glucose from UDP-glucose into a lipid fraction, and the incorporation of glucose into glucolipid was increased by the presence of exogenous Dol-P. No information was presented on the number of glucolipids formed or their properties. Definitive evidence for the reversible formation of GPD by white matter membranes from calf brain appeared later (Scher *et al.*, 1977). Incubation of white matter membranes with UDP-glucose in the presence of Mg^{2+} resulted in the enzymatic transfer of glucose into two glucolipids. Based on the mild acid lability, as well as other hydrolytic and chromatographic criteria, one of the glucolipids was identified as GPD. Consistent with this characterization, the synthesis of the mild-acid-labile glucolipid was stimulated by the addition of exogenous Dol-P. The second glucolipid synthesized was stable to mild acid treatment and tentatively identified as glucosylceramide on the basis of several chemical and chromatographic criteria. The synthesis of GPD was stimulated by divalent metal ions and inhibited by EDTA. An apparent K_m of 0.52 μM was calculated for UDP-glucose, and at a saturating concentration of sugar nucleotide, the pH optimum was 8.0. Additional *in vitro* experiments showed that GPD was not a glucosyl donor in calf brain glucosylceramide biosynthesis. Earlier studies on the synthesis and function of GPD in embryonic chick brain (Behrens *et al.*, 1971*a*) led to the same conclusion.

The synthesis of dolichol-linked mono- and disaccharide derivatives containing GlcNAc was also documented for calf brain (Waechter and Harford, 1977). Incubation of white or gray matter membranes with UDP-GlcNAc and AMP resulted in the incorporation of GlcNAc into two major glycolipids and a minor glycolipid. The rationale for including AMP in the reaction mixtures was to inhibit a membrane-associated sugar nucleotide pyrophosphatase that actively degraded UDP-GlcNAc (Pattabiraman *et al.*, 1964; Waechter and Harford, 1977). The two major glycolipids were characterized as GlcNAc-P-P-Dol and $(GlcNAc)_2$-P-P-Dol. The structure of $(GlcNAc)_2$-P-P-Dol is illustrated in Fig. 3. The presence of a pyrophosphate bridge in the GlcNAc-lipids is indicated by their chromatographic properties on DEAE–cellulose, and the observation that UMP is the uridine nucleotide end-product formed during synthesis of GlcNAc-P-P-Dol. The minor [^{14}C]-GlcNAc-labeled lipid formed by brain membranes has the chemical and chromatographic properties that have been observed for Man $(GlcNAc)_2$-P-P-Dol (Levy *et al.*, 1974; Chen and Lennarz, 1976), and Man $\xrightarrow{\beta}$ $(GlcNAc)_2$ is released by mild acid treatment (Waechter and Harford, 1979).

Due to the technical problems in obtaining sufficient amounts of polyisoprenoid glycolipid intermediates formed by brain membranes, rigorous chemical analyses of the lipid moiety have not been performed. Consequently, the dolichol derivatives have been characterized primarily on the basis of hydrolytic and chromatographic properties of the glycolipids labeled *in vitro* by incubating brain membranes with the appropriate sugar nucleotide, radio-labeled in the glycose unit. Demonstrating that exogenous Dol-P stimulates glycolipid biosynthesis has provided corroborative evidence for the nature of the lipid moiety. The general hydrolytic and chromatographic properties can be summarized as follows: (1) the glycose units are released by mild acid treatment, distinguishing them from glycosphingolipids; (2) the glycolipids are stable to mild alkaline methanolysis, under conditions that deacylate glycosyl diglycerides; (3) they have a characteristic mobility when chromatographed on silica gel plates or silica-gel-impregnated paper; and (4) they are retained by DEAE–cellulose.

The functions of the dolichol-linked mono- and disaccharides in the assembly of the oligosaccharide lipid intermediates are described in the next section (also see the reactions illustrated in Fig. 2).

3.2. Biosynthesis and Structure of Dolichol-Linked Oligosaccharides

The most important early development in understanding the role of lipid intermediates in glycoprotein biosynthesis was probably the discovery of dolichol-linked oligosaccharides in rat liver by Leloir and his co-workers (Behrens *et al.*, 1971*b*). The oligosaccharide lipid intermediates formed by animal tissues consist of oligosaccharide chains bound to dolichol by a pyrophosphate bridge. Because of their amphipathic nature, they have the unusual solubility properties of being insoluble in H_2O or $CHCl_3$–CH_3OH mixtures, but can be readily extracted from the delipidated membrane residue with $CHCl_3$–CH_3OH–H_2O (10:10:3).

The first indication that dolichol-linked oligosaccharides were formed in brain came from a brief survey of several animal tissues by Parodi *et al.* (1973) in which they found that a glucosylated oligosaccharide lipid was labeled by incubating rat brain microsomes with UDP-[^{14}C]glucose. Further proof for the biosynthesis of dolichol-linked oligosaccharides by brain was obtained from *in vitro* studies with membrane preparations from calf (Waechter *et al.*, 1976; Scher *et al.*, 1977) or pig white matter (Harford *et al.*, 1977). When white matter membranes from calf or pig brain are incubated with GDP-[^{14}C]mannose, [^{14}C]mannose is incorporated not only into MPD, but also into a mannosylated oligosaccharide lipid and several membrane-bound mannoproteins. The mannose-containing oligosaccharide lipid had the hydrolytic and chromatographic properties of the oligosaccharide lipid intermediates in which the oligosaccharide chains are attached to dolichol by a pyrophosphate linkage. *In vitro* studies showed that the biosynthesis of the mannosylated oligosaccharide lipid occurs by the mannosylation of *N,N'*-diacetylchitobiosylpyrophosphoryldolichol (Waechter and Harford, 1977). On the basis of chromatographic size estimations and the observation that

$(GlcNAc)_2$-P-P-Dol is a precursor of the mannosylated oligosasccharide lipid, the structure $(Man)_6(GlcNAc)_2$-P-P-Dol was proposed for the mannosylated oligosaccharide lipid formed *in vitro* by white matter membranes. Direct evidence for the role of MPD as a mannosyl donor in the biosynthesis of the dolichol-linked oligosaccharide by brain was obtained by demonstrating that when white matter membranes were incubated with exogenous [^{14}C]-MPD, [^{14}C]mannose was transferred into oligosaccharide lipid Waechter *et al.*, 1976). Moreover, all the [^{14}C]mannose in the lipid-linked oligosaccharide can be released by an α-mannosidase (Harford and Waechter, 1978), indicating that an inversion of anomeric configuration of the mannosyl residue occurs during transfer, since the mannosyl unit is believed to be β-linked in MPD. Considering the similarity between the white matter lipid-bound oligosaccharide and the dolichol-bound oligosaccharides formed by other animal tissues, it is likely that the first mannose residue attached to $(GlcNAc)_2$-P-P-Dol is β-linked.

As mentioned above, white matter membranes catalyze the transfer of glucose from UDP-glucose into GPD (Scher *et al.*, 1977). Under the same conditions, glucose was also actively incorporated into a glucosylated oligosaccharide lipid. The intact glucosylated oligosaccharide lipid exhibited the chromatographic behavior of a pyrophosphate-containing glycolipid on paper and DEAE–cellulose chromatography. The glucosylated oligosaccharide, released by mild acid hydrolysis, was judged to contain approximately 10 glycose units by paper chromatographic and gel filtration analysis. There is indirect evidence for the presence of two glycosamines and an undetermined number of mannose residues (Scher and Waechter, 1978*b*). It should be emphasized that the size of the white matter glucosylated oligosaccharide may have been underestimated due to the presence of branched glycosyl units. As a working hypothesis, the structure $(Glc)_2$-$(Man)_n(GlcNAc)_2$-P-P-Dol was proposed for the glucosylated oligosaccharide lipid formed *in vitro* by white matter membrane preparations. It is quite likely that the glucosylated oligosaccharide lipid is formed in brain by the series of biosynthetic steps illustrated in Fig. 2. Proof that at least one of the glucose units is derived from GPD was based on the observation that white matter membranes catalyzed the transfer of [^{14}C]glucose from exogenous [^{14}C]-GPD into an endogenous oligosaccharide lipid acceptor (Waechter and Scher, 1978). The exact number of glucose residues, as well as the anomeric linkages, remain to be elucidated. In related studies, it was demonstrated that thyroid slices synthesize a dolichol-linked oligosaccharide containing 11 mannose, 1–2 glucose, and 2 *N*-acetylglucosamine residues (Spiro *et al.*, 1976). Thus, it is possible that the lipid-linked glucosylated oligosaccharide formed *in vitro* by white matter membranes is incomplete, and that additional sugars are added *in vivo*. Further studies with brain membrane preparations will be required to resolve this question.

3.3. Mode of Action of Tunicamycin as a Glycoprotein Biosynthesis Inhibitor

Tunicamycin (TM), a glucosamine-containing glycoprotein biosynthesis inhibitor produced by *Streptomyces lysosuperificus*, was first isolated by Tamura and his co-workers (Takatsuki *et al.*, 1971). The molecular weight was estimated to be 870 by the vapor pressure equilibrium method. Recently, the gross structure of the antibiotic was reported, and it was found to be a homologous mixture varying in the chain length of a fatty acid constituent (Takatsuki *et al.*, 1977).

While TM was originally detected by its antiviral activity, it was later shown to block the incorporation of glucosamine into glycoproteins of chick embryo fibroblasts (Takatsuki and Tamura, 1971). Studies with yeast also showed that TM blocks the incorporation of glucosamine into glycoproteins and mannan glycopeptides, but has no effect on the incorporation of glucosamine into chitin (Kuo and Lampen, 1974). On the basis of results obtained with chick embryo fibroblasts (Takatsuki and Tamura, 1971), yeast (Kuo and Lampen, 1974), and hen oviduct slices (Struck and Lennarz, 1977), TM has little, if any, effect on the biosynthesis of polypeptide chains. *In vitro* experiments performed with membrane preparations from yeast (Lehle and Tanner, 1976) and a variety of animal tissues including brain (Tkacz and Lampen, 1975; Takatsuki *et al.*, 1975; Struck and Lennarz, 1977; Waechter and Harford, 1977) revealed that TM blocks the enzymatic transfer of GlcNAc-1-phosphate from UDP-GlcNAc to Dol-P, preventing the formation of GlcNAc-P-P-Dol (see topmost reaction in Fig. 2). The enzymatic studies conducted with white matter membranes from calf brain showed that over 90% inhibition of GlcNAc-lipid synthesis is observed at a level of TM of only 100 ng/ml (Table I). Under the same conditions that inhibit GlcNAc-lipid biosynthesis, TM has virtually no effect on the direct transfer of GlcNAc residues from the nucleotide derivative to nonreducing termini of oligosaccharide chains of several endogenous glycoproteins

Table I. Effect of Tunicamycin on the Enzymatic Transfer of [^{14}C]-GlcNAc from UDP-[^{14}C]-GlcNAc into Glycolipids and Glycoproteins of Calf White Matter Membranes[a]

Tunicamycin in assay mixture (ng)	[^{14}C]-GlcNAc incorporated (cpm)	
	GlcNAc-lipids	Glycoprotein
None	4115	751
5	2993	660
25	695	733
100	314	740

[a] Data from Waechter and Harford (1977).

associated with white matter membranes. Thus, the mode of action of TM is the selective blocking of the transfer of the phosphorylated form of GlcNAc from the nucleotide sugar to the acceptor lipid, thereby inhibiting the synthesis of GlcNAc-P-P-Dol, a precursor of the dolichol-linked oligosaccharide intermediates. By arresting the formation of this important precursor of the oligosaccharide lipids, the assembly of the mannosylated *N,N'*-diacetylchitobiose inner core found *N*-glycosidically bound to glycoprotein is blocked. The enzymatic synthesis of MPD and GPD by white matter membranes is not affected by TM, indicating that the antibiotic is not a broad-range inhibitor of the synthesis of all dolichol-linked sugars. It is possible, however, that TM analogues exist in nature that block the lipid intermediate pathway at other stages of the assembly process.

In nonneural systems, TM has been used as a means of gaining information about the role of protein glycosylation in the synthesis and secretion of hen ovalbumin (Struck and Lennarz, 1977), the conversion of procollagen to collagen (Duksin and Bornstein, 1977), the formation of Semliki forest and fowl plague virus particles (Schwarz *et al.*, 1976), and the biosynthesis of Sindbis envelope glycoproteins (Krag and Robbins, 1977; Leavitt *et al.*, 1977). In future studies, TM should prove to be a useful experimental tool for learning more about the function of the mannose- and *N*-acetylglucosamine-containing oligosaccharide chains of membrane glycoproteins in nervous tissue.

3.4. Possible Regulation of the Lipid Intermediate Pathway by Modulation of the Level of the Carrier Lipid, Dolichyl Monophosphate

As discussed in Section 3.1, several enzymatic studies with membrane preparations from brain tissue showed that the *in vitro* synthesis of GPD, MPD, and GlcNAc-P-P-Dol can be stimulated by the addition of exogenous Dol-P. These results suggest that in the presence of saturating amounts of sugar nucleotide, the level of Dol-P could be rate-limiting for the biosynthesis of the dolichol-linked monosaccharides. Thus, it is plausible that changes in the membrane content of Dol-P could affect the rate of synthesis of the oligosaccharide lipid intermediates by modulating the formation of dolichol-linked monosaccharides, and consequently influence the rate of assembly of certain membrane glycoproteins in nervous tissue. Consistent with this hypothetical control mechanism, it was found that the addition of exogenous Dol-P to reaction mixtures not only stimulated the transfer of mannose from GDP-mannose into MPD, but also resulted in an enhanced incorporation of mannose into a mannosylated oligosaccharide lipid and membrane-associated mannoproteins (Harford *et al.*, 1977). It was also observed that in the presence of a saturating level of GDP-mannose, the initial rate of synthesis of MPD by white matter membranes from actively myelinating pigs was approximately 3 times greater than the initial rate of mannolipid synthesis by corresponding preparations from adult pigs. Two possible explanations for this apparent developmental difference are: (1) that the amount of MPD synthase is higher in white matter membranes from the actively myelinating pigs or (2) that the level of endogenous acceptor lipid, Dol-P, is higher. To try to distinguish between these two possibilities, the initial rates of MPD synthesis were assayed in the presence of amounts of exogenous Dol-P that approached saturation. This comparison showed that while exogenus Dol-P stimulated mannolipid biosynthesis in both preparations, the relative enhancement in adult membranes was higher than in membranes from young animals. The result of this differential enhancement was that the ratio of initial rates in young/adult was significantly reduced. Thus, it is quite possible that the apparent developmental difference in the synthesis of MPD, an important mannosyl donor, is due, at least in part, to a smaller utilizable pool or a lower content of Dol-P in the white matter membranes prepared from adult animals. Results were also obtained with chick oviduct membranes indicating that changes in the level of Dol-P can influence the rate of MPD synthesis during diethylstilbestrol-induced differentiation (Lucas and Levine, 1977).

Before the possible regulatory functions of fluctuating levels of Dol-P can be fully assessed, it will be necessary to know more about the enzymatic properties of the individual lipid-mediated glycosyltransferases catalyzing the mono- and oligosaccharide transfers in the lipid intermediate pathway for glycoprotein assembly. It will also be of critical importance to understand the pathways of *de novo* synthesis of Dol-P, the potential conversion of dolichyl esters to Dol-P, and the dephosphorylation of the Dol-P-P discharged during the transfer of the lipid-bound oligosaccharide to protein (see the postulated reaction scheme in Fig. 2). Recent work has shown that calf brain membranes catalyze the phosphorylation of dolichol with CTP serving as the phosphoryl donor (Burton, Scher, and Waechter, unpublished observation).

3.5. Speculation on the Function of Dolichol-Linked Oligosaccharides and the Possible Nature of the Brain Glycoproteins Glycosylated via Lipophilic Glycosyl Donors

Substantial experimental evidence has shown that the biosynthesis of *O*-glycosidically linked oligosaccharide chains and the terminal trisaccharide units (NeuAc-Gal-GlcNAc) of complex asparagine-linked oligosaccharide chains proceeds by the stepwise addition of single sugar residues with the sugar nucleotide derivatives serving as the direct glycosyl donors. However, until recently, little information has been available on the mechanism for the biosynthesis of the mannosylated *N,N′*-diacetylchitobiose inner-core region found attached to asparagine residues through an *N*-glycosidic bond. Because of the structural resemblance of the dolichol-linked oligosaccharides and the mannosylated *N,N′*-diacetylchitobiose core, speculation that polyisoprenoid glycolipids are involved in the assembly of the inner core became a popular hypothesis. Extensive enzymatic studies carried out with many nonneural tissues have strongly implicated oligosaccharide lipid intermediates in the assembly of asparagine-linked oligosaccharide chains. It is quite likely that the lipid-mediated mechanism for glycosylation is operative in the assembly of soluble and membrane-bound glycoproteins containing the mannosylated *N,N'*-diacetylchitobiose inner core in central nervous tissue.

There is firm evidence for the biosynthesis of dolichol-linked oligosaccharides by brain membranes (see Section 3.2). Several types of experiments have indicated that these oligosaccharide lipids function as intermediates in the assembly of glycoproteins tightly associated with membrane preparations from central nervous tissue. Early enzymatic investigations with GDP-[^{14}C]mannose serving as the isotopic precursor yielded a kinetic pattern of labeling of MPD, the mannosylated oligosaccharide lipid, and glycoprotein that was consistent with an intermediary role for the lipophilic mannosyl carriers (Waechter *et al.*, 1976). The white matter mannoproteins labeled *in vitro* were estimated to have apparent molecular weights of 105,000, 59,000, and 19,000. An investigation of the effect of exogenous Dol-P on the transfer of [^{14}C]mannose from GDP-[^{14}C]mannose into endogenous acceptors yielded evidence for the participation of a mannosylated oligosaccharide lipid in the biosynthesis of membrane-bound mannoproteins in white and gray matter membranes from pig brain (Harford *et al.*, 1977). This work showed that when labeling of MPD and a mannosylated oligosaccharide lipid was stimulated by adding exogenous Dol-P, there was a corresponding increase in the incorporation of label into mannoprotein. These data provide indirect enzymatic evidence for the *en bloc* transfer of the oligosaccharide unit, $Man_nGlcNAc_2$, to polypeptide acceptors, as depicted in Fig. 4 (Reaction A).

This oligosaccharide transfer reaction has also been demonstrated for retinal membranes (Kean, 1977), and there is also evidence that the direct transfer of $Man(GlcNAc)_2$ and $(GlcNAc)_2$ from the dolichol derivatives to protein is catalyzed by hen oviduct membranes (Chen and Lennarz, 1976, 1977).

SDS-PAGE analysis of the polypeptides labeled by incubating pig brain membranes with GDP-[^{14}C]mannose in the presence and absence of exogenous Dol-P showed that the major labeled polypeptide in both white and gray matter membranes has an apparent molecular weight of 60,000. Another glycoprotein with an apparent molecular weight of 100,000 was labeled only by the white matter membranes. Since white matter is relatively enriched in oligodendroglial cells, it is conceivable that this high-molecular-weight mannoprotein is related to the glycoprotein associated with myelin that has a similar molecular weight (see Chapter 11). The stimulation of

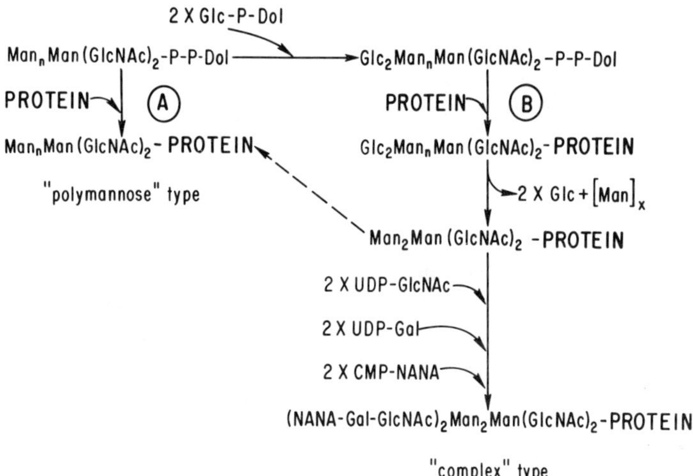

Fig. 4. Possible reaction sequence involving lipid intermediates in the assembly of glycoproteins containing "polymannose" and "complex" asparagine-linked oligosaccharide units.

labeling of this high-molecular-weight glycoprotein, under conditions that result in enhanced labeling of the putative mannosylated oligosaccharide lipid intermediate, indicates that at least one oligosaccharide unit in this glycoprotein is synthesized by the lipid-mediated process.

Three different experimental approaches have provided evidence that a glucosylated oligosaccharide lipid can serve as an oligosaccharide donor in the biosynthesis of calf brain glycoprotein, as shown in Fig. 2 and in Fig. 4 (reaction B). First, the kinetic pattern of labeling of GPD, the glucosylated oligosaccharide lipid, and glycoprotein, when white matter membranes were incubated with UDP-[^{14}C]glucose, was consistent with a precursor–product relationship existing between the lipophilic glycosyl carriers and glycoprotein (Scher *et al.*, 1977). Furthermore, when Ca^{2+} was added to calf brain membranes containing prelabeled endogenous [^{14}C]-Glc-labeled oligosaccharide lipid, label was lost from the oligosaccharide lipid fraction, and there was a concurrent incorporation of a virtually stoichiometric amount of label into glycoprotein (Waechter and Scher, 1978). More convincing data for the intermediary role of the glucosylated oligosaccharide lipid in glycoprotein synthesis in brain was obtained by demonstrating that when calf brain membranes were incubated with exogenous [^{14}C]-Glc-labeled oligosaccharide lipid in the presence of Ca^{2+}, label was enzymatically transferred to glycoprotein. While the exact number and molecular weights of the proteins glycosylated *via* the oligosaccharide lipid are not yet known, the labeled glycopeptide produced by pronase digestion of the product formed *in vitro* is quite similar in size to the lipid-linked [^{14}C]-Glc-labeled oligosaccharide (Scher and Waechter, 1978*a*). The nature of the glycoproteins that appear to be glycosylated via the glucosylated oligosaccharide lipid is still uncertain.

Since there is little evidence for the occurrence of glucoproteins in neural membranes, the ultimate fate of the glucosyl residues remains a controversial question. It is possible that these glucose units represent a regulatory signal for further processing of the glycoprotein, and are subsequently removed by endo- or exoglycosidases. The excision of the glucosyl and probably some mannosyl residues may be intimately involved in determining whether the mannosylated *N,N′*-diacetylchitobiose inner core is to be subsequently elaborated into a "complex" type (Fig. 4) or a "polymannose" type (Fig. 4, dashed arrow) asparagine-linked oligosaccharide unit (see structures below):

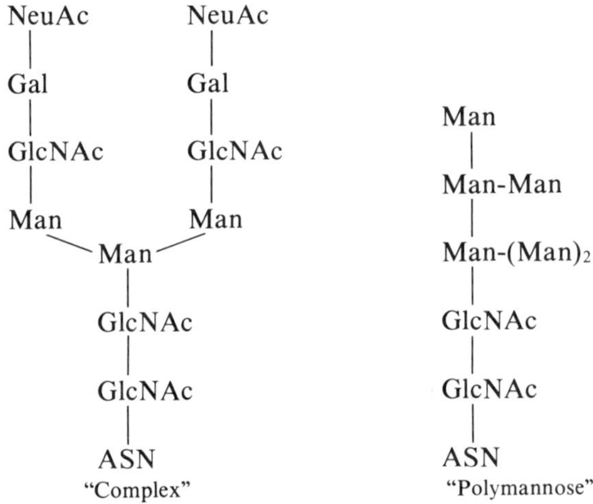

An alternative hypothesis is also illustrated in Fig. 4. In this mechanism, the neutral "polymannose"-type chain would be transferred directly from dolichyl diphosphate prior to the addition of the glucosyl residues (Fig. 4, reaction A). The insertion of the glucose units into the lipid-bound oligosaccharide would then serve as a regulatory signal to mark the oligosaccharide chain for further processing leading ultimately to the completion of an acidic "complex" type of asparagine-bound oligosaccharide chain.

Recent studies (Scher and Waechter, 1978*b*) on lipid-mediated glycoprotein biosynthesis by calf brain membranes indicated that glucosyl residues are excised subsequent to the attachment of the glucosylated oligosaccharide chain to membrane-bound polypeptide acceptors. Furthermore, the presence of a glucosidase activity associated with calf brain membranes that might be involved in the excision process has been documented. Many details of this excision–revision scheme will probably be elucidated by the time this volume appears.

Presumably, the sugar nucleotide glycosyltransferases illustrated in Fig. 1 would participate in the completion of the "complex" type of *N*-glycosidically linked oligosaccharides. The possibility that there are small amounts of membrane-associated glycoproteins containing oligosaccharide chains consisting of mannose, glucose, and GlcNAc, however, cannot yet be excluded. In this context, it was reported that there are protein-bound forms of glycogen in rat liver (Butler *et al.*, 1977), and evidence for a protein primer in glycogen biosynthesis was presented by Krisman and Barengo (1975). These findings warrant further investigation into the possibility that the polysaccharide chains of at least one form of glycogen might be joined to a polypeptide primer by a linkage region formed via lipid intermediates.

4. Subcellular Site(s) of Sugar-Nucleotide- and Lipid-Mediated Glycosyltransferases

Determining the subcellular location(s) of glycosyltransferases in nervous tissue is clearly of critical importance to understanding the possible neurological functions served by these enzymes. Due to the complex organization of nervous tissue, obtaining homogenous membrane fractions has presented a difficult obstacle to unequivocally ascertaining the sites of enzymatic activity. Measurement of glycosyltransferase activity in various subcellular fractions has in most cases resulted in the determination that glycosyltransferase activity is associated with synaptosomes. Many of the *in vivo* and *in vitro* studies on glycosyltransferases associated with synaptosomal preparations have been critically assessed elsewhere (Barondes, 1974). However, the issue remains unresolved.

Bosmann localized *N*-acetylgalactosaminyl-, *N*-acetylglucosaminyl-, glucosyl-, and galactosyltransferase activities in the plasma membrane and mitochondrial subfractions isolated from the synaptosome fraction of guinea pig cerebral cortex (Bosmann, 1972; Bosmann and Hemsworth, 1970). Electron microscopy correlated with enzymatic analysis established that galactosyl- and mannosyltransferase activities were associated with synaptic vesicles from rat brain, although *N*-acetylglucosaminyltransferase was distributed more broadly in other neuronal membrane fractions (Broquet and Louisot, 1971).

Den and Kaufman (1968) reported that the primary location of several glycoprotein glycosyltransferases in embryonic chicken brain was the synaptosomal fraction. Although it is unusual to find a soluble glycosyltransferase involved in glycoprotein synthesis, 60–70% of embryonic chicken brain galactosyltransferase exists in the soluble form (Den *et al.*, 1970). On the basis of specific activity measurements in the chicken serum, vitreous humor, and fluid surrounding the brain, it was suggested that this enzyme might be synthesized in brain cells. The abundance in brain fluids suggests a role for the galactosyltransferase in brain function or development. In addition, when detergent was included in the assay mixture, it was determined that 20–40% of the galactosyltransferase activity was associated with the synaptosome-rich pellet. This also contained *N*-acetylneuraminyl- and *N*-acetylglucosaminyltransferase activities.

However, contradictory reports have appeared stating that galactosyltransferase activity was not enriched in the synaptic vesicles. Morgan and coworkers found that the specific activity of UDP-galactose: *N*-acetylglucosamine galactosyltransferase was higher in a Golgi-enriched fraction from rat brain than in the synaptosomal fractions (Morgan *et al.*, 1972; Reith *et al.*, 1972). They suggested that the activity reported by other laboratories in the synaptic vesicles and synaptosomal plasma membranes was due to contamination with membranes of the Golgi apparatus or endoplasmic reticulum. Ko and Raghupathy (1971) reported that the specific activity for UDP-galactose:glycoprotein galactosyltransferase was higher in the microsomal fraction than in the synaptosomes or mitochondria prepared from rat brains. Subsequently, a comparison of the methods of preparing synaptosomes was performed (Raghupathy *et al.*, 1972). In some preparations, significant contamination of synaptosomes with polysomes was observed. Reduction in the amount of microsomal contamination by repeated washings was accompanied by a reduction in the specific activity of galactosyl-, *N*-acetylgalactosaminyl-, and *N*-acetylneuraminyltransferases.

Results of rigorous enzymatic analysis and quantitation, as well as electron microscopy, of subcellular fractions were recently reported by Den *et al.* (1975). Of the particulate galactosyltransferase activity, 70% was associated with the crude mitochondrial fraction, from which synaptosomes were prepared. Of the activity associated with the crude mitochondrial fraction, 80% was then found in the synaptosomal fraction. *N*-acetylneuraminyl- and *N*-acetylglucosaminyltransferases behaved similarly. Electron microscopy revealed that this fraction was rich in synaptosomes (and also contained membranes of unknown origin), whereas other fractions showed almost no synaptosomes. Synaptic membrane fractions prepared in two ways from osmotically shocked mitochondria also contained the highest specific and total activities of galactosyltransferase. Comparing fucosyltransferase activity in various subcellular fractions from mouse brain, Zatz and Barondes (1971) found that the activity was highest in microsomes. The synaptosomal preparation was virtually inactive.

N-Acetylneuraminyltransferase activity was also isolated from synaptosomes of guinea pig cortex and purified 64-fold (Bosmann, 1973). In contrast, Van den Eijnden and Van Dijk (1974) reported that the distribution of *N*-acetylneuraminyltransferase in frontal gray cortex and frontal white cortex as well as other calf brain areas suggested localization of this enzyme in membranes other than those of synaptic complexes. Van den Eijnden *et al.* (1975) then reported that at least at certain developmental stages, *N*-acetylneuraminyltransferase must be localized at a site other than synaptic membranes, since early in postnatal development the occurrence of rat brain synapses is negligible. *N*-acetylneuraminyltransferase activity was found in neuronal and glial cell bodies (Gielen and Hinzen, 1974). Recent studies with rat cerebra (Ng and Dain, 1977*b*) also indicate that the CMP-*N*-acetylneuraminic acid : glycoprotein *N*-acetylneuraminyltransferase activity is not associated with synaptosomes, but rather appears to be localized in smooth microsomal membranes and Golgi complex.

Unfortunately, there is very little information on the subcellular distribution of lipid-mediated glycosyltransferases in central nervous tissue. Wolfe *et al.* (1974) compared three subcellular fractions from embryonic chick brain for their ability to synthesize MPD from GDP-mannose utilizing endogenous acceptor lipid. Their results showed that the relative rates of MPD synthesis were in the order crude mitochondria > microsomes > nuclei. It is not clear from these studies whether the differential rates were due to variable amounts of the mannosyltransferase or the level of the acceptor lipid.

Evidence was presented for the synthesis of MPD and GlcNAc-P-P-Dol by axolemma-enriched membrane preparations from bovine brain (Harford *et al.*, 1979). These findings may prove to be significant in relation to the postulated role of glycosyltransferases in myelination. Since it is unlikely that sugar nucleotides are available extracellularly, the formation of dolichol-linked sugars in axolemma would provide activated glycosyl groups at the axon surface.

5. Summary and Concluding Remarks

There is now firm experimental evidence that two distinct enzymatic mechanisms for the glycosylation of glycoproteins are operative in nervous tissue. The most familiar mechanism involves the sequential transfer of single sugar residues to several types of nascent oligosaccharide chains with the sugar nucleotide derivatives serving as the direct glycosyl donors. These sugar nucleotide : glycoprotein glycosyltransferases are usually found in particulate preparations, and enzymatic activity is frequently stimulated by detergent. Some have absolute detergent requirements. Each preparation may transfer a sugar to one or more glycoprotein acceptors that have the required glycose residue exposed. However, there is little evidence as to the nature of the endogenous acceptors. Divalent metal ions are required for most of the enzyme preparations, with the notable exception of *N*-acetylneuraminyltransferase.

In the other mechanism, dolichol-bound carbohydrate groups function as activated glycosyl carriers in the biosynthesis of a mannosylated *N,N'*-

diacetylchitobiose common core found *N*-glycosidically linked to the amide group of asparagine residues. In the lipid-mediated pathway, the oligosaccharide unit is preassembled while bound to the polyisoprenol carrier lipid by a pyrophosphate bridge. The covalent attachment of a glucosylated oligosaccharide unit to the acceptor polypeptide occurs subsequently by an *en bloc* transfer. It is quite likely that further processing of the carbohydrate chain occurs while attached to the polypeptide. A picture is beginning to emerge in which the dolichol-bound oligosaccharide containing glucose, mannose, and *N*-acetylglucosamine (Figs. 2 and 4) may serve as a common intermediate in the assembly of both the "polymannose" and the "complex" type of asparagine-linked oligosaccharide units (Robbins *et al.*, 1977). The possibility that the "polymannose" oligosaccharides are transferred directly to the polypeptide acceptor before the unit is modified by glucosylation (as shown in Fig. 4, reaction A) has not yet been eliminated.

Many of the enzymatic properties of sugar-nucleotide- and lipid-mediated glycosyltransferases involved in these two mechanistically different biosynthetic pathways in brain are compiled in this chapter. While there is much information about the enzymatic parameters and specificity of sugar nucleotide glycosyltransferases in nervous tissue, many aspects of these important biosynthetic enzymes need to be clarified. It is to be hoped that progress in developing procedures for obtaining well-characterized membrane fractions will provide more knowledge of the subcellular location(s) and possible neurological functions of sugar nucleotide glycosyltransferases. Recent technical developments in solubilizing and purifying membrane proteins should also lead to more detailed studies on the regulation of these enzymes.

A prospectus for future investigations on the lipid intermediate pathway in nervous tissue should include studies aimed at: (1) elucidating the exact relationship between the glucosylated oligosaccharide lipid intermediate and the "polymannose" and "complex" types of oligosaccharide units found *N*-glycosidically linked to asparagine residues; (2) establishing the subcellular site(s) of the lipid-mediated glycosyltransferases; (3) examining the various aspects of dolichol, dolichyl ester, and dolichyl phosphate metabolism that might influence the regulation of protein glycosylation via lipophilic glycosyl donors; and (4) clarifying the structure and function of glycoproteins glycosylated by lipid intermediates in nervous tissue.

ACKNOWLEDGMENTS. The authors would like to express their appreciation to Miss Cheryl Smith for her invaluable help with the preparation of the manuscript. The studies that were performed by the authors were supported by USPHS Grant No. NS-12296 and by a Basil O'Connor Starter Research Grant from the National Foundation of the March of Dimes.

References

Barondes, S. H., 1970, Brain glycomacromolecules and intraneuronal recognition, in: *Brain Glycoconjugates and Intraneuronal Recognition* (F. O. Schmitt, ed.), pp. 747–760, Rockefeller University Press, New York.

Barondes, S. H., 1974, Synaptic macromolecules: Identification and metabolism, *Annu. Rev. Biochem.* **43:**147.

Behrens, N. H., Parodi, A. J., Leloir, L. F., and Krisman, C. R., 1971*a*, The role of dolichol monophosphate in sugar transfer, *Arch. Biochem. Biophys.* **143:**375.

Behrens, N. H., Parodi, A. J., and Leloir, L. F., 1971*b*, Glucose transfer from dolichol monophosphate glucose: The product formed with endogenous microsomal acceptor, *Proc. Natl. Acad. Sci. U.S.A.* **68:**2857.

Belon, P., Broquet, P., Guidollet, J., Guillaumond, M., Levrat, C., Martin, A., Neveu, F., Richard, M., and Louisot, P., 1975, Séparation des isoenzymes des glucosyltransférases par électrofocalisation sur colonne, *C. R. Acad. Sci. Ser. D* **280:**767.

Bosmann, H. B., 1972, Synthesis of glycoproteins in brain: Identification, purification and properties of glycosyltransferases from purified synaptosomes of guinea pig cerebral cortex, *J. Neurochem.* **19:**763.

Bosmann, H. B., 1973, Synthesis of glycoproteins in brain: Identification, purification and properties of a synaptosomal sialyltransferase utilizing endogenous and exogenous acceptors, *J. Neurochem.* **20:**1037.

Bosmann, H. B., and Hemsworth, B. A., 1970, Intraneural mitochondria, incorporation of amino acids and monosaccharides into macromolecules by isolated synaptosomes and synaptosomal mitochondria, *J. Biol. Chem.* **245:**363.

Brady, R. O., and Quarles, R. H., 1973, The enzymology of myelination, *Mol. Cell. Biochem.* **2:**23.

Breckenridge, W. C., and Wolfe, L. S., 1973, The effect of dolichol phosphate on the synthesis of lipid bound sugars in embryonic chick brain, *FEBS Lett.* **29:**66.

Breckenridge, W. C., Wolfe, L. S. and Ng Ying Kin, N. M. K., 1973, The structure of brain polyisoprenols, *J. Neurochem.* **21:**1311.

Broquet, P., and Louisot, P., 1971, Biosynthèse des glycoprotéines cérébrales. II. Localisation subcellulaire des transglycosylases cérébrales, *Biochimie* **53:**921.

Broquet, P., Richard, M., and Louisot, P., 1971. Biosynthèse des glycoprotéines cérébrales: Étude de l'activité mannosyltransferase particulee du cerveaym *J. Neurochem.* **18:**2291.

Broquet, P., Morelis, R., and Louisot, P., 1975*a*, Biosynthèse des glycoprotéines cérébrales: Étude de l'activité mannosyltransferase mitochondriale, *J. Neurochem.* **24:**989.

Broquet, P., Morelis, R., and Louisot, P., 1975*b*, Evidence for the existence of a cerebral mitochondrial mannosyltransferase, *Biochimie* **57:**983.

Butler, N. A., Lee, E. Y. C., and Whelan, W. J., 1977, A protein-bound glycogen component of rat liver, *Carbohydr. Res.* **55:**73.

Chen, W. W., and Lennarz, W. J., 1976, Participation of a trisaccharide-lipid in glycosylation of oviduct membrane glycoproteins, *J. Biol. Chem.* **251:**7802.

Chen, W. W., and Lennarz, W. J., 1977, Metabolism of lipid-linked *N*-acetylglucosamine intermediates, *J. Biol. Chem.* **252:**3473.

Den, H., and Kaufman, B., 1968, Ganglioside and glycoprotein glycosyltransferases in synaptosomes, *Fed. Proc. Fed. Am. Soc. Exp. Biol.* **27:**346.

Den, H., Kaufman, B., and Roseman, S., 1970, Properties of some glycosyltransferases in embryonic chicken brain, *J. Biol. Chem.* **245:**6607.

Den, H., Kaufman, B., McGuire, E. J., and Roseman, S., 1975, The sialic acids. XVIII. Subcellular distribution of seven glycosyltransferases in embryonic chicken brain, *J. Biol. Chem.* **250:**739.

Duksin, D., and Bornstein, P., 1977, Impaired conversion of procollagen to collagen by fibroblasts and bone treated with tunicamycin, an inhibitor of protein glycosylation, *J. Biol. Chem.* **252:**955.

Garfield, S., and Ilan, J., 1976, Galactosyltransferase activities during embryonic development of chick neural tissue, *Biochim. Biophys. Acta* **444:**154.

Gielen, W., and Hinzen, D. H., 1974, Acetylneuraminat-Cytidyltransferase und Sialyltransferase in isolierten neuronal und Gliazellen des Rattengehirns, *Z. Physiol. Chem.* **355:**895.

Gottschalk, A. (ed.), 1972, *Glycoproteins,* Elsevier, Amsterdam.

Harford, J. B., and Waechter, C. J., 1978, Mannosylphosphoryldolichol as a mannosyl donor in pig brain white matter, *Trans. Am. Soc. Neurochem.* **9:**170.

Harford, J. B., Waechter, C. J., and Earl, F. L., 1977, Effect of exogenous dolichyl monophosphate on a developmental change in mannosylphosphoryldolichol biosynthesis, *Biochem. Biophys. Res. Commun.* **76:**1036.

Harford, J. B., Waechter, C. J., Saul, R., and DeVries, G. H., 1979, Evidence for the biosynthesis of mannosylphosphoryldolichol and *N*-acetylglucosaminyl-pyrophosphoryldolichol by an axolemma-enriched membrane preparation from bovine white matter, *J. Neurochem.* **32:**91.

Hemming, F. W., 1977, Dolichol phosphate, a coenzyme in the glycosylation of animal membrane-bound glycoproteins, *Biochem. Soc. Trans.* **5:**1223.

Jankowski, W., and Chojnacki, T., 1972, Formation of lipid-linked sugars in rat liver and brain microsomes, *Biochim. Biophys. Acta* **260:**93.

Jato-Rodriguez, J. J., and Mookerjea, S., 1974, UDP-galactose: glycoprotein galactosyltransferase activity in tissues of developing rat, *Arch. Biochem. Biophys.* **162:**281.

Kean, E. L., 1977, GDP-mannose-polyprenyl phosphate mannosyltransferases of the retina, *J. Biol. Chem.* **252:**5622.

Ko, G. K. W., and Raghupathy, E., 1971, Glycoprotein biosynthesis in the developing rat brain, *Biochim. Biophys. Acta* **244:**396.

Ko, G. K. W., and Raghupathy, E., 1972, Glycoprotein biosynthesis in the developing rat brain. II. Microsomal galactosyltransferase utilizing endogenous and exogenous protein acceptors, *Biochim. Biophys. Acta* **264:**129.

Ko, G. K. W., and Raghupathy, E., 1973, Glycoprotein biosynthesis in the developing rat brain. IV. Effects of guanosine nucleotides on soluble glycoproteins galactosyl- and *N*-acetylgalactosaminyltransferases, *Biochim. Biophys. Acta* **313:**277.

Kornfeld, R., and Kornfeld, S., 1976, Comparative aspects of glycoprotein structure, *Annu. Rev. Biochem.* **45:**217.

Krag, S. S., and Robbins, P. W., 1977, Sindbis envelope proteins as endogenous acceptors in reactions of guanosine diphosphate [^{14}C]mannose with preparations of infected chicken embryo fibroblasts, *J. Biol. Chem.* **252:**2621.

Krisman, C. P., and Barengo, R., 1975, A precursor of glycogen biosynthesis: α-4-Glucan protein, *Eur. J. Biochem.* **52:**117.

Kuo, S.-C., and Lampen, J. O., 1974, Tunicamycin—an inhibitor of yeast glycoprotein biosynthesis, *Biochem. Biophys. Res. Commun.* **58:**287.

Leavitt, R., Schlesinger, S., and Kornfeld, S., 1977, Tumicamycin inhibits glycosylation and multiplication of Sindbis and vesicular stomatitis viruses, *J. Virol.* **21:**375.

Lehle, L., and Tanner, W., 1976, The specific site of tunicamycin inhibition in the formation of dolichol-bound *N*-acetylglucosamine derivatives, *FEBS Lett.* **71:**167.

Lennarz, W. J., 1975, Lipid-linked sugars in glycoprotein synthesis, *Science* **188:**986.

Lennarz, W. J., and Scher, M. G., 1972, Metabolism and function of polyisoprenol sugar intermediates in membrane-associated reactions, *Biochim. Biophys. Acta* **265:**417.

Levy, J. A., Carminatti, H., Cantarella, A. I., Behrens, N. H., Leloir, L. F., and Tábora, E., 1974, Mannose transfer to lipid linked di-*N*-acetylchitobiose, *Biochem. Biophys. Res. Commun.* **60:**118.

Louisot, P., Belon, P., and Broquet, P., 1976, Cinétiques des réactions de glycosylation dans la biosynthese des glycoproteines: Mécanisme Bi–Bi ordonné des glycosyltransférases et coopérativés, *C. R. Acad. Sci. Ser. D* **283**:401.

Lucas, J. J., and Levine, E., 1977, Increase in the lipid intermediate pathway of protein glycosylation during hen oviduct differentiation, *J. Biol. Chem.* **252**:4330.

Lucas, J. J., and Waechter, C. J., 1976, Polyisoprenoid glycolipids involved in glycoprotein biosynthesis, *Mol. Cell. Biochem.* **11**:67.

Mookerjea, S., and Schimmer, B. P., 1975, UDP-galactose: glycoprotein galactosyltransferase activity in a clonal line of rat glial tumor cells and in rat brain, *Biochim. Biophys. Acta* **384**:381.

Morgan, I. G., Reith, M., Marinari, U., Breckenridge, W. C., and Gombos, G., 1972, The isolation and characterization of synaptosomal plasma membranes, *Adv. Exp. Med. Biol.* **25**:209.

Ng, S. S., and Dain, 1977*a*, Sialyltransferases in rat brain: Reaction kinetics, product analyses, and multiplicities of enzyme species, *J. Neurochem.* **29**:1075.

Ng, S. S., and Dain, J. A., 1977*b*, Sialyltransferases in rat brain: Intracellular localization and some membrane properties, *J. Neurochem.* **29**:1085.

Parodi, A., and Leloir, L. F., 1976, Lipid intermediates in protein glycosylation, *Trends Biochem. Sci.* **1**:58.

Parodi, A. J., Staneloni, R., Cantarella, A. I., Leloir, L. F., Behrens, N. H., Carminatti, H., and Levy, J. A., 1973, Further studies on a glycolipid formed from dolichyl-D-glucosyl monophosphate, *Carbohydr. Res.* **26**:393.

Pattabiraman, T. N., Sekhara Varma, T. N., and Bachhawat, B. K., 1964, Enzymic degradation of uridine diphosphoacetylglucosamine, *Biochim. Biophys. Acta* **83**:74.

Raghupathy, E., Ko, G. K. W., and Peterson, N. A., 1972, Glycoprotein biosynthesis in the developing rat brain. III. Glycoprotein glycosyltransferases present in synaptosomes, *Biochim. Biophys. Acta* **286**:339.

Reith, M., Morgan, J. G., Gombos, G., Breckenridge, W. C., and Vincendon, G., 1972, Synthesis of synaptic glycoproteins. I. The distribution of UDP-galactose: *N*-acetylglucosamine galactosyltransferase and thiamine diphosphate in adult rat brain subcellular fractions, *Neurobiology* **2**:169.

Robbins, P. W., Hubbard, S. C., Turco, S. J., and Wirth, D. F., 1977, Proposal for a common oligosaccharide intermediate in the synthesis of membrane glycoproteins, *Cell* **12**:893.

Roseman, S., 1970, The synthesis of complex carbohydrates by multiglucosyltransferase systems and their potential function in intercellular adhesion, *Chem. Phys. Lipids* **5**:270.

Scher, M. G., and Waechter, C. J., 1978*a*, Glucolipid intermediates in calf brain glycoprotein assembly, *Trans. Am. Soc. Neurochem.* **9**:170.

Scher, M. G., and Waechter, C. J., 1978*b*, Possible role of membrane-bound glucosidase in the processing of calf brain glycoproteins, in: *Abstracts, Second Meeting of the European Society of Neurochemistry*, Vol. 1, p. 559, Verlag Chemie, Weinheim.

Scher, M. G., Jochen, A., and Waechter, C. J., 1977, Biosynthesis of glucosylated derivatives of dolichol: Possible intermediates in the assembly of white matter glycoproteins, *Biochemistry* **16**:5037.

Schwarz, R. T., Rohrschneider, J. M., and Schmidt, M. R. G., 1976, Suppression of glycoprotein formation of Semliki forest, influenza, and avian sarcoma virus by tunicamycin, *J. Virol.* **19**:782.

Spiro, R. G., Spiro, M. J., and Bhoyroo, V. D., 1976, Lipid-saccharide intermediates in glycoprotein biosynthesis. II. Studies on the structure of an oligosaccharide-lipid from thyroid, *J. Biol. Chem.* **251**:6409.

Struck, D. K., and Lennarz, W. J., 1977, Evidence for the participation of saccharide-lipids in the synthesis of the oligosaccharide chain of ovalbumin, *J. Biol. Chem.* **252**:1007.

Takatsuki, A., and Tamura, G., 1971, Effect of tunicamycin on the synthesis of macromolecules in cultures of chick embryo fibroblasts infected with Newcastle disease virus, *J. Antibiot.* **24**:785.

Takatsuki, A., Arima, K., and Tamura, G., 1971, Tunicamycin, a new antibiotic. I. Isolation and characterization of tunicamycin, *J. Antibiot.* **24**:215.

Takatsuki, A., Kohno, K., and Tamura, G., 1975, Inhibition of biosynthesis of polyisoprenol sugars in chick embryo microsomes by tunicamycin, *Agric. Biol. Chem.* **39**:2089.

Takatsuki, A., Kawamura, K., Okina, M., Kodama, T. I., and Tamura, G., 1977, The structure of tunicamycin, *Agric. Biol. Chem.* **41**:2307.

Tkacz, J. S., and Lampen, J. O., 1975, Tunicamycin inhibition of polyisoprenyl *N*-acetylglucosaminyl pyrophosphate formation in calf-liver microsomes, *Biochem. Biophys. Res. Commun.* **65**:248.

Van den Eijnden, D. H., and Van Dijk, W., 1974, Properties and regional distribution of cerebral CMP-*N*-acetylneuraminic acid: glycoprotein sialytransferase, *Biochim. Biophys. Acta* **362**:136.

Van den Eijnden, D. H., Van Dijk, W., and Roukema, P. A., 1975, Sialoglycoprotein synthesis in developing rat brain, *Neurobiology* **5**:221.

Waechter, C. J., and Harford, J. B., 1977, Evidence for the enzymatic transfer of *N*-acetylglucosamine from UDP-*N*-acetylglucosamine into dolichol derivatives and glycoproteins by calf membranes, *Arch. Biochem. Biophys.* **181**:185.

Waechter, C. J., and Harford, J. B., 1979, A dolichol-linked trisaccharide from central nervous tissue: Structure and biosynthesis, *Arch. Biochem. Biophys.* **192**:380.

Waechter, C. J., and Lennarz, W. J., 1976, The role of polyprenol-linked sugars in glycoprotein synthesis, *Annu. Rev. Biochem.* **45**:95.

Waechter, C. J., and Scher, M. G., 1978, Glucosylphosphoryldolichol: Role as a glucosyl donor in the biosynthesis of an oligosaccharide lipid intermediate by calf brain membranes, *Arch. Biochem. Biophys.* **188**:385.

Waechter, C. J., Kennedy, J. L., and Harford, J. B., 1976, Lipid intermediates involved in the assembly of membrane-associated glycoproteins in calf brain white matter, *Arch. Biochem. Biophys.* **174**:726.

Warren, C. D., Liu, I. Y., Herscovics, A., and Jeanloz, R. W., 1975, The synthesis and chemical properties of polyisoprenyl β-D-mannopyranosyl phosphates, *J. Biol. Chem.* **250**:8069.

Wolfe, L. S., Breckenridge, W. C., and Skelton, P. P. C., 1974, Involvement of mannosylphosphoryldolichols in mannose transfer to brain glycoproteins, *J. Neurochem.* **23**:175.

Zatz, M., and Barondes, S. H., 1969, Incorporation of mannose into mouse brain lipid, *Biochem. Biophys. Res. Commun.* **36**:511.

Zatz, M., and Barondes, S. H., 1971, Particulate and solubilized fucosyltransferases from mouse brain, *J. Neurochem.* **18**:1625.

5

Biosynthesis of Glycosaminoglycans

George W. Jourdian

1. Introduction

The low levels of glycosaminoglycans present in nervous tissue have in the past deterred the study of their chemistry and metabolism in this tissue. Now, however, the use of radioisotopes, the recent development of sensitive, specific colorimetric and enzyme methodologies, and the development of procedures for the isolation of the various cell types and subcellular fractions unique to nervous tissue have prompted renewed interest as to the role of glycosaminoglycans in the development and function of this tissue in health and disease. This chapter reviews recent progress concerning the biosynthesis of glycosaminoglycans in nervous tissue.

Several aspects of the chemistry and metabolism of glycosaminoglycans in nervous tissue have been reviewed, including: (1) a cytochemical survey of the distribution of glycosaminoglycans in nervous tissue (Friede, 1966); (2) sulfate metabolism in brain tissue (Balasubramanian and Bachhawat, 1970; Bachhawat *et al.*, 1972); and (3) the metabolism of glycosphingolipids and glycosaminoglycans in cultured cells derived from nervous tissue (Stoolmiller *et al.*, 1973). For background information concerning the chemistry, distribution, and metabolism of glycosaminoglycans in other vertebrate tissues, the reader is referred to the following review articles: Rodén (1970), Rodén *et al.* (1972*a*,*b*), Kimata *et al.* (1973), and Kennedy (1976).

George W. Jourdian • Rackham Arthritis Research Unit and Department of Biological Chemistry, University of Michigan, Ann Arbor, Michigan.

2. Metabolism of Glycosaminoglycans in Nervous Tissue

2.1. *In Vivo* Studies

Odeblad and Boström (1952), Bostrom and Odeblad (1953), and Dziewiatkowski (1953) pioneered the application of ^{35}S-labeled inorganic sulfate to studies of the synthesis of glycosaminoglycans in nervous tissue. Using this methodology, Ringertz (1956) established that ^{35}S-labeled material present in the gray matter of mouse brain after prior administration of inorganic $^{35}SO_4^{2-}$ was largely glycosaminoglycan, and in white matter largely ^{35}S-labeled sulfolipid. Subsequently, similar results were obtained by chemical analyses in human brain (Stary *et al.*, 1964) and in bovine brain (Margolis, R. U., 1967). However, Singh and Bachhawat (1965) found no difference in the glycosaminoglycan content of white and gray matter in sheep brain. Variation in the content and distribution of glycosaminoglycans in nervous tissue from different animal species and at differing developmental stages is not unexpected and is described elsewhere in this chapter and volume.

The results of [^{35}S]sulfate incorporation studies in rat brain (Guha *et al.*, 1960; Robinson and Green, 1962) suggested that the brain synthesizes a spectrum of sulfated glycosaminoglycans. These findings were confirmed and extended by Goldberg and Cunningham (1970), who reported in an abstract that ^{35}S-labeled rat brain contained chondroitin sulfate(s) (71%), heparan sulfate (14%), dermatan sulfate (14%), and unlabeled hyaluronic acid; heparin and keratan sulfate were not detected. The results of chemical analyses of brain tissue from this and other animal species are in general agreement with these findings (Singh *et al.*, 1969; Margolis, R. U., 1967). Vitello *et al.* (1978) reported the presence of a keratan-sulfate-like glycosaminoglycan in the cerebral cortex. Whether this material is keratan sulfate or is a member of a class of sulfated glycoproteins from brain lacking uronic acid (Margolis, R. K., and Margolis, R. U., 1970; Simpson *et al.*, 1976) remains to be established.

The results of a number of studies show that although the change (decrease) in total glycosaminoglycan concentration during brain development is approximately 2.5-fold, the relative amounts of the individual glycosaminoglycans vary by only 40–70%. However, striking changes in the absolute amounts of individual glycosaminoglycans occur with increasing age. In newborn rat brain, their content increases rapidly, reaching a maximum level at about 8 days after birth. After this time, distinct changes in glycosaminoglycan content occur. For example, hyaluronic acid and heparan sulfate decrease rapidly, approximately 70%, and attain adult levels by day 16 after birth (Robinson and Green, 1962; Singh and Bachhawat, 1968; Dziewiatkowski, 1970; Young and Custod, 1972; Margolis, R.U., *et al.*, 1975). A similar change in hyaluronic acid content occurs in developing chick embryo brain (Polansky *et al.*, 1974). Singh and Bachhawat (1968) and R. U. Margolis *et al.* (1975) found that chondroitin sulfate(s) was metabolized in a manner distinct from other glycosaminoglycans; after day 8 of development, the chondroitin sulfate(s) content slowly declined (approximately 30%) and attained adult levels by day 28 after birth. Using microbial chondroitinases, enzymes that discriminate between chondroitin 4- and 6-sulfate, Sampaio *et al.* (1977) found that chondroitin 6-sulfate decreased in a manner similar to that reported above for heparan sulfate and hyaluronic acid; however, the level of chondroitin 4-sulfate continued to increase until day 30 after birth. Heparan sulfate turns over more rapidly in young and adult rat brain than other glycosaminoglycans (Margolis, R. U., and Margolis, R. K., 1972; Margolis, R. U., *et al.*, 1975). A 4-fold increase in the relative specific activity of heparan sulfate as compared to hyaluronic acid occurs between day 7 and adulthood, while the specific activity of chondroitin sulfate(s) decreases only slightly. (For additional data on developmental changes in brain glycosaminoglycans, see Appendix I.)

2.2. Metabolism of Glycosaminoglycans in Cultured Cells

Dorfman and Ho (1970) demonstrated that a clonal strain of glial cells derived from a rat glial tumor induced by *N*-nitrosomethylurea synthesized hyaluronic acid, chondroitin 4-sulfate and heparan sulfate. Subsequently, Stoolmiller (1972) (also see Stoolmiller *et al.*, 1973) isolated small amounts of the same glycosaminoglycans from a clonal cell strain of neuroblastoma cells NB41A.

Organized cultures of fetal mouse dorsal root ganglia, cerebellum, and sympathetic ganglia grown in the presence of [^{35}S]sulfate synthesized a spectrum of glycosaminoglycans similar to that found in whole brain (Uzman *et al.*, 1973). Chondroitin 4-sulfate accounted for 50–60% of the total glycosaminoglycan content; chondroitinase-ABC-resistant material (presumably heparan sulfate), for 13–30%; chondroitin 6-sulfate, for 5–10%; and dermatan sulfate, for 1–3%. These values are in close agreement with those found in whole rat brain. The glycosaminoglycan content of dorsal root and sympathetic chain ganglia exceeded that of cerebellum cultures. However, the cerebellum cultures secreted a higher amount of chondroitin 6-sulfate and a lower amount of dermatan sulfate than either of the other cell types. Whether this "compartmentalized" labeling pattern is characteristic of whole brain or was the result of preponderant contributions of stromal elements in the ganglia cultures to the glycosaminoglycan content observed was not ascertained. Wiesmann *et al.* (1975) found that dissociated cultures of glial and neuronal cells from newborn albino mouse brains secreted a similar spectrum of glycosaminoglycans. Interestingly, and perhaps fortuitously, maximum secretion occurred after 8 days in culture, paralleling the results found *in vivo*. Heparan sulfate constituted the major glycosaminoglycan in cultured glial cells from a patient with Sanfilippo A syndrome, and was present in approximately 8-fold higher concentration than that found in normal glial cells and approximately 3-fold higher than that found in skin fibroblasts from the same patient (Fluharty *et al.*, 1975).

While the studies cited above suggest that several types of cultured cells derived from nervous tissue provide convenient models for the study of glycosaminoglycan metabolism, it has not been established whether the types and amounts of glycosaminoglycans synthesized accurately reflect their occurrence in normal tissue. In this regard, evidence has accumulated that transformed cell lines exhibit qualitative and quantitative differences in glycosaminoglycan content from normal cells (Satoh *et al.*, 1973). Furthermore, scant consideration has been given to the observations of Dorfman and Ho (1970) that fetal calf serum and related culture media supplements contain significant quantities of glycosaminoglycans, or to the observation of Lie *et al.* (1972) that incubation of fibroblasts at pH values between 6 and 8 results in abnormal intracellular accumulations of glycosaminoglycans (see also Chapter 14).

2.3. Metabolism of Glycosaminoglycans in Cellular and Subcellular Fractions of Brain

The recent development of (1) methodology for the preparative isolation of brain cell types including neuronal perikarya, astrocytes, and oligodendroglia (Norton and Poduslo, 1970, 1971; Sellinger *et al.*, 1971; Poduslo and Norton, 1972) of a high degree of purity and (2) methodology for preparation of specific subcellular fractions including nerve endings, synaptic vesicles, and nerve-ending membranes (Gray and Whittaker, 1962; Whittaker and Sheridan, 1965; De Robertis, 1967; Morgan *et al.*, 1971; Cotman *et al.*, 1974) offers a promising new approach to the study of glycosaminoglycan metabolism in nervous tissue.

2.3.1. Glycosaminoglycan Metabolism in Whole Cells

R. U. Margolis and R. K. Margolis (1974) found distinct differences in the specific activities of labeled glycosaminoglycans associated with several cell types isolated from 25-day-old rat brains after exposure to intraperitoneally administered [1-^{3}H]glucosamine and [^{35}S]sulfate. The specific activity of hyaluronic acid in the neuronal cell body fraction approximated twice that of the astrocyte fraction, while the specific activity of the hexosamine and sulfate contained in heparan sulfate was 20–30% less in neuronal cell bodies than in the astrocyte-enriched fraction; the specific activity of chondroitin sulfate(s) in the neuronal cell body and astrocyte fractions was quite similar. These results indicated that the metabolic turnover of hyaluronic acid is considerably more rapid in neurons than in astrocytes, and prompted the hypothesis that a metabolically active pool of hyaluronic acid may have a "specific functional association" with the neuronal cell body, while chondroitin sulfate occurs as a single pool throughout the brain.

2.3.2. *Glycosaminoglycan Metabolism in Subcellular Fractions*

Barondes (1968) found that ^{14}C-labeled glucosamine was rapidly incorporated into macromolecules in all the subcellular fractions of mouse brain including the soluble and particulate fractions of isolated nerve endings. About 10% of the total radioactivity contained in each of these fractions was released on treatment with hyaluronidase. Vos *et al.* (1969) reported that a synaptic vesicle fraction from mouse brain contained approximately twice as much glycosaminoglycan (largely chondroitin sulfate and hyaluronic acid) as the whole brain particulate fraction and about four times the glycosaminoglycan content found in synaptosomes; about 34% of the total glycosaminoglycan present in whole brain was found in a soluble state. Brandt *et al.* (1975) examined the glycosyltransferase activity of the synaptic vesicle fraction of developing chick embryo brain and found that less than 6% of each of the transferases assayed in the parent enriched synaptosome fraction was present in the synaptic vesicle fraction. Taken together, these studies suggest that synaptic vesicles may serve as a storage depot for chondroitin sulfate (and perhaps hyaluronic acid), or alternatively that during isolation, glycosaminoglycans present in the cytosol preferentially adsorb to the vesicles.

2.3.3. *Sulfation of Glycosaminoglycans in Subcellular Fractions*

In developing mouse brain, incorporation of $^{35}SO_4^{2-}$ decreases with increasing age; the highest ^{35}S incorporation occurs at the premyelination stage (Kuriyama and Okada, 1971). Examination of the subcellular fractions of 2-day-old mouse brain revealed that the major sites of [^{35}S]sulfate fixation were the cytosol (39.2%); the crude mitochondrial pellet (29.2%), which contains mitochondria and synaptosomes; and the nuclear fraction (19.7%). On subfractionation of the crude mitochondrial pellet, 53% of the ^{35}S-labeled material was found in the synaptosomal fraction and 20.8% in the mitochondrial fraction of whole mouse brain. The labeled fractions were characterized as glycosaminoglycans by cellulose acetate electrophoresis. Saxena *et al.* (1971) examined the subcellular distribution of sulfotransferase activity(ies) in developing rat brain utilizing dermatan sulfate as an acceptor. The results obtained paralleled those of Kuriyama and Okada (1971) and suggested that sulfate fixation occurs in several subcellular compartments including the cytosol, synaptosomes, microsomes, and perhaps in the nuclei. The latter fraction also contains unbroken cells and cell debris, but recent experiments of R. K. Margolis *et al.* (1976) convincingly demonstrated that a spectrum of glycosaminoglycans are associated with nuclei and that the association is not the result of nonspecific adsorption that occurs during the subcellular fractionation.

The studies described above document the ability of nervous tissue to metabolize a spectrum of glycosaminoglycans, and further suggest that their metabolism differs in specific cell types and compartments. This information has served as a reference point for the initiation of the studies described below, which are concerned with the biosynthesis of glycosaminoglycans in nervous tissue.

3. Biosynthesis of Nucleotide Monosaccharides and Sulfate Precursors of Nervous Tissue Glycosaminoglycans

3.1. Biosynthesis of Nucleotide Monosaccharides

In contrast to our knowledge of other vetebrate tissues, there is a paucity of information regarding the conversion of monosaccharide units to nucleotide monosaccharides that serve as the immediate precursors of the glycosaminoglycans common to nervous tissue. The currently accepted pathway(s) for the formation of the prerequisite nucleotide monosaccharides from glucose are presented in Fig. 1. By and large, these reactions are common to other vertebrate tissues, and their isolation and properties are described in detail in recent review articles (Schachter and Rodén, 1973; Kennedy, 1976).

The phosphorylation of glucose to glucose-6-phosphate (Fig. 1, reaction 1) provides entry into phosphorylated pathways that lead to the formation of the nucleotide derivatives of glucose and *N*-acetylglucosamine. Synthesis of each of these nucleotides is achieved by a parallel series of reactions and includes the formation of the 1-phosphate derivatives of glucose (reaction 2) and *N*-acetylglucosamine (reaction 11), conversion to their corresponding uridine-5′-diphospho derivatives, and subsequent 4′-epimerization at the nucleotide level to form UDP-galactose (reaction 5) and UDP-*N*-acetylgalactosamine (reaction 14), respectively. Several of the nucleotide monosaccharides and monosaccharide derivatives synthesized by these pathways serve as intermediates in the biosynthesis of glycoproteins and glycosphingolipids present in nervous tissue. The individual reactions demonstrated in brain tissue that lead to the biosynthesis of nucleotide precursors of glycosaminoglycans are presented in Table I.

The conversion of fructose-6-phosphate to glucosamine-6-phosphate (Fig. 1, reaction 9) warrants further discussion. While it is commonly accepted that this reaction is mediated by fructose-6-phosphate : L-glutamine transaminase (Roseman, 1959), to the author's knowledge this enzyme has not been demonstrated in nervous tissue. However, Pattabiraman and Bachhawat (1961*a*) and Faulkner and Quastel (1956) demonstrated the presence of a glucosamine-6-phosphate deaminase in brain; the reaction is reversible and

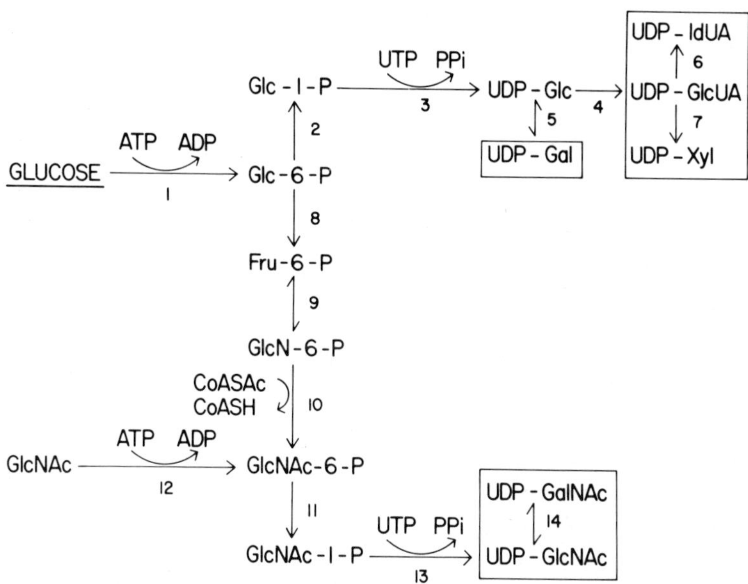

Fig. 1. Pathways for the biosynthesis of nucleotide monosaccharides. Enzymes: (1) hexokinase; (2) phosphoglucomutase; (3) UDP-glucose pyrophosphorylase; (4) UDP-glucose dehydrogenase; (5) UDP-glucose 4′-epimerase; (6) UDP-glucuronate 5′-epimerase; (7) UDP-glucuronate decarboxylase; (8) glucose-6-phosphate isomerase; (9) glucosamine phosphate isomerase; (10) glucosamine phosphate acetyltransferase; (11) phosphoacetylglucosamine mutase; (12) *N*-acetyl-D-glucosamine kinase; (13) UDP-acetylglucosamine pyrophosphorylase; (14) UDP-acetylglucosamine 4-epimerase.

requires ammonia as the nitrogen donor. Comb and Roseman (1958) found that a combination of fructose-6-phosphate, ammonia, and acetyl-CoA when incubated in the presence of fructose-6-phosphate deaminase and *N*-acetyltransferase (reaction 10) resulted in the rapid production of *N*-acetylglucosamine-6-phosphate, and suggested on the basis of the relative activities of these enzymes that the ammonia pathway is a potentially better source of glucosamine derivatives than the glutamine pathway.

3.2. Lipid-Linked Oligosaccharides: Alternative Intermediates in Glycosaminoglycan Biosynthesis?

Lipid-linked oligosaccharides participate in the synthesis of glycoproteins in vertebrate tissues (Hemming, 1974; Lennarz, 1975; Waechter and Lennarz, 1976; Parodi and Leloir, 1976) including chick brain glycoproteins (Breckenridge, W. C., and Wolfe, 1973). Recently, two lipid-linked carbohydrate units common to glycosaminoglycans were isolated, a xylose-containing lipid from hen oviduct (Waechter *et al.*, 1974) and a glucuronosyl-*N*-acetylglucosaminyl pyrophosphodolichol from human lung fibroblasts

Table I. Biosynthesis of Nucleotide Sugars: Enzymes Demonstrated in Brain Tissue

Reaction No.	Enzyme	Brain source	Reference
1	Hexokinase	Mouse	Maker and Lehrer (1972)
2	Phosphoglucomutase	Rabbit	B. M. Breckenridge and Crawford (1961)
3	UDP-glucose pyrophosphorylase	Rabbit	B. M. Breckenridge and Crawford (1961)
		Rat	Maxwell *et al.* (1955)
		Human	Basu and Bachhawat (1961)
4	UDP-glucose dehydrogenase	Rat	Koransky (1958)[a]
5	UDP-glucose 4′-epimerase	Goat	Chandra and Bhaduri (1976)
8	Glucose-6-phosphate isomerase	Mouse	Maker and Lehrer (1972)
9	Glucosamine phosphate deaminase	Human	Pattabiraman and Bachhawat (1961*a*)
		Rat	Faulkner and Quastel (1956)
10	Glucosamine phosphate acetyltransferase	Sheep	Pattabiraman and Bachhawat (1962*a*)
11	Phospho-*N*-acetylglucosamine mutase	Rat	Pattabiraman and Bachhawat (1962*b*)
12	*N*-Acetyl-D-glucosamine kinase	Sheep	Pattabiraman and Bachhawat (1961*b*)
13	UDP-acetylglucosamine pyrophosphorylase	Sheep	Pattabiraman and Bachhawat (1961*c*)

[a] Indirect evidence.

(Turco and Heath, 1977). While not definitively established, it seems possible that lipid-linked oligosaccharides may participate in the biosynthesis of nervous tissue glycosaminoglycans.

3.3. Biosynthesis of 3-Phosphoadenosine 5′-Phosphosulfate

The fixation of sulfate into glycosaminoglycans in vertebrate tissues is mediated by 3-phosphoadenosine 5′-phosphosulfate (PAPS). The metabolic pathway for the biosynthesis of this compound, initially established in yeast (Robbins and Lipmann, 1956, Bandurski *et al.*, 1956), proceeds in the following manner:

$$\mathrm{ATP} + \mathrm{SO_4^{2-}} \rightleftharpoons \mathrm{APS} + \mathrm{PP_i} \tag{1}$$

$$\mathrm{APS} + \mathrm{ATP} \longrightarrow \mathrm{PAPS} + \mathrm{ADP} \tag{2}$$

Sulfate adenyltransferase (reaction 1) catalyzes the formation of adenosine 5′-phosphosulfate (APS); reaction 2, catalyzed by adenylylsulfate kinase, yields 3′-phospho 5′-phosphosulfate. That similar reactions occur in rabbit brain was suggested from the results of studies of Gregory and Lipmann (1957) and subsequently confirmed in rat brain by Balasubramanian and Bachhawat

(1961). The latter investigators observed age-dependent variations in the biosynthesis of PAPS with brain development. Enhanced synthesis of PAPS occurred at two time periods, in 1-day-old brain and at day 12 after birth; the first increase was hypothesized to be associated with glycosaminoglycan biosynthesis and the second with sulfatide synthesis.

4. Biosynthesis of Glycosaminoglycans in Nervous Tissue

4.1. Biosynthesis of Chondroitin Sulfate

The structures of the several glycosaminoglycans common to nervous tissue are presented in Chapter 3. Present information suggests that the sulfated glycosaminoglycans found in nervous tissue occur as proteoglycan complexes (Margolis, R. U., *et al.*, 1976). With the exception of hyaluronic acid (and keratan sulfate), the polysaccharide chains of each glycosaminoglycan are linked to their respective polypeptide cores by a common linkage region consisting of glucuronosyl-galactosyl-galactosyl-xylose. The xylose unit that initiates the polysaccharide chains is glycosidically attached to the polypeptide core via the hydroxyl groups of serine units contained in the polypeptide core (Rodén *et al.*, 1972*b*).

Of the several glycosaminoglycans associated with nervous tissue, chondroitin sulfate is the only one whose biosynthesis has been studied in detail. The assembly of the polysaccharide chains of chondroitin sulfate in brain parallels that found in other vertebrate tissues and involves the transfer of monosaccharides, or monosaccharide derivatives, from their respective uridine 5′-diphospho derivatives to the ends of growing carbohydrate chains in the stepwise manner depicted in Fig. 2. Each reaction is catalyzed by the action of a specific glycosyltransferase, and each product serves as an acceptor substrate for the addition of the next carbohydrate derivative. Glycosyltransferases (Fig. 2, reactions 1, 2, 4, 5, and 6) have been demonstrated in 13-day chick embryo brain preparations (Brandt *et al.*, 1975). With the exception of the xylosyltransferase that initiates the growing polysaccharide chains, each of the glycosyltransferases requires Mn^{2+} as cofactor, and each glycosyltransferase is largely membrane-associated (i.e., sediments at 100,000*g*).

The biosynthesis of glycosaminoglycans in brain does not seem to be restricted to any single brain area. For example, glucuronosyltransferase activity (Fig. 2, reaction 4) is present in the cerebral hemisphere, diencephalon, optic lobes, and cerebellum of 13-day-old chick embryo brain; of these areas, the optic lobes exhibit the highest total enzyme activity and highest specific activity (Brandt, 1971). A marked increase in the total activity of three glycosyltransferases associated with chondroitin sulfate synthesis (reactions 1, 4, and 6) was noted between day 7 of development and hatching at day 21; after hatching, a precipitous decrease in each glycosyltransferase activity

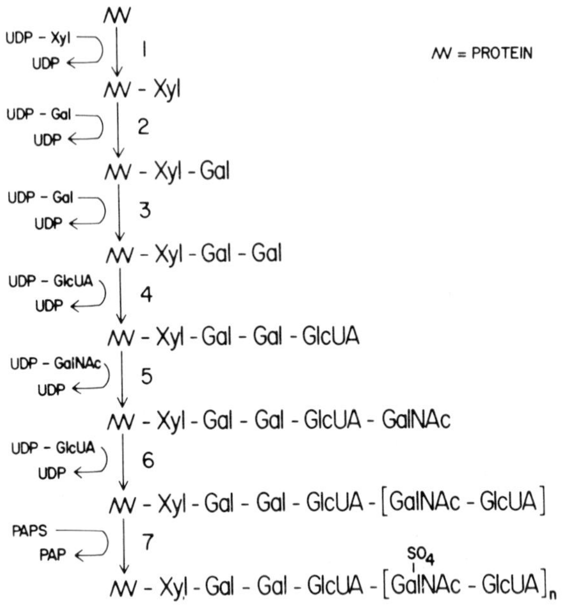

ᐱᐯ = PROTEIN

ᐱᐯ
UDP-Xyl → UDP (1)
ᐱᐯ - Xyl
UDP-Gal → UDP (2)
ᐱᐯ - Xyl - Gal
UDP-Gal → UDP (3)
ᐱᐯ - Xyl - Gal - Gal
UDP-GlcUA → UDP (4)
ᐱᐯ - Xyl - Gal - Gal - GlcUA
UDP-GalNAc → UDP (5)
ᐱᐯ - Xyl - Gal - Gal - GlcUA - GalNAc
UDP-GlcUA → UDP (6)
ᐱᐯ - Xyl - Gal - Gal - GlcUA - [GalNAc - GlcUA]
PAPS → PAP (7)
ᐱᐯ - Xyl - Gal - Gal - GlcUA - [GalNAc(SO_4) - GlcUA]$_n$

Fig. 2. Biosynthesis of chondroitin sulfate proteoglycan. Enzymes: (1) UDP-xylosyltransferase; (2,3) UDP-galactosyltransferase(s); (4) UDP-glucuronosyltransferase I; (5) UDP-*N*-acetylgalactosamine transferase; (6) UDP-glucuronosyltransferase II; (7) 3′-phosphoadenylylsulfate chondroitin 4′-sulfotransferase.

occurred. In contrast, the specific activity of each of the glycosyltransferases remained nearly constant from day 7 through day 21, and after hatching a sharp drop in specific activity was noted (Brandt *et al.*, 1975).

4.1.1. Xylosyltransferase

The results of differential isotope labeling and inhibitor studies in several vertebrate tissues suggest that protein synthesis precedes polysaccharide elaboration (Kennedy, 1976). It has been hypothesized that the polypeptide core of the chondroitin sulfate proteoglycan complex may contain specific amino acid sequences that predetermine the attachment of xylose units to specific serine units that serve as initiation sites for the assembly of the polysaccharide chains. In support of this hypothesis and in agreement with the results obtained in other vertebrate tissues, we found that bovine nasal chondroitin sulfate proteoglycan and hyaluronidase-treated chondroitin sulfate proteoglycan, compounds known to contain unsubstituted serine units, were poor acceptors for the transfer of xylose from UDP-xylose (Fig. 2, reaction 1) in chick brain homogenates (unpublished results). However, Smith-degraded bovine nasal proteoglycan, a material from which the polysaccharide chains are removed to expose serine hydroxyl groups, was a good acceptor (Brandt *et al.*, 1975). Of interest in this regard is the finding of Baker *et al.* (1972) that of a number of chemically defined serine-containing peptides assayed as acceptors for xylose units, only serylglycylglycine served as an acceptor in reaction mixtures with chick epiphyseal cartilage extracts.

4.1.2. Galactosyltransferases

Detergent-"solubilized" (150-fold-purified) preparations from 13-day chick embryo brain catalyzed transfer of galactose from UDP-galactose (Fig. 2, reaction 2) to degraded chondroitin sulfate proteoglycan (sequentially treated with hyaluronidase, β-glucuronidase, and β-galactosidase to expose terminal xylose units), 3-*O*-β-xylosyl-L-serine, and xylose (Brandt, 1971; Brandt *et al.*, 1975). When xylose served as acceptor, the glycosidic linkage formed between galactose and xylose was β1⟶4, the linkage found in the chondroitin sulfate proteoglycan linkage region (unpublished results). Small amounts of digalactosylxylosylserine were formed in reaction mixtures containing 3-*O*-β-xylosylserine as acceptor and crude brain extracts; this product was not detected in reaction mixtures containing the "solubilized" purified enzyme. These results support the existence of a second galactosyltransferase (reaction 3) in developing chick brain. Such an enzyme was demonstrated in chick epiphyseal cartilage by Helting and Rodén (1968) and partially purified from mouse mastocytoma preparations (Helting, 1971). While not demonstrated, it seems probable that the brain enzyme catalyzes formation of a β,1⟶3 glycosidic bond between the two galactosyl units, the linkage synthesized by mastocytoma preparations. In contrast to the galactosyltransferase that catalyzes reaction 2, the mastocytoma galactosyltransferase (reaction 3), and presumably the brain galactosyltransferase, transfer galactose units only to acceptor molecules that terminate in galactose and contain more than a single monosaccharide unit. Mastocytoma preparations are also affected by the nature of the pentultimate monosaccharide. For example, 3-*O*-β-D-galactosylgalactose does not serve as a galactose acceptor. The high degree of specificity exhibited by this transferase may be required in order to limit to 2 the number of galactosyl residues added to the growing polysaccharide chains.

4.1.3. Glucuronosyltransferase I

Two distinct glucuronosyltransferases participate in the biosynthesis of chondroitin sulfate chains: glucuronosyltransferase I (Fig. 2, reaction 4) terminates the linkage region and glucuronosyltransferase II (reaction 6) completes the repeating disaccharide unit common to the polysaccharide chains of chondroitin sulfate. The transferase that catalyzes reaction 4 was solubilized and purified approximately 65-fold from 13-day chick embryo brain extracts (Brandt *et al.*, 1969). That this transferase is distinct from the glucuronosyltransferase that catalyzes transfer of glucuronic acid from UDP-glucuronic acid to terminal *N*-acetylgalactosamine residues (reaction 6) was suggested by the results of substrate competition studies using lactose and a trisaccharide derived from chondroitin sulfate {1-*O*-methyl-[*O*-β-*N*-acetyl-D-galactosaminyl (1 ⟶ 4)-*O*-β-D-glucuronosyl (1 ⟶ 3)-*N*-acetylgalactosaminide]}. Analysis of the reaction product where lactose served as acceptor revealed the occurrence of a β,1 ⟶ 3 glycosidic bond between glucuronic acid and galactose. That this transferase participates in the biosynthesis of the linkage region was suggested by the efficient transfer of glucuronic acid to degraded chondroitin sulfate proteoglycan treated with glycosidases to expose terminal galactose residues, $K_m = 7 \times 10^{-5}$ M. The acceptor specificity of chick embryo brain glucuronosyltransferase differs from that isolated from chick epiphyseal cartilage (Helting and Rodén, 1969) and mouse mastocytoma (Helting, 1972). In chick embryo brain preparations, the ability of lactose to serve as an acceptor is of the same order of magnitude as that observed for position isomers of *O*-β-galactosyl-D-galactose. Surprisingly, lactose is a poor acceptor of glucuronic acid residues when incubated with chick embryo epiphyseal cartilage glucuronosyltransferase, and is inactive as an acceptor in reaction mixtures containing solubilized mouse mastocytoma transferase. However, the apparent lack of specificity of this enzyme in brain with regard to the pentultimate carbohydrate unit may be of little consequence in the synthesis of the chondroitin sulfate linkage region, since the structure of the natural acceptor is defined by the preceding glycosyltransferases. Hence, *in vivo*, the acceptor substrate that the glucuronosyltransferase normally encounters possesses the correct structure. Of a number of chick embryo tissues tested for this transferase, brain preparations exhibited the highest specific activity. In view of the finding that most of the glycosaminoglycan content of brain occurs in gray matter, it is of interest to note that in adult human brain, the specific activity of glucuronosyltransferase I is 7-fold higher in gray matter than in white matter (Brandt *et al.*, 1969).

4.1.4. *N-Acetylgalactosaminyltransferase*

The presence of *N*-acetylglucosaminyltransferase activity in rat brain extracts was described by Louisot and Broquet (1973). Whether this transferase participates in glycosaminoglycan biosynthesis was not established. The transfer of *N*-acetylgalactosamine from UDP-*N*-acetylgalactosamine to glucuronosyl units (Fig. 2, reaction 5) terminating a tetrasaccharide derived from chondroitin sulfate {1-*O*-methyl-[β-D-glucuronosyl (1 ⟶ 3)-*O*-β-*N*-acetyl-D-galactosaminyl (1 ⟶ 4)-*O*-β-D-glucuronosyl (1 ⟶ 3)-*N*-acetyl-D-galactosaminide]} was also demonstrated in developing chick embryo brain. The position of substitution of the glycosidic bond formed was not established, but is presumed to be $\beta 1 \rightarrow 4$ (Brandt *et al.*, 1975). The specificity of the brain *N*-acetylgalactosamine transferase toward the penultimate sugar of the acceptor oligosaccharide is probably not crucial for this reaction to proceed. An *N*-acetylgalactosamine transferase from chick epiphyseal cartilage catalyzes the transfer of *N*-acetylgalactosamine units to the aforementioned acceptor and to a hexasaccharide derived from hyaluronic acid that contains an *N*-acetylglucosamine unit in the pentultimate position (Telser *et al.*, 1966).

4.1.5. Glucuronosyltransferase II

A second glucuronosyltransferase present in the chick embryo brain (Fig. 2, reaction 6) catalyzes the transfer of [^{14}C]glucuronic acid from UDP-glucuronic acid to terminal nonreducing *N*-acetylgalactosaminyl units of growing chondroitin sulfate chains. As anticipated, the ^{14}C-labeled glucuronosyl unit is released by the action of β-glucuronidase, as is found in chondroitin sulfate. No attempt has been made to ascertain the specificity of the brain enzyme. However, a similar preparation from chick epiphyseal cartilage catalyzes the transfer of glucuronic acid to nonsulfated and 6-sulfated terminal *N*-acetylgalactosamine units, but not to 4-sulfated *N*-acetylgalactosamine units (Rodén and Schwartz, 1973). The concerted action of the *N*-acetylgalactosamine and glucuronic acid glycosyltransferases (reactions 5 and 6) results in the formation of the repeating disaccharide units and elongation of the polysaccharide chains of chondroitin sulfate.

4.2. Sulfate Fixation

The transfer of ester sulfate groups to appropriate hydroxyl groups of the monosaccharide units of glycosaminoglycans is mediated via PAPS (Fig. 2, reaction 7). Sulfotransferase activity was first demonstrated in chick embryo cartilage (D'Abramo and Lipmann, 1957) and subsequently in nervous tissue, including rat brain (Balasubramanian and Bachhawat, 1964; George *et al.*, 1970; Saxena *et al.*, 1971), human brain (George *et al.*, 1970; George and Bachhawat, 1970), and pig brain (Miller and Waechter, 1978).

The early studies of Balasubramanian and Bachhawat (1964) indicated that the sulfotransferase(s) in 1- to 10-day-old rat brain is largely soluble, and in contrast to the glycosyltransferases does not require divalent ions for activity. A decrease in sulfotransferase activity occurs with brain development, a finding not unexpected in view of the results of earlier ^{35}S-incorporation studies described above. Partially purified (protamine-treated) extracts of young rat brain catalyzed transfer of $^{35}SO_4^{2-}$ from [^{35}S]-PAP to several glycosaminoglycans (listed in decreasing order of activity): heparan sulfate, 100; dermatan sulfate, 59; chondroitin 4-sulfate, 27; keratan sulfate, 14; chondroitin 6-sulfate and hyaluronic acid, <0.6. The relative rate of sulfation of each glycosaminoglycan paralleled its ability to serve as a sulfate acceptor. The same order of sulfotransferase acceptor activity was subsequently observed with extracts of normal human brain in which "low"-sulfated glycosaminoglycans served as acceptors; similar results were obtained with extracts of 1-day-old rat, human fetal, rabbit, and guinea pig brain tissue (George *et al.*, 1970). In rat and human brain, the ratio of *N*-sulfate to *O*-sulfation decreased with brain maturation. Preliminary experiments with partially purified enzyme preparations suggested that heparan sulfate and chondroitin sulfate were sulfated by distinct sulfotransferases, probably similar to those previously demonstrated in hen oviduct preparations by Suzuki and Strominger (1960*a–c*) and Suzuki *et al.* (1961). Recently, Miller and Waechter (1978) demonstrated the transfer of [^{35}S]sulfate from [^{35}S]-PAP into two endogenous fractions in membrane preparations from pig brain. Approximately 80% of the [^{35}S]-sulfate in one fraction was released on selective *N*-desulfation; the second fraction was resistant to this treatment. These findings parallel and extend the earlier results of George *et al.* (1970) and provide additional support for the existence of multiple sulfotransferases in the brain. The results of Sampaio *et al.* (1977) indicated that the C-4 hydroxyl of the *N*-acetylgalactosamine unit of chondroitin sulfate is the major site of sulfate esterification. While *in vitro* studies showed that sulfate fixation can occur at the macromolecular level (Balasubramanian and Bachhawat, 1964; George *et al.*, 1970), the results of other studies suggested that sulfation can also take place in proximity to glycosaminoglycan polymerization (Silbert and DeLuca, 1969; Höök *et al.*, 1974).

The finding of George and Bachhawat (1970) that brain extracts from two patients with Sanfilippo syndrome (a mucopolysaccharidosis) lacked the ability to sulfate chondroitin 4-sulfate is of particular interest. To the author's knowledge, this represents the only instance in which a disease classified as a lysosomal storage disease has been shown to lack a specific biosynthetic enzyme.

4.3. Biosynthesis of Other Glycosaminoglycans

The observation that heparan sulfate and hyaluronic acid together comprise a major portion of the glycosaminoglycan content of brain, and that the content and metabolism of these glycosaminoglycans undergo marked changes with increasing age, prompt the following brief review of several salient features concerning their biosynthesis in vertebrate tissues. No information is available as to the individual steps required for the biosynthesis of these glycosaminoglycans in nervous tissue.

The substitution of heparan sulfate (and dermatan sulfate) with α-L-iduronic acid units in place of glucuronic acid units distinguishes these from other glycosaminoglycans. Until recently, UDP-L-iduronic acid was believed to be the immediate precursor of L-iduronic units in these glycosaminoglycans (Jacobson and Davidson, 1962; Fransson, 1970). Höök *et al.* (1974), studying the biosynthesis of heparin in mouse mastocytoma preparations, and Malmström *et al.* (1975), studying the biosynthesis of dermatan sulfate in human skin fibroblasts, demonstrated that the 5′-epimerization of D-glucuronic acid to L-iduronic acid occurs at the polymeric level. In each tissue, the inversion process was enhanced by sulfation on an adjacent acetylhexosamine unit. Furthermore, differences in the acceptor specificity, in the extent of epimerization in the absence of sulfation, and in the position of the glucuronosyl units along the polysaccharide chains that undergo inversion (external vs. internal location) in each tissue preparation suggest that separate 5′-epimerases are involved in the biosynthesis of dermatan sulfate and heparan sulfate.

With the exception of heparan sulfate and heparin, transfer of acetylhexosamine units from the appropriate UDP-*N*-acetylhexosamine donor to the uronic acid units terminating the growing polysaccharide chains of glycosaminoglycans involves inversion around the anomeric carbon and the formation of a β-glycosidic bond. The addition of *N*-acetylglucosamine to the growing chains of heparan sulfate is mediated by a distinct *N*-acetylglucosamine transferase that catalyzes the formation of an α-linked product. The similarity in structure of heparan sulfate and heparin suggests that the transfer of the acetylhexosamine unit may be mediated by the same glycosyltransferase (Helting and Lindahl, 1972).

The metabolic steps leading to the biosynthesis of hyaluronic acid were reviewed by Jacobson (1970) and Stoolmiller and Dorfman (1970). A lingering controversy continues as to the existence of a protein–hyaluronic acid complex. Hyaluronic acid preparations isolated from a number of vertebrate sources invariably contain small amounts of protein and neutral sugars. Several investigators reported the presence of arabinose in hyaluronic acid preparations from brain (Wardi *et al.*, 1966; Margolis, R. U., 1967). This finding was disputed by Katzman (1971). However, analyses of brain hyaluronic acid preparations by chromatographic, enzymatic, and chemical ionization–mass spectrometric methodologies provided additional support for the presence of arabinose in these preparations (Varma *et al.*, 1977).

Distler and Jourdian (1972) solubilized a glucosyltransferase (from the 100,000*g* pellet of 13-day chick embryo brain) that transfers glucose from UDP-glucose to L-arabinosylserine and xylosylserine. The transferase activity was shown to be distinct from glucosyltransferases associated with the synthesis of glycogen, collagen, and glycosphingolipids. Whether this glycosyltransferase participates in the biosynthesis of brain proteoglycans, such as a linkage region associated with a protein–hyaluronic acid complex, remains to be established.

5. Subcellular Distribution of Glycosyltransferases and Sulfotransferases in Nervous Tissue

The subcellular distribution of several glycosyltransferases involved in the biosynthesis of chondroitin sulfate (Fig. 2, reactions 1, 2, 4, 5, and 6) in 13-day chick embryo brain was determined (Brandt *et al.*, 1975). Subcellular fractions were prepared by the procedure of Gray and Whittaker (1962) and Whittaker and Sheridan (1965). Identity of each subcellular fraction was established by use of specific marker enzymes and compounds and by electron microscopy. A summary of the results obtained is presented in Table II. Three of the five enzymes examined [a galactosyltransferase (reaction 2), an *N*-acetylgalactosaminyltransferase (reaction 5), and a glucuronosyltransferase (reaction 6)] were found associated with membranous fractions. The highest total activity and specific activity for each transferase were found in the crude mitochondrial fraction, a fraction known to contain synaptosomes and mitochondria; at this stage of development, myelination has not occurred and consequently myelin fragments normally found in this fraction are absent. In agreement with studies using chick epiphyseal cartilage preparations (Stool-

Table II. Distribution of Protein and Glycosyltransferases in Subcellular Fractions from 13-Day-Old Chick Brain Homogenates

		Glycosyltransferases (%)						
		6	5	4	3	2	1	
Fraction	Protein (%)	[GlcUA-	GalNAc(SO₄)]-	GlcUA —	Gal —	Gal —	Xyl —	Ser
Nuclear	14.1	13.4	20.4	12.2		14.2	8.5	
Crude mitochondria	29.9	67.0	56.7	33.8		73.1	21.9	
Microsome	19.6	12.0	16.1	13.8		12.7	4.9	
Cytosol	35.4	7.6	6.9	40.1		0	64.6	
Recovery of material in initial brain homogenate (%)	92.3	70.4	152.0	78.5		80.2	118.4	
Myelin	6.1	1.2	2.4	5.4		1.9	9.4	
Synaptosome	47.9	73.5	77.0	77.0		85.4	59.4	
Mitochondria	46.0	24.2	19.6	15.8		11.6	31.3	
Recovery of material in crude mitochondrial fraction	77.0	56.2	102.8	124.0		95.2	78.0	
Synaptic vesicles	9.9	2.4	1.0	5.0		1.0	6.1	
Membranous material	90.1	97.6	99.0	95.0		99.0	93.9	
Recovery of material in synaptosome fraction	48	21.8	78.7	64.6		75.1	67.0	

miller *et al.*, 1972), a majority (65%) of the xylosyltransferase was found in the cytosol. Unexpectedly, in 13-day chick embryo brain, approximately one third of the total *N*-acetylgalactosamine transferase (reaction 4), involved in the formation of the repeating disaccharide unit, was found in the same fraction. Den *et al.* (1970) had previously reported a shift in the subcellular distribution of a UDP-galactose:glycoprotein galactosyltransferase present in chick embryo brain from the cytosol to the particulate synaptosome-enriched fraction with embryo development. It is possible that the xylosyltransferases and the *N*-acetylgalactosaminyltransferase (reactions 1 and 4) may undergo a similar transition during development of the brain. Whether the association of these transferases with the synaptosome-enriched fraction may be attributed to ionic bonding, or possibly to differential "affinity" binding of the transferases to the synaptosome membrane, perhaps in a manner similar to that described for the xylosyl and galactosyltransferases in chick embryo epiphyseal preparations (Schwartz *et al.*, 1974), has not been established.

When the crude mitochondrial fraction (17,000*g* pellet) was subfractionated into synaptosome- and mitochondria-enriched fractions, 59% of the xylosyltransferase and more than 75% of each of the other transferases associated with the 17,000*g* pellet were found in the synaptosome-enriched fraction. After hypotonic lysis, more than 93% of each glycosyltransferase was associated with the synaptosome membrane fraction and less than 6% of each transferase with the synaptic vesicle fraction. The latter finding was unexpected, since Vos *et al.* (1969) found that the synaptic vesicle fraction contained the highest concentration of glycosaminoglycans (largely hyaluronic acid and chondroitin sulfate) found in brain. At present, it is open to conjecture whether these vesicles serve as storage sites for glycosaminoglycans that in turn are involved in the binding of neurotransmitter substances (Uvnäs, 1973), or whether this finding is attributable to a charge interaction between soluble glycosaminoglycans and the synaptic vesicles.

As can be seen in Table II, significant amounts of glycosyltransferase activities are associated with fractions enriched in microsomes (5–14%) and mitochondria (12–31%). These transferase activities are probably not attributable solely to cross-contamination of subcellular fractions. Saxena *et al.* (1971) found that sulfotransferase activity (Fig. 2, reaction 7) in rat brain was nearly equally distributed among the synaptosomal (40%), the microsomal (30%), and the soluble (32%) fractions. Furthermore, the distribution of ^{35}S-labeled glycosaminoglycans in subcellular fractions from mouse brain of different ages, prepared by the same methodology (Kuriyama and Okada, 1971), closely paralleled the subcellular distribution of the sulfotransferase activity.

6. Summary and Conclusions

Limited information is available concerning the biosynthesis of glycosaminoglycans in nervous tissue. The results of isotope-incorporation studies show that enzyme activities responsible for the biosynthesis of glycosaminoglycans are present in both neurons and glia. Of the biosynthesis of the several glycosaminoglycans found in this tissue, only that of chondroitin sulfate proteoglycan has been studied in detail. To date, no major differences have been observed that differentiate the pathway for its biosynthesis in nervous tissue from that found in other tissues. The core polypeptide is synthesized by routes that are now well established, and in turn the carbohydrate chains are formed by the stepwise transfer of the several monosaccharide units from the corresponding nucleotide sugars. The transfer of each monosaccharide unit is determined by the specificity exhibited by each glycosyltransferase toward characteristic donor and acceptor groups. Sulfate fixation is mediated by 3-phosphoadenosine 5′-phosphosulfate. Whether fixation occurs *in vivo* during or after chain elongation has not been firmly established.

Present evidence suggests that marked changes occur in the biosynthesis of glycosaminoglycans during brain development. Several (if not all) of the glycosyltransferases associated with chondroitin sulfate synthesis change with increasing age from a soluble to a particulate form, resulting in the formation of a multiglycosyltransferase complex that is tightly associated with the synaptosomal membrane.

Glycosaminoglycans are postulated to play an important role in the conduction of nerve impulses (Szabo and Roboz-Einstein, 1962), in tissue hydration (Margolis, R. U., *et al.*, 1975; Polansky *et al.*, 1974), in neuronal migration and differentiation (Margolis, R. U., *et al.*, 1975), and in modifying cellular expression (Hay and Meier, 1974). It is readily apparent that the studies described in this chapter provide at best a limited perception of the mechanisms and controls for the biosynthesis of glycosaminoglycans in nervous tissue. In view of their potential importance and the increasing availability of sensitive and specific methodology, study of the metabolism of glycosaminoglycans in nervous tissue provides a fertile area for new investigations, and offers great promise for enhancement of our current knowledge and understanding of the biochemical processes responsible for the development and function of nervous tissue in health and disease.

References

Bachhawat, B. K., Balasubramanian, K. A., Balasubramanian, A. S., Singh, M., George, E., and Chandrasekaran, E. V., 1972, Chemistry and metabolism of glycosaminoglycans of the nervous system, *Adv. Exp. Med. Biol.* **25**:51.

Baker, J. R., Rodén, L., and Stoolmiller, A. C., 1972, Biosynthesis of chondroitin sulfate proteoglycan and other exogenous acceptors, *J. Biol. Chem.* **247**:3838.

Balasubramanian, A. S., and Bachhawat, B. K., 1961, Formation of active sulfate in rat brain, *J. Sci. Ind. Res.* **20C**:202.

Balasubramanian, A. S., and Bachhawat, B. K., 1964, Enzymic transfer of sulphate from 3′-phosphoadenosine 5′-phosphosulphate to mucopolysaccharides in rat brain, *J. Neurochem.* **11**:877.

Balasubramanian, A. S., and Bachhawat, B. K., 1970, Sulfate metabolism in brain, *Brain Res.* **20**:341.

Bandurski, R. S., Wilson, L. G., and Squires, C. L., 1956, The mechanism of active sulfate formation, *J. Am. Chem. Soc.* **78**:6408.

Barondes, S. H., 1968, Incorporation of radioactive glucosamine into macromolecules at nerve endings, *J. Neurochem.* **15**:699.

Basu, D. K., and Bachhawat, B. K., 1961, Purification of uridine diphosphoglucose pyrophosphorylase from human brain, *J. Neurochem.* **7**:174.

Boström, H., and Odeblad, E., 1953, Autoradiographic observations on the uptake of S^{35}-labelled sodium sulphate in the nervous system of the adult rat, *Acta Psychiatr. Neurol. Scand.* **28**:5.

Brandt, A. E., 1971, Studies on the biosynthesis of chondroitin sulfate, Doctoral dissertation, The University of Michigan, Ann Arbor.

Brandt, A. E., Distler, J., and Jourdian, G. W., 1969, Biosynthesis of the chondroitin sulfate-protein linkage region: Purification and properties of a glucuronosyltransferase from embryonic chick brain, *Proc. Natl. Acad. Sci. U.S.A.* **64**:374.

Brandt, A. E., Distler, J. J., and Jourdian, G. W., 1975, Biosynthesis of chondroitin sulfate proteoglycan, *J. Biol. Chem.* **250**:3996.

Breckenridge, B. M., and Crawford, E. J., 1961, The quantitative histochemistry of the brain, *J. Neurochem.* **7**:234.

Breckenridge, W. C., and Wolfe, L. S., 1973, The effect of dolichol phosphate on the synthesis of lipid bound sugars in embryonic chick brain, *FEBS Lett.* **29**:66.

Chandra, M., and Bhaduri, A. N., 1976, Purification and partial characterization of UDP-glucose-4-epimerase from goat brain, *J. Neurochem.* **27**:641.

Comb, D. G., and Roseman, S., 1958, Glucosamine metabolism. IV. Glucosamine-6-phosphate deaminase, *J. Biol. Chem.* **232**:807.

Cotman, C. W., Banker, G., Churchill, L., and Taylor, D., 1974, Isolation of postsynaptic densities from rat brain, *J. Cell Biol.* **63**:441.

D'Abramo, F., and Lipmann, F., 1957, The formation of adenosine-3′-phosphate-5′-phosphosulfate in extracts of chick embryo cartilage and its conversion into chondroitin sulfate, *Biochim. Biophys. Acta* **25**:211.

Den, H., Kaufman, B., and Roseman, S., 1970, Properties of some glycosyltransferases in embryonic chicken brain, *J. Biol. Chem.* **245**:6607.

De Robertis, E., 1967, Ultrastructure and cytochemistry of the synaptic region, *Science* **156**:907.

Distler, J., and Jourdian, G. W., 1972, UDP-glucose: β-Xyloside transglucosylase from embryonic chicken brain, *Methods Enzymol.* **28**:482.

Dorfman, A., and Ho, P.-L., 1970, Synthesis of acid mucopolysaccharides by glial tumor cells in tissue culture, *Proc. Natl. Acad. Sci. U.S.A.* **66**:495.

Dziewiatkowski, D. D., 1953, Sulfate–sulfur metabolism in the rat fetus as indicated by sulfur-35, *J. Exp. Med.* **96**:119.

Dziewiatkowski, D. D., 1970, Metabolism of sulfate esters, in: *Symposium: Sulfur in Nutrition* (O. H. Muth and J. E. Oldfield, eds.), pp. 97–125, Avi Publishing Co., Westport, Connecticut.

Faulkner, P., and Quastel, J. H., 1956, Anaerobic deamination of D-glucosamine by bacterial and brain extracts, *Nature* (*London*) **177**:1216.

Fluharty, A. L., Davis, M. L., Trammell, J. L., Stevens, R. L., and Kihara, H., 1975, Mucopolysaccharides synthesized by cultured glial cells derived from a patient with Sanfilippo syndrome, *J. Neurochem.* **25**:429.

Fransson, L.-Å., 1970, Structure and metabolism of the proteoglycans of dermatan sulfate, in: *Chemistry and Molecular Biology of the Intercellular Matrix* (E. A. Balazs, ed.), Vol. 2, pp. 823–842, Academic Press, New York.

Friede, R. L., 1966, *Topographic Brain Chemistry*, p. 444, Academic Press, New York.

George, E., and Bachhawat, B. K., 1970, Brain glycosaminoglycans and glyocosaminoglycan sulfotransferase in Sanfilippo syndrome, *Clin. Chim. Acta* **30**:317.

George, E., Singh, M., and Bachhawat, B. K., 1970, The nature of sulfation catalyzed by brain sulfotransferase of uronic acid containing glycosaminoglycans, *J. Neurochem.* **17**:189.

Goldberg, J. M., and Cunningham, W. L., 1970, Incorporation of [^{35}S]sulfate into the glycosaminoglycans of rat brain, *Biochem. J.* **120**:15P.

Gray, E. G., and Whittaker, V. P., 1962, The isolation of nerve endings from brain: An electron-microscopic study of cell fragments derived by homogenization and centrifugation, *J. Anat.* (*London*) **96**:79.

Gregory, J. D., and Lipmann, F., 1957, The transfer of sulfate among phenolic compound with 3′5′-diphosphoadenosine as coenzyme, *J. Biol. Chem.* **229**:1081.

Guha, A., Northover, B. J., and Bachhawat, B. K., 1960, Incorporation of radioactive sulfate into chondroitin sulfate in the developing brain of rats, *J. Sci. Ind. Res.* **19**:287.

Hay, E. D., and Meier, S., 1974, Glycosaminoglycan synthesis by embryonic inductors: Neural tube, notochord, and lens, *J. Cell Biol.* **62**:889.

Helting, T., 1971, Biosynthesis of heparin—Solubilization, partial separation and purification of uridine diphosphate-galactose: Acceptor galactosyltransferases from mouse mastocytoma, *J. Biol. Chem.* **246**:815.

Helting, T., 1972, Biosynthesis of heparin—Solubilization and partial purification of uridine diphosphate glucuronic acid: Acceptor glucuronosyltransferase from mouse mastocytoma, *J. Biol. Chem.* **247**:4327.

Helting, T., and Lindahl, U., 1972, Biosynthesis of heparin. I. Transfer of *N*-acetylglucosamine and glucuronic acid to low molecular weight heparin fragments, *Acta Chem. Scand.* **26**:3515.

Helting, T., and Rodén, L., 1968, The carbohydrate–protein linkage region of chondroitin-6-sulfate, *Biochim. Biophys. Acta* **170**:301.

Helting, T., and Rodén, L., 1969, Biosynthesis of chondroitin sulfate. II. Glucuronosyl transfer in the formation of the carbohydrate–protein linkage region, *J. Biol. Chem.* **244**:2799.

Hemming, F. W., 1974, Lipids in glycan biosynthesis, in: *MTP International Review of Science: Biochemistry Series One* (T. W. Goodwin, ed.), Vol. 4, pp. 39–97, University Park Press, Baltimore.

Höök, M., Lindahl, U., Bäckström, G., Malmström, A., and Fransson, L.-Å., 1974, Biosynthesis of heparin. III. Formation of iduronic acid residues, *J. Biol. Chem.* **249**:3908.

Jacobson, B., 1970, The biosynthesis of hyaluronic acid, in: *Chemistry and Molecular Biology of the Intercellular Matrix* (E. A. Balazs, ed.), Vol. 2, pp. 763–781, Academic Press, New York.

Jacobson, B., and Davidson, E. A., 1962, Biosynthesis of uronic acids by skin enzymes. II. Uridine diphosphate D-glucuronic acid-5-epimerase, *J. Biol. Chem.* **237**:638.

Katzman, R. L., 1971, On arabinose as a constituent of hyaluronic acid from bovine brain, *J. Neurochem.* **18**:1187.

Kennedy, J. F., 1976, Chemical and biochemical aspects of the glycosaminoglycans and proteoglycans in health and disease, in: *Advances in Clinical Chemistry* (O. Bodansky and A. L. Latner, eds.), Vol. 18, pp. 1–76, Academic Press, New York.

Kimata, K., Okayama, M., Oohira, A., and Suzuki, S., 1973, Cytodifferentiation and proteoglycan biosynthesis, *Mol. Cell. Biochem.* **1**:211.

Koransky, W., 1958, Trennung und Bestimmung der Nucleotide des Gehirns, *Naunyn-Schmiedebergs Arch. Pharmakol.* **234**:46.

Kuriyama, K., and Okada, T. A., 1971, Incorporatiion of ^{35}S-sulfate into developing mouse brain: Subcellular fractionation and electron microscopic studies, *Exp. Neurol.* **30**:18.

Lennarz, W. J., 1975, Lipid linked sugars in glycoprotein synthesis, *Science* **188**:986.

Lie, S. O., McKusik, V. A., and Neufeld, E. F., 1972, Simulation of genetic mucopolysaccharidoses in normal human fibroblasts by alteration of pH of the medium, *Proc. Natl. Acad. Sci. U.S.A.* **69**:2361.

Louisot, P., and Broquet, P., 1973, Subcellular localization of glycosyltransferases in synaptosomes and mitochondria of brain, in: *Central Nervous System—Studies on Metabolic Regulation and Function* (E. Genazzani and H. Herken, eds.), pp. 164–166, Springer-Verlag, New York.

Maker, H. S., and Lehrer, G. M., 1972, Carbohydrate chemistry of brain, in: *Basic Neurochemistry* (R. W. Albers, G. J. Siegel, R. Kataman, and B. W. Agranoff, eds.), pp. 169–190, Little, Brown, Boston.

Malmström, A., Fransson, L.-Å., Höök, M., and Lindahl, U., 1975, Biosynthesis of dermatan sulfate. I. Formation of L-iduronic acid residues, *J. Biol. Chem.* **250**:3419.

Margolis, R. K., and Margolis, R. U., 1970, Sulfated glycopeptides from rat brain glycoproteins, *Biochemistry U.S.A.* **9**:4389.

Margolis, R. K., Crockett, C. P., Kiang, W.-L., and Margolis, R. U., 1976, Glycosaminoglycans and glycoproteins associated with rat brain nuclei, *Biochim. Biophys. Acta* **451**:465.

Margolis, R. U., 1967, Acid mucopolysaccharides and proteins of bovine whole brain, white matter and myelin, *Biochim. Biophys. Acta* **141**:91.

Margolis, R. U., and Margolis, R. K., 1972, Sulfate turnover in mucopolysaccharides and glycoproteins of brain, *Biochim. Biophys. Acta* **264**:426.

Margolis, R. U., and Margolis, R. K., 1974, Distribution and metabolism of mucopolysaccharides and glycoproteins in neuronal perikarya, astrocytes, and oligodendroglia, *Biochemistry U.S.A.* **13**:2849.

Margolis, R. U., Margolis, R. K., Chang, L. B., and Preti, C., 1975, Glycosaminoglycans of brain during development, *Biochemistry U.S.A.* **14**:85.

Margolis, R. U., Lalley, K., Kiang, W.-L., Crockett, C., and Margolis, R. K., 1976, Isolation and properties of a soluble chondroitin sulfate proteoglycan from brain, *Biochem. Biophys. Res. Commun.* **73**:1018.

Maxwell, E. S., Kalckar, H. M., and Burton, R. M., 1955, Galacto-waldenase and the enzymatic incorporation of galactose-1-phosphate in mammalian tissue, *Biochim. Biophys. Acta* **18**:444.

Miller, R. R., and Waechter, C. J., 1978, Enzymatic sulfation of membrane-associated proteoglycans by pig brain, *Trans. Am. Soc. Neurochem.* **9**:158.

Morgan, I. G., Wolfe, L. S., Mandel, P., and Gombos, G., 1971, Isolation of plasma membranes from rat brain, *Biochim. Biophys. Acta* **241**:737.

Norton, W. T., and Poduslo, S. E., 1970, Neuronal soma and whole neuroglia of rat brain: A new isolation technique, *Science* **167**:1144.

Norton, W. T., and Poduslo, S. E., 1971, Neuronal perikarya and astroglia of rat brain: Chemical composition during myelination, *J. Lipid Res.* **12**:84.

Odeblad, E., and Boström, H., 1952, An autoradiographic study of the incorporation of ^{35}S-labeled sodium sulfate in different organs of adult rats and rabbits, *Acta Pathol. Microbiol. Scand.* **31**:339.

Parodi, A. J., and Leloir, L. F., 1976, Lipid intermediates in protein glycosylation, *Trends Biochem. Sci.* **1**:58.

Pattabiraman, T. N., and Bachhawat, B. K., 1961*a*, Purification of glucosamine 6-phosphate deaminase from human brain, *Biochim. Biophys. Acta* **54**:273.

Pattabiraman, T. N., and Bachhawat, B. K., 1961*b*, Purification and properties of acetyl glucosamine kinase from sheep brain, *J. Sci. Ind. Res.* **20C**:14.

Pattabiraman, T. N., and Bachhawat, B. K., 1961*c*, Purification of uridine diphospho acetyl glucosamine pyrophosphorylase from sheep brain, *Biochim. Biophys. Acta* **50**:129.

Pattabiraman, T. N., and Bachhawat, B. K., 1962*a*, Purification of glucosamine-6-phosphate *N*-acetylase from sheep brain, *Biochim. Biophys. Acta* **59**:681.

Pattabiraman, T. N., and Bachhawat, B. K., 1962*b*, Interconversion of *N*-acetyl glucosamine 6-phosphate and *N*-acetyl glucosamine 1-phosphate in rat brain, *J. Sci. Ind. Res.* **21C**:352.

Poduslo, S. E., and Norton, W. T., 1972, Isolation and some chemical properties of oligodendroglia from calf brain, *J. Neurochem.* **19**:727.

Polansky, J. R., Toole, B. P., and Gross, J., 1974, Brain hyaluronidase: Changes in activity during chick development, *Science* **183**:862.

Ringertz, N. R., 1956, On the sulphate metabolism of the mouse brain, *Exp. Cell Res.* **10**:230.

Robbins, P. W., and Lipmann, F., 1956, The enzymatic sequence in the biosynthesis of active sulfate, *J. Am. Chem. Soc.* **78**:6409.

Robinson, J. D., and Green, J. P., 1962, Sulfomucopolysaccharides in brain, *Yale J. Biol. Med.* **35**:248.

Rodén, L., 1970, Biosynthesis of acidic glycosaminoglycans (Mucopolysaccharides), in: *Metabolic Conjugation and Metabolic Hydrolysis* (W. H. Fishman, ed.), Vol. II, pp. 346–432, Academic Press, New York.

Rodén, L., and Schwartz, N. B., 1973, The biosynthesis of chondroitin sulfate, *Biochem. Soc. Trans.* **1**:227.

Rodén, L., Baker, J. R., Cifonelli, A., and Mathews, M. B., 1972*a*, Isolation and characterization of connective tissue polysaccharides, *Methods Enzymol.* **28B**:73.

Rodén, L., Baker, J. R., Helting, T., Schwartz, N. B., Stoolmiller, A. C., Yamagata, S., and Yamagata, T., 1972*b*, Biosynthesis of chondroitin sulfate, *Methods Enzymol.* **28B**:638.

Roseman, S., 1959, Metabolism of connective tissue, *Annu. Rev. Biochem.* **28**:545.

Sampaio, L. O., Dietrich, C. P., and Filko, O. G., 1977, Changes in sulfated mucopolysaccharide composition of mammalian tissues during growth and in cancer tissues, *Biochim. Biophys. Acta* **498**:123.

Satoh, C., Duff, R., Ropp, F., and Davidson, E. A., 1973, Production of mucopolysaccharides by normal and transformed cells, *Proc. Natl. Acad. Sci. U.S.A.* **70**:54.

Saxena, S., George, E., Kokrady, S., and Bachhawat, B. K., 1971, Sulphate metabolism in developing rat brain: A study with subcellular fractions, *Indian J. Biochem. Biophys.* **8**:1.

Schachter, H., and Rodén, L., 1973, The biosynthesis of animal glycoproteins, in: *Metabolic Conjugation and Metabolic Hydrolysis* (W. H. Fishman, ed.), Vol. III, pp. 2–149, Academic Press, New York.

Schwartz, N. B., Rodén, L., and Dorfman, A., 1974, Biosynthesis of chondroitin sulfate: Interaction between xylosyltransferase and galactosyltransferase, *Biochem. Biophys. Res. Commun.* **56**:717.

Sellinger, O. Z., Azcurra, J. M., Johnson, D. E., Ohlsson, W. G., and Lodin, Z., 1971, Independence of protein synthesis and drug uptake in nerve cell bodies and glial cells isolated by a new technique, *Nature (London) New Biol.* **230**:253.

Silbert, J. E., and DeLuca, S., 1969, Biosynthesis of chondroitin sulfate. III. Formation of a sulfated glycosaminoglycan with a microsomal preparation from chick embryo cartilage, *J. Biol. Chem.* **244**:876.

Simpson, D. L., Thorne, D. R., and Loh, H. H., 1976, Sulfated glycoproteins, glycolipids, and glycosaminoglycans from synaptic plasma and myelin membranes: Isolation and characterization of sulfated glycopeptides, *Biochemistry U.S.A.* **15**:5449.

Singh, M., and Bachhawat, B. K., 1965, The distribution and variation with age of different uronic acid–containing mucopolysaccharides in brain, *J. Neurochem.* **12**:519.

Singh, M., and Bachhawat, B. K., 1968, Isolation and characterization of glycosaminoglycans in human brain of different age groups, *J. Neurochem.* **15**:249.

Singh, M., Chandrasekaran, E. V., Cherian, R., and Bachhawat, B. K., 1969, Isolation and characterization of glycosaminoglycans in brain of different species, *J. Neurochem.* **16**:1157.

Stary, Z., Wardi, A., and Turner, D., 1964, Galacturonic acid and hydrolysates of defatted human brain, *Biochim. Biophys. Acta* **83**:242.

Stoolmiller, A. C., 1972, Biosynthesis of mucopolysaccharides by neuroblastoma cells in tissue culture, *Fed. Proc. Fed. Am. Soc. Exp. Biol.* **31**:910 (abstract).

Stoolmiller, A. C., and Dorfman, A., 1970, The biosynthesis of hyaluronic acid in group A streptococci, in: *Chemistry and Molecular Biology of the Intercellular Matrix* (E. A. Balazs, ed.), Vol. 2, pp. 783–794, Academic Press, New York.

Stoolmiller, A. C., Horwitz, A. L., and Dorfman, A., 1972, Biosynthesis of the chondroitin sulfate proteoglycan: Purification and properties of xylosyltransferase, *J. Biol. Chem.* **247**:3525.

Stoolmiller, A. C., Dawson, G., and Dorfman, A., 1973, The metabolism of glycosphingolipids and glycosaminoglycans, in: *Tissue Culture of the Nervous System* (G. Sato, ed.), pp. 247–280, Plenum Press, New York.

Suzuki, S., and Strominger, J. L., 1960*a*, Enzymatic sulfation of mucopolysaccharides in hen oviduct. I. Transfer of sulfate from 3′-phosphoadenosine 5′-phosphosulfate to mucopolysaccharides, *J. Biol. Chem.* **235**:257.

Suzuki, S., and Strominger, J. L., 1960*b*, Enzymatic sulfation of mucopolysaccharides in hen oviduct. II. Mechanism of the reaction studied with oligosaccharides and monosaccharides as acceptors, *J. Biol. Chem.* **235**:267.

Suzuki, S., and Strominger, J. L., 1960*c*, Enzymatic sulfation of mucopolysaccharides in hen oviduct. III. Mechanism of sulfation of chondroitin and chondroitin sulfate A, *J. Biol. Chem.* **235**:274.

Suzuki, S., Trenn, R. H., and Strominger, J. L., 1961, Separation of specific mucopolysaccharide sulfotransferases, *Biochim. Biophys. Acta* **50**:169.

Szabo, M. M., and Roboz-Einstein, E., 1962, Acidic polysaccharides in the central nervous system, *Arch. Biochem. Biophys.* **98**:406.

Telser, A., Robinson, H. C., and Dorfman, A., 1966, The biosynthesis of chondroitin sulfate, *Arch. Biochem. Biophys.* **116**:458.

Turco, S. J., and Heath, E. C., 1977, Glucuronosyl-*N*-acetylglucosaminyl pyrophosphoryl dolichol formation in SV_{40}-transformed human lung fibroblasts and biosynthesis in rat lung microsomal preparations, *J. Biol. Chem.* **252**:918.

Uvnäs, B., 1973, An attempt to explain nervous transmitter release as due to nerve impulse–induced cation exchange, *Acta Physiol. Scand.* **87**:168.

Uzman, B. G., Murray, M. R., and Saito, H., 1973, Incorporation of [^{35}S]sulfate into chondroitin sulfates by organized cultures of murine peripheral, sympathetic and central nervous tissue, *J. Neurobiol.* **4**:429.

Varma, R., Vercellotti, J. R., and Varma, R. S., 1977, On arabinose as a component of brain hyaluronate: Confirmation by chromatographic, enzymatic and chemical ionization–mass spectrometric analyses, *Biochim. Biophys. Acta* **497**:608.

Vitello, L., Breen, M., Weinstein, H. G., Sittig, R. A., and Blacik, L. J., 1978, Keratan sulfate–like glycosaminoglycan in the cerebral cortex of the brain and its variation with age, *Biochim. Biophys. Acta* **539**:305.

Vos, J., Kuriyama, K., and Roberts, E., 1969, Distribution of acid mucopolysaccharides in subcellular fractions of mouse brain, *Brain Res.* **12**:172.

Waechter, C. J., and Lennarz, W. J., 1976, The role of polyprenol-linked sugars in glycoprotein synthesis, *Annu. Rev. Biochem.* **45**:95.

Waechter, C. J., Lucas, J. J., and Lennarz, W. J., 1974, Evidence for xylosyl lipids as intermediates in xylosyl transfers in hen oviduct membranes, *Biochem. Biophys. Res. Commun.* **56**:343.

Wardi, A. H., Allen, W. S., Turner, D. L., and Stary, Z., 1966, Isolation of arabinose-containing hyaluronate peptides and xylose-containing chondroitin sulfate peptides from protease-digested brain tissue, *Arch. Biochem. Biophys.* **117**:44.

Whittaker, V. P., and Sheridan, M. N., 1965, The morphology and acetylcholine content of isolated cerebral cortical synaptic vesicles, *J. Neurochem.* **12**:363.

Wiesmann, U. N., Hofmann, K., Burkhart, T., and Herschkowitz, N., 1975, Dissociated cultures of newborn mouse brain. I. Metabolism of sulfated lipids and mucopolysaccharides, *Neurobiology* **5**:305.

Young, I. J., and Custod, J. T., 1972, Isolation of glycosaminoglycans and variation with age in the feline brain, *J. Neurochem.* **19**:923.

6

Brain Glycosidases

Robert R. Townsend, Yu-Teh Li, and Su-Chen Li

1. Introduction

In keeping with the complexity of the CNS, the characteristics and biological roles of the brain glycosidases have been more difficult to define than in visceral organs. The nature of subcellular fractions isolated from brain tissue is not well defined. Therefore, assignment of a given glycosidase to a particular organelle has been more difficult than in other tissues. As with glycosidases of other tissues, brain glycosidases exist in multiple forms, complicating their characterization with respect to substrate specificity, interrelationships, and biological function.

A major part of the investigation of brain glycosidases has received its impetus and has been in the context of a particular glycosidase deficiency resulting in the accumulation of partially degraded glycoconjugates. The major storage material in these disorders is glycosphingolipids, except for fucosidosis and mannosidosis, in which the accumulating materials are derived from glycoproteins. The degradation of glycosphingolipids by brain glycosidases is the subject of several recent reviews (Ledeen and Yu, 1973; Neufeld *et al.*, 1975; Suzuki, 1976). This chapter gives an overview of aspects of brain glycosidases other than their role in glycosphingolipid catabolism.

Although there are numerous reports on the composition of brain glycoproteins, their oligosaccharide structures have not been well characterized (see Chapter 3). However, in some storage disorders, the accumulation of partially degraded glycoprotein derivatives in brain has provided sufficient material for detailed structural analysis (Tsay and Dawson, 1976; Tsay *et al.*, 1976; Öckerman, 1969). Compositional analyses of brain glycoproteins demonstrate that they contain sialic acid, fucose, galactose, *N*-acetylglucosamine

Robert R. Townsend, Yu-Teh Li, and Su-Chen Li • Department of Biochemistry and Delta Primate Research Center, Tulane Medical Center, New Orleans, Louisiana.

(GlcNAc), *N*-acetylgalactosamine (GalNAc), and mannose. Therefore, brain should contain glycosidases to hydrolyze these sugar units.

2. Sialidase

Since sialic acid occurs only in the terminal (or penultimate to sialic acid) position of a saccharide chain, the degradation of the sialoglycoproteins by exoglycosidases is presumably initiated by the action of sialidase. Brain sialidases from different sources have been studied (Morgan and Laurell, 1963; Leibovitz and Gatt, 1968; Tettamanti *et al.*, 1970; Gielen and Harprecht, 1969; Öhman *et al.*, 1970). With the use of sialyllactose and gangliosides as substrates, brain sialidases were shown to occur in soluble (Tettamanti and Zambotti, 1968), membranous (Leibovitz and Gatt, 1968), and lysosomal (Tallman and Brady, 1972) forms. Most of the studies demonstrating the release of sialic acid from brain glycoproteins were carried out with crude enzymes and ill-defined substrates. In general, the enzymic release of sialic acid from endogenous gangliosides occurs more rapidly than from the endogenous glycoproteins. Tettamanti *et al.* (1975) showed that the membrane-bound sialidases from brain of different species released sialic acid from endogenous membrane-bound gangliosides as well as from glycoproteins. The hydrolysis of sialic acid from gangliosides, under the conditions of their studies, was in the range of 40–60% in all animals tested except pig, where it was lower (15%); the hydrolysis of glycoprotein-bound sialic acid was considerably less, from 1.5–3% in human, rabbit, and rat brain, up to a maximum of 12–15% in chicken and calf brain. These investigators also found that the membrane-bound sialidase could simultaneously hydrolyze endogenous sialoglycoprotein and exogenous sialoglycopeptides isolated from calf brain. Heijlman and Roukema (1972) demonstrated that the crude homogenate of calf brain was able to hydrolyze sialic acid from both endogenous and exogenous sialoglycoproteins. They further found that the calf brain sialoglycopeptides could be separated by chromatography on DEAE–Sephadex A-50 into fractions with different susceptibilities to sialidase of brain whole homogenate. Due to the lack of information concerning the structure of brain glycoproteins, very little can be said about the substrate specificity of brain sialidase at this time. Future work should endeavor to examine the action of purified sialidases on well-characterized brain sialoglycoproteins.

3. β-Galactosidase

Interest in generalized gangliosidoses and Krabbe's disease has prompted numerous investigations of brain β-galactosidase from various species (Gatt and Rapport, 1966*a,b;* Gatt, 1967; Miyatake and Suzuki, 1975; Callahan and Gerrie, 1975; Radin *et al.*, 1969; Tanaka and Suzuki, 1977). In general, β-galactosidases from mammalian tissues can be differentiated into two types: those capable of hydrolyzing G_{M1} ganglioside and those active toward galactosylceramide (Suzuki, 1976). In contrast to the knowledge regarding the catabolism of galactose-containing sphingoglycolipids, there is no report demonstrating the enzymic liberation of galactose from brain glycoproteins by a β-galactosidase isolated from the CNS. Accumulation of both G_{M1} ganglioside and galactose-rich glycoproteins in generalized gangliosidoses (O'Brien, 1972; Tsay and Dawson, 1976) implies that the same galactosidase can hydrolyze the β-galactosyl units in both G_{M1} ganglioside and glycoprotein.

4. β-*N*-Acetylhexosaminidase

The name β-*N*-acetylhexosaminidase is derived from the ability of an enzyme to hydrolyze both β-linked *N*-acetylglucosaminide and *N*-acetylgalactosaminide (Li and Li, 1970). All the β-*N*-acetylhexosaminidases isolated from visceral organs hydrolyze both β-*N*-acetylglucosaminide and β-*N*-acetylgalactosaminide. However, this dual specificity may not apply to brain β-*N*-acetylhexosaminidases. Frohwein and Gatt (1967) isolated, from calf brain, three enzymes, one particle-associated and two soluble, with different specificities toward β-*N*-acetylglucosaminide and β-*N*-acetylgalactosaminide. The particulate enzyme hydrolyzed *p*-nitrophenyl β-*N*-acetylglucosaminide and *p*-nitrophenyl β-*N*-acetylgalactosaminide. One of the soluble enzymes hydrolyzed *p*-nitrophenyl β-*N*-acetylglucosaminide 80 times faster than the corresponding galactosaminide. The other soluble enzyme hydrolyzed *p*-nitrophenyl β-*N*-acetylgalactosaminide 11 times faster than the corresponding glucosaminide. Using the synthetic substrates, Overdijk *et al.* (1975) showed that the β-*N*-acetylhexosaminidase activity of bovine brain tissue was composed of four fractions: two fractions contained both glucosaminidase and galactosaminidase activities, a third fraction showed only glucosaminidase activity, and a fourth fraction had specificity toward the galactosaminide moiety. More work is necessary to ascertain whether the aforementioned specificities of β-*N*-acetylhexosaminidases are applicable to natural substrates. Since the deficiency of β-*N*-acetylhexosaminidases is the molecular etiology of G_{M2} gangliosidoses, these enzymes in neural and visceral tissues have been intensively studied (Sloan and Fredrickson, 1972; Robinson *et al.*, 1972; Aruna and Basu, 1976; Tallman *et al.*, 1974). Despite this investigative effort, the role of β-*N*-acetylhexosaminidases in brain glycoprotein catabolism is not well defined. In the brain of patients with Type O G_{M2} gangliosidosis (Sandhoff disease), there is an accumulation of oligosaccharides derived from glycoproteins that contain terminal GalNAc. This suggests that β-*N*-acetylhexosaminidases (Tsay and Dawson, 1976) also participate in the catabolism of brain glycoproteins.

5. α-L-Fucosidase

Only a few studies have been described with respect to the purification and properties of brain α-L-fucosidase (Quarles and Brady, 1970; Bossman and Hemsworth, 1971; Alhadeff *et al.*, 1975; Alhadeff and Janowsky, 1977; Dawson and Tsay, 1977). There is now strong evidence that there is no correlation between the ability of a given fucosidase to hydrolyze synthetic substrates such as *p*-nitrophenyl α-L-fucoside and its ability to hydrolyze natural substrates (Arakawa-Ogata *et al.*, 1977). Whether this is also true for brain α-L-fucosidase remains to be elucidated.

Studies of the characteristics of α-L-fucosidase in brain are necessary for understanding the pathogenesis of fucosidosis. Moreover, fucose-containing glycoproteins are known to occur in normal brain (Dutton and Barondes, 1970; Zatz and Barondes, 1970; Margolis, R. K., and Margolis, R. U., 1972; Glasgow *et al.*, 1972), indicating an active role for α-L-fucosidases in their catabolism.

In rat brain, there is convincing evidence that α-L-fucosidase activity toward *p*-nitrophenyl-α-fucoside is represented by two enzymes. These two enzymes differ in their pH optima, one having an optimum at pH 4 and the other at pH 6 (Quarles and Brady, 1970), and in their different degree of inhibition by L-fucose (Margolis, R. K., and Margolis, R. U., 1972). Bossman and Hemsworth (1971) were able to purify rat brain α-L-fucosidase active against porcine submaxillary mucin and asialo porcine submaxillary mucin, but the enzyme was not able to cleave the 1,3-linked fucose of orosomucoid.

Soluble α-L-fucosidase, active toward *p*-nitrophenyl-α-L-fucoside, was highly purified from human brain using an agarose-ϵ-aminocaproyl-fucosamine column by Alhadeff *et al.* (1975); however, it specificity for natural substrates was not tested (Alhadeff and Janowsky, 1977). Dawson and Tsay (1977) purified α-L-fucosidase from human brain according to the affinity chromatographic procedure described above. They were able to demonstrate the release of fucose from the following substrates: *p*-nitrophenyl α-L-fucopyranoside, 2′-fucosyllactose (Fucα1→2Gal linkage), a fucodecasaccharide from fucosidosis patients (Fucα1→2Gal linkage), lacto-*N*-fucopentaose (Fucα1→4GlcNAc linkage), lacto-*N*-fucopentaose III (Fucα1→3GlcNAc), and Fucα1→6GlcNAc disaccharide.

6. α-Mannosidase

With the use of chromogenic substrates, α-mannosidase was partially purified from human (Chester *et al.*, 1975), rat (Bossman and Hemsworth, 1971), rabbit (Clark, 1974), and bovine (Hocking *et al.*, 1972) brain. α-Mannosidase from human brain was shown to occur in three molecular species by

DEAE–cellulose chromatography: one neutral and two acidic (A and B) enzymes. Forms A and B appeared to be related to each other, differing only in sialic acid content, since form B was found to be converted into form A by neuraminidase treatment (Chester *et al.*, 1975). The neutral activity found in human brain appeared to be more stable than the corresponding activity studied in other tissues (Chester *et al.*, 1975). The existence of three forms of α-mannosidase was also demonstrated with the use of cellulose acetate electrophoresis (Poenaru and Dreyfus, 1973). By using *p*-nitrophenyl-α-D-mannoside, Bossman and Hemsworth (1971) purified from rat brain an α-mannosidase that was capable of hydrolyzing mannose from ovalbumin. The acidic α-mannosidase from normal human brain is a Zn-enzyme, activated by Zn^{2+} and inhibited by Co^{2+} (Hultberg and Masson, 1975). In contrast, the residual acidic α-mannosidase activity in the brain of mannosidosis patients was found to be significantly activated by both Zn^{2+} and Co^{2+} (Hultberg and Masson, 1975).

The involvement of α-mannosidase in the normal catabolism of glycoproteins is evident from studies of subjects lacking the acidic activity of this enzyme. This results in the accumulation in brain tissue of mannose-rich oligosaccharides, which are assumed to be derived from glycoproteins (Lundblad *et al.*, 1976).

7. Enzymes That Degrade Brain Glycosaminoglycans, and Other Glycohydrolases

The polysaccharide chains of the glycosaminoglycans are degraded by the concerted action of hexosaminidases, glucuronidases, and sulfatases. For hyaluronic acid and chondroitin sulfate, the initial catabolic steps are probably performed by hyaluronidase, which is the collective name for a group of endo-β-hexosaminidases which hydrolyze β-(1 $\longrightarrow$ 4)-hexosaminidoglucuronic acid bonds of hyaluronic acid and chondroitin 4- and 6-sulfates to yield at first relatively high-molecular-weight oligosaccharides, and later a tetrasaccharide as the major end-product. This is then further degraded to monosaccharides by β-glucuronidase and an *N*-acetylhexosaminidase. (Heparan sulfate is not a substrate for hyaluronidase.) The most extensively studied of the animal hyaluronidases is the enzyme isolated from testis (where it is produced by sperm) and having a broad pH optimum of approximately 4–6. Hyaluronidases with more acid pH optima (usually below pH 4) have been demonstrated in numerous mammalian tissues and body fluids, and have also been shown to be lysosomal enzymes in a number of cases.

Hyaluronidase activity was demonstrated in rat and bovine brain, where in both cases it had a pH optimum of 3.7 (Margolis, R. U., *et al.*, 1972). The enzyme was enriched 5- to 6-fold in a crude lysosomal fraction of rat brain or bovine cerebral cortex. It degraded hyaluronic acid and, at a slower rate, chondroitin sulfate to a mixture of higher oligosaccharides with *N*-acetylhexsosamine at the reducing end. The hyaluronidase activity in gray matter is more than twice that present in white matter, which parallels the distribution of glycosaminoglycans in brain. This enzyme probably has an important role in the turnover of brain glycosaminoglycans, since the level of hyaluronidase activity in rat brain was found to be considerably greater than that required to account for the rate of catabolism of hyaluronic acid and chondroitin sulfate measured *in vivo*.

Much less information is available concerning the degradation of heparan sulfate (in any animal tissue), and very little is known about the sulfatases active toward brain glycosaminoglycans and glycoproteins. Although specific eliminases capable of degrading heparin and heparan sulfate have been isolated and characterized in bacterial systems, similar enzymes are not known to be present in brain. A sulfamidase capable of removing the *N*-sulfate groups from heparin (and presumably also heparan sulfate) was demonstrated in brain and other tissues, and shown to differ in several respects from the various arylsulfatases that are also present (Friedman and Arsenis, 1972). Up to now, this is the only sulfatase in brain that has been reported to be active toward glycosaminoglycans, and it is obvious that more work is required in the area of brain sulfatases and heparan-sulfate-depolymerizing enzymes.

Other glycohydrolases such as β-xylosidase, β-mannosidase, β-fucosidase (Bossman and Hemsworth, 1971), and α-galactosidase (Lusis and Paigen, 1976) were also reported to be present in nervous tissue, but their role in brain glycoconjugate metabolism is not clear. In addition to the exoglycosidases discussed above, there is also indirect evidence for the presence in brain of endoglycosidases active toward glycoprotein oligosaccharides (see Chapters 3 and 16). However, these enzymes have not yet been purified from nervous tissue or characterized.

8. Biology of Glycosidases

Attempts have been made to localize acid hydrolases in specific anatomical regions of brain, and in cellular and subcellular fractions. This has led to much speculation as to the role of glycohydrolases in physiological and pathological states.

Robins *et al.* (1968) found that depending on the anatomical site sampled, the ratios of β-galactosidase and β-glucuronidase to β-glucosidase in monkey brain are 6.8 and 2.4, respectively, in gray matter, and 4.5 and 3.7, respectively, in white matter. In monkey and rat cerebellum, β-galactosidase, β-glucuronidase (Robins *et al.*, 1968), and β-hexosaminidase (Shuter *et al.*, 1970) are present predominantly in the cellular layers, suggesting an association with neuronal cell bodies. The distribution of β-hexosaminidases A and B in the layers of human cerebellum was determined by Hirsch (1972). The proportion of the heat-stable hexosaminidase B was greater in the granular layer than in the molecular layer or underlying white matter.

Sinha and Rose (1972) found significant differences in the distribution of glycosidases in the neuropil and neurons of rat cortex. They were able to divide the glycosidases into three groups on the basis of neuron/neuropil activity ratios: Group I (ratio 1.4–3.4), containing β-glucosidase, β-glucuronidase, and arabinosidase; Group II (ratio 5.8), β-*N*-acetylglucosaminidase; and Group III (ratio 9.9), β-galactosidase. On the basis of these results, they proposed that β-galactosidase could serve as a neuron marker. Furthermore, Raghaven *et al.* (1972) and Vargas-Idoyaga *et al.* (1972) reported no enrichment of glycohydrolases in glial as compared to neuronal fractions. Verty *et al.* (1973) demonstrated that β-glucosidase and β-glucuronidase were associated with the synaptosomal complex, while β-galactosidase and β-*N*-acetylglucosaminidase were associated with lysosomal-like particles. Brain sialidase activity was also found to be concentrated in the synaptosomal complex (Schengrund and Rosenberg, 1970).

The localization of brain glycosidases in one discrete subcellular particle, similar to that described in liver, has not been demonstrated. Although all brain glycosidases are considered lysosomal, there are convincing reports that more than one population of lysosomes exist in brain tissue)Sellinger and Hiatt, 1968; Sinha and Rose, 1972; Millson and Bountiff, 1973; Sellinger *et al.*, 1973). The role of lysosomal heterogeneity in the metabolism of brain macromolecules remains to be elucidated.

Studies of scrapie mouse brain indicated a characteristic alteration in glycoconjugate metabolism evidenced by a 2- to 3-fold increase in β-glucuronidase and *N*-acetyl β-D-glucosaminidase, and no change in β-galactosidase, α-glucosidase, or β-glucosidase (Hunter *et al.*, 1967; Millson and Bountiff, 1973). Further studies revealed that α-mannosidase and β-glucosidase, both having two pH optima, were elevated only in the isozyme with a lower pH optimum. Bowen *et al.* (1974) correlated the histopathological changes in various forms of degenerative nervous disorders with changes in the activity of acid hydrolases.

Attempts to link the appearance of natural substrates with their respective glycohydrolases have prompted studies of the pre- and postnatal levels of various glycosidases. Robins *et al.* (1961) showed a positive correlation between cellular layers of cerebellum and β-galactosidase activity in rats from birth through adulthood. Roukema *et al.* (1970) found that β-galactosidase and β-galactosaminidase activities increased 2.5- and 2-fold, respectively, from postnatal day 4 to day 20 in rat brain. This was similar to the concurrent development of increased levels of gangliosides, sialoglycoproteins, and sialidase. Traurig *et al.* (1973) demonstrated a decrease in β-glucuronidase activity and an increase in N-acetyl-β-D-glucosaminidase activity in postnatal rat brain. Clark (1974) found a decrease in β-galactosidase, β-glucuronidase, and α-galactosidase activities and an increase in N-acetyl-β-galactosaminidase and N-acetyl-β-glucosaminidase activities throughout prenatal development of rabbit brain. In addition, he found that α-mannosidase was at a low level until day 12 of fetal development, at which time there was a sharp increase up to birth. Schengrund and Rosenberg (1971) were able to detect increasing β-hexosaminidase and β-galactosidase activities in developing chicken brain from 5-day-old embryos to adulthood; however, β-glucosidase activity was not found.

Although the studies cited above have raised interesting questions concerning complex carbohydrate metabolism in normal and diseased nervous tissue, the function of glycosidases in brain glycoconjugate catabolism *in vivo* is difficult to ascertain, since all studies used artificial substrates.

From the information derived by studying various storage disorders, a catabolic pathway for the degradation of the saccharide portion of a hypothetical brain glycoprotein was proposed by Dawson (see Chapter 16).

Following the elucidation of the primary structure of brain glycosphingolipids, major advances were made in identifying the enzymes responsible for their catabolism. In contrast, our lack of knowledge of the saccharide structure of brain glycoproteins has led to confusion as to the role of glycosidases in the degradation of glycoproteins. Future investigations should characterize the glycosidases on the basis of their activity toward brain glycoproteins of known structure. Only then can significant progress be made in elucidating the specific role of a given glycosidase in the catabolism of brain glycoproteins, and its relationship to brain functions.

References

Alhadeff, J. A., and Janowsky, A. J., 1977, Purification and properties of human brain α-L-fucosidase, *J. Neurochem.* **28**:423–427.

Alhadeff, J. A., Miller, A. L., Wenaas, H., Vedvick, T., and O'Brien, J. S., 1975, Human liver α-L-fucosidase, *J. Biol. Chem.* **250**:7106–7113.

Arakawa-Ogata, M., Muramatsu, T., and Kobata, A., 1977, α-L-Fucosidases from almond emulsin: Characterization of the two enzymes with different specificities, *Arch. Biochem. Biophys.* **181**:353–358.

Aruna, R. M., and Basu, D., 1976, Purification and properties of β-hexosaminidase B from monkey brain, *J. Neurochem.* **27**:337–339.

Bossman, H. B., and Hemsworth, B. A., 1971, Intraneuronal glycosidases. II. Purification and properties of α-fucosidase, β-fucosidase, α-mannosidase, and β-xylosidase of rat cerebral cortex, *Biochim. Biophys. Acta* **242**:152–171.

Bowen, D. M., Flack, R. H. A., Martin, R. O., Smith, C. B., White, P., and Davison, A. N., 1974, Biochemical studies on degenerative neurological disorders. 1. Acute experimental encephalitis, *J. Neurochem.* **22**:1099–1107.

Callahan, J. W., and Gerrie, J., 1975, Purification of GMI ganglioside and ceramide lactoside β-galactosidase from rabbit brain, *Biochim. Biophys. Acta* **391**:141–153.

Chester, A. M., Laundblad, A., and Masson, P. K., 1975, The relationship between different forms of human α-mannosidase, *Biochim. Biophys. Acta* **391**:341–348.

Clark, J. T. R., 1974, Studies on acid hydrolysis in embryonic rabbit brain: Prenatal ontogeny of cerebral glycosidases, *Can. J. Biochem.* **52**:294–303.

Dawson, G., and Tsay, G., 1977, Substrate specificity of human α-L-fucosidase, *Arch. Biochem.* **184**:12–23.

Dutton, G. R., and Barondes, S. H., 1970, Glycoprotein metabolism in developing mouse brain, *J. Neurochem.* **17**:913-920.

Friedman, Y., and Arsenis, C., 1972, The resolution of aryl sulfatase and heparin sulfamidase activities from various rat tissues, *Biochem. Biophys. Res. Commun.* **48**:1133–1139.

Frohwein, Y. Z., and Gatt, S., 1967, Isolation of β-*N*-acetylhexosaminidase, β-*N*-acetylglucosaminidase and β-*N*-acetylgalactosaminidase from calf brain, *Biochemistry* **9**:2775–2782.

Gatt, S., 1967, Enzymatic hydrolysis of sphingolipids. V. Hydrolysis of monosialoganglioside and hexosylceramides by rat brain β-galactosidase, *Biochim. Biophys. Acta* **137**:192–195.

Gatt, S., and Rapport, M. M., 1966*a*, Isolation of β-galactosidase and β-glucosidase from brain, *Biochim. Biophys. Acta* **113**:567–576.

Gatt, S., and Rapport, M. M., 1966*b*, Enzymic hydrolysis of sphingolipids: Hydrolysis of ceramide lactoside by an enzyme from rat brain, *Biochem. J.* **101**:680–686.

Gielen, W., and Harprecht, V., 1969, Die Neuraminidase-Activität in einigen Regionen des Rindergehirns, *Hoppe-Seyler's Z. Physiol. Chem.* **350**:201–206.

Glasglow, M. W., Quarles, R. H., and Grallman, S., 1972, Metabolism of fucoglycoproteins in the developing rat brain, *Brain Res.* **42**:129–137.

Heijlman, J., and Roukema, P. A., 1972, The action of calf brain sialidase on gangliosides, sialoglycoproteins, and sialoglycopeptides, *J. Neurochem.* **19**:2567–2575.

Hirsch, H. E., 1972, Differential determination of hexosaminidases A and B and of two forms of β-galactosidase, in the layers of the human cerebellum, *J. Neurochem.* **19**:1513–1517.

Hocking, J. D., Jolly, R. D., and Batt, R. D., 1972, Deficiency of α-mannosidase in Angus cattle: An inherited lysosomal storage disease, *Biochem. J.* **128**:69–78.

Hultberg, B., and Masson, P. K., 1975, Activation of residual acidic mannosidase activity in mannosidosis tissue by metal ions, *Biochim. Biophys. Res. Commum.* **67**:1473–1479.

Hunter, G. D., Millson, G. C., and Vockins, M. D., 1967, Lysosomal enzymes and scrapie, *Biochem. J.* **102**:43–44p (abstract).

Ledeen, R. W., and Yu, R. K., 1973, Structure and enzymic degradation of sphingolipids in lysosomes and storage diseases, in: *Lysosomes and Storage Diseases* (H. G. Hers and F. Van Hoff, eds.), pp. 105–145, Academic Press, New York.

Leibovitz, Z., and Gatt, S., 1968, Enzymatic hydrolysis of sphingolipids. VII. Hydrolysis of gangliosides by a neuraminidase from calf brain, *Biochim. Biophys. Acta* **152**:136–143.

Li, S.-C., and Li, Y.-T., 1970, Studies on the glycosidases of jack bean meal. III. Crystallization and properties of β-*N*-acetylhexosaminidase, *J. Biol. Chem.* **245**:5153–5160.

Lundblad, A., Masson, P., Norden, N. E., Svensson, S., and Ockerman, P. S., 1976, Mannosidosis: Storage material, α-mannosidase specificity and diagnostic methods, *Adv. Exp. Med. Biol.* **68**:301–312.

Lusis, A. J., and Paigen, K., 1976, Properties of mouse α-galactosidase, *Biochim. Biophys. Acta* **437**:487–497.

Margolis, R. K., and Margolis, R. U., 1972, Disposition of fucose in brain, *J. Neurochem.* **19**:1023–1030.

Margolis, R. U., Margolis, R. K., Santella, R., and Atherton, D. M., 1972, The hyaluronidase of brain, *J. Neurochem.* **19**:2325–2332.

Millson, G. C., and Bountiff, L., 1973, Glycosidases in normal and scrapie mouse brain, *J. Neurochem.* **20**:541–546.

Miyatake, T., and Suzuki, N., 1975, Partial purification and characterization of β-galactosidase from rat brain hydrolyzing glycosphingolipids, *J. Biol. Chem.* **250**:585–592.

Morgan, E. H., and Laurell, C. B., 1963, Neuraminidase in mammalian brain, *Nature* (*London*) **197**:921–922.

Neufeld, E. F., Lim, T. W., and Shapiro, L. J., 1975, Inherited disorders of lysosomal metabolism, *Annu. Rev. Biochem.* **44**:357–376.

O'Brien, J. S., 1972, GM1 gangliosidoses, in: *The Metabolic Basis of Inherited Disease* (J. B. Stanbury, J. B. Wynngarden, and D. S. Fredrickson, eds.), pp. 639–662, McGraw-Hill New York.

Ockerman, P. A., 1969, Mannosidosis: Isolation of oligosaccharide storage material from brain, *J. Pediatr.* **75**:360–369.

Öhman, R., Rosenberg, A., and Svennerholm, L., 1970, Human brain sialidase, *Biochemistry* **9**:3774–3782.

Overdijk, B., Van derKroef, W. M., Veltkamp, W. A., and Hooghwinkel, G. J. M., 1975, The separation of bovine brain β-*N*-acetyl-D-hexosaminidases: Abnormal gel-filtration behaviour of β-*N*-acetyl-D-glucosaminidase C, *Biochem. J.* **151**:257–261.

Poenaru, L., and Dreyfus, J. C., 1973, Electrophoretic heterogeneity of human α-mannosidase, *Biochim. Biophys. Acta* **303**:171–174.

Quarles, R. H., and Brady, R. D., 1970, Sialoglycoproteins and several glycosidases in developing rat brain, *J. Neurochem.* **17**:801–807.

Radin, N. S., Hof, L., Bradley, R. M., and Brady, R. O., 1969, Lactosylceramide galactosidase: Comparison with other sphingolipid hydrolases in developing rat brain, *Brain Res.* **14**:497–505.

Raghaven, S. S., Rhoads, D. B., and Kanfer, J. N., 1972, Acid hydrolases in neuronal and glial enriched fractions of rat brains, *Biochim. Biophys. Acta* **268**:755–760.

Robins, E., Fisher, K., and Lowe, I. P., 1961, Quantitative histochemical studies of the morphogenesis of the cerebellum. II. Two β-glycosidases, *J. Neurochem.* **8**:96–104.

Robins, E., Hilde, H. E., and Emmons, S. S., 1968, Glycosidases in the nervous system. I. Assay, some properties, and distribution of β-galactosidase, β-glucuronidase, and β-glucosidase, *J. Biol. Chem.* **243**:4246–4252.

Robinson, D., Jordan, T. W., and Horsburgh, T., 1972, The *N*-acetyl-β-D-hexosaminidases of calf and human brain, *J. Neurochem.* **19**:1975–1985.

Roukema, P. A., Van den Eijnden, D. H., Heijlman, J., and Van Der Berg, G., 1970, Sialoglycoproteins, gangliosides, and related enzymes in developing rat brain, *FEBS Lett.* **9**:267–270.

Schengrund, C.-L., and Rosenberg, A., 1970, Intracellular location and properties of bovine brain sialidase, *J. Biol. Chem.* **245**:6196-6200.

Schengrund, C.-L., and Rosenberg, A., 1971, Gangliosides, glycosidases, and sialidase in the brain and eyes of developing chickens, *Biochemistry* **12**:2424–2428.

Sellinger, O. Z., and Hiatt, R. A., 1968, Cerebral lysosomes. IV. The regional and intracellular distribution of arylsulfatase and evidence for two populations of lysosomes in rat brain, *Brain Res.* **7**:191–200.

Sellinger, O. Z., Johnson, D. E., Santiago, J. C., and Vargas-Idoyaga, V., 1973, A study of the biochemical differentiation of neurons and glia in the rat cerebral cortex, in: *Neurobiological Aspects of Maturation and Aging* (D. Ford, ed.), pp. 331–347, Elsevier, New York.

Shuter, E. R., Robins, E., Freeman, M. L., and Jungalawala, F. B., 1970, β-Hexosaminidase in the nervous system: The quantitative histochemistry of β-glucosaminidase and β-galactosaminidase in the cerebellar cortex and subjacent white matter, *J. Histochem. Cytochem.* **18**:271–277.

Sinha, A. K., and Rose, S. P. R., 1972, Compartmentation of lysosomes in neurons and neuropil and a new neuronal marker, *Brain Res.* **39**:181–196.

Sloan, H. R., and Fredrickson, D. S., 1972, GM2 gangliosidoses: Tay–Sachs disease, in: *The Metabolic Basis of Inherited Disease* (J. B. Stanbury, J. B. Wynngarden, and D. S. Fredrickson, eds.), pp. 615–638, McGraw-Hill, New York.

Suzuki, L., 1976, Catabolism of sialyl compounds in nature, in: *Biological Roles of Sialic Acid* (A. Rosenberg and C. Schengrund, eds.), pp. 159–182, Plenum Press, New York.

Tallman, J. F., and Brady, R. O., 1972, The catabolism of Tay-Sachs ganglioside in rat brain lysosomes, *J. Biol. Chem.* **247**:7570–7575.

Tallman, J. F., Brady, R. O., Quirk, J. M., Villalba, M., and Gal, A. E., 1974, Isolation and relationship of human hexosaminidases, *J. Biol. Chem.* **249**:3489–3499.

Tanaka, H., and Suzuki, K., 1977, Substrate specificites of the two genetically distinct human brain β-galactosidases, *Brain Res.* **122**:325-335.

Tettamanti, G., and Zambotti, V., 1968, Purification of neuraminidase from pig brain and its action on different gangliosides, *Enzymologia* **35**:61–74.

Tettamanti, G., Lombardo, A., Preti, A., and Zambotti, V., 1970, Effect of temperature and Triton X-100 on the activity of particulate neuraminidase from rabbit brain, *Enzymologia* **39**:65–71.

Tettamanti, G., Preti, A., Lombardo, T., Suman, T., and Zambotti, V., 1975, Membrane bound neuraminidase in the brain of different animals: Behavior of the enzyme on endogenous sialo derivatives and rationale for its assay, *J. Neurochem.* **25**:451–456.

Traurig, H. H., Clendenon, N. R., Swenberg, J. A., and Allen, N., 1973, Lysosomal acid hydrolases in neocortical rat brain, *J. Neurobiol.* **4**:105–115.

Tsay, G. C., and Dawson, G., 1976, Oligosaccharide storage in brains from patients with fucosidosis, GM1-gangliosidosis and GM2-gangliosidosis (Sandhoff's disease), *J. Neurochem.* **27**:733–740.

Tsay, G. C., Dawson, G., and Sung, S. S. J., 1976, Structure of the accumulating oligosaccharide in fucosidosis, *J. Biol. Chem.* **251**:5852–5859.

Vargas-Idoyaga, V., Santiago, J. C., Petiet, P. D., and Sellinger, O. Z., 1972, The early postnatal development of the neuronal lysosome, *J. Neurochem.* **19**:2533–2545.

Verity, M. A., Gade, G. F., and Brown, W. J., 1973, Characterization and localization of acid hydrolase activity in the synaptosomal fraction from rat cerebral cortex, *J. Neurochem.* **20**:1635–1648.

Zatz, M., and Barondes, S. H., 1970, Fucose incorporation of glycoproteins of mouse brain, *J. Neurochem.* **17**:157–163.

7

Histochemistry and Cytochemistry of Glycoproteins and Glycosaminoglycans

John G. Wood and Barbara J. McLaughlin

Introduction

One of the most intriguing areas of research in neurobiology is the role of carbohydrates in cell structure and function. Carbohydrates on macromolecules of neurons and glia are expected to play a role in cellular "recognition" and sorting out into histological arrangements (Barondes, 1970, 1975; Barbera *et al.*, 1973; Hughes, 1973; Moscona, 1974; Roseman, 1974), synaptogenesis (Barondes, 1970; Vaughn *et al.*, 1976, McLaughlin and Wood, 1977; Cotman and Taylor, 1974), myelination (Brady and Quarles, 1973; Wood and Engel, 1976), cell permeability (Rambourg, 1971), and cell immunity (Rambourg, 1971; Winzler, 1970). There is much interest and considerable research in each of these areas, but direct evidence linking carbohydrates in cellular macromolecules to specific functions of the nervous system is fragmented and incomplete. It may be that the serotonin receptor in brain is a carbohydrate (Wooley and Gommi, 1965; Vaccari *et al.*, 1971), and there is evidence that the cholera toxin receptor involves a ganglioside (Cuatrecasas, 1973; Holmgren *et al.*, 1973; Van Heyningen W. E., *et al.*, 1971; Van Heyningen, S., 1974). The nicotinic acetylcholine receptor also contains carbohydrate (Meunier *et al.*, 1974; Raftery *et al.*, 1976; Salvaterra *et al.*, 1977), as does acetylcholinesterase (Carlsen and Svensmark, 1970; Taylor *et al.*, 1974; Wiedmer *et al.*, 1974). The major glycoprotein of peripheral nerve myelin and a minor glycoprotein associated with central myelin are glyco-

Abbreviations used in this chapter: (Con A) concanavalin A; (Con A–FT conjugate) Con A–ferritin conjugate; (EM or LM) electron or light microscope (-microscopic, microscopy); (PAS) periodic acid–Schiff's reagent; (WGA) wheat germ agglutinin.

John G. Wood and Barbara J. McLaughlin • Department of Anatomy, University of Tennessee Center for the Health Sciences, Memphis, Tennessee.

sylated (Quarles *et al.*, 1973; Everly *et al.*, 1973; Wood and Dawson, 1973). Several of the proteins of the isolated synaptic membrane are glycoproteins that may be isolated by adsorption onto lectin-matrix columns followed by elution with appropriate sugars (Gurd and Mahler, 1974; Zanetta *et al.*, 1975). In most cases, it is not known how the carbohydrate moieties contribute to the function of these various macromolecules, although in the case of ganglioside receptors specificity must involve the carbohydrate portion of the molecule, since the nonglycosylated portions of different gangliosides are very similar (Svennerholm, 1970).

One indication of a possible functional involvement of carbohydrates in nervous system development is the observation that embryonic brain cells are differentially agglutinated by different lectins, depending on the age of the tissue from which the cells are derived (Moscona, 1974). These results suggest a structural change or reorganization of membrane carbohydrates during the critical early stages of nervous system differentiation, but it is not known whether these changes directly affect the differentiation process.

There have been isolated from cell cultures glycoprotein factors that enhance the property of other cells in rotating culture to aggregate into organized masses resembling in some respects the parent tissue from which the cells are derived (Moscona, 1974, 1976). Other factors that induce morphological transformation of glial cells have been isolated (Lim and Mitsunobu, 1975).

Several workers have presented evidence for the presence in brain and other tissues of developmentally regulated molecules that cause the agglutination of formalized erythrocytes in a manner that suggests that the molecules possess lectin-like activity (Teichberg *et al.*, 1975; Simpson *et al.*, 1977). The endogenous brain lectin activity reaches a peak during development and then declines as the nervous system approaches more mature stages (Simpson *et al.*, 1977). It has been suggested that aggregation-promoting factors and endogenous lectin activity may play a role in the processes by which embryonic cells recognize appropriate environments to establish cellular relationships and undergo histogenesis (Moscona, 1974; Barondes and Rosen, 1976). These suggestions await advances in experimental design to establish the validity of the hypothesis.

One approach to understanding the function of carbohydrates in the nervous system is to determine the localization of specific carbohydrate-containing macromolecules, in the mature brain as well as at various stages of development. Histochemical and cytochemical methods have been successfully employed for several years to determine the general localization of carbohydrates [for reviews, see Revel and Ito (1967) Bennett (1969*a,b*), Martinez-Palomo (1970), Winzler (1970), Rambourg (1971), and Luft (1976)], but only recently have reliable methods for localizing specific carbohydrates become generally available. In this chapter, we will briefly consider historical aspects of the field of histochemistry of glycoproteins and glycosaminoglycans and then examine in greater depth the techniques that are currently being developed.

2. Historical Background

A number of staining techniques have been utilized to localize carbohydrates in cells at both the light-microscopic (LM) and electron-microscopic (EM) levels. Since the general use of the stains and results of a number of studies have been extensively reviewed (Rambourg *et al.*, 1966; Revel and Ito, 1967; Bennett, 1969*a,b*; Martinez-Palomo, 1970; Rambourg, 1971; Pfenninger, 1973; Luft, 1976), the subject will be covered only briefly here, and emphasis will be placed on cytochemical studies of nervous system carbohydrates.

2.1. Periodic Acid–Schiff

The periodic acid–Schiff's reagent (PAS) method (McManus, 1946; Lillie, 1947; Hotchkiss, 1948) has been effectively used to demonstrate carbohydrates in the nervous system (Rambourg *et al.*, 1966; Doshi *et al.*, 1974; Young and Abood, 1960; Sulkin, 1960; Saxena, 1969; Arseni *et al.*, 1967). The reaction itself is remarkably specific and depends on oxidative cleavage of bonds between adjacent carbon atoms containing hydroxyl groups followed by reaction of the resulting dialdehyde with Schiff's reagent. Since many glycoproteins and glycolipids contain sugars with adjacent hydroxyl groups on carbon atoms, the PAS method as normally used cannot distinguish specific carbohydrate-containing molecules. Glycosaminoglycans do not stain by routine PAS procedures, possibly because their sugars are shielded by electrostatic repulsion from the periodate molecule (Scott and Harbinson, 1969; Scott and Dorling, 1969), which only slowly hydrolyzes uronic acid (a major component of many glycosaminoglycans). Supporting the hypothesis that normal PAS staining does not localize glycosaminoglycans are the observations of Young and Abood (1960) that hyaluronidase treatment does not diminish PAS staining in the nervous system of several mammalian species. When tissues are treated with periodate in two steps and the dialdehydes created in the first step (from glycoprotein and glycoplipid polysaccharides) are reduced with sodium borohydride, the PAS method is said to stain glycosaminoglycans (Scott and Dorling, 1969), except perhaps for keratan sulfate (which contains no uronic acid).

The PAS reaction has been modified by several workers to produce an electron-dense product consisting of large crystals that may obscure membrane detail. The aldehydes produced by periodic acid oxidative cleavage of 1,2-glycols will reduce silver methenamine to metallic silver (cf. Rambourg,

1971), which may be viewed in the EM. Nonspecific results can occur with this method, but these are largely due to the presence of endogenous free aldehyde groups or the introduction of aldehyde fixatives into the tissue (cf. Rambourg, 1971). Chromic acid may be employed after the periodate step to abolish some of the endogenous reducing groups and improve the specificity of the silver methenamine technique (Hernandez *et al.*, 1968). The reaction of thiosemicarbazide (Seligman *et al.*, 1965) with periodate-induced aldehyde groups generates a more powerful reducing agent that can react either with osmium vapor (Seligman *et al.*, 1965) or with silver proteinate (Thiery, 1967) to produce an electron-dense product. In one of the few EM studies of the nervous system using techniques based on the PAS reaction, Feeney (1973) showed that periodic acid–silver methenamine staining of the interphotoreceptor matrix of albino mice was sparse in 10-day-old mice and somewhat heavier in the adult mouse retina. Tasso and Rua (1975) distinguished two classes of fibers in the rat hypothalamo–neurohypophyseal system on the basis of staining of the neurosecretory granules with methods based on the PAS reaction. The secretory granules of some fibers stained, whereas the granules of other fibers did not.

2.2 Iron-Based Stains

One of the oldest methods to demonstrate polysaccharides in tissue is the use of colloidal solutions of ferric oxide, which was introduced by Hale (1946). Many variations of this method have been published (cf. Luft, 1976). In all cases, at low pH these metal stains are said to be specific for "acid mucopolysaccharides" [for references, see Luft (1976)]. It is known, however, that colloidal ferric oxide staining of isolated membranes is markedly diminished by prior treatment with neuraminidase (Benedetti and Emmelot, 1967, 1968), suggesting that the method stains glycoproteins as well. Ferric oxide staining of kidney membranes is also diminished by prior treatment with neuraminidase (Groniowski *et al.*, 1969; Jones, 1969) or hyaluronidase (Jones, 1969), and some differentiation of the stain on the basis of neuraminidase or hyaluronidase sensitivity is possible (Jones, 1969). Synaptosomal fractions incubated in neuraminidase show diminished staining with positively charged ferric iron (Feria-Velasco *et al.*, 1976). Rahmann and Katusic (1975) demonstrated colloidal iron hydroxide staining of nerve fibers in teleost brain that was reduced up to 55% after treatment with neuraminidase. Excellent results of labeling axolemmal membranes with ferric compounds at the node of Ranvier (Langley, 1971; Landon and Langley, 1971; Quick and Waxman, 1977) are discussed in Chapter 10. Although it is unlikely that these low-pH heavy-metal stains reliably distinguish glycoproteins from glycosaminoglycans, the methods can be valuable adjuncts to the PAS and other methods to localize carbohydrate-containing macromolecules in cells. Results similar to those obtained with iron-based stains have been obtained with acidic colloidal thorium dioxide (Revel, 1964; Revel and Ito, 1967; Rambourg and Leblond, 1967; Groniowski *et al.*, 1969); Bonneville and Weinstock, 1970; Zacks *et al.*, 1973; Andrews and Porter, 1973).

2.3. Phosphotungstic Acid

In 1966, Pease (1966) explored the use of phosphotungstic acid (PTA) (Hall *et al.*, 1945) and silicotungstic acid at various pH values to stain a variety of tissues, including nervous tissue. At low pH, the method demonstrates the plasma membrane as a negative image by staining "polysaccharide" material in the extracellular space. Rambourg (1969, 1971) used the PTA method to study a variety of cell types and concluded that structures that were PAS-positive were also PTA-positive. In nervous tissue, the PTA method stains plasma membranes of neurons and glia, and regions of apposition between the various cell types of the brain, including the synaptic cleft, are particularly prominent (Pease, 1966; Rambourg, 1969, 1971). The method has been questioned as to specificity (Silverman and Glick, 1969*a,b*; Pease, 1970; Glick and Scott, 1970; Scott and Glick, 1971) [for a résumé of the dispute, see Luft (1976)], and unfortunately the mechanism of PTA staining is still not known. The interpretation of results using PTA is complicated by the fact that some of the most elegant micrographs demonstrating presumptive carbohydrates were obtained with PTA solutions at pH 4.5 (Peterson and Pease, 1972; Pease and Peterson, 1972). Supporting the results of these latter studies, however, is the observation that the major protein in peripheral nerve myelin is a glycoprotein (Everly *et al.*, 1973; Wood and Dawson, 1973) that appears to be located in the interperiod line of myelin (Wood and McLaughlin, 1975; Peterson and Sea, 1975). The interperiod line of peripheral nerve myelin is the region most strongly stained by the PTA method (Peterson and Pease, 1972). Since PTA staining is diminished by proteolytic enzyme treatment as well as by neuraminidase and hyaluronidase, it is unlikely that the method distinguishes glycoproteins from glycosaminoglycans.

2.4. Ruthenium Red

Luft (1976) adapted the ruthenium red (RR) method for plant pectin (Jensen, 1962) and introduced a method primarily to stain cell coats. Although there are some problems with penetration of the reagent into tissues, the method stains most cell surfaces nicely, presumably through electrostatic interaction of the cationic RR with anionic sites on cellular sialoglycoproteins and glycosaminoglycans with possible enhancement of specificity for acid glycosaminoglycans (Luft, 1976). Several workers have investigated the staining of nervous tissue by ruthenium red. Bondareff (1967) demonstrated RR labeling in proximity to the outer surface of cell membranes in cerebral cortex and an accumulation of RR in the synaptic cleft region. These results were confirmed by Tani and Ametani (1971). Other workers showed RR staining of membranes by ependymal cells (Westergaard, 1971) and retina inner limiting membrane (Matsusaka, 1971). Singer *et al.* (1972) and Nordlander *et al.* (1975) showed that RR applied to living peripheral nerves stained the periaxonal space with some penetration into the axon.

2.5. Alcian Blue

The cationic dye alcian blue is a stain with affinity for acid glycosaminoglycans as seen at the LM level (Quintarelli *et al.*, 1964; Scott *et al.*, 1964; Scott, 1972). An interesting modification of the stain is the application of the critical electrolyte principle (Scott, 1960; Scott and Dorling, 1969) to progressively neutralize anionic charges on acid glycosaminoglycans so that, at high electrolyte concentration, only the most anionic glycosaminoglycans are stained. Using critical electrolyte concentration in combination with enzyme digestions, it is apparently possible to distinguish keratan sulfate from chondrotin sulfate (Stockwell and Scott, 1965; Scott and Stockwell, 1967) in human cartilage. Alcian blue staining of nervous tissue has been attempted by several workers. Rakic and Mrsulja (1975) demonstrated an alcian-blue-positive reaction in cerebellar Purkinje cells and claimed circadian rhythm alterations in the staining intensity. Jirge (1971) presented results indicating a staining of the nuclear layers of retina that was enhanced by prior pepsin digestion. B. Smith and Butler (1973) studied the distribution of alcian blue staining (as well as iron diamine) in nervous-tissue-derived tumor cells. They found that cell borders of oligodendrogliomas were positive, whereas cell borders of astrocytomas were negative. Lampert and Lewis (1974) used alcian blue in combination with critical electrolyte methods and hyaluronidase digestion to show an age-dependent acquisition of positive staining by oligodendroglial cells just before the period of rapid onset of myelination. These workers suggested that positive staining may be due to the presence of chondroitin sulfate C (chondroitin 6-sulfate), although the myelin-associated lipid, cerebroside sulfate, might also be expected to stain under these conditions. Lai *et al.* (1975) reported that the staining of oligodendroglia by alcian blue is not removed by prior chloroform–methanol extraction, suggesting that sulfatide is not the stainable component. Martinez-Rodriquez *et al.* (1976) investigated the alcian blue staining of hypothalamic nervous tissue and reported that a population of presumptive neurosecretory cells were positively stained under conditions where sulfated glycosaminoglycans were most likely to stain. Moss (1973) indicated positive alcian blue staining of spongioblasts (neuron and glia precursors) of disaggregated fetal mouse brain cells and little or no staining of astrocytes. Landon and Langley (1971) and Langley (1971) showed positive alcian blue staining at the node of Ranvier of peripheral nerves. H. V. Castejón (1970*a–c*, 1971) presented results indicating the presence of alcian blue staining in the cytoplasm of CNS neurons of several vertebrates. These results have been extended to the EM level (see Section 3). It appears from an examination of the various results that carefully controlled application of the alcian blue method may produce a relatively specific staining of glycosaminoglycans. It should be pointed out, however, that the distribution of stain with regard to neurons as compared to oligodendroglia and astrocytes is not in complete accord with biochemical measurements of cellular and subcellular fractions [for references, see Margolis and Margolis (1977)], and the use of alcian blue staining must still be considered equivocal as far as histochemistry of the nervous system is concerned.

2.6. Cationic Ferritin

When free aldehyde groups on ferritin are blocked, the molecule becomes cationic and will react with cell coats (Nachmias and Marshall, 1961). Various forms of cationic ferritin were recently used to investigate cell surfaces (Danon *et al.*, 1972; Pinto da Silva *et al.*, 1973). Using nervous tissue, Wessells *et al.* (1976) found that cationic ferritin binds to all surfaces of ciliary ganglion neurons in cultures, except for "mounds" and "veils" that are areas thought to represent sites of membrane addition or retrieval. This binding of cationic ferritin is electrostatic and is not inhibited by prior digestion with neuraminidase or enzymes that degrade glycosaminoglycans; hence, its significance is not yet understood. Cationized ferritin will also react with synaptosomal membranes and myelin (Bittiger and Heid, 1977).

In all the histochemical methods discussed thus far, inherent weaknesses are that (1) the method of binding reagents to cellular components is understood only in general chemical terms (e.g., electrostatic interactions), if at all and (2) most of the stains are generally capable of staining carbohydrate with minimal capacity to distinguish glycoproteins from glycosaminoglycans or to distinguish classes of glycoproteins and glycosaminoglycans. It is obvious that the next step in development of histochemical methods for carbohydrate localization must be the acquisition of reagents with increased specificity for individual glycoproteins, glycolipids, and glycosaminoglycans and with the capacity for visualization at the EM level of resolution. Promising reagents that can be used with EM labeling techniques are lectins, toxins, and antibodies. These reagents have been used almost exclusively for localization of nervous system glycoproteins and glycolipids, so they will be considered separately after a discussion of recent advances in the histochemical localization of glycosaminoglycans.

3. Glycosaminoglycans*

The most extensive study of EM localization of glycosaminoglycans in the nervous system has been the work of H. V. and O. J. Castejón (Castejón, H. V., 1970*a–c*; Castejón, H. V., and Castejón, O. J., 1972, 1976; Castejón, O. J., and Castejón, H. V., 1972*a,b*). In the work from this group, mouse cerebella were processed using alcian blue staining or treatment with an osmium coordination compound (Castejón, H. V., and Castejón, O. J., 1976). Attention was directed to the mossy fiber presynaptic terminal of the cerebellar glomerulus, where a moderate electron density was apparent in the axoplasm and surrounding synaptic vesicles and mitochondria after either staining procedure. The presynaptic dense projections also showed increased electron density after staining. The authors present micrographs (Castejón, H. V., and Castejón, O. J., 1976) that show a diminution of staining after hyaluronidase treatment, but relatively little change after neuraminidase treatment. After methylation of carboxy groups, the axoplasmic and vesicular rimming of stain was somewhat diminished, but the staining associated with the synaptic cleft was relatively unchanged. Although these results may represent the localization of glycosaminoglycans at the presynaptic terminal, it should be noted that the changes observed are quite subjective, and upon close comparison of experimental tissue with the appropriate control sections, the differences observed are subtle. Additional studies will be required to confirm these results. Similarly, the findings of Moran and Rice (1975) using alcian blue in combination with lanthanum nitrate might demonstrate changes in glycosaminoglycan localization during amphibian neurulation, but control experiments will be required to support this conclusion. It seems clear that EM localization of glycosaminoglycans of the nervous system is a field that is very much in its infancy, and one that will require many additional studies before a meaningful review can be attempted.

The localization of glycoproteins and glycolipids in nervous tissue has received much more experimental attention than the localization of glycosaminoglycans. The recent methodology has primarily utilized lectins, toxins, and antibodies together with EM labeling techniques. Even though there are many methodological similarities in these techniques, for simplicity they will be considered under separate headings.

4. Glycoproteins: Lectin Cytochemistry

The property of lectins to bind sugars has been known for some time [for references, see Sharon and Lis (1972, 1975)], but only recently have these

*For a discussion of glycosaminoglycans in nonnervous tissue, the reader is referred to Toole (1976).

molecules been used as EM probes for membrane-bound carbohydrates. The lectins are proteins that have been isolated from a variety of sources, but primarily from plants, and that possess different capacities to bind closely related sugars relatively specifically (Sharon and Lis, 1972, 1975; Nicolson, 1974, 1976). Methods have been developed to visualize lectin receptors at the EM level (Bernhard and Avrameas, 1971; Nicolson and Singer, 1971; Gonatas and Avrameas, 1973; Smith, S. B., and Revel, 1972), and these methods have been widely used to map lectin receptors on a variety of cell types [for references, see Nicolson (1974, 1976)].

4.1. Subcellular Fractions

Relatively few studies have been performed on the histochemical detection of lectin receptors in nerve tissue. Stoddart and Kiernan (1973), using fluorescein-labeled concanavalin A (Con A), reported an intense staining in the neuropil of cerebral gray matter just within or on the various cellular processes. Matus *et al.* (1973) examined the ultrastructural localization of Con A–ferritin (Con A–FT) conjugates in unfixed and glutaraldehyde-prefixed synaptosomes, myelin, and brain mitochondria. Brain mitochondria were not labeled with the Con A–FT conjugate. Prefixed synaptosomes were labeled on the external surfaces of both the pre- and postsynaptic membranes, but not within the synaptic cleft. The label was apparently randomly distributed. Synaptosomes exposed to Con A–FT conjugates prior to fixation were also labeled on the external surfaces of the pre- and postsynaptic membranes, but the label was gathered into discrete patches. Under these conditions, the synaptosomes also contained label within the synaptic cleft, which may indicate a movement of label into the synaptic cleft or it may indicate that this region is more accessible to the lectin–ferritin conjugate in unfixed compared to fixed tissue. Wood and McLaughlin (1976*b*) showed that the synaptic cleft of aldehyde-fixed intact adult rat cerebellar slices is not accessible to sequentially applied Con A and peroxidase unless the tissue slice is first subjected to mild proteolytic digestion. In this case, the appearance of label within the cleft is extremely unlikely to be due to movement of receptors since the tissue is fixed, but it may represent "unmasking" of already-present receptors by removal of surrounding peptides, or, alternatively, the proteolytic treatment may render the cleft region more accessible to reagents. The pattern of lectin staining of synaptosomes demonstrated by Matus *et al.* (1973) was confirmed by others (Bittiger and Schnebli, 1974; Cotman and Taylor, 1974; Kelly *et al.*, 1976). More recently, Matus and Walters (1976) suggested that morphologically distinct types of synaptic complexes may be characterized by different patterns of lectin cytochemical staining.

4.2. Myelin

Matus *et al.* (1973) showed that myelin fractions isolated from the CNS bind Con A–FT conjugates on the interperiod line, which corresponds to the point of coming together of the outer surfaces of the oligodendroglial plasma membrane. Similar to the synaptosomal membrane labeling shown by these workers, unfixed myelin shows a patchy distribution of label, whereas fixed myelin shows an even distribution of ferritin. In the cases of synaptosomal and myelin membrane, these results suggest that *in situ* lectin membrane receptors are relatively free to move laterally in the membrane (Matus *et al.* 1973). The cytochemical demonstration of Con A–FT receptors in myelin is of interest in light of the work of Quarles *et al.* (1973) (reviewed in Chapter 11), which yielded evidence of a glycoprotein fraction that is intimately associated with the CNS myelin sheath.

Peripheral nervous system (PNS) myelin also contains a glycoprotein that is the major protein component of the sheath (Everly *et al.*, 1973; Wood and Dawson, 1973). The application of lectin cytochemistry to the PNS was made by Wood and McLaughlin (1975). These workers developed a method to stain polyacrylamide gels with lectin–peroxidase markers (Wood and Sarinana, 1975) and demonstrated that the major myelin glycoprotein present in whole sciatic nerve (Wood and Sarinana, 1975) or isolated myelin fractions (Wood and McLaughlin, 1975) stained with Con A–peroxidase. When fixed slices of rat sciatic nerve were stained with Con A–peroxidase (Wood and McLaughlin, 1975) (Fig. 1), labeling within the myelin sheath was restricted to the interperiod lines of certain lamellae. These results strongly suggested that the major glycoprotein was located on the interperiod line of myelin. Other indications of a localization of polysaccharide on the interperiod line came from the work of Peterson and Pease (1972) and Peterson and Sea (1975). If these results are correct, they represent one of the first examples of the use of lectin cytochemistry to localize a specific glycoprotein in the nervous system. This would indicate the potential advantages of some of the newer cytochemical techniques over methods that rely on general chemical interaction of reagents with carbohydrates.

4.3. Cells and Tissue

The interaction of lectins with dissociated neural retina cells has been studied by several workers (cf. Moscona, 1974; McDonough and Lilien, 1975). McDonough and Lilien (1975) reported the localization of fluorescein-labeled Con A and wheat germ agglutinin (WGA) in embryonic retinal slices or retinal cells dissociated purely by mechanical means or by mechanical means following brief trypsinization. The lectin receptors of dissociated cells are able to redistribute within the plane of the membrane under a variety of conditions, but interestingly, lectin receptors on cells in tissue slices are not capable of redistributing into patches or caps unless the tissue has been trypsinized. This suggests that cell-surface lectin receptors in intact tissue

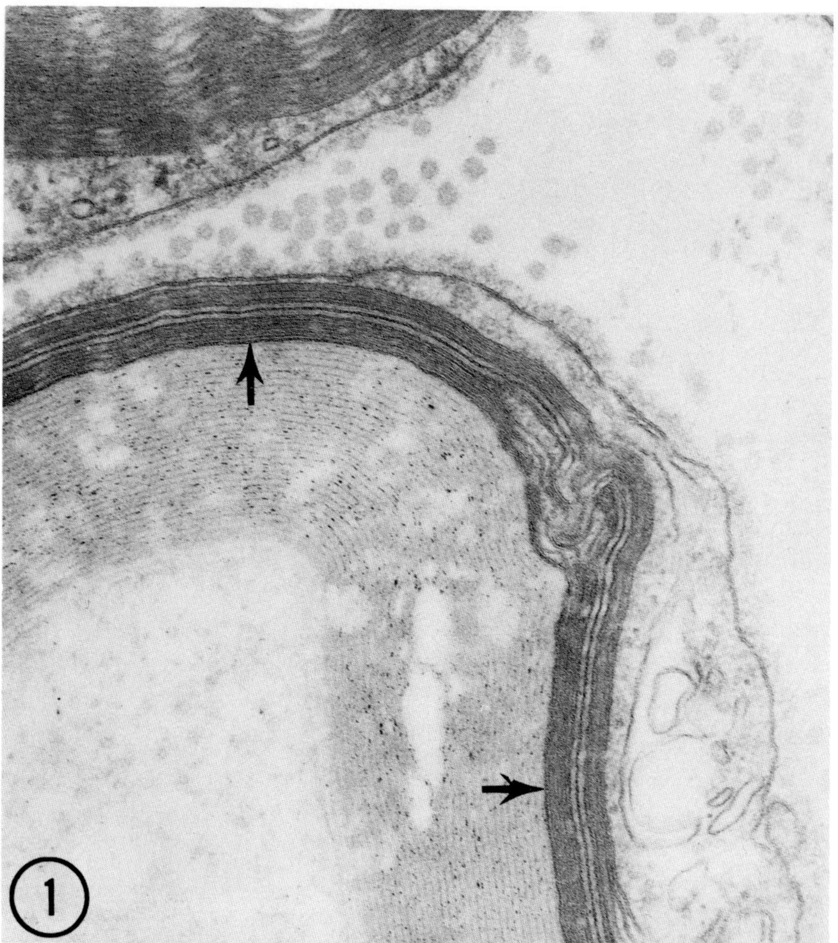

Fig. 1. Electron micrograph showing Con A–peroxidase labeling of the interperiod line (arrows) of several lamellae of rat sciatic nerve myelin. Restriction of label to a few lamellae is probably due to failure of penetration of reagents throughout the rest of the myelin.

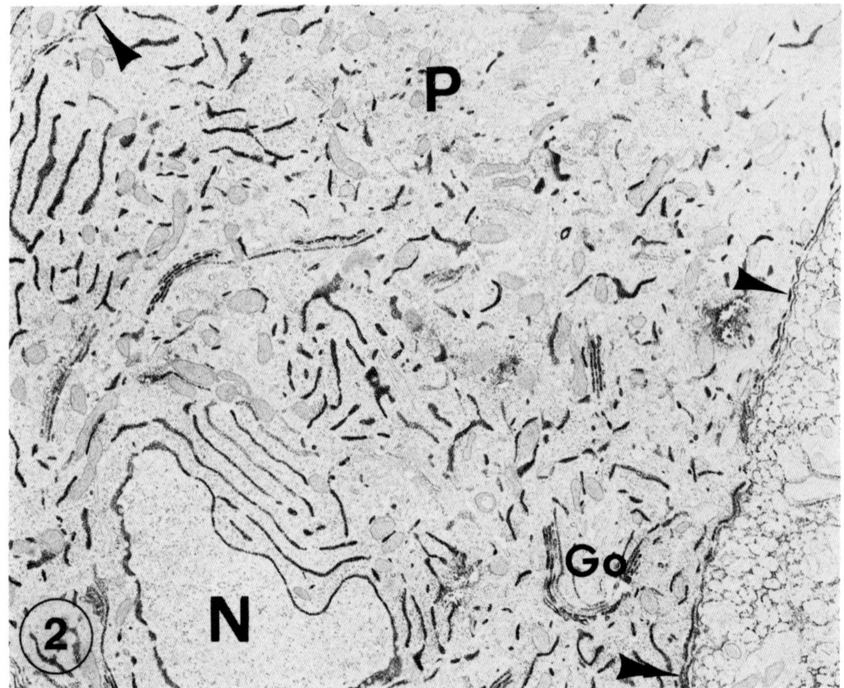

Fig. 2. Electron micrograph showing Con A–peroxidase labeling of a Purkinje (P) cell soma of rat cerebellum. The label is seen lining the nuclear envelope surrounding the nucleus (N) and lining the rough and smooth endoplasmic reticulum and elements of the Golgi (Go) apparatus. The "hypolemmal" cisternae (arrows) at the cell periphery are also labeled.

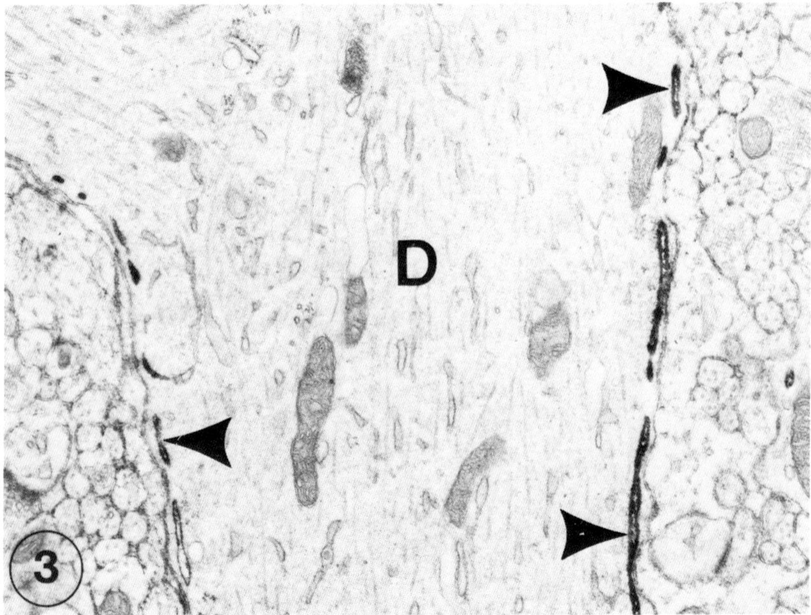

Fig. 3. Electron micrograph showing Con A–peroxidase label lining the "hypolemmal" cisternae (arrows) that lie adjacent to the plasma membrane of a Purkinje cell dendrite (D).

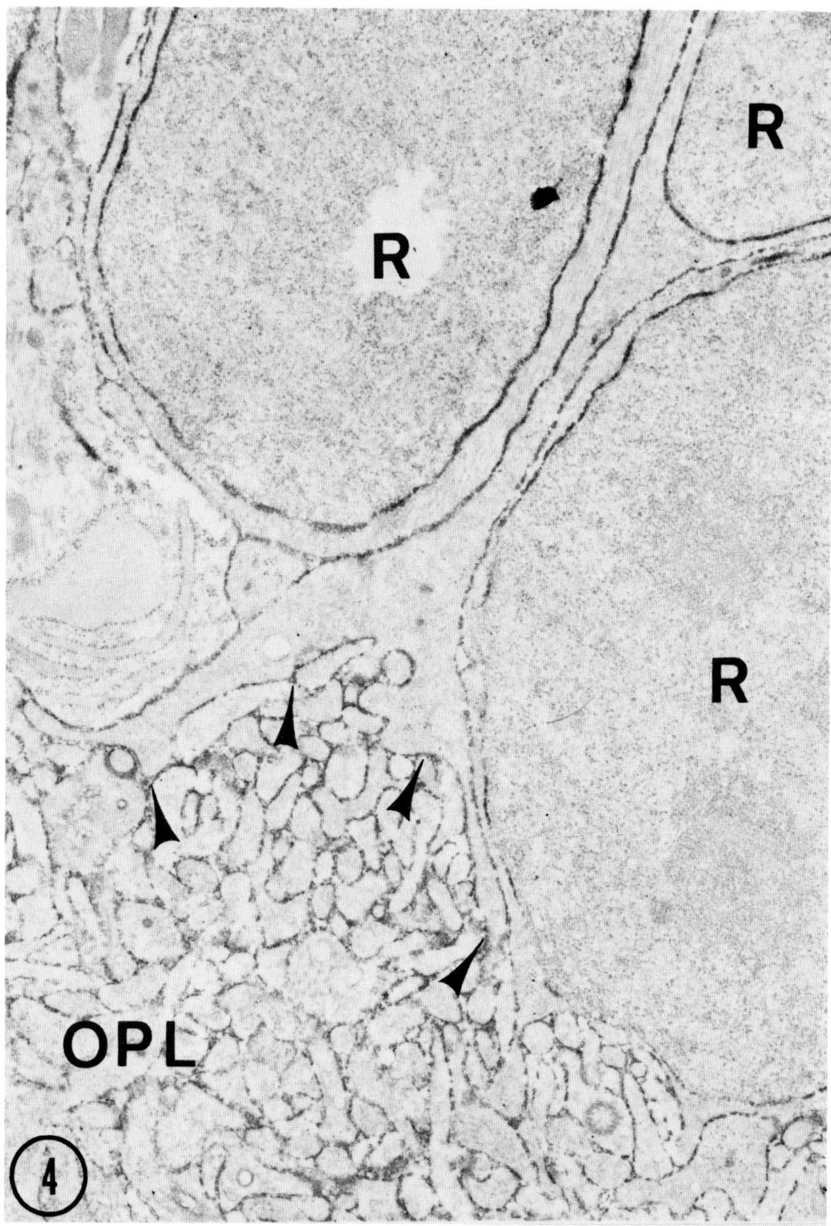

Fig. 4. Electron micrograph showing Con A–peroxidase labeling in the chick retina at 16 embryonic days. Con-A-binding sites are localized along the membrane surfaces of each developing photoreceptor (R) pedicle (arrows) and on the membrane surfaces of profiles in the outer plexiform layer (OPL).

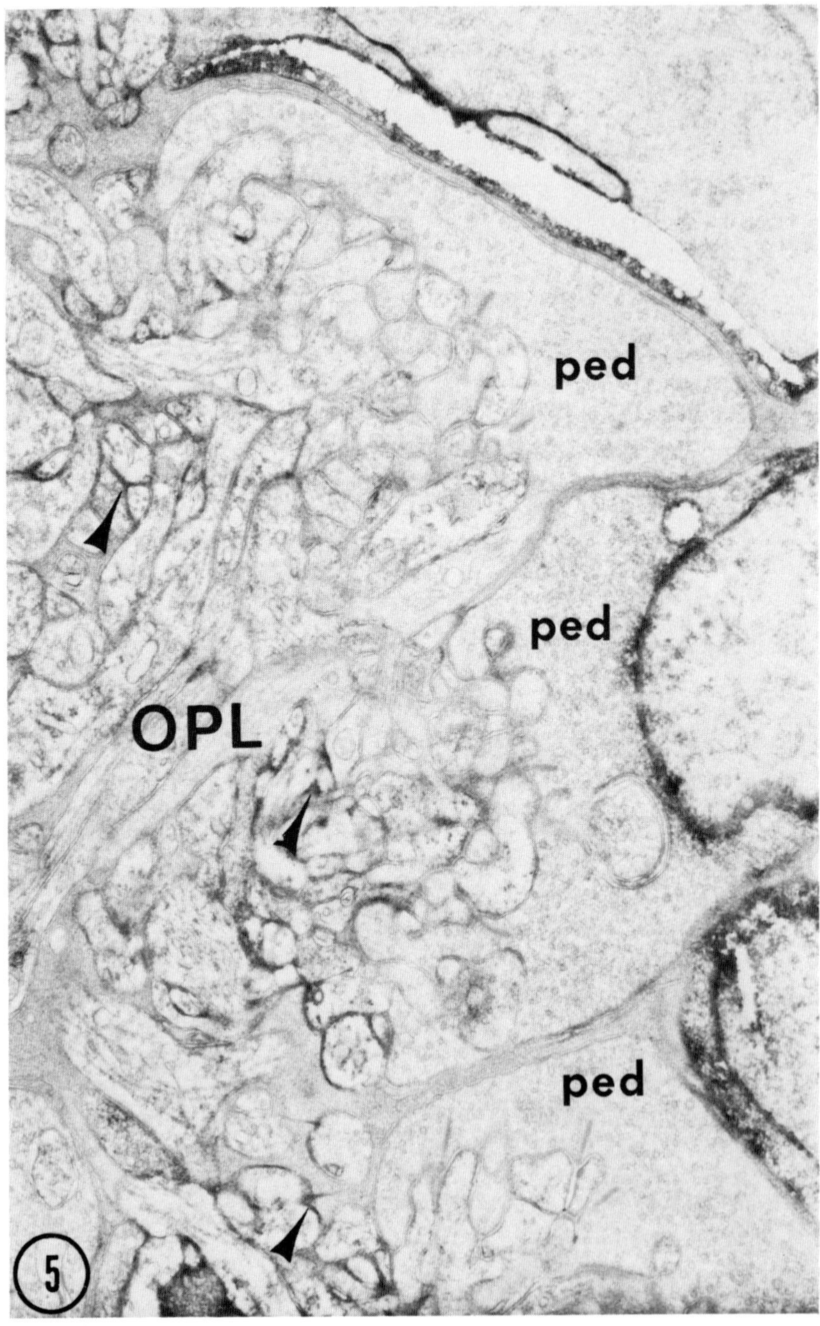

Fig. 5. Electron micrograph showing Con A–peroxidase labeling in the chick retina at 2 days posthatching. Con-A-binding sites are localized on membrane surfaces of some profiles in the OPL (arrows) not in synaptic apposition to the receptor pedicles (ped).

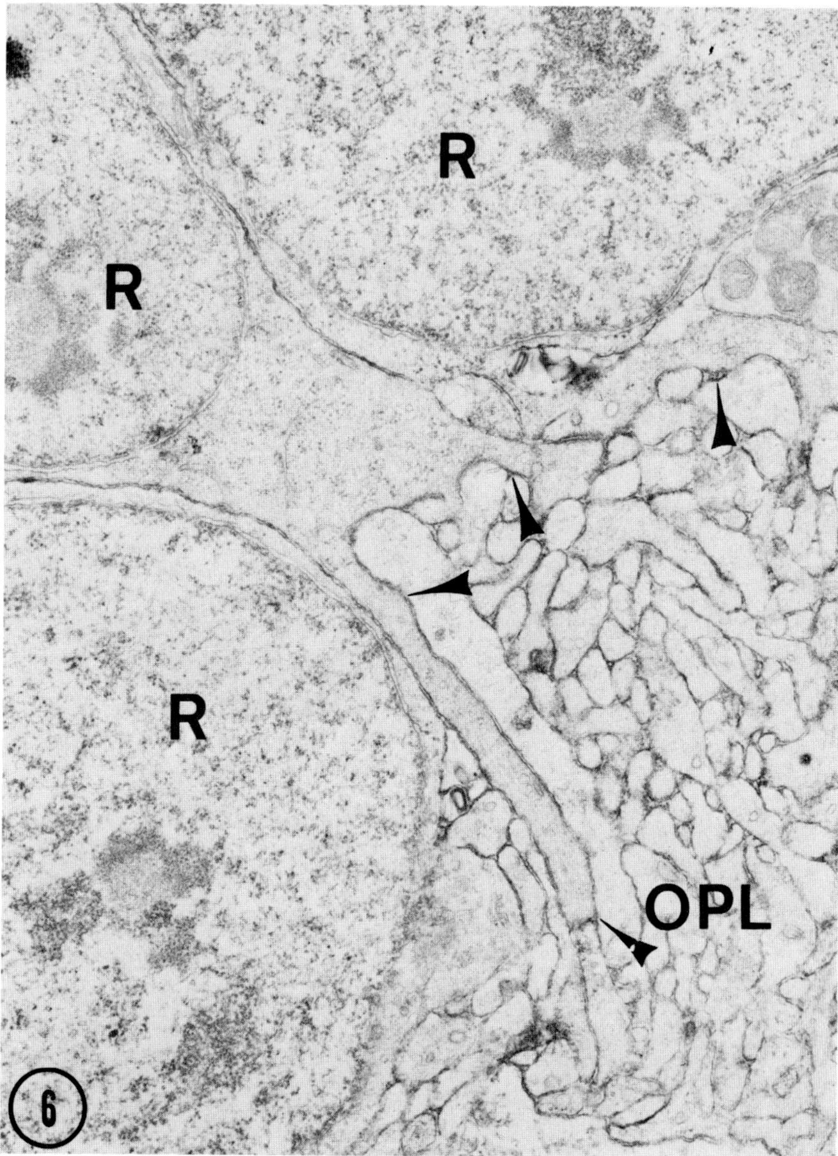

Fig. 6. Electron micrograph showing WGA–peroxidase labeling in the chick retina at 16 embryonic days. WGA-binding sites are localized along the membrane surfaces of each developing photoreceptor (R) pedicle (arrows) and on the membrane surfaces of profiles in the OPL.

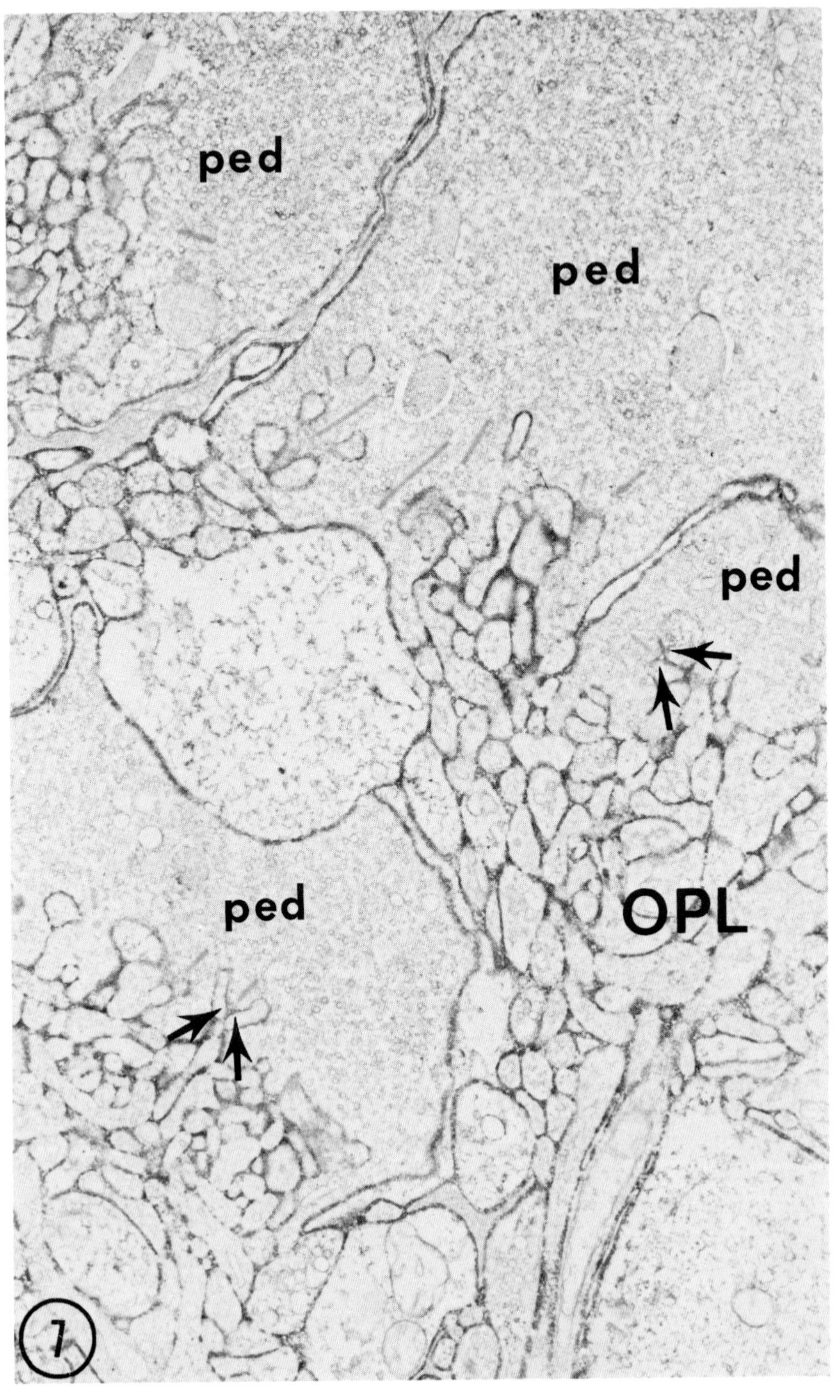

Fig. 7. Electron micrograph showing WGA–peroxidase labeling in the chick retina at 2 days posthatching. WGA-binding sites are localized along the membrane surfaces of each photoreceptor pedicle (ped) in both nonsynaptic and synaptic (arrows) regions and on the membrane surfaces of profiles in the OPL.

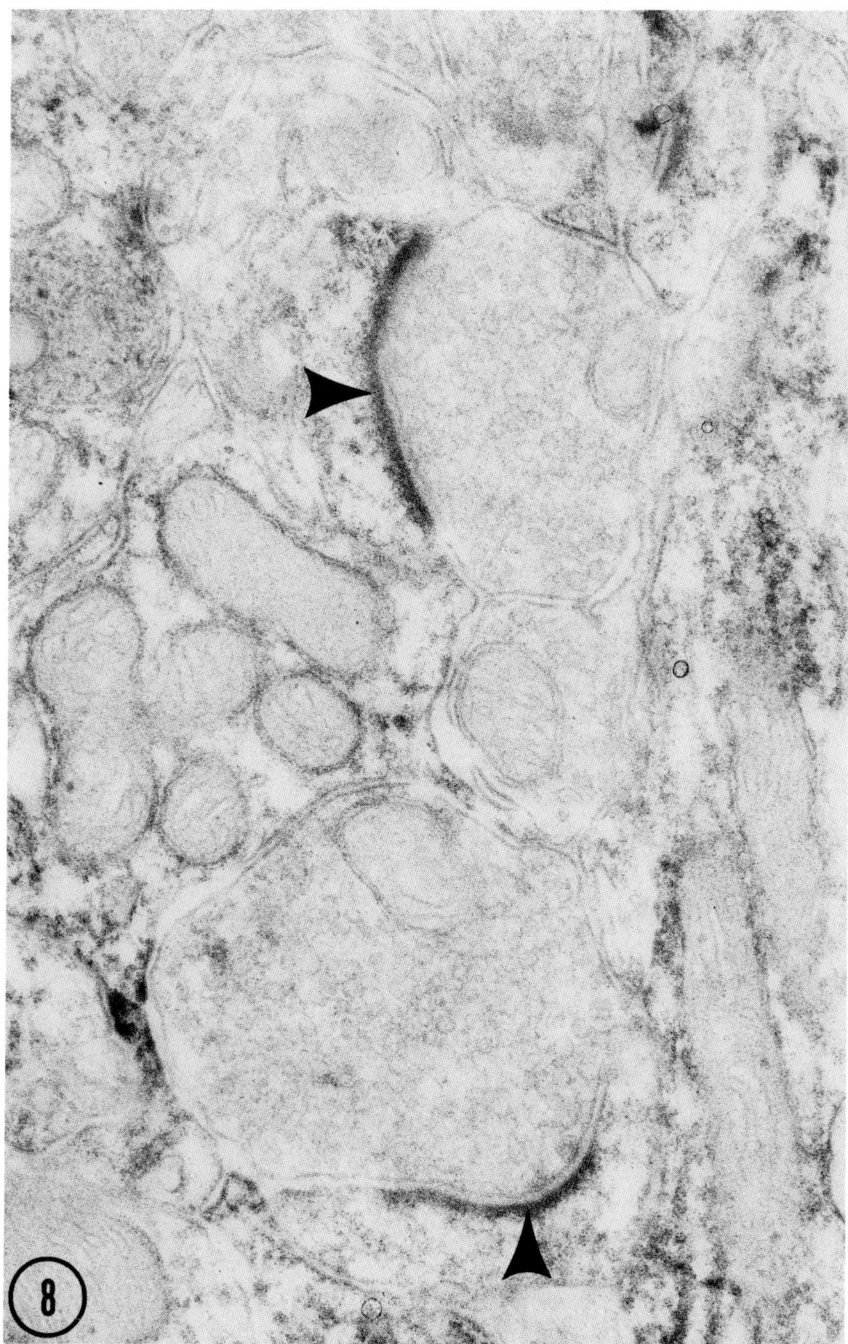

Fig. 8. Electron micrograph showing α-bungarotoxin–peroxidase labeling of synapses in the chick optic tectum. The label is found postsynaptic (arrows) to presynaptic terminals containing round, clear vesicles.

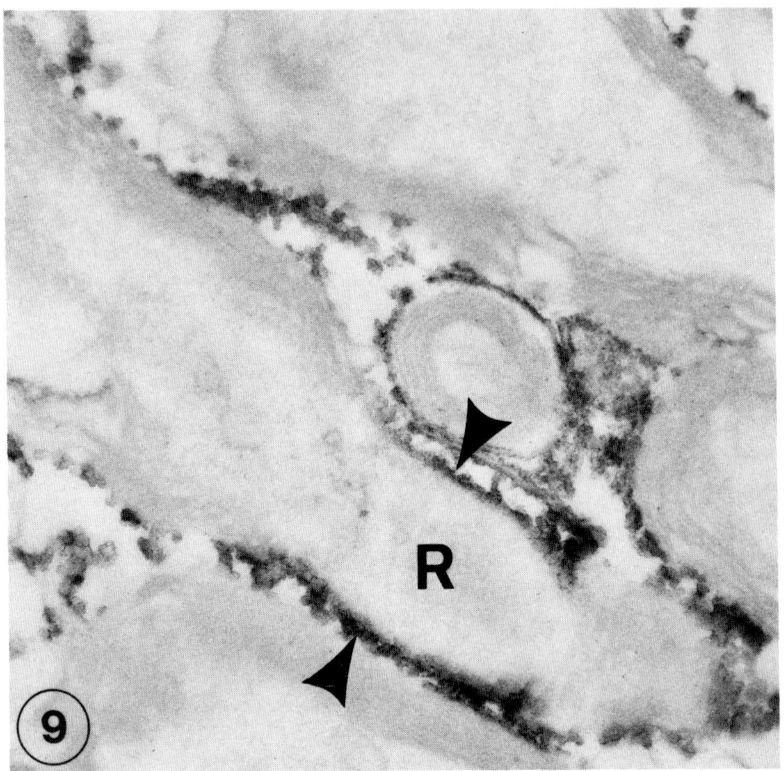

Fig. 9. Electron micrograph showing immunocytochemical staining for the Na^+–K^+-stimulated ATPase in knifefish brain. The label is found on the axolemma (arrows) at the node of Ranvier (R), but is not found along the internodal axolemma.

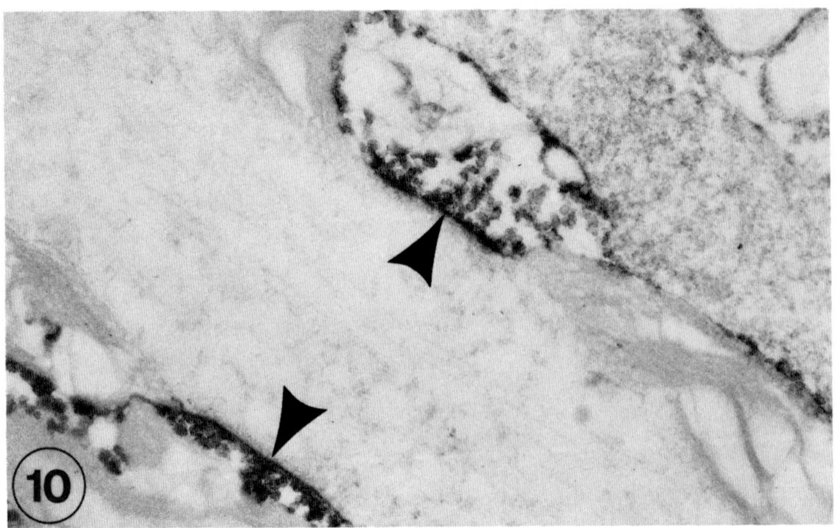

Fig. 10. Another electron micrograph showing the distribution of the [Na^+–K^+]ATPase restricted to the axolemma (arrows) at the node of Ranvier.

may not normally be mobile within the membrane (McDonough and Lilien, 1975).

4.4. Smooth Membrane Cisternae

The application of lectins, together with markers for LM and EM, to aldehyde-fixed slices of nervous tissue allows the histochemical and cytochemical mapping of lectin-binding sites within the brain. Since the initial tissue slicing passes through some cells and their processes, the intercellular as well as extracellular lectin-binding sites may be localized. Wood *et al.* (1974) and Wood and McLaughlin (1976*a*) used this approach to localize Con-A-binding sites in Purkinje cells of rat cerebellum. Con-A-binding sites are found lining the nuclear envelope and within the cisternal space of rough and smooth endoplasmic reticulum as well as lining the cisternae of the Golgi apparatus (Fig. 2*). At the periphery of Purkinje cells, smooth membrane cisternae lie in close apposition to the plasma membrane in an arrangement called the "hypolemmal" cisternae (Kaiserman-Abramoff and Palay, 1969; Palay and Chan-Palay, 1974). The hypolemmal cisternae also contains Con-A-binding sites on the side of the membrane facing the cisternal space. This cisternal system extends into the axons and dendrites of Purkinje cells, where it continues to be labeled with Con A (Wood *et al.*, 1974; Wood and McLaughlin, 1976*a*) (Fig. 3). Other smooth membrane profiles within the dendrites and axons of Purkinje neurons are not seen to contain label, presumably because the membrane macromolecules comprising these cisternae do not contain Con-A-binding sugars in appropriate positions within their oligosaccharide chains. These results suggest a potential specialization of function among different classes of smooth membrane profiles within neurons, a concept elaborated further by Holtzman (1977). The presence of a specialized smooth membrane "channel" lined with carbohydrates just beneath the plasma membranes of neurons is of further interest because recent evidence suggests that smooth membrane profiles lying beneath the axolemma of chicken ciliary ganglion neurons are a part of the morphological substrate for fast axoplasmic transport (Droz *et al.*, 1973, 1975).

4.5. Developing Synapses

Another application of the *in situ* lectin labeling technique has been to localize changes in carbohydrate-containing macromolecules on developing synaptic membranes. Using the developing chick retina as a model system for synaptogenesis, McLaughlin and Wood (1977) showed that Con A, which binds primarily to mannosyl and glucosyl sugar residues, is localized along developing photoreceptor terminals and on pre- and postsynaptic membrane surfaces at early-forming ribbon synapses in the embryonic retina. Con A

*Illustrations for this chapter will be found following page 148.

labeling changes, however, in the hatchling retina after functional differentiation of the photoreceptors and after ribbon synapse maturation, in that Con A no longer labels synaptic membranes but is restricted to nonsynaptic membrane regions of photoreceptor terminals (Figs. 4 and 5). These developmental changes in Con A staining of photoreceptor terminal membranes during synapse formation suggest that Con-A-binding sites on immature synaptic membranes are relatively exposed or terminal in the oligosaccharide chain, whereas their inaccessibility to labeling after synaptic membrane maturation may indicate that other sugars have been added to the chain, thus masking the mannose residues from the lectin label.

McLaughlin *et al.* (1977) showed more recently that two other lectins, WGA and castor bean (CB), which are thought to bind, respectively, to *N*-acetylglucosamine and to galactose *N*-acetylgalactosamine sugar residues or both, also label developing photoreceptor synaptic membranes in the embryonic chick retina. After photoreceptor synapse formation in the hatchling retina, the CB lectin no longer labels photoreceptor synaptic membranes and is greatly reduced on nonsynaptic membrane surfaces. WGA, however, continues to label pre- and postsynaptic membranes at ribbon synapses in addition to nonsynaptic membranes in the hatchling (Figs. 6 and 7). These contrasting results concerning the accessibility of CB- and WGA-binding sites on mature synaptic membranes in the chick retina suggests that like Con A receptors, CB receptors on synaptic membranes are somehow masked or have disappeared after synapse formation and maturation, but that WGA receptors remain accessible.

To further characterize the differences in the accessibility of CB and WGA labeling during synaptogenesis, McLaughlin *et al.* (1977) also studied the effects of enzymatic digestion of CB and WGA staining after prior treatment with either trypsin or neuraminidase. Their results show that CB staining of developing synaptic and nonsynaptic membranes in the embryo remains the same after pretreatment with either enzyme, but in the hatchling, CB staining of nonsynaptic membranes is enhanced by both enzymes. On the other hand, WGA staining of developing synaptic and nonsynaptic membranes in the embryo is diminished after trypsin and is greatly reduced after neuraminidase treatment. WGA staining in the hatchling, however, is not altered after trypsin treatment, but after neuraminidase pretreatment, photoreceptor synaptic membranes are no longer labeled and nonsynaptic membrane staining is diminished. These differences in lectin labeling before and after enzyme pretreatment may reflect basic developmental changes occurring in the oligosaccharide chains on developing synaptic and nonsynaptic membrane surfaces during synapse formation and maturation. For example, the enhancement of CB labeling on nonsynaptic membranes in the hatchling after trypsin treatment may be due to an unmasking of galactose or *N*-acetylgalactosamine residues by the proteolytic digestion. The same could be true of neuraminidase pretreatment, which may be acting to expose galactose to the CB lectin after first removing the terminal sialic acid residues. The dramatic reduction of WGA staining in the embryonic retina after neuraminidase digestion suggests that much of the membrane labeling at this developmental time is due to sialic acid residues. On the other hand, disappearance of WGA staining (after neuraminidase treatment) on mature synaptic membranes in the hatchling and the continued presence of label on nonsynaptic membranes suggests that after synapse formation, synaptic membrane staining is primarily due to sialic acid, whereas most of the nonsynaptic staining is to *N*-acetylglucosamine residues.

4.6. Growth Cone Region

Pfenninger and Maylié-Pfenninger (1975) studied the distribution of several lectin–ferritin conjugates on the growth cone region of neurites of cultured superior cervical ganglion explants. WGA and ricin evenly label the entire neuronal surface, including the perikaryon and growth cone. Con A labels the same surfaces, but in a patchy manner. It may be that in pulse–chase experiments lectin label is cleared from the growth cone region more rapidly than from the rest of the neurite, a phenomenon that may reflect localized insertion of new lectin receptors into the plasmalemma by exocytosis of cytoplasmic vesicle profiles (Pfenninger and Maylié-Pfenninger, 1977).

4.7. Internalization into Neurons

Gonatas *et al.* (1975, 1977) studied distribution of ricin agglutinin and phytohemagglutinin lectin-binding sites on cultured neurons. The bound lectins are initially evenly distributed along the plasma membrane, but with increasing time after incubation lectin-binding sites are seen in a patchy distribution along the plasma membrane. As the plasma membrane staining goes from being evenly distributed to becoming patchy, internalized label appears within the neuron (Gonatas *et al.*, 1975, 1977). The internalized label is seen either within smooth membrane profiles or within elements of the Golgi apparatus, results that suggest that as lectin-binding sites are internalized they are carried directly to the Golgi apparatus. This may mean that some of the glycoproteins or glycolipids that comprise the neuronal plasma membrane lectin receptor are reutilized after modification within the Golgi without necessarily being degraded and resynthesized *de novo* (Gonatas *et al.*, 1975). The gradual transition of plasma membrane lectin receptors into patches could be due to rearrangement of the receptors after lectin binding, or, more likely, the unlabeled regions may represent plasma membrane that had been internalized and then replaced by addition of unlabeled membrane components.

4.8. Lectin Cytochemistry: Summary and Conclusions

The use of lectin cytochemistry has provided valuable information regarding the EM distribution of lectin-binding sites in nervous tissue. Although the lectins are capable of binding different sugars with relative specificity, it is impossible in most instances to identify individual macromolecules that contain the appropriate sugars because many glycoproteins and some glycolipids contain the common sugars that serve as lectin receptors. Since one ultimate goal of cytochemistry is to localize specific macromolecules within the nervous system, other methods are needed that will distinguish features of macromolecules more elaborate than the presence or absence of a given sugar.

5. Toxin and Antibody Cytochemistry

Reagents that offer the necessary increased specificity include (1) various toxin molecules that bind the known receptors and (2) antibodies directed toward a specific macromolecule. Toxin and antibody cytochemistry are just beginning to be exploited in studies of nervous system glycoproteins and glycolipids, and the results indicate considerable promise. To our knowledge, these methods have not yet been employed in cytochemical studies of glycosaminoglycans, but this is an area that merits future attention.

5.1. Toxin Cytochemistry

Several toxins are available that bind to receptor macromolecules that are known to contain carbohydrate residues. α-Bungarotoxin binds tightly to the nicotinic acetylcholine receptor, which is a glycoprotein (Meunier *et al.*, 1974; Raftery *et al.*, 1976; Salvaterra *et al.*, 1977). Several authors have utilized α-bungarotoxin-horseradish peroxidase cytochemistry to localize this glycoprotein at the neuromuscular junction (Lentz *et al.*, 1977) or in brain tissue (Lentz and Chester, 1977; Vogel *et al.*, 1977; Wood and McLaughlin, unpublished). Although results of these studies are preliminary, they indicate that the receptor may be located on both the pre- and postsynaptic membranes at the neuromuscular junction and probably at the postsynaptic membrane in synapses of chick neural retina (Vogel *et al.*, 1977) or rat midbrain and hypothalamic preoptic nucleus (Lentz and Chester, 1977). The α-bungarotoxin cytochemical methods seem to result in preferential staining of presynaptic terminals containing small clear vesicles, while terminals containing dense-cored vesicles are not stained (Lentz and Chester, 1977) (Fig. 8).

Another toxin that has been useful in the cytochemical analysis of membrane carbohydrates is cholera toxin, which apparently binds specifically to the ganglioside G_{M1} (Van Heyningen, S., 1974). Manuelidis and Manuelidis (1976) studied the localization of cholera toxin–peroxidase conjugates in dissociated cell cultures of mouse brain, in human glioblastoma cultures, and in isolated synaptosomal fractions. This marker is found on the plasma membranes of neurons and glia in dissociated cell culture, but it is not found on the plasma membranes of serially propagated glioblastoma cells. These results suggest that long-term culture may alter the glycolipid content of glioblastoma cells. The failure of long-term glioma cultures to bind cholera toxin is apparently not a peculiarity of the membrane unrelated to ganglioside content, since exogenously applied G_{M1} is taken into the membranes of these cells renders them capable of binding cholera toxin. Manuelidis and Manuelidis (1976) also showed that isolated synaptosomal fractions bind cholera toxin–peroxidase conjugates on the presynaptic axolemma and postsynaptic density. Engel *et al.* (1977) localized cholera toxin–peroxidase conjugates in developing rat cerebellum and found the marker on neuronal and glial cell soma and on neurites, including the growth cone plasma membrane. The label was associated with pre- and postsynaptic membranes of immature and mature synapses, and the synaptic cleft was heavily labeled. The localization of cholera toxin–peroxidase along the growth cone membrane indicates that at least one glycolipid, G_{M1}, is present on the neurite membrane prior to the time that membrane is engaged in synaptic formation.

5.2. Antibody Cytochemistry

5.2.1. Rationale and Methods

Antibody techniques to localize tissue antigens (Coons, 1956) have been extensively used in recent years to localize macromolecules of the nervous system, some of which are known to be glycoproteins. In general, the methodology of this approach is similar regardless of the antigen of interest. The antigen must be obtained in highly purified form to ensure the production of a monospecific antibody. When the antibody is obtained in sufficient titer, several choices of immunocytochemical procedures are available to render the antigenic site in the tissue visible in the microscope. The visualization is obtained by attaching markers to any of several antibody reagents used in the localization procedures. If LM results are desired, the marker is usually a fluorescent molecule. If a combination of LM and EM data is needed, the marker is usually horseradish peroxidase, which may be visualized by a standard enzymatic reaction (Graham and Karnovsky, 1966). If EM localization alone provides sufficient data, the marker may be relatively large electron-opaque molecules such as ferritin or hemocyanin. Whatever cytochemical marker is used may be either attached directly to the primary antiserum (directed against the tissue antigen) or attached to a second antiserum directed against the primary antiserum. In the latter case, the immunocytochemical incubations require two incubations to form a "sandwich" of primary antibody and marker-labeled second antibody. A useful method was developed by Sternberger (1974) to avoid attaching the marker molecule to any of the antibodies used in the localization procedures. The tissue is treated with primary and secondary antibodies as in the "sandwich" method, except that neither antibody contains a marker. The tissue is then treated with a peroxidase–antiperoxidase complex that binds to unoccupied sites on the second antibody. The peroxidase is then visualized for LM and EM study.

Several of the brain antigens that have been localized by immunocytochemical procedures are known to be glycoproteins, so this powerful technique allows the localization of individual glycoproteins (or a family of closely related glycoproteins with immunological cross-reactivity) within the nervous system. This specificity offers considerable promise with regard to the elucidation of function of nervous system carbohydrates, since the morphological localization of a given molecule, especially enzymes or hormones, can provide strong hints as to the functional role of that protein in the overall harmony of nerve cells. Although many previous histochemical methods for carbohydrate localization, such as the PAS reaction, are elegant chemically, they do not possess this specificity to localize individual molecules.

5.2.2. Glycoprotein Hormones

The pituitary contains several glycoprotein hormones, including follicle-stimulating hormone (FSH), luteinizing hormone (LH), and thyroid-stimulating hormone (TSH). There have been several attempts to localize these hormones within certain cells of the pituitary using immunocytochemical techniques (for references, see Moriarty, 1976). The results of these studies indicate that TSH is found in a morphologically distinct cell type, and that while LH and FSH are sometimes found in the same cell type, there are present in the pituitary cell types that contain only LH or FSH.

5.2.3. Specific Glycoproteins of Nervous Tissue

A nervous-system-specific sialoglycoprotein, GP-350, was recently described by Van Nieuw Amerongen and Roukema (1974). Immunofluorescence studies of the distribution of this protein (Van Nieuw Amerongen *et al.*, 1974) indicated that it is located almost exclusively in neuronal somata and processes. The protein does not appear to be located in glial cells of the calf corpus callosum or in human glioma cell lines (Van Nieuw Amerongen *et al.*, 1974).

The immunocytochemical localization of the visual pigment rhodopsin, another glycoprotein, was studied by Dewey *et al.* (1969) and by Jan and Revel (1974). In the mouse and cow retinas, the results of Jan and Revel (1974) indicated that the pigment is located on the disk membranes and plasma membranes of the rod outer segment as well as on the connecting cilium and some portions of the rod inner segment.

Delpech *et al.* (1976) studied the distribution of a high-molecular-weight glycoprotein from human brain using immunofluorescence methods. The localization results suggested that the glycoprotein was present in adult brain and perhaps localized to oligodendroglia.

5.2.4. Sodium–Potassium Adenosine Triphosphatase

One of the major membrane-associated proteins of brain is the $[Na^+–K^+]$ATPase (E.C.3.6.1.3) that is an active transport unit for Na^+ and K^+ ions (cf. Swanson and Stahl, 1976) [for references, see Albers (1976)]. The $[Na^+–K^+]$ATPase consists of at least two types of polypeptide subunits, one of which is a glycoprotein (Hokin, 1974; Kyte, 1972; Swann *et al.*, 1975). It is known that lectins will interact with the $[Na^+–K^+]$ATPase (Swann *et al.*, 1975), so it is likely that one of the membrane components localized in brain utilizing lectin cytochemistry (cf. McLaughlin and Wood, 1977) is this enzyme. To specifically localize the enzyme at the EM level in brain tissue, Wood *et al.* (1977) employed immunocytochemical methods utilizing antisera raised in rabbits against eel electric organ $[Na^+–K^+]$ATPase to label this antigen in the Black Ghost knifefish (*Sternarchus albifrons*). In general, the results (Wood *et al.*, 1977) showed that the $[Na^+–K^+]$ATPase is restricted to membrane profiles. At the EM level, the enzyme is associated with the plasma membranes of neurons, oligodendroglia, and astrocytes, although the pattern of distribution varies for the different cell types. In neurons, the enzyme appears to be relatively even in distribution over the somal and dendritic membrane. Along the axolemma of myelinated axons, the enzyme appears to be restricted to the axolemma at the node of Ranvier, and internodal axolemma is not observed to be labeled using these methods (Figs. 9 and 10). This observation suggests that the neuron has the capacity to restrict a functionally important protein to sites along the axolemma where that protein is required to perform a specific task. The capacity to detect this apparent restricted localization of a major membrane glycoprotein, the $[Na^+–K^+]$ATPase, is an illustration of the potential power of immunocytochemical methodology to contribute meaningful results in the cytochemical localization of glycoproteins (and glycosaminoglycans). It must be remembered, however, that the immunocytochemical methods localize enzyme protein, not enzyme activity, and information regarding this activity must be obtained by the elegant histochemical methods being employed [for references, see Stahl and Broderson (1976)].

6. Summary

We have attempted to review the general cytochemical methods that have been used to demonstrate carbohydrates in the nervous system. Most of these methods rely either on chemical reactions with sugars that are common to many glycoproteins and glycosaminoglycans or on electrostatic interaction between generally cationic "stain" molecules and anionic groups on glycoproteins and glycosaminoglycans. Although there is evidence from other tissues that several of these staining procedures, when rigorously controlled, may distinguish glycoproteins from glycosaminoglycans and perhaps distinguish among the glycosaminoglycans, there is very little evidence that any distinction has been achieved in cytochemical studies of the nervous system.

The most extensive recent work on localizing identified carbohydrates of the nervous system has been achieved utilizing lectin, toxin, and antibody cytochemistry. For example, using lectin cytochemistry, a major peripheral nerve glycoprotein has been localized to the interperiod line of the myelin sheath. Several lines of investigation have indicated the presence of specific carbohydrates in the synaptic complex, and there are changes that occur in the carbohydrates during synaptogenesis that may be monitored using lectin cytochemistry.

The use of toxin molecules that bind to specific carbohydrate receptors, together with EM labeling techniques, has led to the localization of specific receptors including the nicotinic acetylcholine receptor and perhaps the ganglioside G_{M1}.

Antibody cytochemistry offers a broad range of possibilities with regard to localizing specific glycoproteins (and potentially glycosaminoglycans) in the nervous system. Results have been obtained indicating the localization of individual glycoprotein hormones in the pituitary and the Na^+–K^+-stimulated ATPase in various regions of the knifefish brain. This field should produce a great deal of new information in the near future.

ACKNOWLEDGMENTS. Original research by the authors has been supported by USPHS Grant NS-12590, the Alfred Sloan Foundation, and Fight for Sight, Inc., New York.

References

Albers, R. W., 1976, The (sodium plus potassium) transport ATPase, in: *The Enzymes of Biological Membranes* (A. Martonosi, ed.), Vol. 3, pp. 283–301, Plenum Press, New York.

Andrews, P. M., and Porter, K. R., 1973, The ultrastructural morphology and possible functional significance of mesothelial microvilli, *Anat. Rec.* **117**:409–426.

Arseni, C., Carp, N., Mestes, E., and Adel, M., 1967, Histochemistry of mucopolysaccharides in brain tumors, *Acta Neuropathol.* **7**:275–284.

Barbera, A. J., Marchase, R. B., and Roth, S., 1973, Adhesive recognition and retinotectal specificity, *Proc. Natl. Acad. Sci. U.S.A.* **60**:2482–2486.

Barondes, S. H., 1970, Brain glycomacromolecules and interneuronal recognition, in: *The Neurosciences, Second Study Program* (F. O. Schmidt, ed.), pp. 747–760, Rockefeller University Press, New York.

Barondes, S. H., 1975, Towards a molecular basis of neuronal recognition, in: *The Nervous System*, Vol. I, *The Basic Neurosciences* (R. Brady, ed.), pp. 129–136, Raven Press, New York.

Barondes, S. H., and Rosen, S. D., 1976, Cell surface carbohydrate-binding proteins: Role in cell recognition, in: *Neuronal Recognition* (S. H., Barondes, ed.), pp. 331–356, Plenum Press, New York.

Benedetti, E. L., and Emmelot, P., 1967, Studies on plasma membranes. IV. The ultrastructural localization and content of sialic acid in plasma membranes isolated from rat liver and hepatoma, *J. Cell Sci.* **2**:499–512.

Benedetti, E. L., and Emmelot, P., 1968, Structure and function of plasma membranes isolated from liver, in: *The Membranes* (A. Dalton and F. Haguenau, eds.), pp. 33–120, Academic Press, New York.

Bennett, H. S., 1969*a*, The cell surface: Components and configurations, in: *Handbook of Molecular Cytology* (A. Lima-de-Faria, ed.), pp. 1261–1293, North-Holland, Amsterdam.

Bennett, H. S., 1969*b*, The cell surface: Movements and recombinations, in: *Handbook of Molecular Cytology* (A. Lima-de-Faria, ed.), pp. 1294–1319, North-Holland, Amsterdam.

Bernard, W., and Avrameas, S., 1971, Ultrastructural visualization of cellular carbohydrate components by means of concanavalin A, *Exp. Cell Res.* **64**:232–236.

Bittiger, H., and Heid, J., 1977, The subcellular distribution of particle-bound negative charges in rat brain, *J. Neurochem.* **28**:917–922.

Bittiger, H., and Schnebli, H. P., 1974, Binding of concanavalin A and ricin to synaptic junctions of rat brain, *Nature (London)* **249**:370–371.

Bondareff, W., 1967, An intercellular substance in rat cerebral cortex: Submicroscopic distribution of ruthenium red, *Anat. Rec.* **157**:527–536.

Bonneville, M., and Weinstock, M., 1970, Brush border development in the intestinal absorptive cells of *Xenopus* during metamorphosis, *J. Cell Biol.* **44**:151–171.

Brady, R. O., and Quarles, R. H., 1973, The enzymology of myelination, *Mol. Cell. Biochem.* **2**:23–29.

Carlsen, J. B., and Svensmark, O., 1970, Multiple forms of soluble butyrylcholinesterase in human brain, *Biochem. Biophys. Acta* **207**:477–484.

Castejón, H. V., 1970*a*, Histochemical demonstration of acid glycosaminoglycans in the nerve cell cytoplasm of mouse central nervous system, *Acta Histochem.* **35**:161–172.

Castejón, H. V., 1970*b*, Histochemical demonstration of sulfated polysaccharides at the surface coat of nerve cells in the mouse central nervous system, *Acta Histochem.* **38**:55–64.

Castejón, H. V., 1970*c*, A sulfated polysaccharide in some nerve cell coats in brains of different species of vertebrates, *J. Histochem. Cytochem.* **18**:685a.

Castejón, H. V., 1971, Histochemical studies of the neuronal acid glycosaminoglycans in the central nervous system of various species of vertebrates, *Acta Cient. Venez.* **22**(Supple. 2): R-24.

Castejón, H. V., and Castejón, O. J., 1972, Application of alcian blue and Os-DMEDA in the electron-histochemical study of the cerebellar cortex. I. Alcian blue staining, *Rev. Microsc. Electron.* **1**:207–226.

Castejón, H. V., and Castejón, O. J., 1976, Electron microscopic demonstration of hyaluronidase sensible proteoglycans at the presynaptic area in mouse cerebellar cortex, *Acta Histochem.* **55**:300–316.

Castejón, O. J., and Castejón, H. V., 1972*a*, The tinctorial potentiality of two basic stains in the electro-histochemical study of polyanionic compounds in nerve tissue. I. Synaptic region, *Acta Histochem.* **43:**153–163.

Castejón, O. J., and Castejón, H. V., 1972*b*, Application of alcian blue and Os-DMEDA in the electron-histochemical study of the cerebellar cortex. II. Os-DMEDA staining, *Rev. Microsc. Electron.* **1:**227–238.

Coons, A. H., 1956, Histochemistry with labeled antibody, *Int. Rev. Cytol.* **5:**1–23.

Cotman, C. W., and Taylor, C., 1974, Localization and characterization of concanavalin A receptors in the synaptic cleft, *J. Cell Biol.* **62:**236–242.

Cuatrecasas, P., 1973, Gangliosides and membrane receptors for cholera toxin, *Biochemistry* **12:**3547–3558.

Danon, D., Goldstein, L., Marikovsky, Y., and Skutelsky, E., 1972, Use of cationized ferritin as a label of negative charges on cell surfaces, *J. Ultrastruct. Res.* **38:**500–510.

Delpech, B., Vidard, M. N., and Delpech, A., 1976, Caracterisation immunochimique et immunohistologigue d'une glycoprotéine asociée au système nerveux, *Immunochemistry* **13:**111–116.

Dewey, M. M., Davis, P. K., Blasie, J. K., and Barr, L., 1969, Localization of rhodopsin antibody in the retina of the frog, *J. Mol. Biol.* **39:**395–405.

Doshi, R., Sandry, S. A., Churchhill,, A. W., and Brownell, B., 1974, The cerebellum in mucopolysaccharidosis: A histological, histochemical and ultrastructural study, *J. Neurol. Neurosurg. Psychiatry* **37:**1133–1138.

Droz, B., Koenig, H. L., and Giamberardino, L., 1973, Axonal migration of protein and glycoprotein to nerve endings. I. Radio-autographic analysis of the renewal of protein in nerve endings of chicken ciliary ganglion after intracerebral injection of [^{3}H]lysine, *Brain Res.* **60:**93–127.

Droz, B., Rambourg, A., and Koenig, H. L., 1975, The smooth endoplasmic reticulum: Structure and role in the renewal of axonal membrane and synaptic vesicles by fast axonal transport, *Brain Res.* **93:**1–13.

Engel, E. L., Wood, J. G., and Byrd, F. I., 1977, Studies of ganglioside patterns and cholera toxin–peroxidase labeling of aggregating cells from the chick optic tectum, Society of Neuroscience 7th Annual Meeting, Anaheim, California (abstract).

Everly, J. L., Brady, R. O., and Quarles, R. H., 1973, Evidence that the major protein in rat sciatic nerve myelin is a glycoprotein, *J. Neurochem.* **21:**329–334.

Feeney, L., 1973, The interphotoreceptor space. II. Histochemistry of the matrix, *Dev. Biol.* **32:**115–128.

Feria-Velasco, A., Sánchez-de-la-Peña, S., and Magdaleno, V., 1976, Labeling of electrical surface charges at synaptosome membrane: An electron cytochemical and biochemical study, *Brain Res.* **112:**214–220.

Glick, D., and Scott, J. E., 1970, Phosphotungstic acid not a stain for carbohydrates, *J. Histochem. Cytochem.* **18:**455.

Gonatas, N. K., and Avrameas, S., 1973, Detection of plasma membrane carbohydrates with lectin–peroxidase conjugates, *J. Cell Biol.* **59:**436–443.

Gonatas, N. K., Steiber, A., Kim, S. U., Graham, D. I., and Avrameas, S., 1975, Internalization of neuronal plasma membrane ricin receptors into the Golgi apparatus, *Exp. Cell Res.* **94:**426–431.

Gonatas, N. K., Kim, S. U., Steiber, A., and Avrameas, S., 1977, Internalization of lectins in neuronal GERL, *J. Cell Biol.* **73:**1–13.

Graham, R. C., and Karnovsky, M. J., 1966, The early stages of absorption of injected horseradish peroxidase in the proximal tubules of mouse kidney: Ultrastructural cytochemistry by a new technique, *J. Histochem. Cytochem.* **22:**320–326.

Groniowski, J., Biczyskowa, W., and Walski, M., 1969, Electron microscope studies on the surface coat of the nephron, *J. Cell Biol.* **40:**585–601.

Gurd, J. W., and Mahler, H. R., 1974, Fractionation of synaptic plasma membrane glycoproteins by lectin affinity, *Biochemistry* **13:**5193–5198.

Hale, C. W., 1946, Histochemical demonstration of acid polysaccharides in animal tissues, *Nature* (*London*) **157:**802.

Hall, C. E., Jakus, M. A., and Schmidt, F. O., 1945, The structure of certain muscle fibrils as revealed by the use of electron stains, *J. Appl. Phys.* **16:**459–465.

Hernandez, W., Rambourg, A., and Leblond, C. P., 1968, Periodic acid–chromic acid–methenamine silver technique for glycoprotein detection in the electron microscope, *J. Histochem. Cytochem.* **16:**507.

Hokin, L. E., 1974, Purification and properties of the (Na + K) activated adenosine triphosphatase and reconstitution of sodium transport. *Ann. N. Y. Acad. Sci.* **242:**12–23.

Holmgren, J., Lonroth, I., and Svennerholm, L., 1973, Fixation and inactivation of cholera toxin by G_{M1} ganglioside, *Scand. J. Infect. Dis.* **5:**77–78.

Holtzman, E., 1977, The origin and fate of secretory packages, especially synaptic vesicles, *Neuroscience* **2:**327–355.

Hotchkiss, R. D., 1948, A microchemical reaction resulting in the staining of polysaccharide structures in fixed tissue preparations, *Arch. Biochem.* **16:**131–141.

Hughes, R. C., 1973, Glycoproteins as components of cellular membranes, *Prog. Biophys. Mol. Biol.* **26:**189–268.

Jan, L. Y., and Revel, J. P., 1974, Ultrastructural localization of rhodopsin in the vertebrate retina, *J. Cell Biol.* **62:**257–273.

Jensen, W. A., 1962, *Botanical Histochemistry*, Freeman Press, San Francisco.

Jirge, S. K., 1971, Mucopolysaccharides in the retinal cells of developing fish eye: A histochemical study, *Acta Histochem.* (*Jena*) **41:**51–61.

Jones, D. B., 1969, Mucosubstances of the glomerulus, *Lab. Invest.* **21:**119–125.

Kaiserman-Abramoff, I. R., and Palay, S. L., 1969, Fine structural studies of the cerebellar cortex in a mormyrid fish, In: *Neurobiology of Cerebellar Evolution and Development* (R. Llinas, ed.), p. 71, AMA-ERF Institute for Biomedical Research, Chicago.

Kelly, P., Cotman, C. W., Gentry, G. L., and Nicolson, G. L., 1976, Distribution and mobility of lectin receptors on synaptic membranes of identified neurons in the central nervous system, *J. Cell Biol.* **71:**487–496.

Kyte, J., 1972, Properties of the two polypeptides of the sodium and potassium–dependent adenosine triphosphatase, *J. Biol. Chem.* **247:**7642–7649.

Lai, M., Lampert, I. A., and Lewis, P. D., 1975, The influence of fixation on staining of glycosaminoglycans in glial cells, *Histochemistry* **41:**275–279.

Lampert, I., and Lewis, P. D., 1974, Demonstration of acidic polyanions in certain glial cells during postnatal rat brain development, *Brain Res.* **73:**356–361.

Landon, D. N., and Langley, O. K., 1971, The local chemical environment of nodes of Ranvier: A study of cation binding, *J. Anat.* **108:**419–432.

Langley, O. K., 1971, A comparison of the binding of alcian blue and inorganic cations to polyanions in peripheral nerve, *Histochem. J.* **3:**251–260.

Lentz, T. L., and Chester, J., 1977, Localization of acetylcholine receptors in central synapses, *J. Cell Biol.* **75:**258–267.

Lentz, T. L., Mazurkiewicz, J. E., and Rosenthal, J., 1977, Cytochemical localization of acetylcholine receptors at the neuromuscular junction by means of horseradish peroxidase-labeled α-bungarotoxin, *Brain Res.* **132:**423–442.

Lillie, R. D., 1947, Reticulum staining with Schiff reagent after oxidation by acidified sodium periodate, *J. Lab. Clin. Med.* **32:**910–912.

Lim, R., and Mitsunobu, K., 1975, Partial purification of a morphological transforming factor from pig brain, *Biochim. Biophys. Acta* **400:**200–207.

Luft, J. H., 1976, The structure and properties of the cell surface coat, *Int. Rev. Cytol.* **45:** 291–382.

Manuelidis, L., and Manuelidis, E. E., 1976, Ultrastructural study of plasma membrane G_{M1} in neuroectodermal cells using cholera–peroxidase, *J. Neurocytol.* **5**:575–589.

Margolis, R. U., and Margolis, R. K., 1977, Metabolism and function of glycoproteins and glycosaminoglycans in nervous tissue, *Int. J. Biochem.* **8**:85–91.

Martinez-Palomo, A., 1970, The surface coats of animal cells, *Int. Rev. Cytol.* **29**:29–75.

Martinez-Rodriguez, R., Toledano, A., Garcia-Segura, L. M., Gonzalez-Eloviaga, M., Gamonal, A., Diaz-Gonzales, P., De Agustin, M., and Rodriguez-Gonzalez, C., 1976, Mucopolysaccharides in hypothalamic neurons of the rat, *J. Anat.* **121**:231–239.

Matsusaka, T., 1971, The fine structure of the inner limiting membrane of the rat retina as revealed by ruthenium red staining, *J. Ultrastruct. Res.* **36**:312–317.

Matus, A. I., and Walters, B. B., 1976, Type 1 and 2 synaptic junctions: Differences in distribution of concanavalin A binding sites and stability of the junctional adhesion, *Brain Res.* **108**:249–256.

Matus, A., Depetris, S., and Raff, M. D., 1973, Mobility of concanavalin A receptors in myelin and synaptic membranes, *Nature* (*London*) **244**:278–279.

McDonough, J., and Lilien, J., 1975, Spontaneous and lectin-induced redistribution of cell surface receptors on embryonic chick neural retina cells, *J. Cell Sci.* **19**:357–367.

McLaughlin, B. J., and Wood, J. G., 1977, The localization of concanavalin A binding sites during photoreceptor synaptogenesis in the chick retina, *Brain Res.* **119**:57–71.

McLaughlin, B. J., Wood, J. G., and Gurd, J. W., 1977, The localization of WGA and CB lectin binding sites at developing photoreceptor synapses before and after enzyme digestion, *J. Cell Biol.* **75**:115a.

McManus, J. F. A., 1946, Histological demonstration of mucin after periodic acid, *Nature* (*London*) **158**:202.

Meunier, J. C., Sealock, R., Olsen, R., and Changeux, J. P., 1974, Purification and properties of the cholinergic receptor protein from *Electrophorus electricus* electric tissue, *Eur. J. Biochem.* **45**:371–394.

Moran, D., and Rice, R. W., 1975, An ultrastructural examination of the role of cell membrane surface coat material during neurulation, *J. Cell Biol.* **64**:172–181.

Moriarty, G. C., 1976, Immunocytochemistry of the pituitary glycoprotein hormones, *J. Histochem. Cytochem.* **24**:846–863.

Moscona, A. A., 1974, Surface specification of embryonic cells: Lectin receptors, cell recognition, and specific cell ligands, in: *The Cell Surface in Development* (A. A. Moscona, ed.), pp. 67–99, John Wiley and Sons, New York.

Moscona, A. A., 1976, Cell recognition in embryonic morphogenesis and the problem of neuronal specificities, in: *Neuronal Recognition* (S. H. Barondes, ed.), pp. 205–226, Plenum Press, New York.

Moss, C. A., 1973, Glycosaminoglycans of disaggregated fetal mouse brain tissue cultures, *Histochem. J.* **5**:547–556.

Nachmias, U. T., and Marshall, J. M., Jr., 1961, Protein uptake by pinocytosis in amoebae: Studies on ferritin and methylated ferritin, in: *Biological Structure and Function* (T. Goodwin and O. Lindberg, eds.), Vol. 2, pp. 605–619, Academic Press, New York.

Nicolson, G. L., 1974, The interaction of lectins with animal cell surfaces, *Int. Rev. Cytol.* **39**:90–190.

Nicolson, G. L., 1976, Trans-membrane control of the receptors on normal and tumor cells. II. Surface changes associated with transformation and malignancy, *Biochim. Biophys. Acta* **458**:1–72.

Nicolson, G. L., and Singer, S. J., 1971, Ferritin-conjugated plant agglutinins as specific saccharide stains for electron microscopy: Application to saccharides bound to cell membranes, *Proc. Natl. Acad. Sci. U.S.A.* **68**:943–945.

Nordlander, R. H., Masnyi, J. A., and Singer, M., 1975, Distribution of ultrastructural tracers in crustacean axons, *J. Comp. Neurol.* **161**:499–514.

Palay, S. L., and Chan-Palay, V., 1974, *Cerebellar Cortex*, p. 28, Springer-Verlag, New York.

Pease, D. C., 1966, Polysaccharides associated with the exterior surface of epithelial cells: Kidney, intestine, brain, *J. Ultrastruct. Res.* **15**:555–588.

Pease, D. C., 1970, Phosphotungstic acid as a specific electron stain for complex carbohydrates, *J. Histochem. Cytochem.* **18**:455–458.

Pease, D. C., and Peterson, R. G., 1972, Polymerizable glutaraldehyde–urea mixtures as polar, water-containing embedding media, *J. Ultrastruct. Res.* **41**:133–159.

Peterson, R. G., and Pease, D. C., 1972, Myelin embedded in polymerized glutaraldehyde–urea, *J. Ultrastruct. Res.* **41**:115–132.

Peterson, R. G., and Sea, C. P., 1975, Localization of the main myelin protein in PNS, *Trans. Am. Soc. Neurochem.* **6**:211a.

Pfenninger, K. H., 1973, Synaptic morphology and cytochemistry, *Prog. Histochem. Cytochem.* **5**:1–86.

Pfenninger, K. H., and Maylié-Pfenninger, M. F., 1975, Distribution and fate of lectin binding sites on the surface of growing neuronal processes, *J. Cell Biol.* **67**:332a.

Pfenninger, K. H., and Maylié-Pfenninger, M. F., 1977, Localized appearance of new lectin receptors on the surface of growing neurites, *J. Cell Biol.* **75**:54a.

Pinto da Silva, P., Moss, P. S., and Fudenberg, H. H., 1973, Anionic sites on the membrane intercalated particles of human erythrocyte ghost membranes: Freeze–etch localization, *Exp. Cell Res.* **81**:127–138.

Quarles, R. H., Everly, J. L., and Brady, R. O., 1973, Evidence for the close association of a glycoprotein with myelin in rat brain, *J. Neurochem.* **21**:1177–1191.

Quick, D. C., and Waxman, S. G., 1977, Specific staining of the axon membrane at nodes of Ranvier wth ferric ion and ferrocyanide, *J. Neurol. Sci.* **31**:1–11.

Quintarelli, G., Scott, J. E., and Dellovo, M. C., 1964, The chemical and histochemical properties of alcian blue. II. Dye binding of tissue polyanions, *Histochemie* **4**:73–85.

Raftery, M. A., Vandlen, R. L., Reed, K. L., and Lee, T., 1976, Characterization of *Torpedo californica* acetylcholine receptor: Its subunit composition and ligand-binding properties, *Cold Spring Harbor Symp. Quant. Biol.* **40**:193–202.

Rahmann, H., and Katusic, J., 1975, Histochemischer Nachweis sialinsäurehaltiger Verbindungen im ZNS von Teleosteern, *Histochemie* **44**:291–302.

Rakic, L. M., and Mrsulja, B. J., 1975, Histochemical distribution of acid mucopolysaccharides during circadian rhythm in the cerebellum of *Serranus scriba*, *Acta Anat.* (*Basel*) **92**:61–71.

Rambourg, A., 1969, Localisation ultrastructurale et nature du materiel coloré au niveau de la surface cellulaire por le mélange chromique–phosphotungstique, *J. Microsc.* (*Paris*) **8**: 325–342.

Rambourg, A., 1971, Morphological and histochemical aspects of glycoproteins at the surface of animal cells, *Int. Rev. Cytol.* **31**:57–114.

Rambourg, A., and Leblond, C. P., 1967, Electron microscope observations on the carbohydrate-rich cell coat present at the surface of cells in the rat, *J. Cell Biol.* **32**:27–53.

Rambourg, A., Neutra, M., and Leblond, C. P., 1966, Presence of a "cell coat" rich in carbohydrate at the surface of cells in the rat, *Anat. Rec.* **154**:41–72.

Revel, J. P., 1964, A stain for the ultrastructural localization of acid mucopolysaccharides, *J. Microsc.* (*Paris*) **3**:535–544.

Revel, J. P., and Ito, S., 1967, The surface components of cells, in: *The Specificity of Cell Surfaces* (B. Davis and L. Warren, eds.), pp. 211–234, Prentice-Hall, Englewood Cliffs, New Jersey.

Roseman, S., 1974, Complex carbohydrates and intercellular adhesion, in: *The Cell Surface in Development* (A. A. Moscona, ed.), pp. 255–271, John Wiley and Sons, New York.

Salvaterra, P. M., Gurd, J. M., and Mahler, H. R., 1977, Interactions of the nicotinic acetylcholine receptor from rat brain with lectins, *J. Neurochem.* **29**:345–348.

Saxena, P. K., 1969, Histochemical studies in the brain of certain teleostean fishes, *Acta Anat.* **73**:569–587.

Scott, J. E., 1960, Aliphatic ammonium salts in the assay of acidic polysaccharides from tissues, *Methods Biochem. Anal.* **8**:145–197.

Scott, J. E., 1972, Histochemistry of alcian blue. III. The molecular biological basis of staining by alcian blue 8GX and analogous phthalocyanins, *Histochemie* **32**:191–212.

Scott, J. E., and Dorling, J., 1969, Periodate oxidation of acid polysaccharides. III. A PAS method for chondroitin sulfates and other glycosamino-glycuronans, *Histochemie* **19**: 295–301.

Scott, J. E., and Glick, D., 1971, The invalidity of "phosphotungstic acid as a specific electron stain for complex carbohydrates," *J. Histochem. Cytochem.* **19**:63–64.

Scott, J. E., and Harbinson, R. J., 1969, Periodate oxidation of acid polysaccharides. II. Rates of oxidation of uronic acids in polyuronides and acid mucopolysaccharides, *Histochemie* **19**:155–161.

Scott, J. E., and Stockwell, R. A., 1967, On the use and abuse of the critical electrolyte concentration approach to the localization of tissue polyanions, *J. Histochem. Cytochem.* **15**:111–113.

Scott, J. E., Quintarelli, G., and Dellovo, M. D., 1964, The chemical and histochemical properties of alcian blue. I. The mechanism of alcian blue staining, *Histochemie* **4**:73–85.

Seligman, A. M., Hankers, J. S., Wasserkrug, H., Dmochowski, H., and Katzoff, L., 1965, Histochemical demonstration of some oxidized macromolecules with thiocarbohydrazide (TCH) or thiosemicarbazide (TSC) and osmium tetroxide, *J. Histochem. Cytochem.* **13**: 629–639.

Sharon, N., and Lis, H., 1972, Lectins: Cell agglutinating and sugar-specific proteins, *Science* **177**:949–959.

Sharon, N., and Lis, H., 1975, Use of lectins for the study of membranes, in: *Methods in Membrane Biology,* Vol. 3 (E. D. Korn, ed.), pp. 147–220, Plenum Press, New York.

Silverman, L., and Glick, D., 1969*a*, The reactivity and staining of tissue proteins with phosphotungstic acid, *J. Cell Biol.* **40**:761–767.

Silverman, L., and Glick, D., 1969*b*, Measurement of protein concentration by quantitative electron microscopy, *J. Cell Biol.* **40**:773–778.

Simpson, D. L., Thorne, D. R., and Loh, H. H., 1977, Developmentally regulated "lectin-like" activity in rat brain, *Trans. Am. Soc. Neurochem.* **8**:234a.

Singer, M., Krishnan, N., and Fyfe, D. A., 1972, Penetration of ruthenium red into peripheral nerve fibers, *Anat. Rec.* **173**:375–390.

Smith, B., and Butler, M., 1973, Acid mucopolysaccharides in tumours of the myelin sheath cells, the oligodendroglioma and the neurilemmona, *Acta Neuropathol.* **23**:181–185.

Smith, S. B., and Revel, J. P., 1972, Mapping of concanavalin A binding sites on the surface of several cell types, *Dev. Biol.* **27**:434–441.

Stahl, W. L., and Broderson, S. H., 1976, Localization of Na^+, K^+-ATPase in brain, *Fed. Proc. Fed. Am. Soc. Exp. Biol.* **35**:1260–1265.

Sternberger, L. A., 1974, *Immunocytochemistry*, Prentice-Hall, Englewood Cliffs, New Jersey.

Stockwell, R. A., and Scott, J. E., 1965, Observations on the acid glycosaminoglycan (mucopolysaccharide) content of the matrix of aging cartilage, *Ann. Rheum. Dis.* **24**:341–350.

Stoddart, R. W., and Kiernan, J. A., 1973, Histochemical detection of the α-D-arabinopyranoside configuration using fluorescent-labelled concanavalin A, *Histochemie* **33**:98–94.

Sulkin, N. M., 1960, The distribution of mucopolysaccharides in the cytoplasm of vertebrate nerve cells, *J. Neurochem.* **5**:231–235.

Svennerholm, L., 1970, Gangliosides, in: *Handbook of Neurochemistry*, Vol. 3 (A. Lajtha, ed.), pp. 425–452, Plenum Press, New York.

Swann, A. C., Daniel, A., Albers, R. W., and Koval, G. J., 1975, Interactions of lectins with ($Na^+ + K^+$)-ATPase of eel electric organ, *Biochim. Biophys. Acta* **401**:299–306.

Swanson, P. D., and Stahl, W. L., 1976, Ion transport, in: *Basic Neurochemistry*, 2nd ed. (G. J. Siegel, R. W. Albers, B. W. Agranoff, and R. Katzman, eds.), pp. 125–147, Little, Brown, Boston.

Tani, E., and Ametani, T., 1971, Extracellular distribution of ruthenium red positive substance in the cerebral cortex, *J. Ultrastruct. Res.* **34**:1–14.

Tasso, F., and Rua, S., 1975, Étude cytochimique ultrastructurale des glycoprotéines dans le complexe hypothalamo-post-hypophysaire du rat, *Arch. Anat. Microsc.* **3**:247–260.

Taylor, P., Jones, J. W., and Jacobs, N. M., 1974, Acetylcholinesterase from torpedo: Characterization of an enzyme species isolated by lytic procedures, *Mol. Pharmacol.* **10**:78–92.

Teichberg, V. I., Silman, I., Beitsch, D. D., and Resheff, G., 1975, A β-D-galactoside binding protein from electric organ tissue of *Electrophorus electricus*, *Proc. Natl. Acad. Sci. U.S.A.* **72**:1383–1387.

Thiery, J. P., 1967, Mise en évidence des polysaccharides sur coupes fines en microscopie électronique, *J. Microsc.* **6**:987–1018.

Toole, B. P., 1976, Morphogenetic role of glycosaminoglycans (acid mucopolysaccharides in brain and other tissues), in: *Neuronal Recognition* (S. H. Barondes, ed.), pp. 275–329, Plenum Press, New York.

Vaccari, A., Vertua, R., and Furlani, A., 1971, Decreased calcium uptake by rat fundal strips after pretreatment with neuraminidase or LSD *in vitro*, *Biochem. Pharmacol.* **20**:2603–2612.

Van Heyningen, S., 1974, Cholera toxin: Interaction of its subunits with ganglioside G_{M1}, *Science* **183**:656–657.

Van Heyningen, W. E., Carpenter, W. B., Pierce, N. F., and Greenough, W. B., III, 1971, Deactivation of cholera toxin by ganglioside, *J. Infect. Dis.* **124**:415–418.

Van Nieuw Amerongen, A., and Roukema, P. A., 1974, GP-350, a sialoglycoprotein from calf brain: Its subcellular localization and occurrence in various brain areas, *J. Neurochem.* **23**: 85–89.

Van Nieuw Amerongen, A., Roukema, P. A., and Van Rossum, A. L., 1974, Immunofluorescence study on the cellular localization of GP-350, a sialoglycoprotein from brain, *Brain Res.* **81**:1–19.

Vaughn, J. E., Henrikson, C. K., and Wood, J. G., 1976, Surface specializations of neurites in embryonic mouse spinal cord, *Brain Res.* **110**:431–445.

Vogel, Z., Maloney, G. J., Ling, A., and Daniels, M. P., 1977, Identification of synaptic acetylcholine receptor sites in retina with peroxidase-labeled α-bungarotoxin, *Proc. Natl. Acad. Sci. U.S.A.* **74**:3268–3272.

Wessells, N. K., Nuttall, R. P., Wrenn, J. T., and Johnson, S., 1976, Differential labeling of the cell surface of single ciliary ganglion neuron *in vitro*, *Proc. Natl. Acad. Sci. U.S.A.* **73**:4100–4104.

Westergaard, E., 1971, Ruthenium red in the ependyma after perfusion with the dye in the fixative, *J. Ultrastruct. Res.* **36**:562a.

Wiedmer, T., Gentinetta, R., and Brodbeck, U., 1974, Binding of acetylcholinesterases to concanavalin A, *FEBS Lett.* **47**:260–263.

Winzler, R. J., 1970, Carbohydrates in cell surfaces, *Int. Rev. Cytol.* **29**:77–125.

Wood, J. G., and Dawson, R. M. C., 1973, A major myelin glycoprotein of sciatic nerve *J. Neurochem.* **21**:717–719.

Wood, J. G., and Engel, E. L., 1976, Peripheral nerve glycoproteins and myelin fine structure during development of rat sciatic nerve, *J. Neurocytol.* **5**:605–615.

Wood, J. G., McLaughlin, B. J., 1975, The visualization of concanavalin A binding sites in the interperiod line of rat sciatic nerve myelin, *J. Neurochem.* **24**:233–235.

Wood, J. G., and McLaughlin, B. J., 1976*a*, Cytochemical studies of lectin binding sites in smooth membrane cisternae of rat brain, *Brain Res.* **118**:15–26.

Wood, J. G., and McLaughlin, B. J., 1976*b*, The effects of mild proteolytic degradation on synaptic lectin binding sites, *J. Neurobiol.* **7**:469–474.

Wood, J. G., and Sarinana, F. O., 1975, The staining of sciatic nerve glycoproteins on polyacrylamide gels with concanavalin A peroxidase, *Anal. Biochem.* **69**:320–322.

Wood, J. G., McLaughlin, B. J., and Barber, R. P. 1974, The visualization of concanavalin A binding sites in Purkinje cell somata and dendrites of rat cerebellum, *J. Cell Biol.* **63**:541–549.

Wood, J. G., Jean, D. H., Whitaker, J. N., McLaughlin, B. J., and Albers, R. W., 1977, Immunocytochemical localization of the sodium, potassium activated ATPase in knifefish brain, *J. Neurocytol.* **6**:571–581.

Wooley, D. W., and Gommi, B. W., 1965, Serotonin receptors. VII. Activities of various pure gangliosides as the receptors, *Proc. Natl. Acad. Sci. U.S.A.* **53**:959–963.

Young, I. J., and Abood, L. G., 1960, Histological demonstration of hyaluronic acid in the central nervous system, *J. Neurochem.* **6**:89–94.

Zacks, S. I., Sheff, M. F., and Saito, A., 1973, Cytochemical and physical properties of myofiber external lamina, *J. Histochem. Cytochem.* **21**:895–910.

Zanetta, J. P., Morgan, I. G., and Gombos, G., 1975, Synaptosomal plasma membrane glycoproteins: Fractionation by affinity chromatography on concanavalin A, *Brain Res.* **83**: 337–348.

8

Glycoproteins of the Synapse

Henry R. Mahler

1. Introduction

A consideration of the current state of our knowledge (or should I say ignorance) of the glycoproteins in CNS synapses is pivotal to an understanding of the function of the complex carbohydrates of nerve tissues. Just as snyapses fulfill the crucial role in the establishment, construction, reconstruction, and reinforcement of neuronal pathways of communication, so do glycoproteins subserve the task of establishing and maintaining contact among cells and provide a framework for information transfer at the molecular level. Thus, they may provide key structural and sensory elements not only in the maintenance of synaptic contacts but also in their establishment during synaptogenesis and in their possible modification as a result of experience.

In this chapter, I plan to be selective rather than exhaustive regarding the voluminous literature surveyed. I shall stress advances in methodology and those experimental findings that have served to provide conceptual progress rather than mere documentation.

2. Current Status

2.1. Methodology

Major advances in this area, as is generally true of experimental disciplines, have been dependent on and greatly aided by the development of novel methodological approaches. These have consisted of both preparative and analytical techniques. The former are concerned principally with the isolation of relatively pure synaptosomal (frequently also called synaptic) plasma membranes (SPMs), retaining morphologically intact synaptic clefts and postsynaptic adhesions, useful and interesting both in their own right and as

Henry R. Mahler • Molecular and Cellular Biology Program and Department of Chemistry, Indiana University, Bloomington, Indiana.

starting materials for the preparation of synaptic junctional complexes (SJCs) [or, simply, synaptic junctions (SJs)] and their characteristic postsynaptic elements, such as the postsynaptic junctional lattice or its most characteristic feature, the postsynaptic density (PSD).

2.1.1. Isolation of Synaptosomal Plasma Membranes and Junctional Membranes

Methods for the isolation and characterization of highly purified SPMs and their properties were recently reviewed by Cotman (1974). Morgan (1976), Morgan *et al.* (1977*b*), Mahler (1977), and Matus and Taff-Jones (1978), and therefore will not be considered here. It should be mentioned, however, that retention of cleft and postsynaptic structures is favored by the inclusion of Ca^{2+} (50 μM) in homogenization and fractionation media and by the substitution of 4-(2-hydroxyethyl-1-piperazine ethane sulfonic acid (HEPES), pH 7.4, for other buffers.

The isolation of SJCs is based on the observation, originally reported by Fiszer and de Robertis (1967), that the nonionic detergent Triton X-100, at relatively low concentrations, selectively disperses membranes not attached to postsynaptic adhesions and can thus be used for the isolation of synaptic complexes. Methods for the isolation of these structures, and for their characterization and properties, which have been, or should be, useful for the studies of synaptic glycoproteins, were described by Davis and Bloom (1973), by Cotman and Taylor (1972), slightly modified by Kelly and Cotman (1977), and by Matus and Taff-Jones (1978).

This process of selective solubilization of membrane constituents can be extended to those of the junction itself, with the presynaptic membrane the most susceptible target. Depending on the nature and the concentration of detergent used, various parts of the cleft and postsynaptic specializations can be removed as well, resulting in preparations that consist almost exclusively of PSDs [sodium lauroyl sarcosinate (Cotman *et al.*, 1974; Kelly and Cotman, 1977, 1978)], PSDs with the subsynaptic web and junctional lattice retained [sodium deoxycholate (Walters and Matus, 1975*a,b*)], and finally, virtually the complete postsynaptic apparatus with only part of its membrane removed [Triton X-100 (Cohen *et al.*, 1977; Blomberg *et al.*, 1977; Matus and Taff-Jones, 1978)]. Matus and Taff-Jones subjected these procedures and the resulting structures to critical examination and demonstrated that 10-min treatments of SPMs with 3% detergents at 4°C results in the retention in the "junctional fraction" of 0.4 ± 0.05, 1.66 ± 0.19, and 6.80 ± 0.42 mg/100 mg of the starting material for Na^+-lauroyl sarcosinate, deoxycholate, and Triton X-100, respectively. The junctional fractions were isolated in each instance by subjecting the total detergent extract to density gradient centrifugation (adding 2 vol. 48 wt. % sucrose and centrifuging at $100{,}000g \times$ 45 min). This treatment pellets junctional material, leaving nonjunctional, insoluble membranes floating on the top. This fraction accounts for 6.6, 6.8, and 14 mg/100 mg of the starting SPMs.

However, although the morphology and purity of these preparations appear satisfactory (e.g., see evaluations by Morgan *et al.*, 1977*b*, Matus and Taff-Jones, 1978) and are probably susceptible to further improvement by additional refinements, the use of detergents at the concentrations required for solubilization of adhering or contaminating membranes carries with it an inherent disadvantage. It simultaneously leads to the solubilization (and often the inactivation) of some of the potentially most interesting integral proteins of the postsynaptic membrane—many of them glycoproteins—such as receptors for neurotransmitters (e.g., see Salvaterra and Mahler, 1976; McQuarrie *et al.*, 1976), proteins or subunits involved in transporting solutes across the membranes, such as the $[Na^+-K^+]$ATPase (Guidotti, 1976), or in the synthesis of cyclic AMP, and other proteins. Attempts are therefore being made in a number of laboratories to circumvent the restriction imposed by the detergent treatments. One such, communicated in abstract form by Bittiger (1976), makes use of mechanical shear at high ionic strength for the separation of components, followed by adsorption of postsynaptic structures to the lectin concanavalin A (Con A) bound to sepharose gels and eluted with the specific hapten α-methyl-D-mannoside (see next section). Another method under exploration in our laboratory (Crawford, in prep.) involves the sequential extraction of SPMs with EGTA, EDTA, and low concentrations of Nonidet P40; the separated pre- and postsynaptic membranes are then isolated on the basis of their different buoyant densities.

2.1.2. Separation and Identification of Glycoproteins: Use of Lectins

The usual method for the separation, analysis, and display of polypeptide subunits of membranes consists of their treatment with sodium dodecyl (lauroyl) sulfate (SDS) under strongly denaturing conditions (relative high concentrations of detergent and elevated temperatures, usually in the presence of a reducing agent), followed by electrophoresis on cross-linked polyacrylamide gels. Particularly when modified to the use of gel slabs with a stacking gel, and gradients (linear or exponential) of acrylamide in the separating gels, excellent resolution of the constituents of the SPM have been achieved (e.g., Wang and Mahler, 1976; Kelly and Cotman, 1977; Cohen *et al.*, 1977; Blomberg *et al.*, 1977; Mahler *et al.*, 1977; Matus and Taff-Jones, 1978). Care must be exercised, however, in the interpretation of the results obtained. Glycoproteins, because of their high negative charge density, do not bind SDS to the same extent as proteins devoid of such side chains, and hence their electrophoretic mobility, unlike that of standard proteins, is not a simple function of the logarithm of their molecular weight (for discussion, see Mahler, 1977).

The most useful and specific tools for the study of synaptic glycoproteins—and for other members of the species as well—are provided by techniques employing lectins [plant (hem)agglutinins, phyto (hem)agglutinins]. These proteins of plant origin, which are frequently though by no means always themselves glycoproteins [the most commonly used lectins such as Con A and wheat germ agglutinin (WGA) are simple proteins], recognize and are recognized by receptor regions of cognate glycoproteins, consisting of mono- or disaccharide units within specific constellations of the glycan moiety. Since these interactions are specifically antagonized by appropriate homo- or oligosaccharides at relatively high concentrations in free solution, they provide exquisite probes for the identification and analysis of glycoproteins. These properties (Table I) have been put to use in two ways: (1) for the identification on polyacrylamide gels of glycoproteins capable of binding specific lectins (Rostas *et al.*, 1977; Gurd, 1977*a*) and (2) for the separation of different glycoproteins solubilized from various populations of synaptic membranes by chromatography on lectin affinity columns prepared by attaching lectins covalently to an inert support medium, usually Sepharose gels. A number of recent publications (Susz *et al.*, 1973; Gurd and Mahler, 1974; Gurd, 1977*a,b*; Gombos and Zanetta, 1977; Zanetta *et al.*, 1975, 1977*b*) describe this application of lectin technology: for maximal resolution, columns of different lectins can be used not only in parallel but also—since the same glycoproteins can carry recognition sites for more than one lectin—in series.

2.1.3. *Quantitative Analysis of Carbohydrate Constituents*

The most sensitive and accurate contemporary methods all make use of gas–liquid chromatography (GLC) of appropriate sugar derivatives, subsequent either to monomerization (and conversion to *O*-methyl glycosides) of the glycan by methanolysis in the presence of 0.5 M HCl at 80° C or to its hydrolysis with 3 M HCl at 100° C. In the first instance, the derivatives used for GLC are the trifluoroacetate esters (Gombos and Zanetta, 1977; Zanetta *et al.*, 1972); in the second, the corresponding acetate derivatives (Churchill *et al.*, 1976).

2.2. Studies on Intact Junctions in Synaptosomal Plasma Membranes

2.2.1. *Localization in Situ*

The most direct demonstration of the presence of glycoproteins in intact CNS synapses would be their unambiguous localization on electron micrographs by means of cytochemical techniques. Such a demonstration has not yet been possible, but two types of inferential evidence are available. The first is based on the use of histochemical stains, such as the periodic acid–Schiff (PAS) reaction, or colloidal iron. Such studies were first performed by Rambourg and LeBlond (1967) in 1967 and confirmed and extended by Bondareff and Sjöstrand (1969) and, in our laboratory, by McBride *et al.* (1970). They

Table I. Useful Lectins [a]

Name		Abbreviation	$M_R \times 10^{-3}$	Specificity (monosaccharide hapten)
Concanavalin A (from *Canavalia ensiformis*)		Con A	32	α-Me-D-mannoside, α-Me-D-glucoside
Lens culinaris hemagglutinin		LcH	23.5	α-Me-D-mannoside, α-Me-D-glucoside
Wheat germ agglutinin		WGA	17	N'-acetylglucosamine
Ricinus communis agglutinin	I	RCA I	120 ($\alpha_2\beta_2$)	D-galactose, D-galactosamine
	II	RCA II	60 (α β')	L-rhamnose, D-galactose; not D-galactosamine
Ulex europeus lectin	I	UEL I	170	L-fucose
	II	UEL II		(*bis-N*-acetyl)chitobiose

[a] Data and references in Lis and Sharon (1973), Gurd (1977*a*), and Gombos and Zanetta (1977).

served to indicate that glycoproteins in the CNS appear to be prominent at synaptic junctions, including those adhering to isolated synaptosomes.

More convincing is the second line of evidence, which has provided striking displays of lectins attached to sites in junctional regions of synaptosomal preparations of rodent brain (derived from cortex, hippocampus, and cerebellum) using ferritin conjugates of the ligand for purposes of visualization.

Such experiments were first described briefly in 1973–1974 by Matus *et al.* (1973), by Bittinger and Schnebli (1974), and by Cotman and Taylor (1974), and more extensive studies were published by Kelly *et al.* (1976) and Matus and Walters (1976) (see also Matus, 1977). They are in essential agreement in demonstrating: (1) The existence of binding sites for Con A and for ricin, the lectin from *Ricinus communis*, on the external membrane of both pre- and postsynaptic, and of extrajunctional as well as partially opened or open, junctional regions, but with a preponderance of Con-A-binding sites concentrated in the junctions. (2) Virtually free lateral mobility of receptors in extrajunctional membranes, with their mobility greatly restricted in junctional, and in particular in postsynaptic, membranes, even when the latter are no longer attached to presynaptic elements. This topographic restriction may be due either to intramolecular (aggregation) or intermolecular interactions (linkage to fibrous elements, intrinsic or attached to the strongly stainable postsynaptic density). (3) These topographies are those characteristic of synaptosomes derived from Type I (excitatory) synapses. Among the other characteristic features of such synapses are the presence of spherical vesicles in the presynaptic space, and of 80- to 90-Å particles and a postsynaptic junctional lattice attached to the postsynaptic membrane. In contrast, Type II (inhibitory) synapses contain flattened storage vesicles and exhibit little structure in and no significant stainable postsynaptic material. However, this whole region appears accessible to and capable of binding Con A–ferritin conjugates.

2.2.2. *Identification of Glycoprotein Constituents of Synaptosomal Membranes*

A number of different methods have been used for the demonstration of the presence of side-chain carbohydrates in the glycoproteins of synaptosomal membranes. Those using lectins have already been described in Section 2.1.2.; briefly, they measure lectin binding either to the intact membrane (Gurd, 1977 *a,b*) or to its constituent proteins or polypeptides. In the case of the intact membrane, membrane proteins are first solubilized by detergent treatment [deoxycholate (Gurd and Mahler, 1974), SDS in relatively low concentration (Zanetta *et al.*, 1975, 1977*b*), or a combination of these two agents (Susz *et al.*, 1973)], and the detergent extracts are then subjected to chromatography on lectin columns. In the case of the subunits, the membrane proteins are first dissociated, separated, and displayed according to their electrophoretic mobility on SDS gels and then reacted with the appropriate lectin, usually in its radioactive form (Gurd, 1977*a*; Kelly and Cotman, 1977; Rostas *et al.*, 1977). Extent of reaction is measured either by counting gel slices or by radioautography.

More classic methods include assays for aldehyde functions generated by periodate oxidation [mostly originating from terminal sialic (*N*-acetylneuraminic) acid (NeuAc) residues] and identified either by virtue of their reaction with the Schiff reagent, or radiochemically, subsequent to their reduction with $Na_3B[^3H]_4$. Free D-galactose and *N*-acetylgalactosamine (GalNAc) termini can be identified in an analogous fashion by sequential oxidation with galactose oxidase and reduction with $Na_3B[^3H]_4$. Finally, the presence of fucosyl residues has been assessed by using this sugar as a precursor *in vivo* followed by isolation and analysis of the labeled membrane. The results of a number of investigations using these various techniques are summarized in Table II. The composition of the glycan moieties of the total SPM glycoproteins is shown in Table III.

The conclusions from the studies summarized in Tables II and III and from related investigations (e.g., Mahadik *et al.*, 1978; Chiu and Babitch, 1978) are as follows: (1) Synaptosomal (frequently called synaptic) plasma membranes (SPMs) contain a large number of glycoprotein subunit classes varying widely in their electrophoretic mobility and hence, presumably, in their molecular weights. Among them, eight, here designated as gp 180, 140–150, 120, 90–100, 65–75, 53, 33, and 26 are the most prominent and consistently identified in different laboratories. (2) Each of these mobility classes of glycoproteins is itself heterodisperse and consists of a number of species differing in their side chain composition. As many as a total of 39 individual species may in fact be present in the usual SPM preparation (Zanetta *et al.*, 1977*b*). The origin of this heterogeneity has been discussed explicitly (Gurd and Mahler, 1974; Zanetta *et al.*, 1977*b*); part of it may be referable to microheterogeneity in the glycan moiety, and part is ascribable to the heterogeneity of the synapses [axodendritic vs. axosomatic, different transmitters, excitatory (Gray's Type I) vs. inhibitory (Gray's Type II)]. Some contribution to the heterogeneity by contaminating structures can also not be ruled out, particularly in the case of minority species. (3) Most species contain mannosyl, galactosyl (and *N*-acetylgalactosaminyl), and *N'*-acetylneuraminic acid residues in their glycan moiety; fucosyl residues are present as well in many, but not all, of these glycoproteins. The proportion of these monomers varies, however, from class to class and species to species. (4) Purified glycoproteins appear to contain separate and distinct oligosaccharide recognition and binding sites (receptors) for different lectins (no competition even when the lectins are capable of reacting with the same monosaccharide such as Con A and LcH). However, one and the same glycan (and glycoprotein) may contain receptors for more than one lectin [e.g., for LcH and WGA (Gurd and Mahler, 1974; Gurd, 1977*a*)]. (5) In contrast, lectins can compete for binding

Table II. Glycoproteins of Purified Synaptic Membranes

Designation[k]	Method[a]: Ref.:	(1) *b*	(1) *c*	(1) *d*	(1) *e–g*	(2) *g*	(3) *g*	(3) *h*	(4) *f*	(5) *i*	(5) *e,f*	(6) *j*	(7) *e,f*	(8) *e,f*	(9) *b*	(10) *b*
gp >200		±	+		±	±	+		±			±			+	
gp 180		++	+			+	+		+		++	+	±	++	++	
gp 140–150		++	+		+	++	+++	+		+	++	++		++	++	++
gp 120		+	++	++	+				++	++		++	++	++	++	++
gp 90–100		±	+		++	++	++	+	++	+++	++	+	++	++	+	
gp 65–75		+	++++	++	++	+++	+++	+++	+++	++	+	++	++++	+++	++	+
gp 60		++										+++		++	+++	
gp 53		++	+++		+++	+++	+++	+++	++	++++	+++	+++		++	+++	++
gp 43		+		++							++	+++	+	+	+++	+
gp 33		±	++	++	+	+++	++++	+	+		++		+	+		+
gp 26		++		+	+			+			+++	+++			+++	+
gp 13								+								++
gp 11																++

[a] Methods: (1) PAS stain; (2) periodic acid, followed by $NaB[^3H]_4$—for sialic acid; (3) galactose oxidase, followed by $NaB[^3H]_4$—for galactose and galactosamine; (4) incorporation of fucose *in vivo*; (5) Con A binding on gels; (6) chromatography on Con A column; (7) WGA binding on gels; (8) RCA binding on gels; (9) chromatography on UEL, also adsorbed on Con A; (10) same as (9), but not adsorbed on Con A.

[b–j] References: [b] Zanetta *et al.* (1977*b*); [c] Banker *et al.* (1972); [d] Morgan *et al.* (1973*a,c*); [e] Gurd (1977*a*); [f] Gurd and Mahler (1974); [g] Wang and Mahler (1977); [h] Rapport and Mahadik (1977); [i] Kelly and Cotman (1977); [j] Zanetta *et al.* (1975)

[k] The designations shown are "gp" for glycoprotein, followed by the ostensible approximate molecular weight—ostensible since glycoproteins exhibit anomalous mobilities in the SDS–acrylamide gels used for these assignments; because of variations in the actual systems used, the molecular-weight assignments are only approximate even for "normal" proteins (for discussion, see Mahler, 1977).

Table III. Carbohydrate Composition of Synaptic Membrane Glycoproteins[a]

Sugar	Prep.: Ref.:	SPM *c*	SPM *d*	SPM *e*	SPM *f*	SPM *g*	SPM-C[b] *c*	SPM-CU[b] *c*	SPM-U[b] *c*
Fuc		26	18	18	8.5	14	0.0	174	67
		(0.29)	(0.28)	(0.17)	(0.29)	(0.42)	(0.00)	(0.46)	(0.36)
Man		98.3	38	43	54	48	325	841	112
		(1.09)	(0.60)	(0.40)	(1.86)	(1.43)	(1.20)	(2.22)	(0.60)
Gal		53	28	29	15	28	230	166	141
		(0.59)	(0.44)	(0.27)	(0.52)	(0.84)	(0.85)	(0.44)	(0.76)
GluNAc		90.2	63	108	29	34	271	379	186
		(1.00)	(1.00)	(1.00)	(1.00)	(1.00)	(1.00)	(1.00)	(1.00)
GalNAC		11	13	23	1.4	2.8	60	49	28
		(0.12)	(0.20)	(0.20)	(0.050)	(0.08)	(0.22)	(0.13)	(0.15)
NeuNAc		44	47	35	9.9	29	135	117	78
		(0.49)	(0.75)	(0.32)	(0.34)	(0.87)	(0.50)	(0.31)	(0.42)
TOTALS:		323	207	256	118	156	1022	1726	612

[a] All values exclusive of glucose, expressed either as nanomoles per milligram protein or, in parentheses, relative to *N*-acetylglucosamine (GlcNAc) set equal to 1.00.
[b] (SPM-C) extracts (0.08% SDS) retained on Con A but not on UEL columns; (SPM-CU) retained on both; (SPM-U) retained on UEL, but not on Con A.
[c] Zanetta *et al.* (1977*b*): membranes prepared according to Morgan *et al.* (1973*c*); analyzed by GLC of trifluoroacetates (see Section 2.1.3).
[d] Churchill *et al.* (1976): membranes prepared according to Cotman and Taylor (1972) as modified by Cotman *et al.* (1974); analyzed by GLC of alditol acetates.
[e] Breckenridge *et al.* (1972): membranes (fractions F+G) prepared according to Whittaker *et al.* (1964); analysis by chromatography on polycarbonate. Values are the averages for the two fractions.
[f] R. K. Margolis *et al.* (1975): membranes prepared according to Gurd *et al.* (1974). Values are expressed as nanomoles sugar per milligram lipid-free dry weight.
[g] Krusius *et al.* (1978): membranes prepared according to Jones and Matus (1974) as modified by Walters and Matus (1975*c*); analyzed by GLC of trimethylsilyl derivatives. Values are expressed as nanomoles sugar per milligram lipid-free dry weight.

sites as long as glycoproteins remain inserted in the membrane. It would be interesting to establish whether this kind of competition is retained in isolated synaptic junctions. Since it has been demonstrated that junctional lectin receptors are relatively immobile (see above), such a finding would indicate the existence of definite topographical constraints between different glycoproteins in the junction and permit inferences concerning the molecular organization of the synapse.

2.2.3. Glycoproteins in Subsynaptosomal Membranes

Synaptosomal membranes may be subfractionated further by a number of techniques, most of them described in Section 2.1. Conversion of synaptic membranes to preparations of synaptic junctions (SJCs) and postsynaptic densities (PSDs) results in a diminution of the amount of membrane-associated glycoproteins and in a considerable simplification of the pattern of these constituents (Tables IV and V). This is a surprising observation and

Table IV. Carbohydrate Content of Subsynaptosomal Structures[a]

Sugar	Struct.: Ref.:	SJC *b*	SJC *c*	SJC *d*	SJC *e*	SV *d*	PSD *c*
Fuc		11	44,19	12	2.4	13	12,21
		(0.38)	(0.98)(0.42)	(0.77)	(0.24)	(0.41)	
Man		35	43	21	40	35	154,106
		(1.17)	(0.95)	(1.3)	(4.0)	(1.07)	
Gal		9.4	44	145(!)	4.5	30	46,16
		(0.31)	(0.98)	(9.2)	(0.45)	(0.94)	
GlcNAc		30	45	16	10	32	41,52
		(1.00)	(1.00)	(1.00)	(1.00)	(1.00)	
GalNAc		10	20	2.4	0.23	4.9	5,19
		(0.33)	(0.44)	(0.19)	(0.023)	(0.15)	
NeuNAc		17	19	3.0	3.1	11	223,391
		(0.57)	(0.42)	(0.19)	(0.31)	(0.35)	
TOTALS:		112	202	(200)	60.2	127	215,258

[a] All values exclusive of glucose, expressed either as nonomoles per milligram protein or, in parentheses, relative to *N*-acetylglucosamine (GalNAc) set equal to 1.00.
[b] Recalculated from Vincendon *et al.* (1973).
[c] See reference *d* of Table III.
[d] Recalculated from Morgan and Gombos (1976); see also Zanetta *et al.* (1975, 1977*b*).
[e] R. K. Margolis *et al.* (1975); see reference *f* of Table III.

Table V. Glycoproteins in Junctional Membranes

SPM glycoproteins (see Table II)	Prep.: Ref.:	SJC *a*	SJC-C *b*	SJC-G *b*	PSD-C *c*	PSD-L *c*	PSD-W *c*	PSD *d*
gp 180		+++	+++	++	++	+++	++	+
gp 140–150		++[e]		+				
gp 120		++++	+++	++	+++	++	+++	+++
gp 90–100		+++	+++	++++	+	++	++	++
gp 65–75		±		±	±	++	+	
gp 53		+	±	±				
gp 43		+		++				
gp 33				++				
gp 26		++		++				

[a] Gurd (1977*a,b*): Con-A-binding proteins.
[b] Kelly and Cotman (1977): (C) Con-A-binding (G) galactose oxidase, $B[^3H]_4^-$ labeling; the M_R's quoted by these authors are 160–165, 123, 108–118, and 95–100 × 10^3 for (C).
[c] Gurd (1977*b*): (C) Con A, (L) LcH, and (W) WGA binding components.
[d] Kelly and Cotman (1977) Con-A-binding; specific binding activity much lower (<30%) than to SJCs.
[e] Two components.

aparently inconsistent with the localization and possible functional significance of glycoproteins in the synaptic cleft proper (see Section 2.1.1). The resolution of this discrepancy is probably to be sought in the mode of preparation of the subsynaptic membranes (see Section 2.1); the detergents used in their isolation have also served to extract some—and perhaps many—of their glycoproteins. A particular case in point is that of the α-bungarotoxin receptor, a glycoprotein (Salvaterra *et al.*, 1977), which can be extracted from synaptic membranes by concentrations of Triton X-100 (Salvaterra and Mahler, 1976; McQuarrie *et al.*, 1976, 1978) lower than those used for the preparation of subsynaptic membranes.

It may be appropriate at this point to mention additional, functionally important, synaptic entities that have been shown to be glycoproteins. These include acetylcholinesterase (Wenthold *et al.*, 1974; Stefanovic *et al.*, 1975; Gurd, 1976); an L-glutamate-binding protein, perhaps identical with the receptor for this amino acid (Michaelis *et al.*, 1974); and $[Na^+–K^+]$ATPase (Dahl and Hokin, 1974; Guidotti, 1976). Another interesting entity with an as yet unknown function is GP-350, a specific synaptosomal sialoglycoprotein that may be a peripheral membrane protein (Van Nieuw Amerongen and Roukema, 1974). Finally, Simpson *et al.* (1976) presented evidence for the presence in synaptic membranes of some species of sulfated glycoproteins of high molecular weight, not identical with the single member of this class found in myelin or with the known glycosaminoglycans (Margolis, R. K., *et al.*, 1975; Margolis, R. U., and Margolis, R. K., 1977).

Turning next to intrasynaptosomal entities, synaptic vesicles (SVs) were shown by Morgan *et al.* (1973*a*) (see also Zanetta and Gombos, 1974) to contain a subset of the glycoproteins of the synaptic membranes—specifically gp 140–150, gp 120, gp 65–75, and gp 43 (Table II). This observation is consistent with the hypothesis of the release of transmitter from vesicles by exocytosis coincident with membrane fusion.* If this hypothesis is correct, it would demand a topology for these constituents opposite that found in synaptic membranes, i.e., facing toward the *interior*. Such a reversal of topology was actually found (Mahler *et al.*, 1977) for some proteins of the vesicle, but remains to be established for its glycoproteins.

Intrasynaptic mitochondria appear to be devoid of glycoproteins (Zanetta *et al.*, 1977*a*).

2.3. Biosynthesis and Plasticity: Biosynthesis of Synaptic Glycoproteins

Most studies are in agreement in suggesting that the bulk, and probably all, of the protein moieties of presynaptic glycoproteins are synthesized in the cell body and transported down the axon by fast axoplasmic flow (see Morgan

*The vesicles actually primed for or engaged in release are those in the active zone (Heuser, 1977; Llinas and Heuser, 1978; Fried and Blaustein, 1978) and may constitute only a minority subpopulation, possibly different in composition from the majority.

et al., 1973*b*, 1977*a*; Barondes, 1974; Margolis, R. U., and Margolis, R. K., 1977; Gurd, 1978). The mode and kinetics of these processes therefore do not differ substantively from those of other unglycosylated synaptic membrane proteins (Gurd, 1978). Evidence for local synthesis is weak and can probably be accounted for either by residual synthetic capacities inherent in, or adherent to, postsynaptic structures, or by slight contamination with elements of the rough endoplasmic reticulum. The question whether the same considerations apply to the glycan moiety is not answered as easily. The bulk of the glycosylation probably occurs at or close to the site of polypeptide synthesis as determined by double or simultaneous labeling techniques using either labeled glucosamine (Morgan *et al.*, 1973*b*) or fucose (Gurd and Mahler, 1974; Gurd, J. W., personal communication) as a tracer. But experiments of this type are not sufficiently precise to address the important problem of possible junctional modification of the side chains of a (restricted) class of synaptic glycoproteins. In principle, it should be possible to answer this question by experiments with isolated synaptic fractions *in vitro*. However, even if such modification of specific glycoproteins is found, it will be very difficult to rule out its origin in catalysis by trace amounts of contaminating enzymes from the Golgi apparatus or other membranes (Reith *et al.*, 1972). The evidence is good for the localization in synaptosomes and SPMs of at least one enzyme potentially capable of modifying its intrinsic glycoproteins (Yohe and Rosenberg, 1977; Tettamanti *et al.*, 1972, 1975). Although the principal targets of this enzyme, the membrane-associated *N*-acetylneuraminidase, appear to be the gangliosides of the membrane, particularly at physiological pH, membrane-bound sialoproteins are susceptible to its attack as well. Since the content of NeuAc may be particularly high in nerve endings of young rats [40 μg/mg protein according to Dicesare and Rapport (1973), but see Yohe and Rosenberg (1977)], the reaction may be of some significance in synapse formation (see Section 2.4). Terminal *N*′-acetylneuraminic acid residues of synaptoplasmic, i.e., soluble, proteins also appear capable of rapid turnover (Margolis, R. K., *et al.*, 1975).

So far, the most important aspect of the whole problem of biosynthesis, the mode and mechanism of construction of the synaptic membrane from its constituent biosynthetic intermediates, is a virtually unknown quantity, as is the nature of these intermediates—but, unfortunately, this holds true for all membrane proteins and not just its glycoproteins. However, the severity of the problem is accentuated in this instance, for glycoproteins are localized on the exterior surface, and the insertion of such proteins is now believed to be tightly coupled to their synthesis (Bretscher and Raff, 1975; Rothman and Lenard, 1977). An interesting lead toward a resolution of this problem is the suggestion by Marchisio *et al.* (1975) that embryonic chick optic nerves transport glycoproteins along the axolemma; thus, there may exist two means for somatosynaptic transport: an axo*lemmal* in addition to the traditional axo*plasmic* mode.

2.4. Glycoproteins in Recognition and Development: General Considerations and Potential Leads

The hypothesis that glycoproteins play a determinative role in cell–cell recognition, adhesion, and formation during synaptogenesis (Brunngraber, 1969; Barondes, 1970; Moscona, 1974; Barondes and Rosen, 1976) is an appealing one. Its conceptual framework rests on the following assumptions: (1) the existence of sets of complementary "recognition molecules," probably of a proteinaceous nature, on the surfaces of participating cells; (2) the participation of cell-surface glycoproteins in such recognition processes in other systems; (3) the presence of glycoproteins on the pre- and postsynaptic surface of the cleft; and (4) the possibility that at least some of these glycoproteins constitute the sets of recognition molecules postulated under (1), the construction of such sets and their specificity having been achieved by virtue of modification of one of the partners by its complement. The conceptual and experimental bases for (1) and (2) were reviewed in several articles in Barondes (1976), as well as by Oseroff *et al.* (1973), Cook and Stoddard (1973), Sharon (1975), and Hughes (1976). Among various aggregation phenomena in neural tissues (e.g., see Seeds, 1973; Schubert *et al.*, 1973) that might serve as clues or models for the ontogenetic events, the most persuasive are specific retinotectal interactions of dissociated embryonic chick brain tissue observable *in vitro* (Garfield *et al.*, 1974; Edds *et al.*, 1973). In this system, a beginning has been made toward the isolation, purification, and characterization of distinct molecular species capable of acting as adhesion molecules; one such molecule, capable of bringing about the adhesion of retinal cells, was purified from this source by immunochemical techniques by Edelman and his collaborators (Edelman, 1976; Brackenbury *et al.*, 1977; Thiery *et al.*, 1977). This is a protein with a molecular weight of 140,000 that probably has to be cleaved proteolytically to render it functional for adhesion (Edelman, 1976). It is not yet known whether this molecule is a glycoprotein, as is the case for the entities on the cell surface of cellular slime molds implicated in analogous processes (Barondes and Rosen, 1976; Gerisch, 1977). However, very recently evidence has been provided for the presence of a set of developmentally regulated glycoproteins in this system (Mintz and Glaser, 1978). Even less is known concerning the structure and function of the lectin activity found in extracts of embryonic chick and rat brain, the nature of which appears to be developmentally regulated (Kobiler and Barondes, 1978; Kobiler *et al.*, 1978; Simpson *et al.*, 1977).

Another system of great potential resolving power is provided by the postnatally developing cerebellar cortex of the rat (Altman, 1972*a–c*; Eccles *et al.*, 1967). In this system, a massive and transient accumulation of glycoproteins, capable of reacting with Con A, was demonstrated in the course of development and shown to be specifically correlated with the maturation of and synapse formation by parallel fibers (Zanetta *et al.*, 1978; Gombos *et al.*,

1977). The experimental foundation for item (4) is unfortunately virtually nonexistent, mainly for the reasons briefly alluded to in Section 2.2.3. The existence of such modification systems, constituted by membrane-bound glycosyltransferases, was first suggested by Roseman (1970) and reviewed by Roth (1973) and Shur and Roth (1975); among the systems studied by Roth and his collaborators was a galactosyltransferase of neural retina. However, critical appraisal of past claims concerning an association of glycosyltransferases specifically with synaptic membranes appears to provide no basis for such localization (Morgan *et al.*, 1977*a*). The evidence for the association of a neuraminidase with, and activity on, SPMs was mentioned in Section 2.3.1.

3. Outlook

From the brief description in Section 2, the reader will perhaps appreciate that research on synaptic glycoproteins is on the verge of coming of age. We are reasonably certain that synaptic junctions contain glycoproteins that are located on both their pre- and postsynaptic membrane, and the glycan side chains of which penetrate into the cleft. We also have some notion just what kind of glycoproteins these are. What we do not know is the nature and disposition of the actual molecules that are to be found in *any* given homogenous population of synapses defined as to source, location, and transmitter type. Nor do we know the structure or topography of *any* single, homogenous glycoprotein molecule. Yet, these are the details that are essential for any advance from the general to the molecular, from following purely descriptive to the design of interventive strategies.

What this will require is probably the development of new approaches and experimental systems, and the application to them of the most incisive and rigorous techniques of carbohydrate and protein chemistry, of membrane biochemistry and biophysics, and particularly of immunology, to the fractionation, characterization, localization, and functional assay of selected members among the class of synaptic glycoproteins.

Only in this manner will we be able to probe the outstanding, fundamental questions concerning the nature and function of these entities that we have postulated in the introduction and throughout the chapter to be essential for the intercommunication of nerve cells: Are presynaptic and postsynaptic glycoproteins capable of mutual recognition and interaction? If so, is the basis of this capability complementariness or identity (the antibody–antigen vs. the oligomeric protein analogy); if not, do they interact with other kinds of receptors on the opposite membrane? What is the mode of transfer of the protein and glycan moiety from their respective sites of synthesis to the synapse; what is their mode of insertion? Are they coupled in an obligatory manner, coordinated, relatively independent? Is there processing of either component prior to, in the course of, or subsequent to the insertion of the molecule in the membrane? Do glycoproteins participate in any of the functional aspects of the synapse? What is their role, if any, during depolarization, exocytosis, and transmitter release; during vesicle retrieval? Do they play any part in the interaction of transmitters with their postsynaptic receptors and the attendant change in membrane permeability? Are they, or their dispositions, altered as a result of repeated firing? Do they participate in recognition and synaptogenesis during ontogeny?

ACKNOWLEDGMENTS. Dr. Mahler is the recipient of Research Career Award K06 05060 from National Institute of General Medical Sciences, National Institutes of Health. Experimental work supported by Research Grant NS 08309 from the National Institute of Neurological and Communicative Disorders and Stroke, National Institutes of Health, Bethesda, Maryland. This is Publication No. 3196 from the Department of Chemistry. I want to thank the large number of colleagues who have contributed to the information summarized in these pages, and in particular Profs. S. Barondes, C. Cotman, G. Gombos, P. Mandel, A. Matus, I. G. Morgan, M. Rapport, and A. Rosenberg for providing me with a collection of their manuscripts, published and some as yet unpublished. My special thanks go to all current and recent members of our brain research group, especially Drs. Paul Salvaterra, Yng-Jiin Wang, and Carole McQuarrie, for their experimental work, and for the lively exchange of ideas made possible by their presence, but above all to Prof. James Gurd, whose all too brief stay in my laboratory served to initiate me and all of us to the challenge, mystery, and frustrations of working with and thinking about glycoproteins.

References

Altman, J., 1972*a*, Postnatal development of the cerebellar cortex in the rat. I. The external germinal layer and the transitional molecular layer, *J. Comp. Neurol.* **145**:353–398.

Altman, J., 1972*b*, Postnatal development of the cerebellar cortex in the rat. II. Phases in the maturation of Purkinje cells and of the molecular layer, *J. Comp. Neurol.* **145**:399–464.

Altman, J., 1972*c*, Postnatal development of the cerebellar cortex in the rat. III. Maturation of the components of the granular layer, *J. Comp. Neurol.* **145**:465–514.

Banker, G., Crain, B., and Cotman, C. W., 1972, Molecular weights of the polypeptide chains of synaptic plasma membranes, *Brain Res.* **42**:508–513.

Barondes, S. H., 1970, Brain glycomacromolecules and interneuronal recognition, in: *The Neurosciences: Second Study Program* (F. O. Schmitt, ed.), pp. 747–760, Rockefeller University Press, New York.

Barondes, S. H., 1974, Synaptic macromolecules: Identification and metabolism, *Annu. Rev. Biochem.* **43**:147–168.

Barondes, S. H., 1976 (ed.), *Neuronal Recognition,* Plenum Press, New York.

Barondes, S. H., and Rosen, S. D., 1976, Cell surface carbohydrate-binding proteins: Role in cell recognition, in: *Neuronal Recognition* (S. H. Barondes, ed.), pp. 331–356, Plenum Press, New York.

Bittiger, H., 1976, Fractionation of rat brain by affinity binding on concanavalin A–sepharose: Separation of mitochondria and postsynaptic densities, *Brain Res.* **24**:5–6.

Bittiger, H., and Schnebli, H. P., 1974, Binding of concanavalin A and ricin to synaptic junctions of rat brain, *Nature* (*London*) **249**:370–371.

Blomberg, F., Cohen, R. S., and Siekevitz, P., 1977, The structure of postsynaptic densities isolated from dog cerebral cortex. II. Characterization and arrangement of some of the major proteins within the structure, *J. Cell Biol.* **74**:204–225.

Bondareff, S. and Sjöstrand, J., 1969, Cytochemistry of synaptosomes, *Exp. Neurol.* **24**:450–458.

Brackenbury, R., Thiery, J.-P., Rutishauser, U., and Edelman, G. M., 1977, Adhesion among neural cells of the chick embryo. I. An immunological assay for molecules involved in cell–cell binding, *J. Biol. Chem.* **252**:6835–6840.

Breckenridge, W. C., Breckenridge, J. E., and Morgan, I. G., 1972, Glycoproteins of the synaptic region, *Adv. Exp. Med. Biol.* **32**:135–153.

Bretscher, M. S., and Raff, M. C., 1975, Mammalian plasma membranes, *Nature* (*London*) **258**:43–49.

Brunngraber, E. G., 1969, Possible role of glycoproteins in neural function, *Perspect. Biol. Med.* **12**:467–470.

Chiu, T.-C., and Babitch, J. A., 1978, Topography of glycoproteins: The chick synaptosomal plasma membrane, *Biochim. Biophys. Acta* **570**:112–123.

Churchill, L., Cotman, C., Banker, G., Kelly, P., and Shannon, L., 1976, Carbohydrate composition of central nervous system synapses: Analysis of isolated synaptic junctional complexes and postsynaptic densities, *Biochim. Biophys. Acta* **448**:57–72.

Cohen, R. S., Blomberg, F., Berzins, K., and Siekevitz, P., 1977, The structure of postsynaptic densities isolated from dog cerebral cortex. I. Overall morphology and protein composition, *J. Cell Biol.* **74**:181–203.

Cook, G. M. W., and Stoddard, R. W., 1973, *Surface Carbohydrates of the Eukaryotic Cell*, pp. 257–270, Academic Press, New York.

Cotman, C. W., 1974, Isolation of synaptosomal and synaptic plasma membrane fractions, *Methods Enzymol.* **31**:445–452.

Cotman, C. W., and Taylor, D., 1972, Isolation and structural studies on synaptic complexes from rat brain, *J. Cell Biol.* **55**:696–711.

Cotman, C. W., and Taylor, D., 1974, Localization and characterization of concanavalin A receptors in the synaptic cleft, *J. Cell Biol.* **62**:226–242.

Cotman, C. W., Banker, G., Churchill, L., and Taylor, D., 1974, Isolation of postsynaptic densities from rat brain, *J. Cell Biol.* **63**:441–455.

Dahl, J. L., and Hokin, L. E., 1974, The sodium–potassium adenosinetriphosphatase, *Annu. Rev. Biochem.* **43**:327–356.

Davis, G. A., and Bloom, F. E., 1973, Isolation of synaptic junctional complexes from rat brain, *Brain Res.* **62**:135–153.

Dicesare, J. L., and Rapport, M. M., 1973, Availability of neuraminidase of gangliosides and sialoglycoproteins in neuronal membranes, *J. Neurochem.* **20**:1781–1783.

Eccles, J. C., Ito, M., and Szentagothai, J., 1967, *The Cerebellum as a Neuronal Machine*, Springer-Verlag, Berlin.

Edds, M. V., Barkley, D. S., and Fambrough, D. M. (eds.), 1973, *Genesis of Neuronal Patterns*, NRP Bulletin, Vol. 10, No. 3.

Edelman, G. M., 1976, Surface modulation in cell recognition and cell growth, *Science* **192**:218–226.

Fiszer, S., and de Robertis, E., 1967, Action of Triton X-100 on ultrastructure and membrane-bound enzymes of isolated nerve endings from rat brain, *Brain Res.* **5**:31–44.

Fried, R. C., and Blaustein, M. P., 1978, Retrieval and recycling of synaptic vesicle membrane in pinched-off nerve terminals (synaptosomes), *J. Cell Biol.* **78**:685–700.

Garfield, S., Hausmann, R. E., and Moscona, A. A., 1974, Embryonic cell aggregation: Absence of galactosyltransferase activity in retina-specific cell-aggregating factor, *Cell Differ.* **3:** 215–219.

Gerisch, G., 1977, Membrane sites implicated in cell adhesion: Their developmental control in *Dictyostelium discoideum,* in: *Cell Biology 1976–77* (B. R. Brinkley and K. P. Porter, eds.), pp. 36–42, Rockefeller University Press, New York.

Gombos, G., and Zanetta, J.-P., 1977, Recent methods for the separation and analysis of central nervous system glycoproteins, *Res. Methods Neurochem.* **4:**307–363.

Gombos, G., Chandour, M. S., Vincendon, G., Reeber, A., and Zanetta, J.-P., 1977, Formation of the neuronal circuitry of rat cerebellum: Plasma membrane modifications, in: *Maturation of Neurotransmission* (E. Giacobini, A. Vernadakis, and S. Karger, eds.), Symposium on Maturation Aspects of Neurotransmission Mechanisms, Saint Vincent (Aosta), Italy.

Guidotti, G., 1976, The structure of membrane transport systems, *Trends Biochem. Sci.* **1:**11–13.

Gurd, J. W., 1976, Fractionation of rat brain acetylcholinesterase by lectin affinity chromatography, *J. Neurochem.* **27:**1257–1259.

Gurd, J. W., 1977*a*, Synaptic plasma membrane glycoproteins: Molecular identification of lectin receptors, *Biochemistry* **16:**369–374.

Gurd, J. W., 1977*b*, Identification of lectin receptors associated with rat brain postsynaptic densities, *Brain Res.* **126:**154–159.

Gurd, J. W., 1978, Biosynthesis of synaptic membranes: Incorporation of [^{3}H]leucine into proteins and glycoproteins of rat brain synaptic plasma membranes, *Brain Res.* **142:**201–204.

Gurd, J. W., and Mahler, H. R., 1974, Fractionation of synaptic plasma membrane glycoproteins by lectin affinity chromatography, *Biochemistry* **13:**5193–5198.

Gurd, J. W., Jones, L. R., Mahler, H. R., and Moore, W. J., 1974, Isolation and partial characterization of rat brain synaptic plasma membranes, *J. Neurochem.* **22:**281–290.

Heuser, J. E., 1977, Synaptic vesicle exocytosis revealed in quick-frozen frog neuromuscular junctions treated with 4-amino-pyridine and given a single electrical shock, in: *Society for Neuroscience Symposia*, Vol. II, *Approaches to the Cell Biology of Neurons* (W. M. Cowan and J. A. Ferrendelli, eds.), pp. 215–239, Society for Neuroscience, Bethesda, Maryland.

Hughes, R. C., 1976, *Membrane Glycoproteins: A Review of Structure and Function*, Butterworths, London.

Jones, D. H., and Matus, A. I., 1974, Isolation of synaptic plasma membrane from brain by combined flotation–sedimentation density gradient centrifugation, *Biochim, Biophys. Acta* **356:**276–287.

Kelly, P. T., and Cotman, C. W., 1977, Identification of glycoproteins, proteins at synapses in the central nervous system, *J. Biol. Chem.* **252:**786–793.

Kelly, P. T., and Cotman, C. W., 1978, Synaptic proteins. Characterization of tubulin and actin and identification of a distinct postsynaptic density polypeptide, *J. Cell Biol.* **79:**173–183.

Kelly, P., Cotman, C. W., Gentry, C., and Nicolson, G. L., 1976, Distribution and mobility of lectin receptors on synaptic membranes of identified neurons in the central nervous system, *J. Cell Biol.* **71:**487–496.

Kobiler, D., and Barondes, S. H., 1978, Lectin activity from embryonic chick brain, heart and liver: Changes with development, *Dev. Biol.* **60:**326–330.

Kobiler, D., Beyer, E. C., and Barondes, S. H., 1978, Developmentally regulated lectins from chick muscle, brain, and liver have similar chemical and immunological properties, *Dev. Biol.* **64:**265–272.

Krusius, T., Finne, J., Margolis, R. U., and Margolis, R. K., 1978, Structural features of microsomal, synaptosomal, mitochondrial and soluble glycoproteins of brain, *Biochemistry* **17:**3849–3854.

Lis, H., and Sharon, N., 1973, The biochemistry of plant lectins (phytohemagglutinins), *Annu. Rev. Biochem.* **42:**541–574.

Llinas, R. R., and Heuser, J. E. (eds.), 1978, *Depolarization–Release Coupling Systems in Neurons*, MIT Press, Boston.

Mahadik, S. P., Hungund, B., and Rapport, M. M., 1978, Topographic studies of glycoproteins of intact synaptosomes from rat brain cortex, *Biochim. Biophys. Acta* **511**:240–250.

Mahler, H. R., 1977, Proteins of the synaptic membrane, *Neurochem. Res.* **2**:119–147.

Mahler, H. R., Wang, Y.-J., De Blas, A., and Crawford, G., 1977, Topography of membrane proteins at vertebrate synapses, in: *Mechanisms, Regulation and Special Functions of Protein Synthesis in the Brain* (S. Roberts, A. Lajtha, and W. H. Gispen, eds.), pp. 205–220, Elsevier/North-Holland Biomedical Press, New York.

Marchisio, P. C., Gremo, F., and Sjöstrand, J., 1975, Axonal transport in embryonic neurons: The possibility of a proximo–distal axolemmal transfer of glycoproteins, *Brain Res.* **85**:281–285.

Margolis, R. K., Margolis, R. U., Preti, C., and Lai, D., 1975, Distribution and metabolism of glycoproteins and glycosaminoglycans in subcellular fractions of brain, *Biochemistry* **14**:4797–4804.

Margolis, R. U., and Margolis, R. K., 1977, Metabolism and function of glycoproteins and glycosaminoglycans in nervous tissue, *Int. J. Biochem.* **8**:85–91.

Matus, A., 1977, The chemical synapse in structure and function, in: *Intracellular Junctions and Synapses in Development* (J. D. Feldman, N. B. Gilula, and J. O. Pitts, eds.), Chapman and Hall, London.

Matus, A. I., and Taff-Jones, D. H., 1978, Morphology and molecular composition of isolated postsynaptic junctional structures, *Proc. R. Soc. London Ser. B* **203**:135–151.

Matus, A. I., and Walters, B. B., 1976, Type 1 and 2 synaptic junctions: Differences in distribution of concanavalin A binding sites and stability of the junctional adhesion, *Brain Res.* **108**: 249–256.

Matus, A., De Petris, S., and Raff, M. C., 1973, Mobility of concanavalin A receptors in myelin and synaptic membranes, *Nature (London) New Biol.* **244**:278–279.

McBride, W. J., Mahler, H. R., Moore, W. J., and White, F. P., 1970, Isolation and characterization of membranes from rat cerebral cortex, *J. Neurobiol.* **2**:73–92.

McQuarrie, C., Salvaterra, P. M., De Blas, A., Routes, J., and Mahler, H. R., 1976, Studies on nicotinic acetylcholine receptors in mammalian brain. V. Preliminary characterization of membrane bound α-bungarotoxin receptors in rat cerebral cortex, *J. Biol. Chem.* **251**:6335–6339.

McQuarrie, C., Salvaterra, P. M., and Mahler, H. R., 1978, Studies on nicotinic acetylcholine acceptors in mammalian brain. Interaction of solubilized protein with cholinergic ligands, *J. Biol. Chem.* **253**:2743–1247.

Michaelis, E. K., Michaelis, M. L., and Boyarski, L. L., 1974, High-affinity glutamic acid binding to brain synaptic membranes, *Biochim. Biophys. Acta* **367**:338–348.

Mintz, G., and Glaser, L., 1978, Specific glycoprotein changes during development of the chick neural retina, *J. Cell Biol.* **79**:132–137.

Morgan, I. G., 1976, Synaptosomes and cell separation, *Neuroscience* **1**:159–165.

Morgan, I. G., and Gombos, G., 1976, Biochemical studies of synaptic macromolecules: Are there specific synaptic components?, in: *Neuronal Recognition* (S. H. Barondes, ed.), pp. 179–202, Plenum Press, New York.

Morgan, I. G., Breckenridge, W. C., Vincendon, G., and Gombos, G., 1973*a*, The proteins of nerve-ending membranes, in: *Proteins of the Nervous System* (D. J. Schneider, R. H. Angeletti, R. A. Bradshaw, A. Grasso, and B. W. Moore, eds.), pp. 171–192, Raven Press, New York.

Morgan, I. G., Marinari, U. M., Dutton, G. R., Vincendon, G., and Gombos, G., 1973*b*, The biosynthesis of synaptic glycoproteins, in: *Methodologie de la Structure et du Metabolisme des Glycoconjugues (Glycoproteins et Glycolipids)*, Vol. 2, pp. 1043–1055, CNRS, Paris.

Morgan, I. G., Zanetta, J.-P., Breckenridge, W. C., Vincendon, G., and Gombos, G., 1973*c*, The chemical structure of synaptic membranes, *Brain Res.* **62**:405–411.

Morgan, I. G., Gombos, G., and Tettamanti, G., 1977*a*, Glycoproteins and glycolipids of the nervous system, in: *The Glycoconjugates*, Vol. 1 (M. I. Horowitz and W. Pigman, eds.), pp. 351–383, Academic Press, New York.

Morgan, I. G., Woolstron, M. E., and Hambley, J. W., 1977*b*, Biochemical studies of synaptic macromolecules: The developing chick visual system, in: *Mechanisms, Regulation and Special Functions of Protein Synthesis in the Brain* (S. Roberts, A. Lajtha, and W. H. Gispen, eds.), pp. 231–246, Elsevier/North-Holland Biomedical Press, New York.

Moscona, A. A., 1974, Surface specification on embryonic cells: Lectin receptors, cell recognition, and specific cell ligands, in: *The Cell Surface in Development* (A. A. Moscona, ed.), pp. 67–99, John Wiley & Sons, New York.

Oseroff, A. R., Robbins, P. W., and Burger, M. M., 1973, The cell surface membrane: Biochemical aspects and biophysical probes, *Annu. Rev. Biochem.* **42**:648–682.

Rambourg, A., and LeBlond, C. P., 1967, Electron microscope observations on the carbohydrate-rich cell coat present at the surface of cells in the rat, *J. Cell Biol.* **32**:27–53.

Rapport, M. M., and Mahadik, S. P., 1977, Topographic studies of glycoproteins and gangliosides in the surface of intact synaptosomes, in: *Mechanisms, Regulation and Special Functions of Protein Synthesis in the Brain* (S. Roberts, A. Lajatha, and W. H. Gispen, eds.), pp. 221–230, Elsevier/North-Holland Biomedical Press, New York.

Reith, M., Morgan, I. G., Gombos, G., Breckenridge, W. C., and Vincendon, G., 1972, Synthesis of synaptic glycoproteins. I. The distribution of UDP-galactose: *N*-acetyl glucosamine galactosyl transferase and thiamine diphosphatase in adult rat brain subcellular fractions, *Neurobiology* **2**:169–175.

Roseman, S., 1970, The synthesis of complex carbohydrates by multiglycosyltransferase systems and their potential function in intercellular adhesion, *Chem. Phys. Lipids* **5**:270–297.

Rostas, J. A., Kelly, P. T., and Cotman, C. W., 1977, The identification of membrane glycocomponents in polyacrylamide gels: A rapid method using ^{125}I-labeled lectins, *Anal. Biochem.* **80**:366–372.

Roth, R. S., 1973, A molecular model for cell interactions, *Q. Rev. Biol.* **48**:541–562.

Rothman, J. E., and Lenard, J., 1977, Membrane asymmetry, *Science* **195**:743–753.

Salvaterra, P. M., and Mahler, H. R., 1976, Nicotinic acetylcholine receptor from rat brain: Solubilization, partial purification and characterization, *J. Biol. Chem.* **251**: 6326–6334.

Salvaterra, P. M., Gurd, J. M., and Mahler, H. R., 1977, Interactions of the nicotinic acetylcholine receptor from rat brain with lectins, *J. Neurochem.* **29**:345–348.

Schubert, D., Harris, A. J., Heinemann, S., Kidokoro, Y., Patrick, J., and Steinbach, J. H., 1973, Differentiation and interaction of clonal cell lines of nerve and muscle, in: *Tissue Culture of the Nervous System* (G. Sato, ed.), pp. 55–85, Plenum Press, New York.

Seeds, N. W., 1973, Differentiation of aggregating brain cell cultures, in: *Tissue Culture of the Nervous System* (G. Sato, ed.), pp. 35–53, Plenum Press, New York.

Sharon, N., 1975, *Complex Carbohydrates: Their Chemistry, Biosynthesis and Function*, pp. 26–29, 177–190, Addison-Wesley, Reading, Massachusetts.

Shur, B. D., and Roth, S., 1975, Cell surface glycosyltransferase, *Biochim. Biophys. Acta* **415**: 473–512.

Simpson, D. L., Thorne, D. R., and Loh, H. H., 1976, Sulfated glycoproteins, glycolipids, and glycosaminoglycans from synaptic plasma and myelin membranes: Isolation and characterization of sulfated glycopeptides, *Biochemistry* **15**:5449–5457.

Simpson, D. L., Thorne, D. R., and Loh, H. H., 1977, Developmentally regulated lectin in neonatal rat brain, *Nature (London)* **226**:367–369.

Stefanovic, V., Mandel, P., and Rosenberg, A., 1975, Activation of acetyl- and butyrylcholinesterase by enzymatic removal of sialic acid from intact neuroblastoma and astroblastoma cells in culture, *Biochemistry* **14**:5257–5260.

Susz, J. P., Hof, H. I., and Brunngraber, E. G., 1973, Isolation of concanavalin A binding glycoproteins from rat brain, *FEBS Lett.* **32**:289–292.

Tettamanti, G., Morgan, I. G., Gombos, G., Vincendon, G., and Mandel, P., 1972, Sub-synaptosomal localization of brain particulate neuraminidase, *Brain Res.* **47**:515–518.

Tettamanti, G., Preti, A., Lombardo, A., Suman, T., and Zambotti, V., 1975, Membrane-bound neuraminidase in the brain of different animals: Behaviour of the enzyme on endogenous sialo derivatives and rationale for its assay, *J. Neurochem.* **25**:451–456.

Thiery, J.-P., Brackenbury, R., Rutishause, U., and Edelman, G. M., 1977, Adhesion among neural cells of the chick embryo. II. Purification and characterization of a cell adhesion molecule from neural retina, *J. Biol. Chem.* **252**:6841–6845.

Van Nieuw Amerongen, A., and Roukema, P. A., 1974, GP-350, a sialoglycoprotein from calf brain: Its subcellular localization and occurrence in various brain areas, *J. Neurochem.* **23**:85–89.

Vincendon, G., Gombos, G., and Morgan, I. G., 1973, The interest of studying glycoproteins in the central nervous system, in: *Methodologie de la Structure et du Metabolisme des Glycoconjugues (Glycoproteins et Glycolipids)*, Vol. 2, CNRS, Paris.

Walters, B. B., and Matus, A. I., 1975*a*, Ultrastructure of the synaptic junctional lattice from mammalian brain, *J. Neurocytol.* **4**:369–375.

Walters, B. B., and Matus, A. I., 1975*b*, Proteins of the synaptic junction, *Biochem. Soc. Trans.* **3**:109–112.

Walters, B. B., and Matus, A. I., 1975*c*, Tubulin in postsynaptic junctional lattice, *Nature (London)* **257**:496–498.

Wang, Y.-J., and Mahler, H. R., 1976, Topography of the synaptosomal plasma membrane, *J. Cell Biol.* **71**:639–658.

Wenthold, R. J., Mahler, H. R., and Moore, W. J., 1974, Properties of rat brain acetylcholinesterase, *J. Neurochem.* **22**:945–949.

Whittaker, V. P., Michaelson, I. A., and Kirkland, R. J. A., 1964, The separation of synaptic vesicles from nerve-ending particles ("synaptosomes"), *Biochem. J.* **90**:293–303.

Yohe, H. C., and Rosenberg, A., 1977, Action of intrinsic sialidase of rat brain synaptic membranes on membrane sialolipid and sialoprotein components *in situ, J. Biol. Chem.* **252**: 2412–2418.

Zanetta, J. P., and Gombos, G., 1974, Affinity chromatography on Con A–Sepharose of synaptic vesicle membrane glycoproteins, *FEBS Lett.* **47**:276–478.

Zanetta, J. P., Breckenridge, W. C., and Vincendon, C., 1972, Analysis of monosaccharides by gas–liquid chromatography of the *O*-methyl glycosides as trifluoroacetate derivatives: Application to glycoprotein and glycolipids, *J. Chromatogr.* **69**:291–304.

Zanetta, J. P., Morgan, I. G., and Gombos, G., 1975, Synaptosomal plasma membrane glycoproteins: Fractionation by affinity chromatography on concanavalin A, *Brain Res.* **83**: 337–348.

Zanetta, J. P., Reeber, A., Ghandour, M. S., Vincendon, G., and Gombos, G., 1977*a*, Glycoproteins of intrasynaptosomal mitochondria, *Brain Res.* **125**:386–389.

Zanetta, J. P., Reeber, A., Vincendon, G. and Gombos, G., 1977*b*, Synaptosomal plasma membrane glycoproteins. II. Isolation of fucosyl-glycoproteins by affinity chromatography on the *Ulex europeus* lectin specific for L-fucose, *Brain Res.* **138**:317–328.

Zanetta, J. P., Roussel, G., Ghandour, M. S., Vincendon, G., and Gombos, G., 1978, Postnatal development of rat cerebellum: Massive and transient accumulation of concanavalin A binding glycoproteins in parallel fiber axolemma, *Brain Res.* **142**:301–319.

9

Surface Glycoconjugates in the Differentiating Neuron

Karl H. Pfenninger and Marie-France Maylié-Pfenninger

1. Introduction

A wide range of data suggests that the cell surface, and especially its carbohydrates, play a major role in developmental mechanisms (e.g., Moscona, 1974). The glycoconjugates involved may provide the cell with one or more surface codes and, perhaps, with specific sensing units that would enable each cell to (1) signal its own identity to its environment and (2) recognize the identity of its neighbors and acquire positional information. Within the nervous system, cellular interactions of this kind must occur during the formation of neuronal assemblies such as brain nuclei and, at still another level, when neurons form long processes—dendrites and axons. These processes are capable of growing into specific regions of the brain or the body ("guidance"), and are then able to recognize a specific target cell for the formation of a synapse. Our understanding of these phenomena in neuronal differentiation can be furthered, for example, by studying components of the plasma membrane that may be involved in relevant cellular mechanisms. However, the complex geometry of the neuron, and its subdivision into functionally different regions, have long interfered with the biochemical analysis of its plasma membrane in both developing and mature nervous systems. Hence the need for mapping techniques that enable us to demonstrate and quantitatively assess at the ultrastructural level the presence or absence of various membrane components. This chapter deals with the results of such mapping experiments, carried out on neurons during process

Karl H. Pfenninger • Department of Anatomy, College of Physicians and Surgeons, Columbia University, New York, New York. **Marie-France Maylié-Pfenninger** • Department of Developmental Genetics, Sloan-Kettering Institute for Cancer Research, New York, New York.

formation, and with lectins utilized as probes for the detection of specific cell-surface carbohydrates. The material is presented here in abbreviated form because it was recently reviewed from the point of view of neuronal growth and recognition (Pfenninger and Maylié-Pfenninger, 1978), from the point of view of synaptogenesis (Pfenninger, 1979), and in the framework of neuronal membrane organization in general (Pfenninger, 1978).

2. An Approach to the Mapping of Cell-Surface Carbohydrates

Lectins form a heterogeneous class of proteins derived mostly from plant seeds, but also from invertebrate hemolymph as well as various vertebrate tissues (endogenous lectins) (e.g., Sharon and Lis, 1975; Den *et al.*, 1976; DeWaard *et al.*, 1976; Nowak *et al.*, 1976; Kabat, 1978). They have in common high and selective affinity for specific carbohydrate residues and therefore have been widely used for the detection of sugars in a great variety of systems (cf. Chapter 7). For mapping of such residues at the ultrastructural level, lectins must be conjugated to an electron-dense marker such as ferritin, which contains an iron core, or horseradish peroxidase, which can be localized by an enzymatic reaction resulting in an electron-dense precipitate. The particulate marker ferritin is especially suitable because, if it is linked to the lectin in a known ratio, it can be used to quantitate directly the number of lectin-binding sites in a specific anatomical region. Maylié-Pfenninger and collaborators (Maylié-Pfenninger *et al.*, 1975; Maylié-Pfenninger and Jamieson, 1979) have made particular efforts to generate lectin–ferritin conjugates that contain the electron-dense marker and the binding protein in a ratio of approximately 1:1, and that retain, at the same time, their original affinity and specificity for certain sugar residues.

The properties of probes for cell-surface components, however, are not the only determinants of the success or failure of mapping studies. Of equal importance are the conditions under which such probes are applied to the cells to be analyzed: (1) cell surfaces must be free of contaminants such as serum glycoproteins; (2) cell surfaces must be freely accessible to the label; (3) the system must permit the removal of unbound excess label; and (4) nonspecific binding, i.e., labeling in the presence of the hapten, must be very low. To fulfill these conditions, we chose to carry out our mapping studies on neurons grown *in vitro* (Pfenninger and Maylié-Pfenninger, 1975, 1976, 1978). This system ensures ready access of the label to cell surfaces and enables us to clear them of serum glycoproteins and of glycoconjugates released by the cultured cells in an extensive washing step preceding the application of the marker. Similarly, excess marker can easily be removed following the labeling step. Furthermore, labeling in the presence of bovine serum albumin dramatically reduces nonspecific binding because of the albumin's ability to interfere with the nonspecific association of ferritin and cell surfaces (Maylié-Pfenninger and Jamieson, 1979). In this system, it has been possible to study lectin receptors of neuronal plasmalemma (1) as a function of the neuronal type, (2) as a function of the neuronal region, and (3) as a function of time during growth of the neurite. Figure 1* shows the results of a labeling experiment in which a culture of rat superior cervical ganglion was exposed live to ferritin conjugate of wheat germ agglutinin (WGA) for 15 min at 36°C. Note the uniform, dense layer of ferritin particles along the cell membrane of the neurites. In a corresponding control experiment, carried out in the presence of 0.2 M *N*-acetylglucosamine (GlcNAc), the amount of surface label is negligible, as shown in Fig. 2.

This approach is not without limitations, however. First, it may not be possible to grow every type of neuron in each phase of differentiation *in vitro*. Second, neurons may exhibit somewhat different cell-surface properties *in vitro* as compared to the situation *in vivo*. Third, the probes that are being used here are relatively bulky, and therefore steric hindrance is an important factor influencing quantitation of lectin receptors if they occur in high density. Last, plant lectins recognize primarily one type of sugar residue (although neighboring sugar residues will influence lectin binding), rather than an entire oligosaccharide, and are therefore relatively limited probes. However, the work carried out so far has provided a substantial amount of important information that could not have been obtained by other means. Future developments, especially the generation of specific antibodies and the perfection of cell culture methods, will undoubtedly allow for an even wider range of significant cell-surface mapping studies.

3. Regional and Type-Specific Lectin Receptor Content in the Plasmalemma of Growing Neurons

The study of regional differences in lectin receptor content is of great importance in the neuron because of the functional specialization of its different parts. During neurite formation, it is particularly interesting to know whether the growth cone, which is capable of finding and recognizing target cells, exhibits surface properties that are different from those in the perikaryon. This question gains further interest in view of the fact that the density of intramembranous particles in growth cone plasmalemma as seen by freeze–fracture is about one order of magnitude lower than that in the perikaryal plasmalemma (Pfenninger and Bunge, 1974). We studied lectin binding to cell surfaces in mechanically or enzymatically dissociated and cultured neurons from the superior cervical ganglion and the anterior horn of the spinal cord of the rat (Pfenninger and Maylié-Pfenninger, 1978). These cultures were first either cooled to 0–4°C or fixed with glutaralde-

*Illustrations for this chapter will be found following page 188.

hyde, then labeled with a saturating concentration of ferritin–lectin conjugate, and finally processed for thin-section electron microscopy. Although it has not been possible to distinguish with certainty dendrites and axons in our culture system, it is clear that receptor sites for various lectins are uniformly distributed, and that, at least in gross terms, their density is equal in neurite and perikaryon. In view of the freeze–fracture data mentioned above, this is an unexpected finding. However, it could be argued that the lectin probes are too crude to detect the more subtle differences in cell-surface and membrane properties that may exist in the various neuronal regions.

It is of equal interest to ask whether there are differences in plasmalemmal lectin receptor content when comparing growth cones originating from different neuronal types. This problem was approached by growing various neurons from fetal or newborn rats *in vitro* and by studying, as before, the density of lectin receptor sites on growth cones labeled at 0–4° C or after aldehyde prefixation. A battery of lectin–ferritin conjugates was utilized for these studies. Results are shown in Figs. 3–5 and listed in Table I (Pfenninger and Maylié-Pfenninger, 1978). As is evident from this material, each neuronal type has its own characteristic set of plasmalemmal lectin receptors (cf. Hatten and Sidman, 1977). In other words, neurons bear a type-specific carbohydrate code, at least at the stage of differentiation when they are forming processes. Our data show, furthermore, that the various surface carbohydrate patterns fall into two major classes: (1) sparse binding of concanavalin A (Con A), WGA, and *Ricinus communis* agglutinins I and II, (RCAs I and II) to neurites from olfactory bulb, anterior spinal cord, and cerebellum; (2) high density of such binding sites in neurites from dorsal root and superior cervical ganglia (cf. Figs. 3 and 4). However, neuraminidase pretreatment of the former group of neurites, i.e., removal of terminal sialic

Table I. Lectin Receptor Densities (Binding Sites/μm^2 Plasma Membrane) on Developing Neurites of Different Origin[a]

Origin[b]	Con A Glc/Man	WGA GlcNAc	RCA I Gal	Neuramin. RCA I	SBA GalNAc	LTA/UEL Fuc
SCG	(1400)	2740	1440	—	(1400)	0
DRG	(1800)	2220	(2200)	—	—	0
SC	(800)	910	210	(2200)	0	0
CBL	—	530	560	2210	0	0
OB	(600)	1830	0	1950	—	0

[a] From Pfenninger and Maylié-Pfenninger (1978). Preliminary data; figures in parentheses are estimated values. *Abbreviations:* (Con A) concanavalin A; (WGA) wheat germ agglutinin; (RCA I) *Ricinus communis* agglutinin I; (Neuramin.) neuraminidase treatment before labeling; (SBA) soybean agglutinin; (UEL) *Ulex europeus* lectin; (LTA) *Lotus tetragonolobus* agglutinin.

[b] (SCG) superior cervical ganglion; (DRG) dorsal root ganglion; (SC) spinal cord; (CBL) cerebellum; (OB) olfactory bulb (fetal rat tissues).

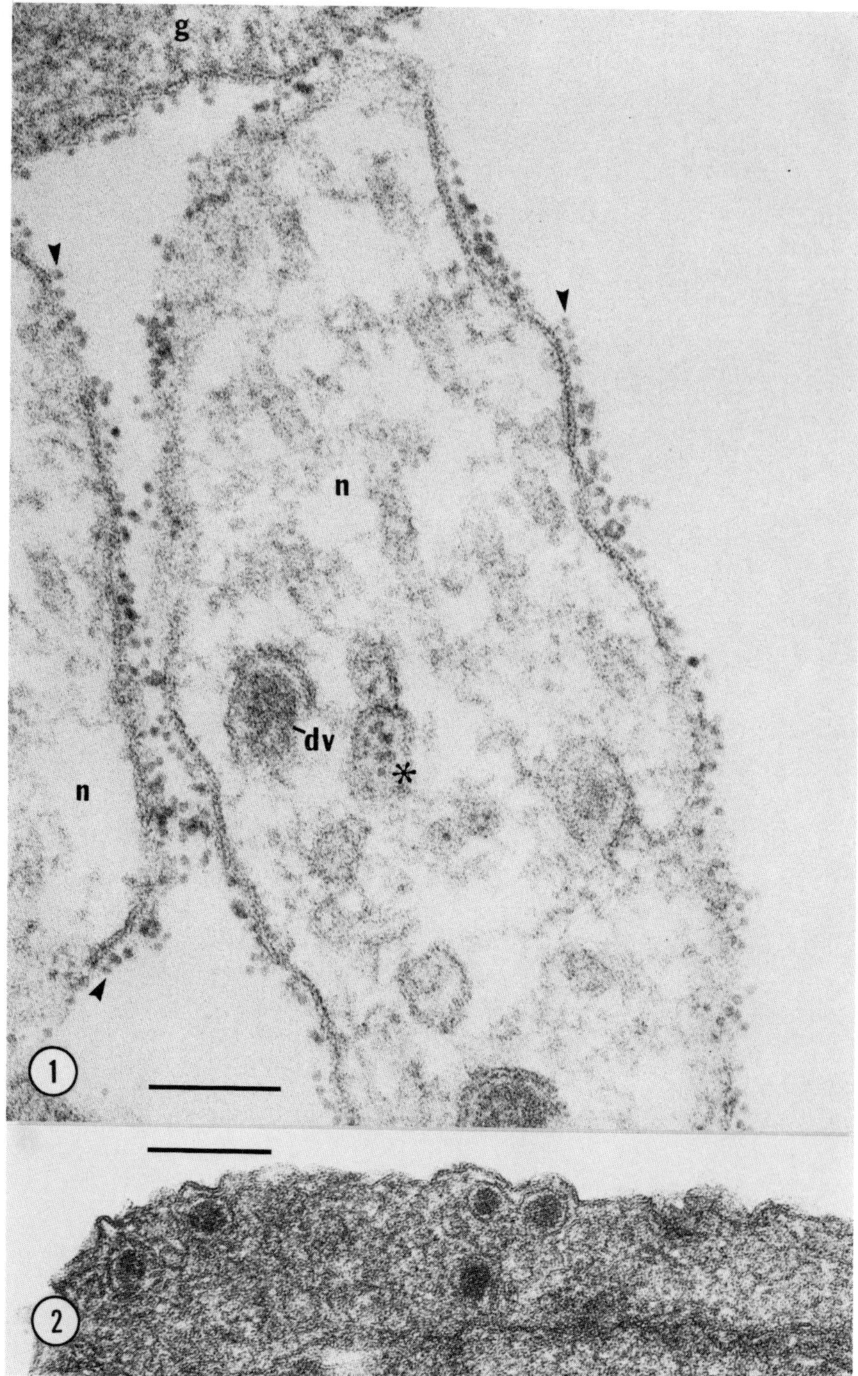

Fig. 1. Neurites (n) of a rat superior cervical ganglion culture labeled live with ferritin–WGA. Note the layer of electron-dense ferritin particles (arrowheads) along the cell surfaces. Some of the label has been internalized (*). (dv) Large dense core vesicle; (g) glial process. Calibration bar: 0.1 μm.

Fig. 2. Control experiment in which a similar neurite was labeled with ferritin–WGA in the presence of 0.2 M GlcNAc. Note the absence of surface label. Calibration bar: 0.2 μm.

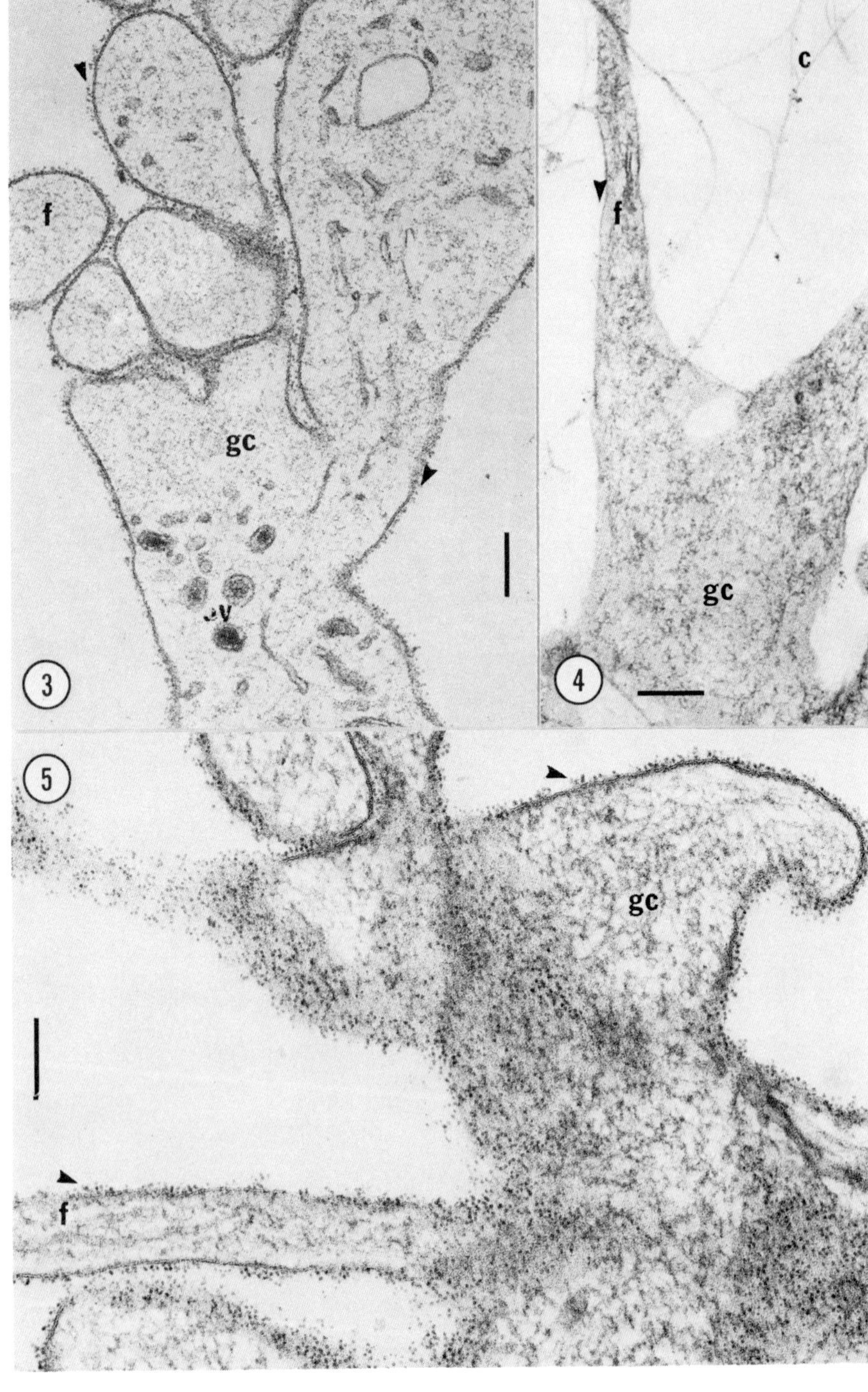
f
gc
ev
3
c
f
gc
4
5
gc
f

←

Figs. 3–5. Differential labeling of nerve growth cones with ferritin conjugate of RCA I. The growth cone (gc) in a rat superior cervical ganglion culture is relatively heavily labeled **(Fig. 3)**, whereas the same structure in a spinal cord culture, treated in the same manner, exhibits hardly any surface marker (**Fig. 4**). However, pretreatment of spinal cord neurites with neuraminidase reveals a high density of binding sites for RCA I (**Fig. 5**). Glutaraldehyde-prefixed preparations. (f) Filopodia; (dv) large dense core vesicles; (c) collagen fibrils of culture substratum. Calibration bars: 0.2 μm.

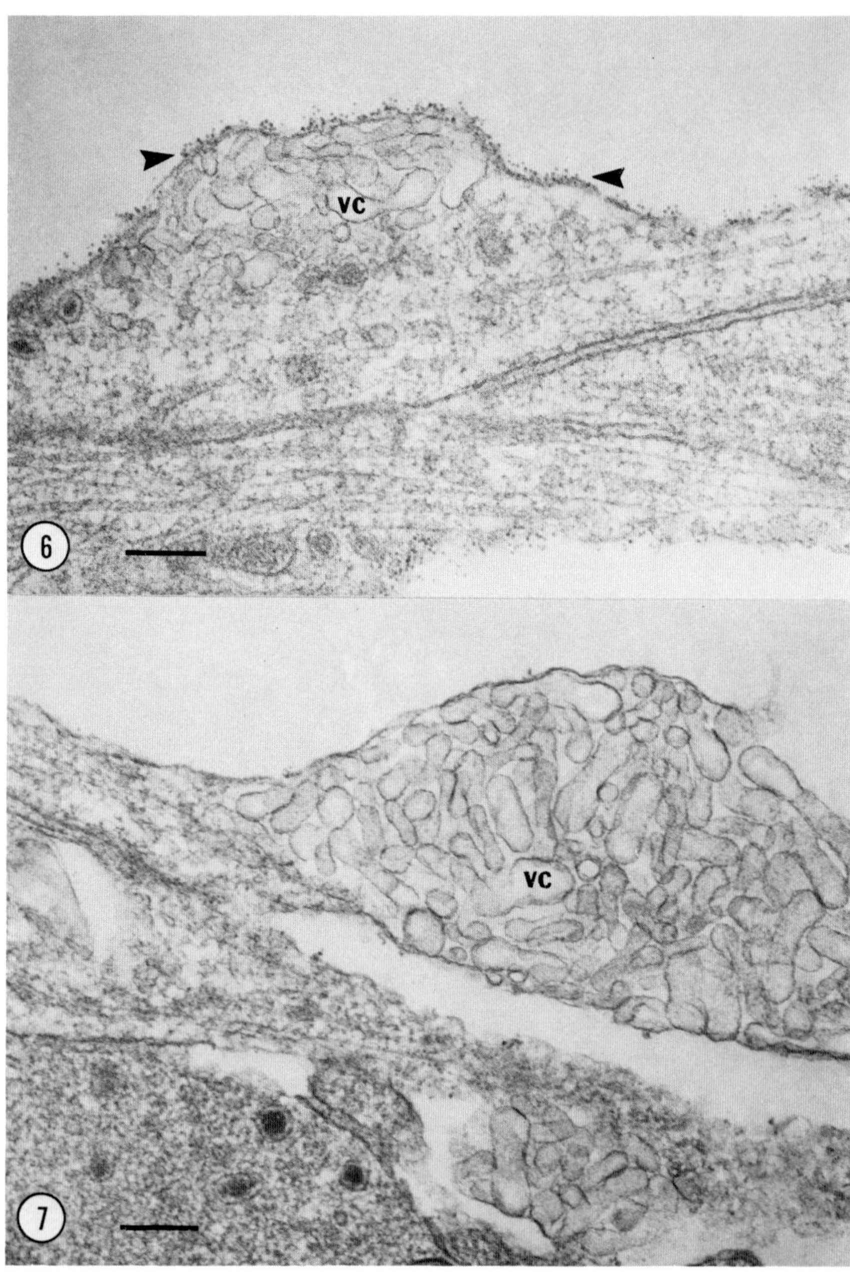

Figs. 6 and 7. Double-label pulse–chase experiments with ferritin–conjugated lectins (RCAs I and II; other lectins yield comparable results). Neurites in cultures of rat superior cervical ganglion were first exposed to unconjugated lectin, then washed and incubated (chased) for 3 min in the absence of lectin, and finally, following aldehyde fixation, labeled with the respective ferritin conjugate. Note the appearance of a relatively dense patch of marker (arrowheads) on the surface of a subplasmalemmal vesicle cluster (vc) in the experiment carried out at 37° C (**Fig. 6**), whereas in the control experiment at 0° C (**Fig. 7**), virtually no surface label is found. For further explanation, see the text. Calibration bars: 0.2 μm.

acid residues, exposes a large number of binding sites for RCA I (galactosyl residues), as shown in Figs. 4 and 5. Interestingly, this distinction of two classes of neuronal membrane by their lectin-binding properties parallels the anatomical central–peripheral division of the nervous system. Thus, these experiments suggest that one important feature distinguishing neural-tube-derived neurons from those originating from the crest may lie in the amount of terminal sialic acid residues on their cell surfaces.

4. Dynamics of Surface Glycoconjugates during Neuritic Elongation

Surface properties of the developing neuron, and especially those of the growing neurite, should be considered, not from a purely static viewpoint, but rather in the light of the fast process of expansion of the plasma membrane that must occur in order to keep up with the rapidly increasing cellular size and surface. The uniform distribution of lectin-binding sites in the various regions of a growing neuron seems to indicate that expansion of the membrane matrix is paralleled by insertion of the appropriate types and numbers of lectin receptor sites. This hypothesis, however, needs to be tested experimentally and the mechanism of insertion must be elucidated. We decided to approach these problems in a series of pulse–chase experiments in which live growing neurites in rat superior cervical ganglion cultures were exposed to lectins for a brief period of time and then allowed to continue to grow in the absence of the surface marker. In this way, it is possible to determine (1) the relative movement of preexisting lectin receptors and (2) the appearance of new lectin receptors during neuritic growth. The results of such experiments have been reported (Pfenninger and Maylié-Pfenninger, 1975, 1977, 1978) and will be described in detail elsewhere. In brief, the data are as follows:

In the most important and significant experiment, all preexisting lectin receptors are saturated with an unconjugated lectin that is invisible in the electron microscope. After a chase period of 3–20 min, with growth continuing in the absence of surface label, the neurites are aldehyde-fixed and then labeled with the ferritin conjugate of the same lectin. In this experiment, one would expect the ferritin-conjugated marker to be attached only to those lectin receptors that were previously inaccessible to the lectin but appeared on the cell surface during the chase period. If this experiment is carried out in the cold, when no growth occurs, the ferritin marker is only very rarely observed on cell surfaces (Fig. 7). However, if the experiment is carried out under growth conditions, at 36°C, receptors for WGA and RCAs I and II appear in patches on the surface of plasmalemmal protrusions filled with a fairly heterogeneous population of large clear vesicles (Fig. 6). These subplasmalemmal vesicle clusters are found particularly frequently in the growth cone region. In this area, furthermore, large patches of ferritin-labeled lectin receptors are seen to extend over filopodia and lamellopodia in the same experiment.

In conjunction with the earlier freeze–fracture findings (Pfenninger and Bunge, 1974; Pfenninger and Rees, 1976), these results strongly suggest that the plasmalemma is expanded by the insertion of preassembled membrane in the form of vesicles that are exported to the cell periphery and then fuse in exocytosis-like fashion with the plasmalemma. Our data seem to indicate, furthermore, that the newly inserted membrane carries the full complement of lectin receptors that is seen throughout the surface of the growing neuron. Because the nerve growth cone appears to be the primary site of insertion of new membrane into the plasmalemma, the properties of newly added membrane seem to be of particular importance for the interactions between the growth cone and its microenvironment.

5. Concluding Remarks

This brief review of our lectin labeling studies in culture is intended to demonstrate that certain aspects of cell-surface biochemistry, i.e., the distribution of membrane glycoconjugates, can now be studied quantitatively and as a function of time at the ultrastructural level. Clearly, the lectins available today are of limited specificity in that they do not permit identification of an individual glycoprotein or glycolipid, and the data obtained in these studies must be interpreted with this limitation in mind. Yet, it has been possible to at least partially define a type-specific carbohydrate signature for the cell surface of the neuron. The significance of this carbohydrate signature is not clear and awaits further investigation. However, it is highly likely that this code plays an important role in certain developmental processes, such as the sorting out of neuronal types during the morphogenesis of nervous tissues and/or recognition and synaptogenesis. One of the most puzzling findings of this work is the discrepancy in the various anatomical regions of the growing neuron between the uneven distribution of intramembranous particles and the uniform pattern of lectin receptors. A relatively high lectin receptor density and a very low density of intramembranous particles are found in the same plasma membrane at the level of the nerve growth cone. In light of our present understanding of membrane organization (e.g., see Pfenninger, 1978), one might suspect that in this case, a large proportion of lectin-binding sites could be contributed by glycolipids rather than glycoproteins. In preliminary experiments, we have found that the vast majority of receptors for RCAs I and II resist proteolytic degradation but can be extracted with chloroform–methanol. If this can be confirmed in further experiments, we would have to conclude that glycolipids are indeed major determinants of cell-surface properties in the growing neurite.

References

Den, H., Malinzak, D. A., and Rosenberg, A., 1976, Lack of evidence for the involvement of β-D-galactosyl-specific lectin in the fusion of chick myoblasts, *Biochem. Biophys. Res. Commun.* **69**:621–627.

DeWaard, A., Hickman, S., and Kornfeld, S., 1976, Isolation and properties of β-galactoside binding lectins of calf heart and lung, *J. Biol. Chem.* **251**:7581–7587.

Hatten, M. E., and Sidman, R. L., 1977, Plant lectins detect age- and region-specific differences in cell surface carbohydrates and cell reassociation behavior of embryonic mouse cerebellar cells, *J. Supramol. Struct.* **7**:267–275.

Kabat, E. A., 1978, Dimensions and specificities of recognition sites on lectins and antibodies, *J. Supramol. Struct.* **8**:79–88.

Maylié-Pfenninger, M.-F., and Jamieson, J. D., 1979, Distribution of cell surface saccharides on pancreatic cells. I. General method for preparation and purification of lectins and lectin–ferritin conjugates, *J. Cell Biol.* **80**:69–76.

Maylié-Pfenninger, M.-F., Palade, G. E., and Jamieson, J. D., 1975, Interaction of lectins with the surface of dispersed pancreatic cells, *J. Cell Biol.* **67**:333a.

Moscona, A. A., (ed.), 1974, *The Cell Surface in Development*, John Wiley & Sons, New York.

Nowak, T. P., Haywood, P. L., and Barondes, S. H., 1976, Developmentally regulated lectin in embryonic chick muscle and a myogenic cell line, *Biochem. Biophys. Res. Commun.* **68**: 650–657.

Pfenninger, K. H., 1978, Neuronal membrane organization, *Annu. Rev. Neurosci.* **1**:445–471.

Pfenninger, K. H., 1979, Synaptic membrane differentiation, in: *The Neurosciences: Fourth Study Program* (F. O. Schmitt, ed.), pp. 779–795, MIT Press, Cambridge.

Pfenninger, K. H., and Bunge, R. P., 1974, Freeze–fracturing of nerve growth cones and young fibers: A study of developing plasma membrane, *J. Cell Biol.* **63**:180–196.

Pfenninger, K. H., and Maylié-Pfenninger, M.-F., 1975, Distribution and fate of lectin binding sites on the surface of growing neuronal processes, *J. Cell Biol.* **67**:322a.

Pfenninger, K. H., and Maylié-Pfenninger, M.-F., 1976, Differential lectin receptor content on the surface of nerve growth cones of different origin, *Neurosci. Abstr.* **II** (Pt. 1):224.

Pfenninger, K. H., and Maylié-Pfenninger, M.-F., 1977, Localized appearance of new lectin receptors on the surface of growing neurites, *J. Cell Biol.* **75**:54a.

Pfenninger, K. H., and Maylié-Pfenninger, M.-F., 1978, Characterization, distribution and appearance of surface carbohydrates on growing neurites, in: *Neuronal Information Transfer* (A. Karlin, H. Vogel, and V. M. Tennyson, eds.), pp. 373-386, Academic Press, New York.

Pfenninger, K. H., and Rees, R. P., 1976, From the growth cone to the synapse: Properties of membranes involved in synapse formation, in: *Neuronal Recognition* (S. H. Barondes, ed.), pp. 131–178, 357–358, Plenum Press, New York.

Sharon, N., and Lis, H., 1975, Use of lectins for the study of membranes, in: *Methods in Membrane Biology*, Vol. 3 (E. D. Korn, ed.), pp. 147–200, Plenum Press, New York.

10

Histochemistry of Polyanions in Peripheral Nerve

O. K. Langley

1. Introduction

Ranvier (1871) in 1871 was the first to describe periodic constrictions of the external contour of peripheral myelinated nerve fibers that coincided with interruptions in the myelin sheath. He also observed that the axon itself was continuous across the "annular constrictions" or nodes, and that only a single Schwann cell nucleus appeared between each pair of nodes along the same fiber. Ranvier was also the first to use heavy metals to stain nodes. He found that after immersion of unfixed nerves in dilute silver nitrate solution and subsequent exposure to sunlight, a "petite croix latine," now known as the "cross of Ranvier," revealed the location of each node (see Fig. 1*). He believed the transverse bar of the cross to be due to the presence of a ring of "cement" substance representing the contact zone between adjacent Schwann cells, while the longitudinal bar was due to staining within the axon. Nemiloff (1908) also described the staining of the nodal cement substance in unfixed nerves with methylene blue. Early attempts to demonstrate other ionic species at nodes of Ranvier by precipitation with heavy metals were also made (Macallum, 1905; Macallum and Menten, 1906). Ranvier's observations have largely been validated in more recent histological studies (Hess and Young, 1952), but it was not until the application of the electron microscope to this structure that the morphology of nodes could be defined in detail (Gasser, 1952; Robertson, 1957; Uzman and Villegas, 1960; Elfvin, 1961; Sjöstrand, 1963; Landon and Williams, (1963). A diagrammatic representation of a node of a large mammalian peripheral nerve fiber is illustrated in Fig. 2. It is seen that the cementing disk described a century ago by Ranvier occupies the region in which microvilli extend radially toward the axon from the cyto-

*Illustrations for this chapter will be found following page 196.

O. K. Langley • Institute of Neurology, London, England. Present address: Institut de Chimie Biologique de la Faculté de Medecine, Strasbourg, France.

plasm of Schwann cells terminating on both sides of the node. These microvillous Schwann cell fingers are embedded in a matrix of moderate electron density (using conventional electron-microscopic preparative techniques) that is segregated from the endoneural space by a continuous Schwann cell basal lamina.

In reviewing the staining properties of the node of Ranvier, Hess and Young (1952) emphasized the lack of available data on the nature of permeability barriers existing around the axon at the node. Such knowledge is essential for a complete understanding of impulse conduction. The work reviewed in this chapter represents an attempt to define the properties of the matrix that surrounds the excitable axon membrane at the node in order to relate the complex morphology and cytochemistry of this region to nerve function, and specifically to the ionic movements occurring during impulse conduction.

2. Histochemistry of Peripheral Nerve

2.1. Colloidal Iron

The most consistent feature of teased preparations of peripheral nerve stained with the colloidal iron technique (Langley and Landon, 1968) is that of a transverse bar filling the region of the cementing disc (Fig. 3). In nerves fixed in glutaraldehyde, axonal staining is also evident, though this is not discerned in nerves fixed in osmium tetroxide. In sectioned material, it is clear that the transverse bar represents a circular collar of reaction product around the nodal axon membrane.

The electron microscope confirms the differences in localization of the stain using the two fixatives, apparent in the light microscope. With glutaraldehyde alone as fixative, Prussian blue crystals are seen inside the axon, and attached to the inside of the axolemma, in addition to their presence in the nodal gap (Langley and Landon, 1968). With osmium tetroxide, a dense deposit appears over the nodal axolemma and in the region immediately beneath it with a few scattered crystals in the gap (Fig. 4).

2.2. Ferric Chloride

A staining pattern, viewed in the light microscope, essentially similar to that obtained with colloidal iron is produced when osmium-fixed nerves are immersed in dilute solutions of ferric chloride and the attached iron is subsequently converted to Prussian blue with potassium ferrocyanide (Langley, 1969). In transverse sections, the pigment is clearly seen to form a ring around the nodal axon (Fig. 5). Electron microscopy, however, reveals that while the reaction product of the Hale stain in osmium-fixed nerve is essentially restricted to a narrow zone spanning the nodal axon membrane, bound ferric ions are found to be wholly extraaxonal. A finely granular precipitate is found distributed throughout the gap substance with coarser crystals often in the adjacent endoneurium, the Schwann cell finger processes appearing in negative contrast (Fig. 6).

More recent studies on the ultrastructural localization of ferric ion bound to nodes of Ranvier indicate that the localization of the reaction product is dependent on the fixative employed (as was found for colloidal iron staining). Quick and Waxman (1971*a*) found chiefly intraaxonal staining and binding to the inner aspect of the nodal axolemma when cacodylate-buffered aldehyde and osmium were used. They found that staining of the gap substance is evinced with fixatives containing phosphate buffer and considered such staining to be the result of phosphate binding at this location (Quick and Waxman, 1977*b*). While it appears that the use of phosphate solutions prior to staining nerves with ferric chloride can enhance the staining of the nodal gap, the studies of Langley and Landon (Langley, 1969, 1970; Landon and Langley, 1971) demonstrate clearly that fixing or soaking nerves in phosphate-buffered solutions is not a prerequisite for nodal gap staining. Furthermore, the reversibility of cation exchange at nodes of Ranvier (see Section 3.3) and the similar nodal binding of other ions is not compatible with an interpretation of nodal staining in terms of binding to phosphate ions themselves bound to nodes.

2.3. Copper

The use of solutions containing copper ions to stain nerve fibers stems from attempts to demonstrate cholinesterase activity in nervous tissue (Gerebtzoff, 1959, 1964; Zenker, 1964, 1969; Grant *et al.*, 1967; Adams *et al.*, 1968). Thiocholine, released by tissue esterases, is precipitated in such reactions by copper ions present in the incubation medium. If fixed nerves are incubated in solutions containing cupric ions, in either the simple or the complexed form, copper is bound to nodes of Ranvier, and may be demonstrated by its conversion to the insoluble pigment Hatchett's brown with potassium ferrocyanide (Fig. 7). Since the ionic composition of cholinesterase incubation media is similar to that of solutions capable of producing nodal staining in the absence of enzyme substrate, a considerable debate has resulted on the significance of the localization of the insoluble end-product of the reaction (Herbst, 1965; Mladenov, 1965; Adams *et al.*, 1968; Langley and Landon, 1969). Such appearances of nodal staining were first suggested by Ramón y Cajal (1955) to be the result of particular affinity of this part of the fiber for the reagents used.

The fine structural localization of copper ferrocyanide was determined in the electron microscope (Fig. 8). Dense amorphous deposits were found to fill the nodal gap substance extending into the extracellular space between the terminating myelin loops on either side of the node and extending also for a short distance across the nodal axon membrane.

The localization of the insoluble copper pigment was found to be dependent both on the fixative employed and in particular on the reagent used to precipitate tissue-bound copper (Krammer *et al.*, 1972; Krammer and Lishka, 1973). When copper sulfide is selected as the end-product, the predominant pattern is one of membrane staining. Axonal membranes including the axolemma and the membranes of intraaxonal organelles, Schwann cell membranes, including the inner and outer plasmalemma, and major and minor dense lines in the region of the paranodal splitting myelin sheath and those of Schwann cell nodal microvilli are found to be stained under such conditions. Both inner and outer aspects of the "unit membrane" are apparently stained. The use of ferrocyanide to precipitate copper produces a contrasting pattern in which the nodal gap substance is the predominantly stained element (Krammer and Lishka, 1973), in agreement with the data reported by Langley and Landon (1969). The nature of ferrocyanide as a precipitating agent for copper was examined in its relationship to the pH of the precipitating medium. Acidified ferrocyanide (in which the copper salt is less soluble) was found to be more effective and may explain the negative results obtained by some workers in control (i.e., without substrate) cholinesterase incubations (Adams *et al.*, 1968). The solubility of the end-product may also account for the different localization of copper resulting from using ammonium sulfide as precipitant.

2.4. Barium

Barium bound to nodes of Ranvier after immersion of nerves in weak solutions of barium chloride may be demonstrated by conversion either to the insoluble pink rhodizonate or to the sulfate. When the former method is adopted, a pink crystalline deposit is seen located as a ring around the nodal axon. In sections, the reaction product appears to fill the nodal gap substance (Fig. 9). When bound barium is precipitated as its sulfate, a highly refractile material may be observed in the same location by phase-contrast microscopy of sectioned nerve. Neither of these methods is best suited for use with the electron microscope, since the crystal size is such that the barium salt is torn from the tissue during sectioning. It is nevertheless evident that barium is precipitated in the region of the nodal gap substance, and in addition, barium sulfate crystals are found attached to the innermost layer of myelin in paranodal regions (Fig. 10) (Landon and Langley, 1971).

2.5. Alcian Blue

Nodes of Ranvier of both fixed and unfixed peripheral nerve display alcianophilia. Nodes are the exclusive sites of staining in unfixed teased nerves (Langley, 1970, 1971*a*), a result presumably of the relative impermeability of cellular membranes to the bulky charged alcian blue molecule. In fixed tissue, staining is more widespread and includes the axon and Schwann cell nuclei in

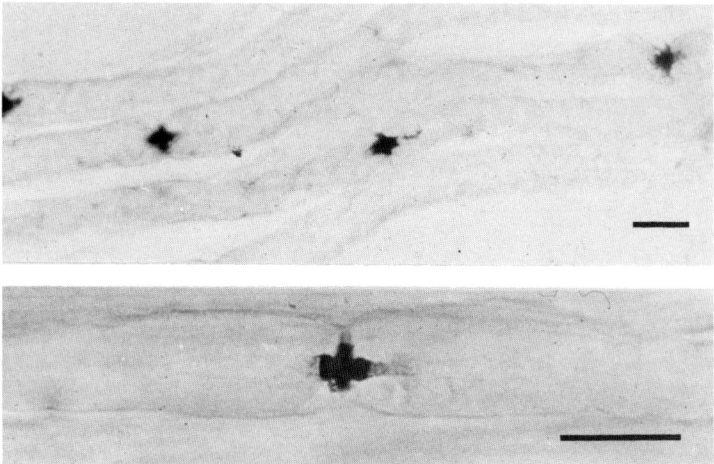

Fig. 1. Demonstration of the cross of Ranvier. Unfixed teased rat sciatic nerve was immersed in dilute silver nitrate solution and the precipitated silver chloride subsequently converted to the sulfide with ammonium sulfide. Low power (*top*) and high power (*bottom*) showing the two members of the cross. The transverse element occupies the region of the "cementing disk" and the longitudinal element is apparently in the axon. Calibration bars: 20 μm.

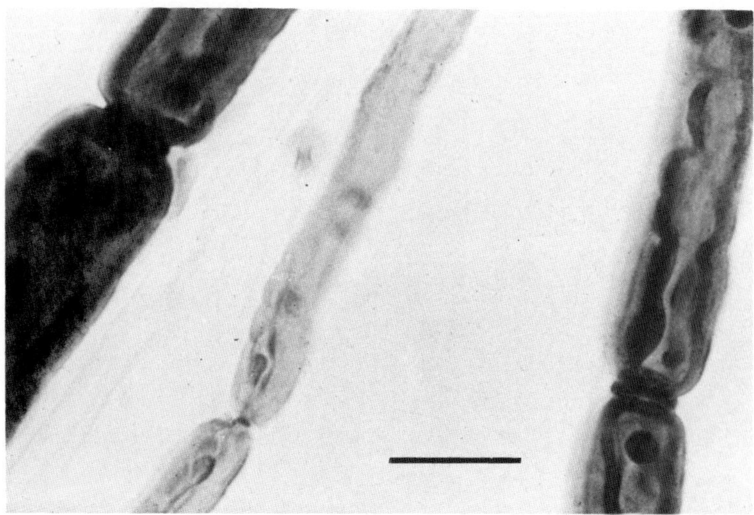

Fig. 3. Teased osmium-tetroxide-fixed rat sciatic nerve stained with the Hale stain. Reaction product appears as a collar at the nodes of three fibers of different diameters. Calibration bar: 20 μm.

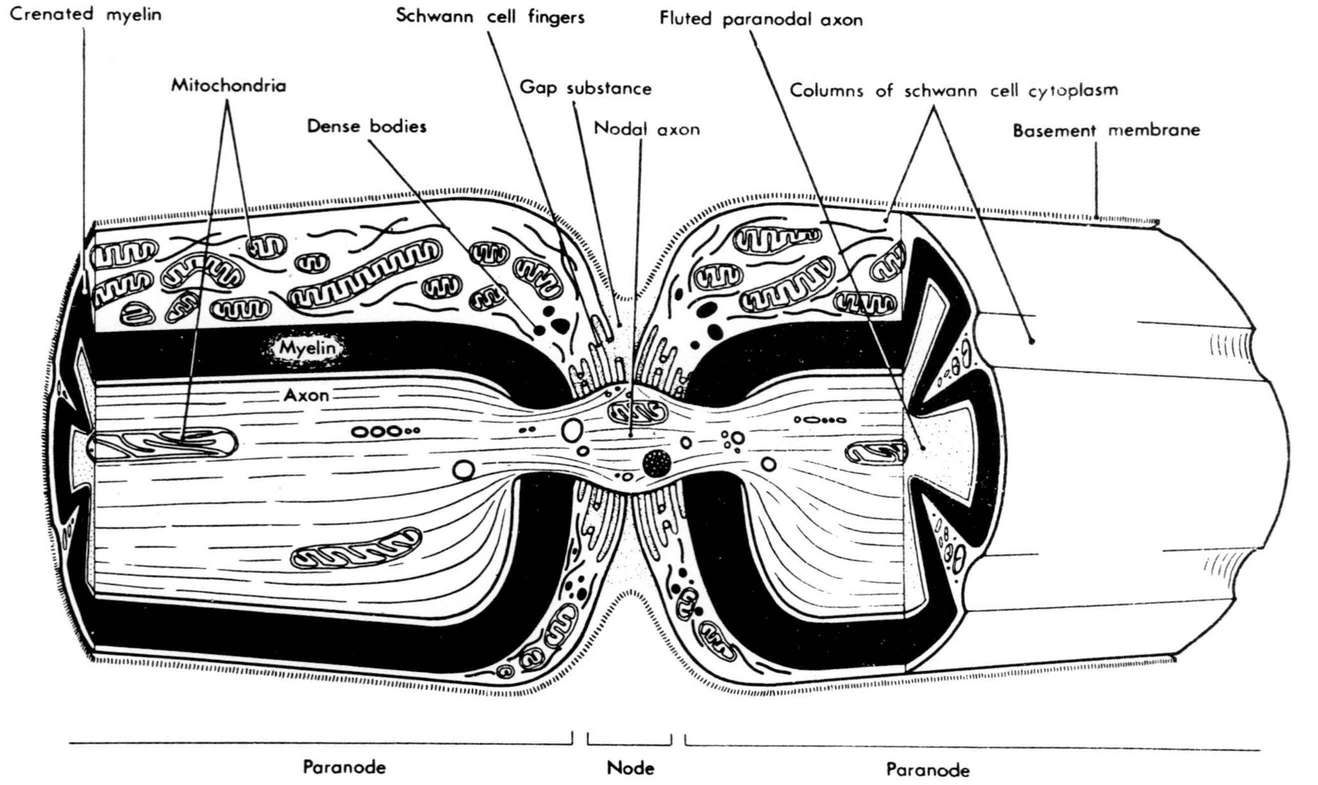

Fig. 2. Diagrammatic representation of a node of Ranvier of a large mammalian peripheral nerve fiber. Reproduced by the courtesy of Dr. D. N. Landon and the *New Scientist*.

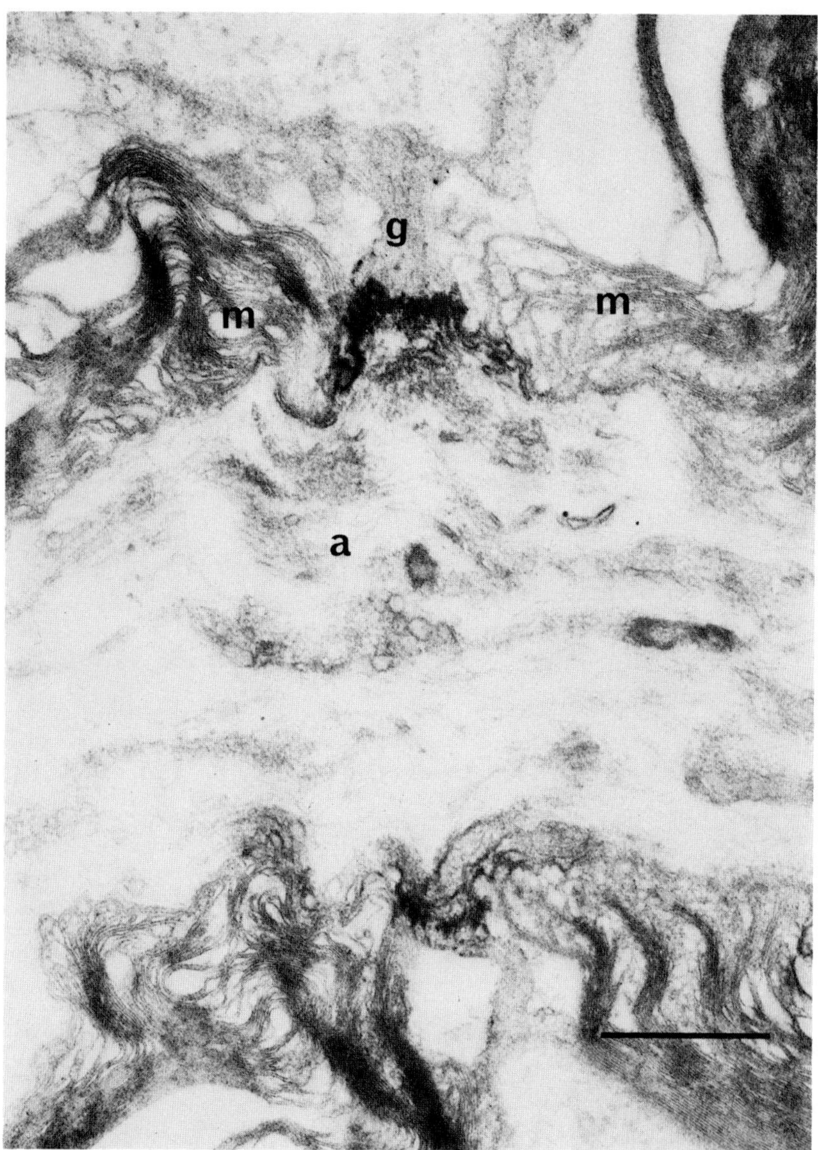

Fig. 4. Electron micrograph of osmium-fixed rat sciatic nerve stained with the Hale stain. Prussian blue deposit is situated over the nodal axon membrane with a few scattered crystals in the overlying gap substance. (a) Axon; (g) gap substance; (m) myelin. Calibration bar: 1 μm. Provided by Dr. D. N. Landon.

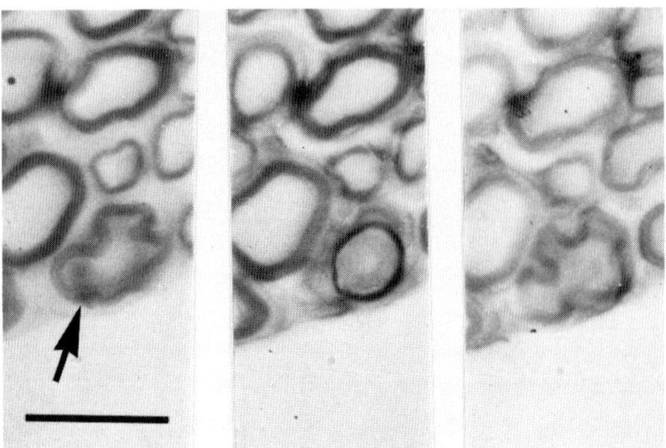

Fig. 5. Through-focus series of 2-μm-thick araldite transverse section of rat sciatic nerve stained with ferric chloride showing paranodal myelin fluting (*left*, arrow), ring of reaction product at the node (*middle*), and paranodal convoluted myelin sheath on the other side of the node (*right*). Calibration bar: 20 μm.

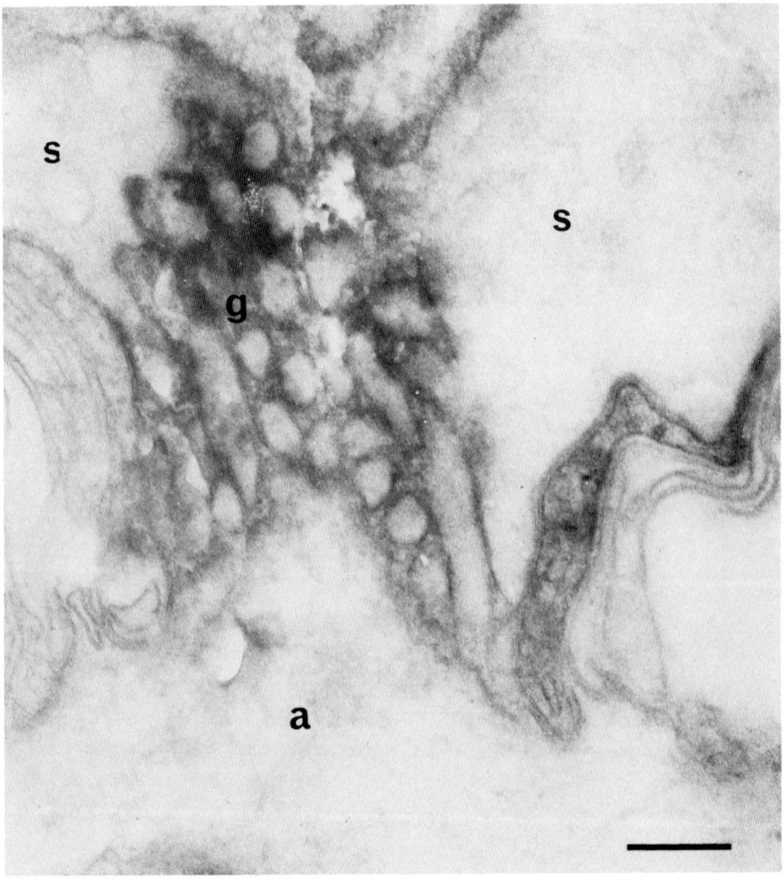

Fig. 6. Electron micrograph of preparation shown in Fig. 5. Prussian blue pigment fills the gap substance, revealing Schwann cell fingers in negative contrast. (a) Axon; (g) gap substance; (s) Schwann cell cytoplasm. Calibration bar: 0.2 μm. Provided by Dr. D. N. Landon.

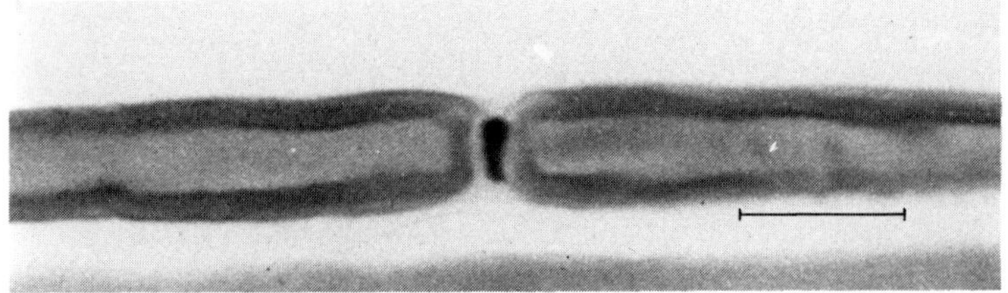

Fig. 7. Teased osmium-tetroxide-fixed rat sciatic nerve stained with copper, visualized as Hatchett's brown. Reaction product appears as a transverse bar at nodes of Ranvier. Calibration bar: 20 μm.

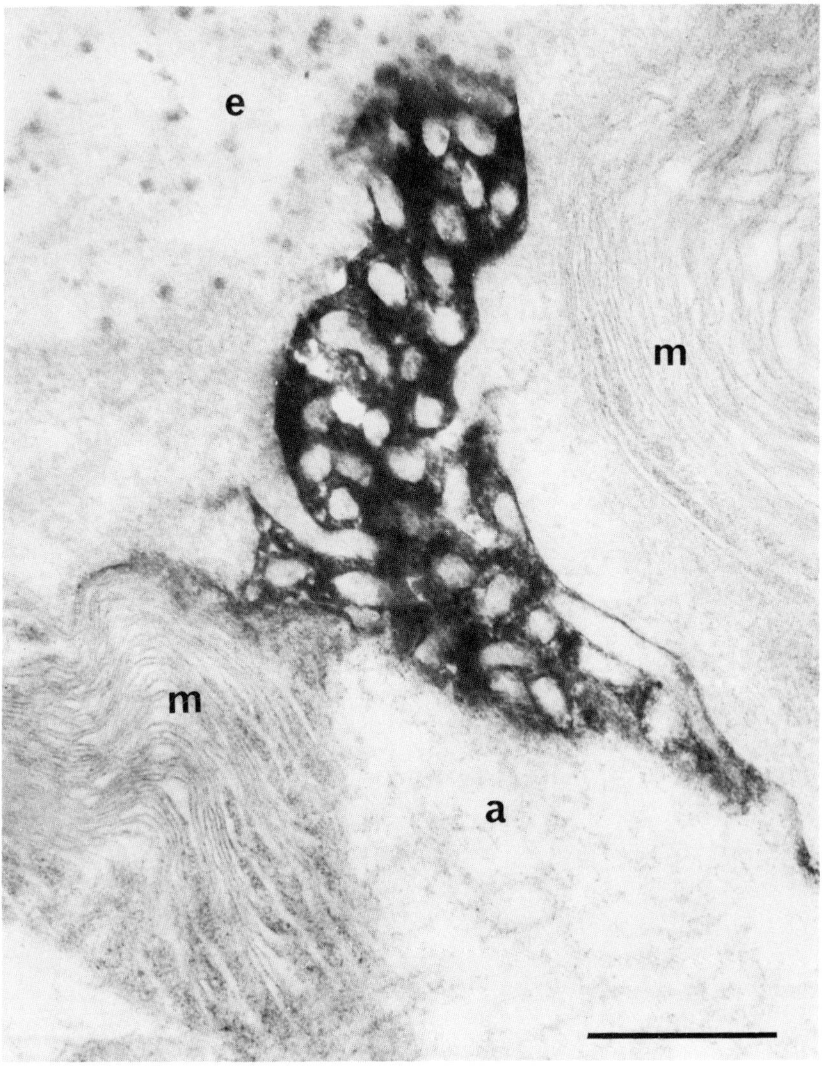

Fig. 8. Electron micrograph of rat sciatic nerve stained with copper. Electron-dense precipitate is largely confined to the "gap substance," with a few scattered crystals in the endoneurium (e). (a) Axon; (m) myelin. Calibration bar: 0.5 μm. Provided by Dr. D. N. Landon.

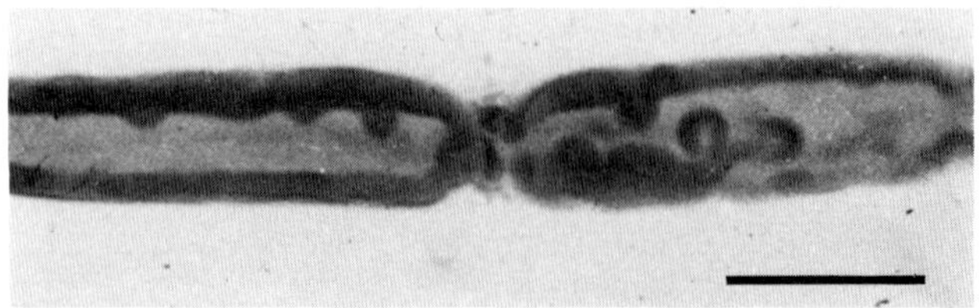

Fig. 9. Teased osmium-fixed rat sciatic nerve stained with barium chloride. Bound barium converted to the insoluble rhodizonate. The reaction product is situated in the nodal gap. Calibration bar: 20 μm.

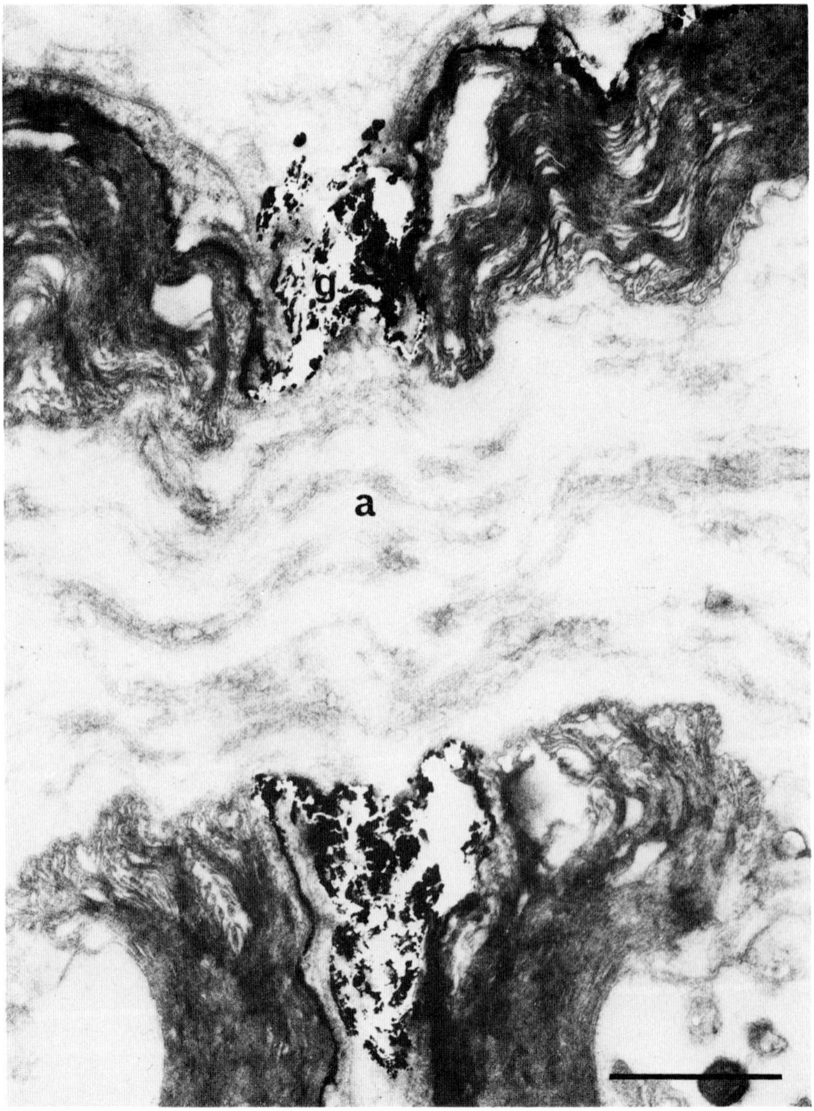

Fig. 10. Electron micrograph of nerve stained with barium chloride. Bound barium was precipitated as the sulfate. Reaction product is present in the region of the "gap substance" (g) and attached to the outermost myelin lamellae. Some crystals have been lost during sectioning due to their large size. (a) Axon. Calibration bar: 1 μm. Provided by Dr. D. N. Landon.

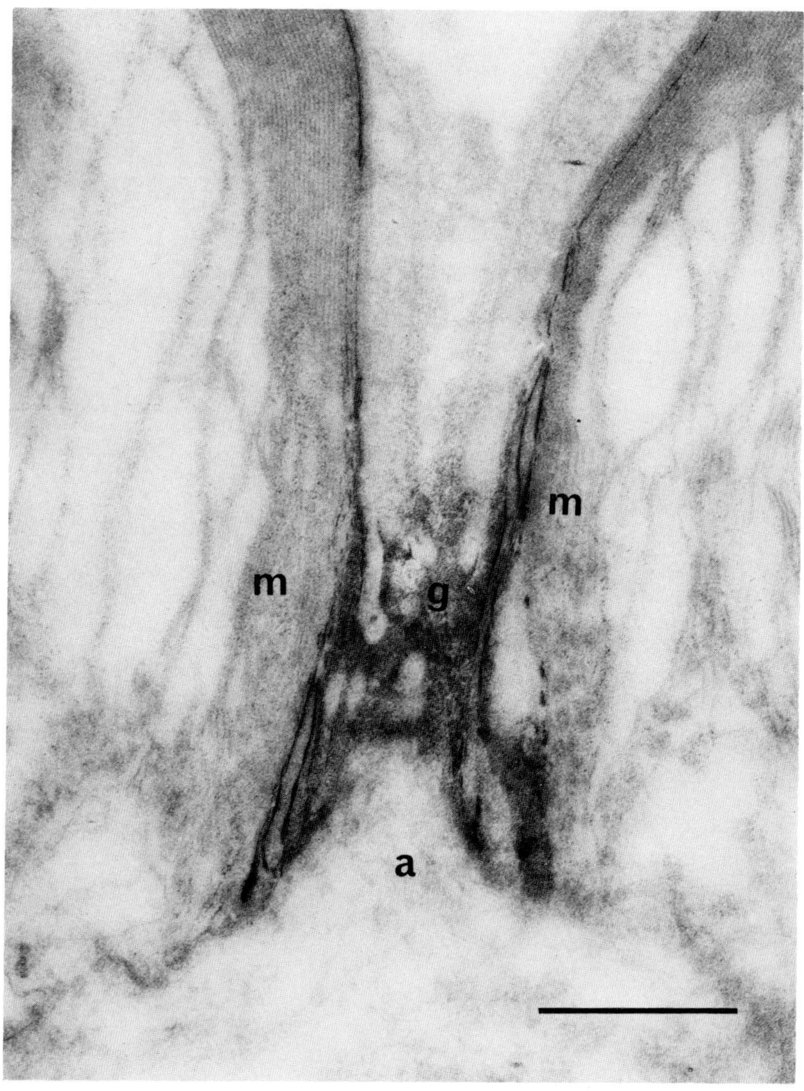

Fig. 11. Electron micrograph of peripheral nerve stained with lanthanum. Electron-dense product fills the "gap substance" (g) and stains the intraperiod lines of the outermost layers of the myelin (m) sheath. (a) Axon. Calibration bar: 0.5 μm. Provided by Dr. D. N. Landon.

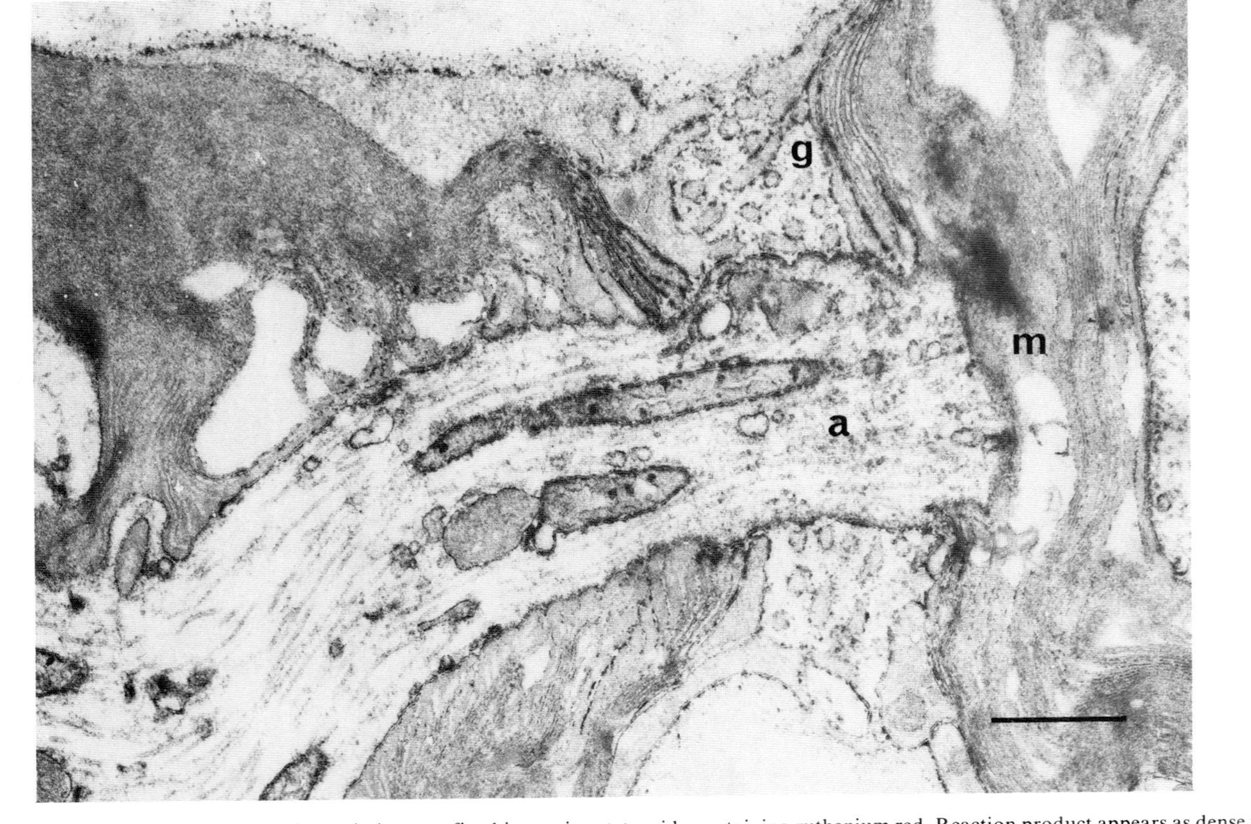

Fig. 12. Off-axis longitudinal section of rat sciatic nerve fixed in osmium tetroxide containing ruthenium red. Reaction product appears as dense flakes attached to membrane structures including those of Schwann cell fingers, the intraperiod lines of terminating myelin loops, and those of intraaxonal organelles. Scattered flakes are present within the "gap substance" (g), and the Schwann cell basement membrane appears as a thick layer in continuity with the plasma membrane of the Schwann cell. (a) Axon; (m) myelin. Calibration bar: 0.5 μm. Provided by Dr. D. N. Landon.

addition to the nodes (for a detailed account of alcianophilia, see Section 3.4). In the electron microscope, alcian blue was shown to enhance the osmiophilic extraneous cell coat of a variety of cells (Behnke and Zelander, 1970). A marginal increase in the electron density of the gap substance of nodes of peripheral nerve was found with alcian blue (Landon and Langley, unpublished results). A thin layer of alcianophilic material on the surface of all intraaxonal structures was also demonstrated in crayfish ventral axons, presumably the ultrastructural correlate of intraaxonal staining visible in sectioned nerve in the light microscope (Hinkley, 1973).

2.6. Lanthanum

In its tissue-staining properties, lanthanum displays a duality that depends largely on the experimental procedure adopted. It has been used to demonstrate extracellular spaces as an inert tracer, in the form of colloidal lanthanum hydroxide, when there is little evidence of true chemical binding to associated structures (Revel and Karnovsky, 1967; Hall and Williams, 1970; Bannister, 1972). In contrast, at slightly lower pHs, lanthanum has been shown to interact with cell-surface components (Doggenweiler and Frenk, 1965; Lesseps, 1967; Behnke, 1968*a,b*; Overton, 1968; Kahn and Overton, 1970) and with polysaccharides of both internal and external cellular matrices (Jong, 1949; Lane and Treherne, 1970). When used in this latter mode, trivalent lanthanum has been shown to produce a highly refractile ring around the nodal axon in sectioned nerve. Electron microscopy showed this to be caused by a dense amorphous lake of finely granular precipitate filling the gap substance and extending beneath the terminal loops and within the intraperiod line of the external three or four lamellae of the myelin sheath (Fig. 11) (Landon and Langley, 1971).

2.7. Ruthenium Red

Ruthenium red has been employed as a classic stain for plant pectin by botanists for a considerable time. In more recent years, its application to cytochemical studies on cell membranes has been explored in detail (Luft, 1966, 1971). In peripheral nerve, the distribution of stain is quite different from that of the other heavy metals discussed above when ruthenium red is included in the osmium tetroxide fixative. The chief characteristic (Fig. 12) is the appearance of electron-dense plaques attached to the plasma membranes of the axon and Schwann cells including the microvillous nodal projections. This deposit is more heavily concentrated on the external aspects of the membranes and also emphasizes the intraperiod line of terminating myelin lamellae at nodes. The zone between the Schwann cell plasma membrane and its basement membrane, electron-lucent in conventional preparations, is filled with similar but paler deposits, as is also the gap substance itself. The surface membranes of intraaxonal organelles such as mitochondria are also similarly enhanced.

3. Chemical Nature of Peripheral Nerve Polyanions

3.1. Chemical Blocking Reactions

From a comparison of the ultrastructural localizations of reaction product of the different stains, it is evident that more than a single entity is responsible for apparently similar staining patterns when observed with the light microscope. While colloidal iron is largely confined to the immediate vicinity of the nodal axolemma, the other metals (with the exception of ruthenium red) are seen to stain the full extent of the nodal gap substance. The use of both histochemical blocking reactions and enzymic digestion of peripheral nerve has confirmed that more than one substance demonstrable with these histochemical techniques is present at nodes, and has emphasized differences in their chemical natures.

Methylation and subsequent saponification have long been used by histochemists to differentiate between the basophilia of carboxyl and sulfate groups in tissues (Fisher and Lillie, 1954; Spicer and Lillie, 1959). Methylation with methanolic hydrochloric acid removes tissue sulfate groups, and in addition esterifies carboxyl groups, a process that is reversed by treatment with alkali (when it is accompanied by a return of basophilia). The effect of such procedures on nodal staining with colloidal iron clearly indicate the involvement of sulfate ester groups (Langley and Landon, 1968). Attempts to block the nodal staining by ferric, barium, and cupric ions by methylation have proved unsuccessful, treatment of nerves with organic solvents being sufficient to eliminate nodal ion-binding. One explanation of this phenomenon is that disruption of tissue components responsible for binding occurs (Langley, 1970), though an alternative suggestion, that lipids are involved in ion-binding, was made by Krammer and Lishka (1973).

3.2. Enzyme Digestion

The action of certain enzymes has been studied to further elucidate the nature of macromolecules responsible for the attachment of colloidal iron at nodes. Neuraminidase was found to have no effect either on the internodal staining after glutaraldehyde fixation or on nodal staining. Trypsin was also found to have no effect on nodal staining and produced only a limited reduction on internodal staining. Testicular hyaluronidase, however, completely eliminated nodal binding. This strongly indicates that sulfated mucopolysaccharides containing β-*N*-acetyl-D-glucosaminidic residues, such as chondroitin 4- and 6-sulfates, are present in the region immediately adjacent to the nodal axon membrane.

The action of enzymes on the ion-binding macromolecules differed markedly from their effect on the Hale-stainable moiety. While no effect on cation-binding was evinced by prolonged incubation of unfixed nerve with phospholipase C or neuraminidase, trypsin completely eliminated the ability of nodes to stain. Hyaluronidase reduced the intensity of nodal staining with ferric chloride, suggesting that part of such staining is due to components that are also responsible for binding colloidal iron. No effect of this enzyme has been observed on the nodal uptake of barium.

In a detailed study of copper-binding employing ammonium sulfide as the precipitant (Krammer *et al.*, 1972), neuraminidase was found to reduce the staining of plasma membranes situated in the nodal region. The involvement of the carboxyl groups of sialic acid residues was suggested, since in model systems the carboxyl groups of both fatty acids and free sialic acid have been shown to interact with copper.

3.3. Ionic Competition and Ion Exchange at the Node

The nature of the bond between various cations and tissue components has been explored by immersing fixed nerves in solutions containing a mixture of cationic species. In such experiments, the selection of the cation bound to nodal sites was found to depend both on the concentration of each cation present and on its charge (Langley, 1969). Trivalent ferric ions were found to be strongly preferred over divalent cations and divalent over monovalent ions. Attempts to replace cations already bound to nodes (but not precipitated for histochemical visualization) with other cations has been examined with regard to the charges of both displacing and displaced ions. Histochemical demonstration of such cation exchange showed that while bound Fe^{3+} could be displaced only by high concentrations of hydronium or trivalent ions such as ruthenium, Ba^{2+} could be exchanged by monovalent, divalent, and trivalent ions. The reversibility of this ion exchange was shown by nodal binding of Ba^{2+}, replacement of this with K^{+} (shown in the same location on parellel nerves by precipitation with sodium cobaltinitrite), and subsequent exchange of this ion with Ba^{2+}.

These experiments verify the contention that the bond between cations and the nodal matrix is in essence electrostatic. While charge appears to be of overriding significance in determining the affinity of nodes for different cations, hydrated ionic radius may also have some importance. The hydronium ion was peculiarly efficient, considering its low charge, in removing bound cations. The binding of cations to nodal components superficially resembles the interaction between complex cobalt ions and chondroitin sulfate (Vouras and Schubert, 1957; Matthews, 1960). Cobalt bound to the sulfate residues of the acid mucopolysaccharide are exchangeable with other ions, and charge determines the effectiveness of such exchange.

3.4. Nodal Alcianophilia

A particular example of ionic competition for nodal anionic sites is that encountered between organic cations and simple inorganic ions. This was shown to occur both with alcian blue (Langley, 1970) and with certain local anesthetics (Langley, 1973). Alcian blue, a copper-containing phthalocyanin, has the advantage over metallic cations in that it does not require conversion to an insoluble visible product, a process that may produce artifacts due to the physical rather than cytochemical nature of the material. The competitive interaction of alcian blue and inorganic cations with biological polyanions was extensively investigated (Scott, 1967*a*,*b*) and developed into a standard method for the demonstration of tissue acidic polysaccharides (Scott and Dorling, 1965). Under standardized staining conditions, a critical electrolyte concentration (CEC) (the salt concentration sufficient to prevent staining) may be determined for each tissue component with different electrolytes. Values obtained are characteristic of the acidic groups of the polyanions.

At low salt concentration, alcian blue was found to complex with a variety of peripheral nerve components. The CEC was assessed for Schwann cell nuclei, axons, the endoneurium, and mast cell granules in addition to nodal constituents. These CECs were found to be independent of the fixative used, if any. The CEC of the axonal alcianophilia reported by Langley (1971*a*) compares exactly with that determined in the electron microscope for the alcianophilic layer surrounding microtubules and the intraaxonal filamentous network also stained with alcian blue (Hinkley, 1973). A feature of nodal staining, in contrast to that of other elements, is that while nodal staining is completely suppressed at salt concentrations above 0.3 M $MgCl_2$ and 1.0 M NaCl, staining intensity abruptly increases below 0.15 M $MgCl_2$ and 0.6 M NaCl. A similar dependence of the intensity of nodal staining on pH was also observed, weak staining occurring at pH 3–4 and more intense staining apparent at lower hydrogen ion concentrations.

The effects of methylation and enzyme digestion were also examined. Testicular hyaluronidase reduced both the CEC and the intensity of nodal staining, and eliminated axonal staining, while not influencing the nuclear and mast cell interactions. Hyaluronidase was reported to have similar effect on the alcianophilia of crayfish nerve axons (Hinkley, 1973). Trypsin reduced the intensity of nodal staining, though the CEC remained unaltered, and neuraminidase was found to have no effect.

Methylation eliminated alcian-blue-binding by all components of peripheral nerve. Subsequent saponification restored the ability of axons and nuclei to stain, but the restored nodal staining was characterised by a lower CEC.

3.5. Conclusions on Nodal Cytochemistry

It is possible to conclude that two types of macromolecules are involved in nodal staining. One, situated close to the axon membrane, is destroyed both by hyaluronidase and by methylation followed by saponification, and binds both colloidal iron and Fe^{3+}. The second fills the annular zone surrounding the nodal axolemma between Schwann cell fingers, is bounded externally by the satellite cell basal lamina, and has a strong affinity toward Fe^{3+}, La^{3+}, Ba^{2+}, Cu^{2+}, and other cations. With alcian blue, it is possible to differentiate among separate classes of tissue polyanions, since the CEC for a particular acidic group depends on its ability to be polarized. In addition, the presence in nervous tissue of easily recognized and well-characterized polyanions (e.g., nucleic acids of Schwann cell nuclei and the heparin of mast cell granules) provides internal markers with which the nodal and axonal constituents may be compared.

The alcianophilia of the axon corresponds to that of a carboxylated polymer such as hyaluronic acid and confirms the earlier work of Abood and Abul-Haj (1956). A similar interpretation was placed on ultrastructural observations (Hinkley, 1973). It is probable, however, that hyaluronic acid, which in the hydrated state occupies a large domain, would be condensed and deposited on axonal structures during heavy metal fixation and tissue dehydration, resulting in an artifactual localization. It is possible that hyaluronic acid carboxyl groups are also responsible for the binding of ruthenium red in this location (Tani and Ametani, 1970).

The reduction of the nodal CEC by methylation followed by alkali treatment is compatible with the removal of sulfate esters, and it is reasonable to conclude that a limited layer of sulfated mucopolysaccharide overlays the axon at this site. This material probably contains both carboxyl and sulfate groups, though by comparison with the staining properties of mast cell granules the degree of sulfation is much lower than that existing in heparin. The nodal material that is insensitive to hyaluronidase and that is apparently somewhat digested by trypsin has a CEC similar to that of the axonal carboxylated polyanions. It is likely that its anionogenic groups are chemically similar. It is of interest to note that the greater affinity of trivalent ions like Fe^{3+} for this nodal component probably reflects their much greater polarizing power compared with that of divalent ions, which would result in their binding more strongly to easily polarized groups such as carboxyl.

The strikingly different picture produced by ruthenium red provides information on other membrane components. While the pale flakes found within the nodal gap substance probably result from interaction with the polyanions also responsible for cation-binding, it is likely that the marked staining of the external surface of membranes is due to other anionogenic groups. Acidic glycoproteins containing sialic acid are prevalent on mammalian cell surfaces, and significant amounts are present in peripheral nerve, where they are considered to be components of the axolemma and satellite cell internal and external membranes (Langley, 1971*b*). Part of the ruthenium red staining of nodal plasma membranes is probably due to their constituent sialoproteins. Ruthenium red was also shown to interact with sulfated mucopolysaccharides on the surface of other cells (Behnke, 1968*b*), and the nodal axolemmal staining may also partly reflect the presence of such substances at this site.

4. Functional Implications

The polyanions associated with the nodal axon membrane should be viewed within the general concept of the "greater membrane unit" (Lehninger, 1968). The carboxyl-rich matrix of the gap substance and the sulfated macromolecules close to the axolemma provide a microenvironment for the axon membrane from which its functional activity may not be isolated. Polysaccharide-rich cell coats are properly considered integral components of the cell membrane, and are thought to confer permeability, enzymatic, antigenic, or adhesion properties on, or to modify such properties in, the cells with which they are associated (Bennet, 1963; Anderson, 1968).

Such concepts are in striking contrast to membrane models employed by electrophysiologists in discussing theories of saltatory conduction, in which a semipermeable membrane is considered to be bathed on either side by dilute electrolyte solutions (Tasaki, 1968; Hodgkin, 1971). While the physicochemical properties of the material surrounding the nodal axon have not been quantitated, with the cytochemical data presently available it is possible to examine ways in which local mechanisms of ion fluxes may operate. The rising phase of the action potential of the nerve impulse results from an influx of Na^+ ions across the axon membrane, and both the steepness of the rise and the maximum amplitude attained are related to the magnitude of the external sodium ion concentration (Hodgkin, 1951). The largest myelinated fibers, which are required to cope with repetitive chains of action potentials, would appear to need a system capable of maintaining high levels of Na^+ ions immediately outside the axon. The histochemical data reviewed herein demonstrate that nodal polyanions may provide the basis for a cation reservoir that satisfies such a requirement. The distribution of ions in tissues is in part a function of the charge density of the polyelectrolytes present. In peripheral nerve, this results in the cation concentration being greatest close to the axon membrane, where the maximum effect on the electrical activity of the fiber may be exerted.

The possible existence of "ionic sinks" at nodes of Ranvier was first mooted by Abood and Abul-Haj (1956); these sinks are also considered to be present in the CNS of insects. Ganglionic fibers of the cockroach are embedded in an electron-dense matrix thought to be acid mucopolysaccharide (Ashurst, 1961). Treherne (1966) suggested that this material plays a role in maintaining a high concentration of sodium inside ganglia by cation-binding to fixed tissue anions. The nodal system described here also bears an unusual resemblance to the cells of salt-secreting glands of many marine vertebrates, which display highly differentiated cytological features (Abel and Ellis, 1966; Bulger, 1963). Microvilli project from the apical (secretory) surfaces of such cells and are embedded in a cap of sulfated mucopolysaccharide.

A further role of nodal polyelectrolytes is envisaged in relation to the potassium ions that flow from the axons during the falling phase of the action potential. While free diffusion of potassium into the endoneurium is prevented in unmyelinated fibers by a relatively closely applied satellite cell layer, no such structural barrier is apparent at the nodes of myelinated fibers. It is likely that the nodal polyanions would prevent loss of potassium at this site so that a return to the resting membrane state, a process involving an interdependent exchange of K^+ and Na^+ across the membrane (Judah and Ahmed, 1964), could easily be effected. The influence of K^+ ion diffusion barriers on the ionic flow across neural membranes has been considered in other systems (Adelman and Palti, 1969). Perineural macromolecules have also been considered to limit the ionic diffusion between hippocampal neurons and CSF (Greengard and Straub, 1958; Zuckerman and Glaser, 1970), and intercellular macromolecules in the CNS are known to share some histochemical properties with nodal polyanions (Bondareff, 1967). A function similar to that ascribed here in limiting the free diffusion of K^+ from the immediate vicinity of the sarcolemma during the propagation of action potentials has been suggested for the "glycocalyx" of striated muscle (Bennet, 1963).

Ionic mechanisms of impulse conduction suggested by Lieberman (1967), based on the concept of high-affinity sites on either side of the excitable membrane for Na^+ and K^+, employ membrane models that have certain features in common with others in which the membrane is lined on both sides by a negatively charged layer with which both Na^+ and K^+ can associate as counterions (Hechter, 1964). It has also been postulated that fixed-charge matrices associated with membranes can alone provide an adequate basis for explaining both the distributions of cations across membranes and the electrical excitation phenomenon (Ling, 1962). The interaction between the membrane-associated fixed anionic sites and the cations has been regarded as essentially electrostatic in nature, as has been found at the node of Ranvier. It is reasonable to conclude that the interactions of nodal polyelectrolytes with physiologically important cations, including sodium and potassium, are intimately involved in regulating the ionic balance during the passage of the action potential. In this regard, it is of interest to note that the affinity of polyanions for different cations is related to their exact chemical constitution and in particular to their content of the less easily polarized sulfate group. Polyanions with the readily polarized carboxyl group prefer divalent over monovalent countercations, while the reverse is true for purely sulfated macromolecules. For mixed polyanions, the affinities are arranged in the order $Ba^{2+} > Ca^{2+} > Mg^{2+} > K^+ \approx Na^+$ (Scott, 1967*a*). It is envisaged that Ca^{2+} or K^+ ions withdrawn electrotonically from the axon membrane (Tobias, 1964) promote the flow of sodium ions (maintained at high levels by polyanions outside the axon membrane in its resting state) into the axon as the action potential passes. It follows that the counterions associated with the polyanions at this site continually change in response to the ionic movements involved in impulse conduction.

Hess and Young (1952) emphasized the relative obscurity of the nature of the barrier existing around the node and remarked that ". . . until the permeability properties of the cementing disc, perinodal space and outer endoneurium are known it cannot be assumed that at the node there is free diffusion between the axon surface and whatever tissue fluids constitute the external environment of the fibre." Ultrastructural and cytochemical studies have now removed much of this obscurity. While the intrinsic limitations of histochemical methods do not permit quantitative evaluation of the permeability properties of tissue components, such methods have shown a specialized structure to exist surrounding the axon at nodes of Ranvier, and have enabled its chemical nature to be partially defined.

References

Abel, J. H., and Ellis, R. A., 1966, Histochemical and electron microscope observations on the salt secreting lachrymal glands of marine turtles, *Am. J. Anat.* **118**:337–358.

Abood, A. L., and Abul-Haj, S. K., 1956, Histochemistry and characteristics of hyaluronic acid in axons of peripheral nerve, *J. Neurochem.* **1**:119–125.

Adams, C. W. M., Bayliss, O. B., and Grant, R. T., 1968, Cholinesterases in peripheral nervous system. III. Validity of localization around the node of Ranvier, *Histochem. J.* **1**:68–77.

Adelman, W. J. Jr., and Palti, Y., 1969, The effects of external potassium and long duration voltage conditioning on the amplitude of sodium currents in the giant axon of the squid *Loligo pealei*, *J. Gen. Physiol.* **54**:589–606.

Anderson, W. A., 1968, Cytochemistry of sea urchin gametes. II. Ruthenium red staining of gamete membranes of sea urchins, *J. Ultrastruct. Res.* **24**:322–333.

Ashurst, D. E., 1961, An acid mucopolysaccharide in cockroach ganglia, *Nature* (*London*) **191**:1224–1225.

Bannister, L. H., 1972, Lanthanum as an intercellular stain, *J. Microsc.* **95**:413–419.

Behnke, O., 1968*a*, Electron microscopical observations on the surface coating of human blood platelets, *J. Ultrastruct. Res.* **24**:51–69.

Behnke, O., 1968*b*, An electron microscope study of the megacaryocyte of the rat bone marrow. I. The development of the demarcation membrane system and the platelet surface coat, *J. Ultrastruct. Res.* **24**:412–433.

Behnke, O., and Zelander, T., 1970, Preservation of intercellular substances by the cationic dye alcian blue in preparative procedures for electron microscopy, *J. Ultrastruct. Res.* **31**: 424–438.

Bennet, H. S., 1963, Morphological aspects of extracellular polysaccharides, *J. Histochem. Cytochem.* **11**:14–23.

Bondareff, W., 1967, An intercellular substance in rat cerebral cortex: Submicroscopic distribution of ruthenium red, *Anat. Rec.* **157**:527–536.

Bulger, R. E., 1963, Fine structure of the rectal (salt secreting) gland of the spiny dogfish, *squalus acanthus*, *Anat. Rec.* **147**:95–127.

Doggenweiler, C. F., and Frenk, S., 1965, Staining properties of lanthanum on cell membranes, *Proc. Natl. Acad. Sci. U.S.A.* **53**:425–430.

Elfvin, L. G., 1961, The ultrastructure of the nodes of Ranvier in cat sympathetic nerve fibers, *J. Ultrastruct. Res.* **5**:374–387.

Fisher, E. R., and Lillie, R. D., 1954, The effect of methylation on basophilia, *J. Histochem. Cytochem.* **2**:81–87.

Gasser, H. S., 1952, Discussion, in: The hypothesis of saltatory conduction, *Cold Spring Harbor Symp. Quant. Biol.* **17**:32–36.

Gerebtzoff, M. A., 1959, *Cholinesterases: A Histochemical Contribution to the Solution of Some Functional Problems,* p. 195, Pergamon Press, New York—London.

Gerebtzoff, M. A., 1964, Histoenzymologie de l'étranglement de Ranvier dans la myélinisée périphérique normale et en dégénérescence, *Ann. Histochem.* **9**:209–216.

Grant, R. T., Bayliss, O. B., and Adams, C. W. M., 1967, Cholinesterases in peripheral nervous system. II. The motor, sensory and sympathetic nerves in the rabbit ear perichondrium and rat cremaster muscle, *Brain Res.* **6**:457–474.

Greengard, P., and Straub, R. W. 1958, After-potentials in mammalian non-myelinated nerve fibers, *J. Physiol.* **144**:442–462.

Hall, S. M., and Williams, P. L., 1970, The distribution of electron dense tracers in peripheral nerve fibers, *J. Cell Sci.* **8**:541–555.

Hechter, O., 1964, Role of water structure in the molecular organization of cell membranes, *Neurosci. Res. Prog. Bull.* **II**:36–66.

Herbst, F., 1965, Untersuchungen über Metallsalzreaktionen an den Ranvierschen Schnürringen, *Acta Histochem.* (*Jena*) **22**:223–233.

Hess, A., and Young, J. Z., 1952, The nodes of Ranvier, *Proc. R. Soc. London Ser. B* **140**: 301–320.

Hinkley, R. E., 1973, Axonal microtubules and associated filaments stained by alcian blue, *J. Cell. Sci.* **13**:753–761.

Hodgkin, A. L., 1951, The ionic basis of electrical activity in nerve and muscle, *Biol. Rev.* **26**: 339–409.

Hodgkin, A. L., 1971, *The Conduction of the Nervous Impulse,* University of Liverpool Press.

Jong, H. G. B. de, 1949, in: *Colloid Science* (H. R. Kruty, ed.), Vol. 2, Chapt. 9, Elsevier, New York.

Judah, J. D., and Ahmed, K., 1964, The biochemistry of sodium transport, *Biol. Rev.* **39**:160–193.

Kahn, T. A., and Overton, J., 1970, Lanthanum staining of developing chick cartilage and reaggregating cartilage cells, *J. Cell Biol.* **44**:433–438.

Krammer, E. B., and Lishka, M. F., 1973, Schwermetallaffine Strukturen des periphen Nerven. I. Potentieller Störfaktor beim cytochemischen AChE-Nachweis, *Histochemie* **36**: 269–282.

Krammer, E. B., Buchinger, W., and Lishka, M., 1972, Über Bindungstellen zweiwertiger Schwermetallionen am Ranvierschen Schnürring, *Histochemie* **29**:97–102.

Landon, D. N., and Langley, O. K., 1971, The local chemical environment of nodes of Ranvier: A study of cation binding, *J. Anat.* **108**:419–432.

Landon, D. N., and Williams, P. L., 1963, Ultrastructure of the node of Ranvier, *Nature* (*London*) **199**:575–577.

Lane, N. J., and Treherne, J. E., 1970, Lanthanum staining of neurotubules in axons from cockroach ganglia, *J. Cell Sci.* **7**:217–231.

Langley, O. K., 1969, Ion-exchange at the node of Ranvier, *Histochem. J.* **1**:295–309.

Langley, O. K., 1970, The interaction between peripheral nerve polyanions and alcian blue, *J. Neurochem.* **17**:1535–1541.

Langley, O. K., 1971*a*, A comparison of the binding of alcian blue and inorganic cations to polyanions in peripheral nerve, *Histochem. J.* **3**:251–260.

Langley, O. K., 1971*b*, Sialic acid and membrane contact relationships in peripheral nerve, *Exp. Cell Res.* **68**:97–105.

Langley, O. K., 1973, Local anaesthetics and nodal polyanions in peripheral nerve, *Histochem. J.* **5**:79–86.

Langley, O. K., and Landon, D. N., 1968, A light and electron histochemical approach to the node of Ranvier of peripheral nerve fibers, *J. Histochem. Cytochem.* **15**:722–731.

Langley, O. K., and Landon, D. N., 1969, Copper binding at nodes of Ranvier: A new electron-histochemical technique for the demonstration of polyanions, *J. Histochem. Cytochem.* **17**:66–69.

Lehninger, A. L., 1968, The neuronal membrane, *Proc. Natl. Acad. Sci. U.S.A.* **60**:1069–1080.

Lesseps, R. J., 1967, The removal by phospholipase C of a layer of La^{3+}-staining material external to the membrane in embryonic chick cells, *J. Cell Biol.* **34**:173–183.

Lieberman, E. M., 1967, Structural and functional sites of action of ultraviolet radiations in crab nerve fibers, *Exp. Cell Res.* **47**:518–535.

Ling, G. N., 1962, *Physical Theory of the Living State: Association–Induction Hypothesis,* Blaisdell, New York.

Luft, J. H., 1966, Fine structure of capillary and endocapillary layer as revealed by ruthenium red, *Fed. Proc. Fed. Am. Soc. Exp. Biol.* **25**:1773–1783.

Luft, J. H., 1971, Ruthenium red and violet. II. Fine structural localization in animal tissues, *Anat. Rec.* **171**:369–416.

Macallum, A. B., 1905, On the distribution of potassium in animal and vegetable cells, *J. Physiol. (London)* **32**:95–128.

Macallum, A. B., and Menten, M. C., 1906, On the distribution of chlorides in nerve cells and fibres, *Proc. R. Soc. London Ser. B* **177**:165–193.

Matthews, M. B., 1960, Trivalent cation binding of acid mucopolysaccharides, *Biochim. Biophys. Acta* **37**:288–295.

Mladenov, S., 1965, Histochemical investigations on the enzyme activity in the nodes of Ranvier, *C. R. Acad. Bulg. Sci.* **18**:1047–1049.

Nemiloff, A., 1908, Einige Beobachtungen über den Bau des Nervengewebes bei Ganoiden und Knochen-fischen. II. Bau der Nervenfasern, *Arch. Mikrosk. Anat.* **72**:575–606.

Overton, J., 1968, Localized lanthanum staining of the intestinal brush border, *J. Cell Biol.* **38**:447–452.

Quick, D. C., and Waxman, S. G., 1977*a*, Specific staining of the axon membrane at nodes of Ranvier with ferric ion and ferrocyanide, *J. Neurol. Sci.* **31**:1–11.

Quick, D. C., and Waxman, S. G., 1977*b*, Ferric ion, ferrocyanide and inorganic phosphate as cytochemical reactants at peripheral nodes of Ranvier, *J. Neurocytol.* **6**:555–570.

Ramón y Cajal, S., 1955, *Histologie du Système Nerveux de l'Homme et des Vertèbres*, Vol. 1 (translated by L. Azonlay), pp. 252–286, Maloine, Madrid.

Ranvier, L. A., 1871, Contributions a l'histologie et à la physiologie des nerfs périphérique, *C. R. Acad. Sci.* **73**:1168–1171.

Revel, J.-P., and Karnovsky, M. J., 1967, Hexagonal array of subunits in intracellular junctions of the mouse heart and liver, *J. Cell Biol.* **33**:C7.

Robertson, J. D., 1957, The ultrastructure of nodes of Ranvier, *J. Physiol. (London)* **137**:8P.

Scott, J. E., 1967*a*, Ion binding in solutions containing acid mucopolysaccharides, in: *The Chemical Physiology of Mucopolysaccharides* (G. Quintarelli, ed.), pp. 171–186, Little, Brown, Boston.

Scott, J. E., 1967*b*, Patterns of specificity in the interaction of organic cations with acid mucopolysaccharides, in: *The Chemical Physiology of Mucopolysaccharides* (G. Quintarelli, ed.) pp. 219–231, Little, Brown, Boston.

Scott, J. E., and Dorling, J., 1965, Differential staining of acid glycosaminoglycans (mucopolysaccharides) by alcian blue in salt solutions, *Histochemie* **5**:221–233.

Sjöstrand, F. S., 1963, The structure and formation of the myelin sheath, in: *Mechanisms of Demyelination*, pp. 1–43, McGraw-Hill, New York.

Spicer, S. S., and Lillie, R. D., 1959, Saponification as a means of selectively reversing the methylation blockade of tissue basophilia, *J. Histochem. Cytochem.* **7**:123–125.

Tani, E., and Ametani, T., 1970, Substructure of microtubules in brain nerve cells as revealed by ruthenium red, *J. Cell Biol.* **46:**159–165.

Tasaki, I., 1968, *Nerve Excitation—A Macromolecular Approach,* Charles C. Thomas, Springfield, Illinois.

Tobias, J. M., 1964, A chemically specified molecular mechanism underlying excitation in nerve: A hypothesis, *Nature* (*London*) **203:**13–17.

Treherne, J. E., 1966, *The Neurochemistry of Arthropods,* Cambridge University Press, Cambridge, England.

Uzman, B. G., and Villegas, G. M., 1960, A comparison of nodes of Ranvier in sciatic nerves with node-like structures in optic nerves of the mouse, *J. Biophys. Biochem. Cytol.* **7:**761–762.

Vouras, M., and Schubert, M., 1957, The outer sphere association of chondroitin sulfate with polyvalent complex cations, *J. Am. Chem. Soc.* **79:**792–795.

Zenker, W., 1964, Über die Aufärbung der Ranvierschen Schnürrige beim Koelle-Verfahren zum histochemischen Nachweis der Cholinesterase, *Acta Histochem.* (*Jena*) **19:**67–72.

Zenker, W., 1969, Cholinesterase und Metallophilie an den Ranvierschen Schnürringen, *Acta Histochem.* (*Jena*) **33:**247–273.

Zuckerman, E. C., and Glaser, G. H., 1970, Activation of experimental epileptogenic foci: Action of increased K^+ in extracellular spaces of the brain, *Arch. Neurol.* **23:**358–364.

11

Glycoproteins in Myelin and Myelin-Related Membranes

Richard H. Quarles

1. Historical Background and Introduction

Glycoproteins are well established as cell-surface components and appear to participate in specific cell–cell interactions. Myelin is formed as an extension of the plasma membrane of the oligodendrocyte in the central nervous system (CNS) and the Schwann cell in the peripheral nervous system (PNS). Therefore, it is reasonable to expect that glycoproteins could be present in the myelin membrane and be involved in the formation of the spiraled myelin sheath around the axon. Histochemical studies at the light-microscopic level by Wolman (1957) and Wolman and Hestrin-Lerner (1960) indicated the presence of polysaccharide in myelin. On the basis of early electron-microscopic studies, Robertson (1959) suggested that the intraperiod region of peripheral myelin contained polysaccharide or glycoprotein material, possibly derived from the surface of the Schwann cell. Later, Peterson and Pease (1972) developed a specialized embedding technique and stained with silicotungstic acid to provide electron-microscopic evidence for the presence of glycoproteins in the intraperiod line of peripheral myelin, supporting Robertson's earlier hypothesis. The first direct biochemical demonstration of glycoproteins in purified myelin was provided by Margolis (1967), who showed that *N*-acetylglucosamine is present in several protein fractions obtained from bovine CNS myelin and that glycosaminoglycans are not present. Nevertheless, the nature of the glycoprotein components in myelin remained poorly defined.

Abbreviations used in this chapter: (Con A) concanavalin A; (CNP) 2′,3′-cyclic nucleotide 3′-phosphohydrolase; (CNS) central nervous system; (EAN) experimental allergic neuritis; (MAG) myelin-associated glycoprotein; (PAS) periodic–Schiff; (PNS) peripheral nervous system; (SDS) sodium dodecyl sulfate.

Richard H. Quarles • Section on Myelin and Brain Development, Developmental and Metabolic Neurology Branch, NINCDS, National Institutes of Health, Bethesda, Maryland.

Our work on myelin began with the observation that the [^{3}H]fucose-labeled glycoproteins in a highly purified preparation of rat brain myelin gave a very different pattern on sodium dodecyl sulfate (SDS)–polyacrylamide gels from those in the whole particulate fraction (Fig. 1) (Quarles *et al.*, 1972, 1973*a*). In particular, there was a prominent fucose-labeled peak with an apparent molecular weight of approximately 100,000. This pattern of labeled glycoproteins was unique for myelin, and after the 100,000-dalton component was detected by the sensitive labeling technique, it was possible to show it by periodic acid–Schiff (PAS) staining (Fig. 2). This glycoprotein is present in purified CNS myelin fractions from all mammalian species that have been examined, including humans (Matthieu *et al.*, 1974*a*; Everly *et al.*, 1977). Although the glycoprotein is quantitatively a minor component of the whole myelin fraction, extensive studies (summarized in Section 3.1) showed that it is in myelin or closely related membranes, or both, and is not in an unrelated contaminant of the myelin fraction. Therefore, we refer to this CNS glycoprotein with a molecular weight of about 100,000 daltons as the "myelin-

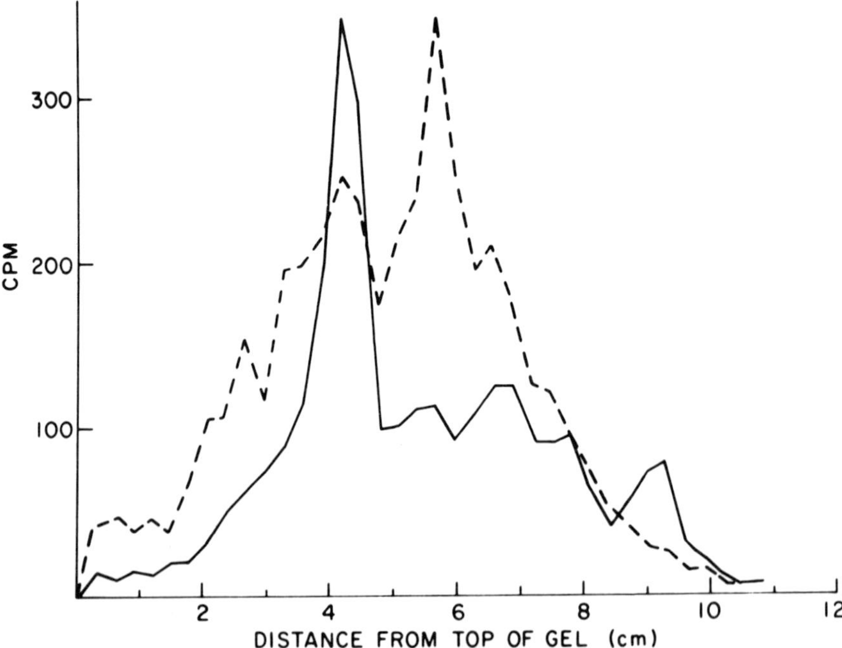

Fig. 1. Polyacrylamide gel electrophoresis of fucose-labeled glycoproteins of CNS myelin in the presence of SDS. In this experiment, 21-day-old rats were injected intracranially with [^{3}H]- or [^{14}C]fucose to label glycoproteins. The ^{14}C-labeled myelin fraction from one group of rats was electrophoresed on the same gel with the ^{3}H-labeled, whole 100,000*g* particulate fraction from the other group. (—) ^{14}C; (----) ^{3}H. The prominent peak in the myelin fraction is about 4.5 cm down the gel. From Quarles *et al.* (1972), with the permission of Academic Press.

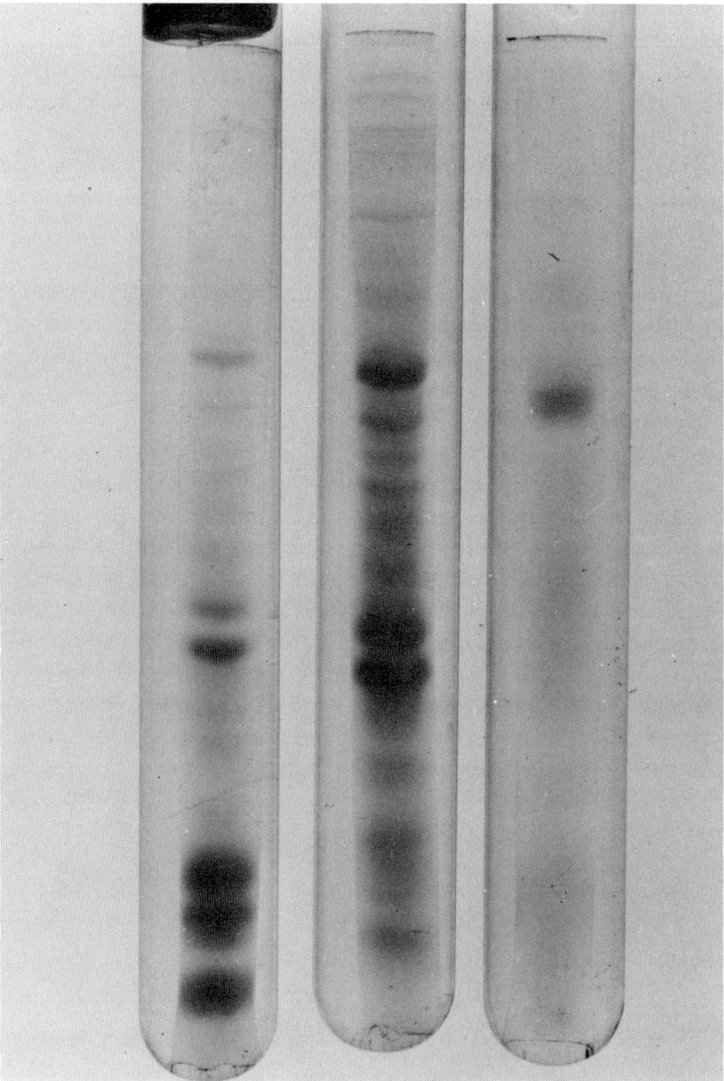

Fig. 2. Staining of CNS myelin proteins on SDS 5% polyacrylamide gels. *Left*: Gel showing total myelin proteins stained with Fast Green. The three bands at the bottom of the gel are the major proteins of CNS myelin; from bottom to top, they are the small basic protein, the large basic protein, and proteolipid protein, respectively. Also, there are a number of high-molecular-weight proteins in the upper two thirds of the gel. These high-molecular-weight proteins are enriched in the insoluble residue obtained when myelin is extracted with chloroform–methanol (2:1, vol./vol.). *Middle*: Gel showing the proteins in this residual fraction after electrophoresis and staining with Fast Green. *Right*: Gel showing the same fraction stained with PAS reagent for carbohydrate. The glycoprotein is clearly seen migrating about one third of the way down the gel. From Quarles (1975), with the permission of Raven Press.

associated glycoprotein," abbreviated MAG. In Section 3.1, I will describe in some detail what we mean by myelin-related membranes.

After MAG was detected in CNS myelin, purified sciatic nerve myelin was examined to see whether the same glycoprotein was present in peripheral myelin (Everly *et al.*, 1973). Neither labeling nor Schiff staining gave any indication of the high-molecular-weight glycoprotein that had been detected in purified brain myelin. Much to our surprise, however, the major protein of peripheral myelin stained with PAS reagents and was labeled *in vivo* by radioactive fucose (Fig. 3), indicating that it was glycosylated. Wood and Dawson (1973) and Savolainen (1972) independently made similar observations at about the same time. This peripheral myelin glycoprotein has now been purified and partially characterized (Brostoff *et al.*, 1975; Kitamura *et al.*, 1976; Roomi *et al.*, 1978*a*).

The studies described above indicated, then, that the glycoproteins in purified CNS myelin are quantitatively minor components, while peripheral myelin has a major protein constituent that is glycosylated. To obtain information about the magnitude of the difference in the glycoprotein-carbohydrate content of the two types of myelin, quantitative gas–liquid chromatographic (GLC) analyses were performed (Quarles and Everly, 1977). The levels of protein-bound carbohydrate in purified rat brain and sciatic nerve myelin were compared (Table I). Peripheral myelin contained 4- to 7-fold more of each of the sugars that are characteristic of glycoproteins than did central myelin. These analytical studies confirmed the concept that glycoproteins are quantitatively more important components of peripheral myelin than of central myelin. The possible significance of this difference for the structures of central and peripheral myelin will be considered in Section 4.

2. Peripheral Myelin

2.1. Introduction

The finding that the major protein of peripheral myelin was a glycoprotein was made shortly after it had been demonstrated in several laboratories that the proteins in peripheral myelin are different from those in central myelin (for a detailed review, see Braun and Brostoff, 1977). Although both types of myelin have basic proteins, PNS myelin does not contain the proteolipid protein that is a major component of CNS myelin. Instead, the major protein of peripheral myelin is insoluble in neutral chloroform–methanol and has a molecular weight of about 30,000. It is this protein, referred to as the P_0 protein in the widely used terminology of Greenfield *et al.* (1973), that is glycosylated. After the initial demonstrations that this protein was glycosylated (Savolainen, 1972; Everly *et al.*, 1973; Wood and Dawson, 1973), this was confirmed by Schiff staining in other

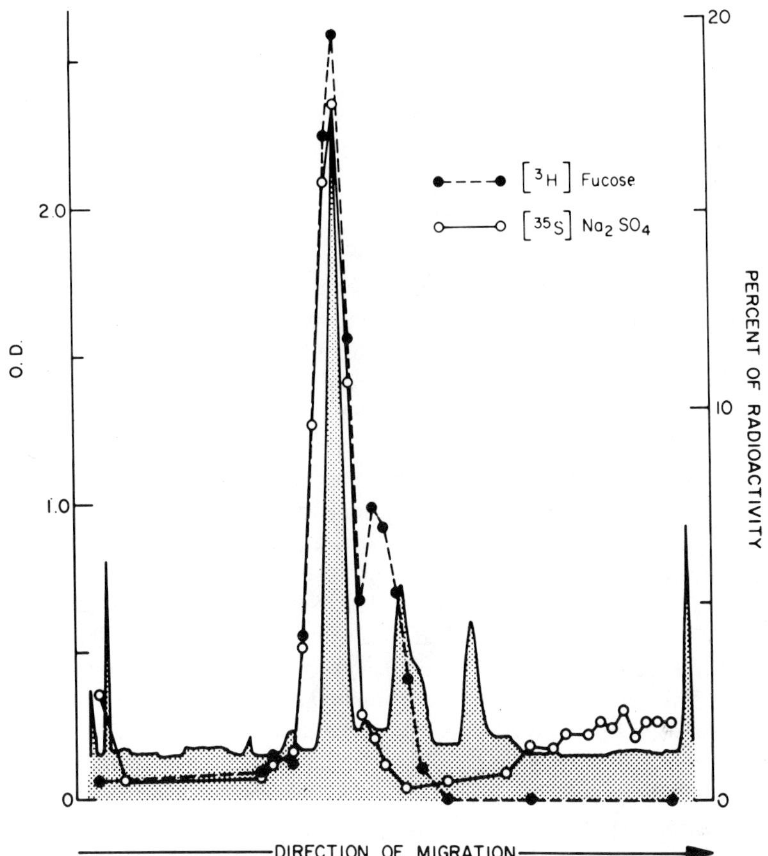

Fig. 3. Polyacrylamide gel electrophoresis of [^{3}H]fucose- and [^{35}S]sulfate-labeled glycoproteins of rat sciatic nerve myelin on a 10% polyacrylamide gel in the presence of SDS. The shaded area is a densitometric scan of Fast-Green-stained proteins. The highest peak is the major P_0 protein and the two peaks below it are the P_1 and P_2 basic proteins. (•) Curve showing that [^{3}H]fucose was incorporated into the P_0 protein and a second smaller component after injection into the nerve. [^{35}S]sulfate (o) was incorporated only into the P_0 protein. There was very little of either isotope in the high-molecular-weight portion of the gel where MAG of CNS myelin would run. From Matthieu *et al.* (1975*a*), with the permission of Elsevier Scientific Publishing Co.

laboratories (Singh and Spritz, 1974; Peterson, 1976). In most investigations, another glycoprotein of lower molecular weight was also shown to be present in smaller amount in peripheral myelin (Everly *et al.*, 1973; Wood and Dawson, 1973; Singh and Spritz, 1974; Matthieu *et al.*, 1975*a*; Kitamura *et al.*, 1976; Quarles *et al.*, 1978; Singh *et al.*, 1978; Roomi *et al.*, 1978*b*). The most recent reports showed the presence of P_0 and other glycoproteins in purified peripheral myelin from a wide variety of species (Koski and Max, 1977; Quarles *et al.*, 1978; Singh *et al.*, 1978; Roomi, *et al.*, 1978*b*). In some of these recent investigations, several minor glycoproteins in both the low- and high-molecular-weight regions of polyacrylamide gels were detected in addition to P_0 (Koski and Max, 1977; Singh *et al.*, 1978; Roomi *et al.*, 1978*b*). However, the very minor glycoproteins could be in contaminants of the purified myelin fraction, and it remains to be established which are true components of the myelin membrane.

2.2. Chemistry

Some of the early investigations qualitatively revealed the presence of fucose, mannose, galactose, glucosamine, and sialic acid in the PNS myelin glycoproteins (Everly *et al.*, 1973; Wood and Dawson, 1974*a*; Brostoff *et al.*, 1975). However, the first detailed carbohydrate analyses on the P_0 glycoprotein were those of Kitamura *et al.* (1976), who isolated it from bovine intradural spinal roots by gel filtration in the presence of SDS, but called it the BR protein in their terminology. They found 2.6 mol *N*-acetylglucosamine, 2.7 mol mannose, 0.8 mol fucose, 1 mol galactose, and 0.8 mol sialic acid per molecule of protein. *N*-acetylgalactosamine was not detected, and also was not found in the glycopeptides isolated from whole rat sciatic nerve myelin (Quarles and Everly, 1977). The NH_2-terminal amino acid of the glycoprotein was isoleucine. More recently, Roomi *et al.* (1978*a*) reported very similar results for the P_0 glycoprotein purified from rabbit sciatic nerve myelin. They conclude that the carbohydrate is in a single nonasaccharide unit consisting of 3 mannose, 3 *N*-acetylglucosamine, 1 sialic acid, 1 galactose, and 1 fucose. They found a high content of nonpolar amino acids in the polypeptide and also identified isoleucine at the NH_2-terminus. It should be emphasized that the sugar content of P_0 is rather low, accounting for only about 5% by weight. The absence of galactosamine suggests that the oligosaccharide unit is linked *N*-glycosidically to asparagine residues. [^{35}S]Sulfate incorporation studies in the rat showed that the P_0 glycoprotein is sulfated (Fig. 3) (Matthieu *et al.*, 1975*a*), probably on sugar residues.

Table I. Glycoprotein-Carbohydrate in PNS and CNS Myelin[a]

Carbohydrate	PNS	CNS
Fucose	8.7 ± 1.4	1.3 ± 0.27
Mannose	37 ± 3.3	5.4 ± 0.78
Galactose	14 ± 0.9	2.7 ± 0.37
N-Acetylglucosamine	28 ± 4.6	5.8 ± 0.88
N-Acetylneuraminic acid	11 ± 1.2	1.7 ± 0.14

[a]The glycoproteins in purified rat brain and sciatic nerve myelin were converted to glycopeptides by exhaustive pronase digestion. The glycopeptides were purified by gel filtration and analyzed for sugars by GLC of trimethylsilyl derivatives. Values are given as nanomoles of each sugar per milligram protein in the whole myelin fraction. Data are from Quarles and Everly (1977), with the permission of Elsevier/North-Holland Biomedical Press.

Treatment of peripheral myelin with trypsin degrades the P_0 glycoprotein to a lower-molecular-weight glycoprotein (Wood and Dawson, 1974*a*; Peterson, 1976; Roomi and Eylar, 1978). Chemical analysis of this smaller glycoprotein (mol.wt. 19,000) revealed the same nonasaccharide as in P_0, a greater content of hydrophobic amino acids, and methionine at the NH_2-terminus (Roomi and Eylar, 1978). The results suggest that about one third of the P_0 glycoprotein at the amino terminus is relatively polar and protrudes above the lipid leaflet of the myelin membrane, where it is susceptible to trypsin. The effects of trypsin treatment also raise the possibility that one or more of the minor glycoproteins with lower molecular weight than P_0 that have been detected in purified myelin (see previous section) are proteolytic degradation products of P_0. However, a smaller glycoprotein purified from bovine myelin by Kitamura *et al.* (1976) had a carbohydrate composition different from that of P_0. Also, [^{35}S]sulfate-incorporation studies showed that in contrast to P_0, there is a smaller glycoprotein that is not sulfated (Fig. 3) (Matthieu *et al.*, 1975*a*). A variety of nomenclatures (e.g., X, Y, PAS-II, P_3) further complicates discussion of these minor glycoproteins, and clarification of their exact number and their possible relationship to P_0 remains a problem for further investigations.

2.3. Localization

Since there is so much P_0 glycoprotein in purified PNS myelin, it must be a major component of the lipid–protein matrix of the myelin membrane along with the basic proteins and lipids. Several lines of evidence indicate that it is at least partially localized in the intraperiod line where the external surfaces of the Schwann cell membranes come together in forming the multilamellar myelin. The evidence includes correlative electron-microscopic and biochemical studies (Sea and Peterson, 1975) and a molecular probe technique with lactoperoxidase-catalyzed iodination (Peterson and Gruener, 1978). Also, electron-microscopic histochemical studies involving staining with silicotungstic acid (Peterson and Pease, 1972) and concanavalin A (Con A)-binding (Wood and McLaughlin, 1975) revealed carbohydrate at the intraperiod line, indicating that the oligosaccharide units are present at this localization. The effect on P_0 when purified myelin is treated with trypsin, as described in the previous section, suggests that the molecule has a hydrophobic domain that is buried in the lipid bilayer, where it is protected from the enzyme (Roomi and Eylar, 1978). The structural role of the glycoprotein in PNS myelin membranes is considered in more detail in Section 4.

2.4. Development

In view of the probable functions of cell-surface glycoproteins in cell–cell interactions, it is interesting to speculate on the possible role of glycoprotein-carbohydrate in Schwann cell–axon interactions or in spiraling of the myelin sheath during development. However, nothing is known about developmental changes in the carbohydrate portion of the P_0 glycoprotein. The amount of the P_0 glycoprotein in peripheral nerve increases during development and correlates well with the deposition of myelin (Wiggins *et al.*, 1975; Wood and Engel, 1976). When the composition of purified peripheral myelin was examined, two studies showed a decrease in P_0 relative to the basic proteins during development (Zgorzalewicz *et al.*, 1974; Koski and Max, personal communication), whereas another investigation revealed little change (Wiggins *et al.*, 1975).

Recently, Gould (1977) conducted very elegant autoradiographic studies with [^{3}H]fucose to obtain information about the incorporation of glycoproteins into developing and mature sciatic nerve myelin. Fucosylation of the P_0 glycoprotein occurs in the Golgi apparatus of the Schwann cell, following which the protein is transferred to the outermost region of the myelin sheath in developing and mature animals. Diffusion from the site where the glycoprotein is added to the myelin membrane to the remainder of the myelin sheath is slow, occurring at about 1/100th the rate of phospholipid diffusion.

2.5. Pathology

Glycoproteins may be involved in demyelinating conditions of the PNS. Early histochemical studies (Wolman, 1957) suggested the involvement of polysaccharide-containing material in Wallerian degeneration. More recently, it was shown that the P_0 glycoprotein decreases more rapidly than other proteins during Wallerian degeneration (Wood and Dawson, 1974*b*; Peterson and Sea, 1975). Diphtheria-toxin-induced demyelination is accompanied by a decreased biosynthesis of the P_0 glycoprotein as well as basic proteins (Pleasure *et al.*, 1973).

In view of the known role of glycoproteins as cell-surface antigens, an involvement of the P_0 glycoprotein in experimental allergic neuritis (EAN) or the similar Landry–Guillain–Barré syndrome in humans would not be surprising. There is one report that P_0 can induce mild EAN (Wood and Dawson, 1974*a*). However, other attempts to induce EAN with purified P_0 were unsuccessful (Brostoff *et al.*, 1975; Kitamura *et al.*, 1976), and there is now considerable evidence that the P_2 basic protein is the principal antigen involved (Brostoff, 1977; Brostoff *et al.*, 1977). Nevertheless, it has generally been more difficult to induce EAN with purified proteins than with whole myelin (Brostoff, 1977), and it is known that guinea pigs with EAN induced by whole myelin have cells sensitized to the P_0 glycoprotein (Carlo *et al.*, 1975). Therefore, immunological roles for P_0 and other glycoproteins in EAN and in Landry–Guillain–Barré syndrome remain possibilities. If so, differences in the glycoproteins between myelin of motor and sensory nerves could be responsible for differences in the susceptibility of various nerves to EAN (Koski and Max, 1977; Hardwicke *et al.*, 1977).

3. Central Myelin

3.1. Introduction and Localization

Unlike the significance of the P_0 glycoprotein of peripheral myelin, which must be a major structural component of the myelin sheath, the significance of the high-molecular-weight glycoprotein that is present in small amount in purified CNS myelin (MAG) has been more difficult to elucidate. The first possibility that had to be considered was that this glycoprotein was not a myelin constituent at all, but was in a contaminant of the myelin fraction. Although the myelin preparations in which it was detected met the usual criteria of purity (Norton, 1977), this did not rule out the possibility that it could be in a small amount of another membrane with a very high content of this particular glycoprotein. Therefore, a considerable amount of effort was spent obtaining information bearing on this question. Although these experiments cannot be reviewed here in detail, they involved regional and developmental studies (Quarles *et al.*, 1973*a*; Druse *et al.*, 1974; Everly *et al.*, 1977) as well as experiments with myelin-deficient mutant mice (Matthieu *et al.*, 1974*b,c*). All results indicated that the glycoprotein was associated with myelin in the CNS and was not in an unrelated contaminant of the myelin fraction. One of the most convincing experiments involved the use of Jimpy mutant mice, which form very little myelin (Matthieu *et al.*, 1974*c*). Double-labeling experiments were done in which Jimpy mice were injected with $[^{14}C]$fucose and age-matched controls were injected with $[^{3}H]$fucose (Fig. 4). These experiments, the details of which are described in the Fig. 4 caption, showed that the glycoprotein found in normal myelin was not obtained from Jimpy brain during the myelin-purification procedure, although a substantial amount of normal myelin was present during the isolation to potentially trap nonmyelin contaminants from the Jimpy brains. Since the principal abnormality in Jimpy mice is an almost complete absence of myelin, the results are strong evidence for the close association of the glycoprotein with myelin in normal brains. This experiment and numerous other experiments provide a composite of strong circumstantial evidence indicating that MAG is in myelin or intimately associated structures. Very recently, we utilized an antiserum prepared to purified MAG in the immunocytochemical procedure previously used to localize myelin basic protein at the light-microscopic level (Sternberger *et al.*, 1978). These experiments give direct confirmation for the specific localization of MAG in the oligodendrocyte–myelin complexes of developing rat brain (Sternberger, Quarles, Itoyama, and Webster, unpublished results).

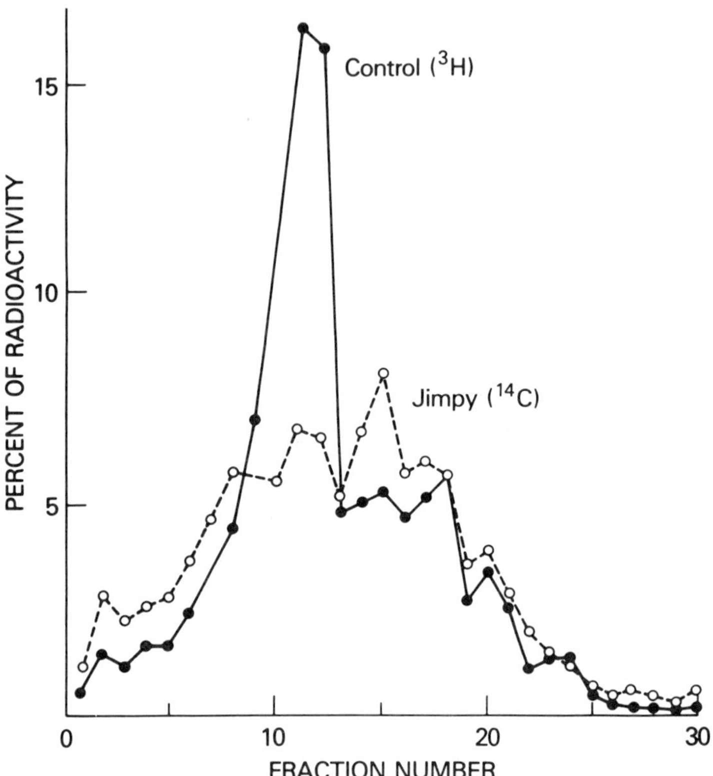

Fig. 4. Double-label polyacrylamide gel showing the absence of MAG in the fraction obtained from Jimpy mice by the procedure for myelin purification. In this experiment, 17-day-old Jimpy mice were injected with [^{14}C]fucose and age-matched controls were injected with [^{3}H]fucose. The brains were combined and homogenized together, and a myelin fraction was purified. The fraction contained 0.7% of the macromolecular [^{3}H]fucose from the control brains, but only 0.06% of the [^{14}C]fucose from the Jimpy brains. When the fraction was electrophoresed, [^{3}H]-MAG was clearly present from the controls (•), but there was little or no [^{14}C]-MAG from the Jimpy mice (o). From Matthieu *et al.* (1974*c*), with the permission of Pergamon Press.

It should be emhasized that only the most prominent 100,000-dalton glycoprotein (MAG) has definitively been shown to be in myelin or closely associated structures. It is clear from Figs. 1 and 2 that there are also a number of other glycoproteins in the myelin fraction, most of which are of lower molecular weight. It has been more difficult to obtain information about the association of these smaller glycoproteins with myelin, because many other membranes of brain contain a large number of glycoproteins in the 30- to 70-kilodalton range (Quarles and Brady, 1971; Quarles *et al.*, 1973*a*; Zanetta *et al.*, 1977*b*). Nevertheless, it seems likely that some of these other glycoproteins are also genuinely associated with myelin in the tissue, and there is some experimental evidence to support this idea (Quarles *et al.*, 1973*a*; Poduslo, J. F., *et al.*, 1977).

One myelin-related structure that was originally considered as a possible source for MAG was the axolemma of myelinated axons, although there were indications that this was not the case (Matthieu *et al.*, 1973; Druse *et al.*, 1974). More recent experiments in which axolemmal glycoproteins in the myelin fraction were labeled with [^{3}H]fucose via axoplasmic transport (Autilio-Gambetti *et al.*, 1975; Monticone and Elam, 1975; Matthieu *et al.*, 1978*a*) and in which a subcellular fraction enriched in axolemma was isolated (Matthieu *et al.*, 1978*a*) argue more strongly that MAG is not in the axolemma. Rather, MAG is a component of myelin itself, or of other "oligodendroglia-derived membranes," or of both.

The suggestion that MAG is in other "oligodendroglia-derived membranes" is based largely on investigation of subfractions of myelin and related membranes (Matthieu *et al.*, 1973; Zimmerman *et al.*, 1975; Waehneldt *et al.*, 1977; McIntyre R. J., *et al.*, 1978). These studies show that MAG, the major Wolfgram protein, and 2′,3′-cyclic nucleotide 3′-phosphohydrolase (CNP) are most concentrated in fractions consisting of single membranes and vesicles, rather than in multilamellar myelin (Table II). The first indication that the glycoprotein and CNP were not uniformly distributed in the myelin fraction was the finding that the heavy myelin subfraction had nearly three times the levels of these components as the light subfraction. More recently, even higher levels of MAG and CNP were found in a membrane fraction ($W_1$3) that is purified from the material released from crude myelin by osmotic shock. The major Wolfgram protein of myelin has a distribution in these fractions similar to that of MAG and CNP. Table II also shows that there is an inverse relationship between the levels of these components and myelin basic protein. Low-power electron microscopy clearly shows that the morphology of the fraction with highest MAG is very

Table II. Myelin and Myelin-Related Fractions

	Basic protein (% of total)	CNP (μmol/hr/mg protein)	MAG (U/mg protein)[a]
Myelin[b]	35	1063	12
Light myelin[c]	47	550	5
Heavy myelin[c]	26	1350	17
Microsome-depleted ($W_1$3)[d]	15	3909	32

[a]One unit is the amount of MAG that stains as intensely as 1 μg fetuin with PAS reagents.
[b]Data are from Matthieu *et al.* (1973) and R. J. McIntyre *et al.* (1978).
[c]Light and heavy subfractions of highly purified myelin obtained by the procedure of Norton and Poduslo (1973) were isolated. Data are from Matthieu *et al.* (1973), with the permission of Elsevier Scientific Publishing Co.
[d]This fraction was purified by density gradient centrifugation of the material released when myelin is osmotically shocked. Data are from R. J. McIntyre *et al.* (1978), with the permission of Pergamon Press. Similar findings were made by Waehneldt *et al.* (1977) for their SN-4 fraction.

different from the whole myelin fraction, which contains many large, multilamellar fragments (Fig. 5). Interpretation of these experiments is difficult, and detailed consideration of the results is beyond the scope of this chapter and has been presented elsewhere (Quarles, 1979). Nevertheless, the results strongly suggest that MAG, CNP, and the major Wolfgram protein, although clearly associated with myelin in the tissue, are most concentrated in membranes other than compact multilamellar myelin. It seems likely that they are in membranes that are part of the continuum between the surface of the oligodendrocyte and compact myelin. This would include the glial processes, the outer and inner surfaces of the myelin sheath, and the paranodal membranes.

A completely different experimental approach leading to similar conclusions involved the use of an impermeant molecular probe for surface structures. When the intact rat spinal cord was treated with galactose oxidase followed by sodium [^{3}H]borohydride, MAG was labeled, indicating that it is at least partially exposed on the surface of the intact myelin sheath or oligodendrocyte (Poduslo, J. F., *et al.*, 1976). For the reasons discussed, it seems likely that MAG is particularly concentrated in oligodendroglia-derived membranes that are adjacent to compact myelin. As these membranes are extended or modified to form multilamellar myelin, MAG would be diluted by the addition of the major myelin proteins, so that it is present in low concentration in the whole myelin fraction. Possibly it is even removed completely so that it is not present at all in compacted myelin (see Section 3.4).

3.2. Development

In view of the considerations discussed above, it seems likely that MAG would be exposed on the surface of the oligodendroglial processes as they are extended in the early stages of myelinogenesis. Since there is a large body of evidence indicating that cell-surface glycoproteins are involved in recognition and contact relationships between cells, it is reasonable to suggest that MAG could function in the recognition of axons to be myelinated or in layering of the first few turns of myelin. Although the amount of MAG per milligram total myelin protein remains nearly constant during development (Druse *et al.*, 1974; Everly *et al.*, 1977), an interesting developmental change in its apparent molecular weight on SDS gels was observed. Double-labeling experiments with radioactive fucose showed that MAG in very immature rat myelin has a slightly higher apparent molecular weight than that in mature myelin (Fig. 6) (Quarles *et al.*, 1973*b*). Also, PAS staining of gels on which mature and immature myelin were mixed showed that the larger glycoprotein is not restricted to the newly synthesized radioactive component, but the bulk of the glycoprotein in immature myelin is larger (Matthieu *et al.*, 1975*b*). Detailed developmental studies showed that the magnitude of this difference

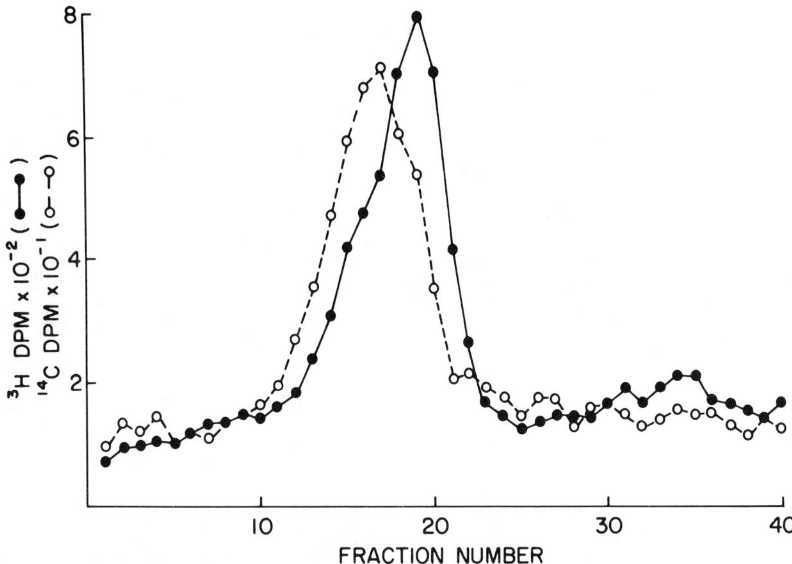

Fig. 6. Developmental change in the apparent molecular weight of MAG from rat brain. In this experiment, 14-day-old rats were injected with [^{14}C]fucose and 22-day-old rats were injected with [^{3}H]fucose. All the brains were homogenized together, and the myelin was purified. The labeled glycoproteins in the mixed myelin fraction were electrophoresed on SDS 5% polyacrylamide gels, which were cut into 1-mm slices. The figure shows only a 4-cm segment of the gel containing MAG. (o) 14-day myelin labeled with ^{14}C; (•) 22-day myelin labeled with ^{3}H. Reproduced from Quarles *et al.* (1973*b*), with the permission of Elsevier Scientific Publishing Co.

decreases gradually with age as the animal matures (Matthieu *et al.*, 1975*b*). The transition from the large to the small glycoprotein also correlates well with the caudal–cephalo progression of myelination (McIntyre L. J., *et al.*, 1978*a*). Indeed, the change in molecular weight of MAG appears to be a sensitive indicator of myelin maturation in the rat, since a small shift of MAG could be detected in the slightly immature myelin of hypothyroid (Matthieu *et al.*, 1975*c*) or copper-deficient rats (Zimmerman *et al.*, 1976). A similar developmental shift toward higher molecular weight was observed in the immature myelin of gerbils and hamsters, but could not be detected in mice or prairie deer mice (Matthieu *et al.*, 1974*a*). The failure to find the developmental differences in all species has put the general significance of the molecular-weight change for the process of myelinogenesis into question. Nevertheless, a shift of MAG toward higher molecular weight was observed in the immature myelin of Quaking murine mutants (Matthieu *et al.*, 1974*b*, 1978*b*), i.e., in one of the species in which the difference could not be detected during normal myelinogenesis. Therefore, it remains possible that the presence of the large glycoprotein is incompatible with normal myelino-

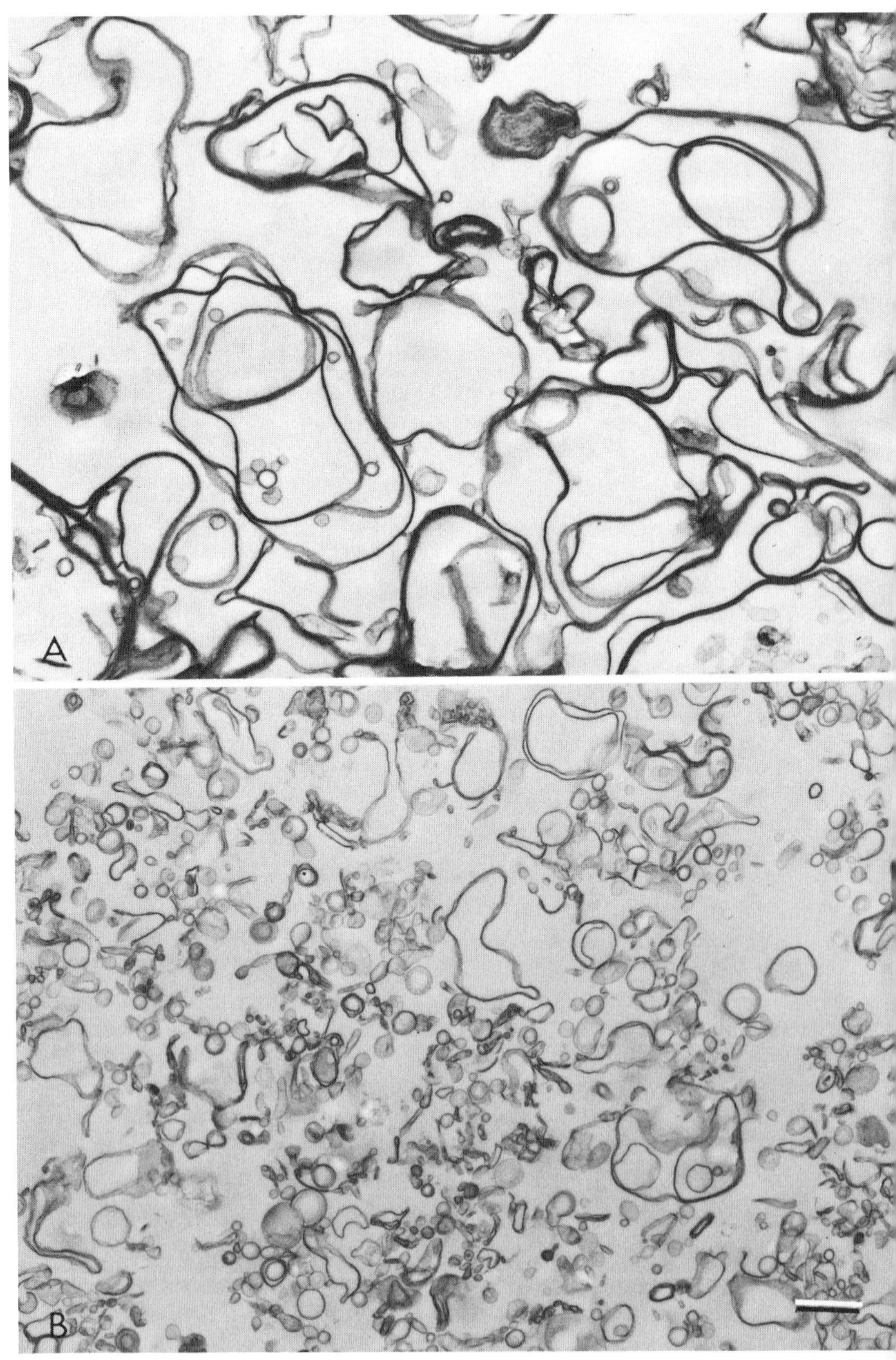
A
B

genesis, and transition to the smaller form is necessary for normal compaction and development.

The chemical reason for the developmental change is currently being investigated in our laboratory. Several possible explanations are a change in the oligosaccharide moieties of the glycoprotein, a shortening of the polypeptide, or a change in the relative amounts of two different glycoproteins running at about the same position on polyacrylamide gels. The possibility that it involves the oligosasccharide moieties now appears less likely (Quarles, 1976; Quarles *et al.*, 1977), and there are indications that it could be produced by cleavage of a peptide bond (McIntyre L. J., *et al.*, 1978*b*). Proteolytic modification of the biological activity of proteins, as in zymogen activation, for example, is now known to be a rather common phenomenon. However, confirmation of this mechanism for the developmental change in MAG must await the outcome of peptide-mapping studies that are in progress.

3.3. Metabolism

Although *in vivo* labeling methods have been used extensively in studying MAG, relatively little is known about the details of its biosynthesis or catabolism. Metabolic experiments in young animals have been complicated by the continuing deposition of myelin and the possible conversion of precursor membranes to compact myelin (Druse *et al.*, 1974). Nevertheless, the glycoprotein does turn over in young adult rats (Druse *et al.*, 1974), and appears to have a greater rate of turnover than some of the lower-molecular-weight glycoproteins in the isolated myelin fraction (Druse *et al.*, 1974; Poduslo, J. F., *et al.*, 1977). Detailed studies by Waechter and colleagues on the biosynthesis of white matter glycoproteins (reviewed in Chapter 4) indicated that dolichol-linked sugars are intermediates in the synthesis of MAG (Waechter *et al.*, 1976; Harford *et al.*, 1977).

3.4. Pathology

An interesting observation was made with regard to MAG in the edematous demyelination caused by hexachlorophene (HCP) intoxication. In this condition, myelin splits at the intraperiod line with the formation of large fluid-containing vacuoles. When this condition was induced in developing rats, the yield of isolated myelin was reduced to nearly one half the normal level, but the difference could be largely accounted for by the appearance of a fraction that floated on 0.32 M sucrose (Matthieu *et al.*, 1974*d*). The protein composition of the "floating fraction" was similar to that of the myelin fraction, except that MAG could not be detected. A possible explanation for

←

Fig. 5. Low-power electron micrographs contrasting the large multilamellar fragments in the purified rat myelin fraction (*top*) with the single membranes, vesicles, and small myelin fragments in the $W_1 3$ fraction (*bottom*). Calibration bar: 0.5 μm. From R. J. McIntyre *et al.* (1978), with the permission of Pergamon Press.

this is that MAG is rapidly destroyed via proteolytic or glycosidic degradation in this pathological situation. Indeed, there is a recent study indicating that MAG is particularly susceptible to degradation during *postmortem* autolysis (Matthieu *et al.*, 1977). However, another possible explanation for the absence of MAG in the floating fraction is that this fraction is derived from a portion of the myelin–oligodendroglia unit that does not contain the glycoprotein. The vacuoles in HCP intoxication form in multilamellar myelin, and there is evidence suggesting that the paranodal membranes and glial–axonal junctions remain relatively undamaged in this condition (Reier *et al.*, 1978). Therefore, it is likely that the floating fraction is derived from compact myelin, and these findings could indicate that MAG is not present in multilamellar myelin, but is restricted entirely to adjacent oligodendroglial membranes. This latter explanation for the absence of MAG in the floating fraction is rendered more likely by the recent isolation from normal CNS tissue of small amounts of a similar floating fraction that also lacks MAG (McIntyre, R. J., *et al.*, 1978). In normal tissue, a catabolic degradation would be less likely to account for the absence of MAG than in the pathological situation.

Although MAG is quantitatively a minor component of the whole myelin fraction, it probably accounts for a considerably higher proportion of the protein molecules exposed on the surface of the myelin–oligodendroglia units. As already indicated, surface-labeling techniques showed that MAG is accessible to nonpermeant surface probes (Poduslo, J. F., *et al.*, 1976), and this is in contrast to myelin basic protein, for example, which was not labeled when similar probes were applied to the intact tissue (Braun, 1977). Therefore, the glycoprotein seems well situated to interact with external agents that could be involved in demyelinating processes. Furthermore, it is well established that membrane glycoproteins function as cell-surface antigens and receptors for viruses. Therefore, a role for MAG in the immune or viral aspects of demyelinating diseases is reasonable. This possibility is enhanced by recent demonstrations that MAG is immunogenic and that rabbits with experimental allergic encephalomyelitis (EAE) induced by innoculation with whole myelin have antibodies directed against this glycoprotein (Poduslo, J. F., and McFarlin, 1978; Quarles and Brady, unpublished results). There is a substantial body of evidence indicating that autoimmune or viral factors or both are involved in the pathogenesis of multiple sclerosis, which is the most significant demyelinating disease in humans (Poser, 1975). Paramyxoviruses have long been suspected to be involved because of increased measles antibodies in the serum of multiple sclerosis patients, and cell-surface receptors for viruses of this class are glycoproteins (Howe and Lee, 1972). It is also known that viral infections often cause alterations in the carbohydrate moieties of cell-surface glycoproteins (Warren *et al.*, 1973). Conceivably, a viral agent involved in multiple sclerosis could induce changes in glycoproteins of myelin or oligodendroglial membranes, which in turn would alter their antigenicity and render them susceptible to autoimmune attack. Although these ideas are only speculative at this time, a principal objective in our laboratory is to elucidate the possible involvement of MAG in the viral or autoimmune aspects of multiple sclerosis and other demyelinating diseases.

3.5. Chemistry

It is clear that a good understanding of the structural role and function of MAG in myelin or oligodendroglial membranes will require a detailed knowledge of its chemical and immunological properties. For this reason, we have put a substantial amount of effort into the purification of MAG. A useful procedure for obtaining a fraction enriched in glycoproteins of myelin is to treat the myelin with chloroform–methanol (2:1, vol./vol.) (Quarles *et al.*, 1973*a*; Zanetta *et al.*, 1977*b*). This solubilizes the lipids, basic proteins, and proteolipid, while leaving the glycoproteins and other high-molecular-weight proteins in the residue, and results in about a 4-fold enrichment of glycoproteins. However, complete purification of MAG is difficult due to problems in working with membrane proteins, and to the small amount of MAG in the myelin fraction. Conventional techniques of protein isolation have not been satisfactory even when applied in the presence of detergents. However, it was recently found that solubilization with the surface active agent, lithium diiodosalicylate, followed by partitioning with phenol (Marchesi and Andrews, 1971), results in a highly selective extraction of MAG from myelin (Quarles and Pasnak, 1977). The procedure takes only 1 day starting from purified myelin and gives a good yield ($\approx$50%) of relatively pure MAG. Using this isolation, we expect to obtain detailed information about the chemistry of MAG in the near future.

It must be emphasized that the carbohydrate analysis of CNS myelin glycopeptides in Table I is for the whole mixture of glycoproteins in isolated CNS myelin and not for the 100,000-dalton component (MAG) alone. Nevertheless, substantial information about the chemistry of MAG in rat brain can be inferred from various studies. The relative intensities of staining with protein stains and PAS reagents suggest a much higher carbohydrate content than the 5% in the P_0 glycoprotein of peripheral myelin. Some membrane glycoproteins such as glycophorin in the red cell contain as much as 60% carbohydrate (Marchesi *et al.*, 1973). *In vivo* labeling with several sugar precursors (Quarles *et al.*, 1973*a*; Quarles and Everly, 1977) and binding to a variety of immobilized lectins (Quarles *et al.*, 1977; Zanetta *et al.*, 1977*a*) indicate that MAG contains fucose, mannose, galactose, *N*-acetylglucosamine, and sialic acid. Gel filtration of labeled glycopeptides that were prepared by exhaustive pronase digestin of MAG purified on preparative gels indicated a substantial heterogeneity and range in the size of the oligosaccharide units (Quarles, 1976). The sialic acid content of MAG is probably rather high because of extensive labeling by *N*-acetylmannosamine (Quarles *et al.*, 1973*a*) and a reduction in apparent molecular weight of about 10,000 daltons when MAG is treated with neuraminidase (Quarles, 1976). Labeling experiments with [^{35}S]sulfate showed that MAG is also sulfated probably on sugar residues (Matthieu *et al.*, 1975*d*; Simpson *et al.*, 1976). It would be expected that the sialic acid and sulfate would contribute a substantial negative charge to the protein, and isoelectric focusing indicates a pI of 3.8 (unpublished results).

In summary, it appears likely that MAG is a highly glycosylated protein with a relatively high negative charge on many of the oligosaccharide groups. By analogy with the red cell glycoprotein (Marchesi *et al.*, 1973), there may be a hydrophobic portion of the polypeptide chain by which MAG interacts with lipids in the membrane. Recent experiments in which purified CNS myelin was treated with trypsin as a membrane probe (McIntyre, L. J., *et al.*, 1978*b*) are consistent with this idea. MAG was degraded to a somewhat smaller glycoprotein (mol.wt. ≈80,000) that was protected from further degradation by its situation in the membrane. These findings are analogous to those described earlier for the P_0 glycoprotein of peripheral myelin (see Section 2.2), and suggest that a substantial portion of the polypeptide of MAG may be buried in the hydrophobic portion of the membrane. However, detailed biochemical information about MAG must await the outcome of peptide-mapping and analytical studies that are currently under way.

4. Relevance for Myelin Structure and Perspectives

There are both qualitative differences and similarities between the proteins of central and peripheral myelin. The results reviewed here clearly show that the glycoproteins in central and peripheral myelin are quite different. Figure 7 is a hypothetical model representing our current ideas about the molecular arrangements of proteins and glycoproteins in central and peripheral myelin and the adjacent oligodendroglial or Schwann cell membranes, respectively. This model is based on our own work on glycoproteins as well as the work of many others.

Basic proteins are present in both central and peripheral myelin, and it seems likely that they have similar structural roles in the two kinds of myelin. They are placed in the major dense line of Fig. 7 where the cytoplasmic surfaces of the membranes of the myelin-forming cells join. This localization is based largely on the results of molecular probe studies (Braun, 1977; Peterson and Gruener, 1978) and should be considered tentative, since there is conflicting evidence from immunohistochemical studies. Although an early immunocytochemical localization suggested that basic protein was in the major dense line (Herndon *et al.*, 1973), more recent studies indicate the

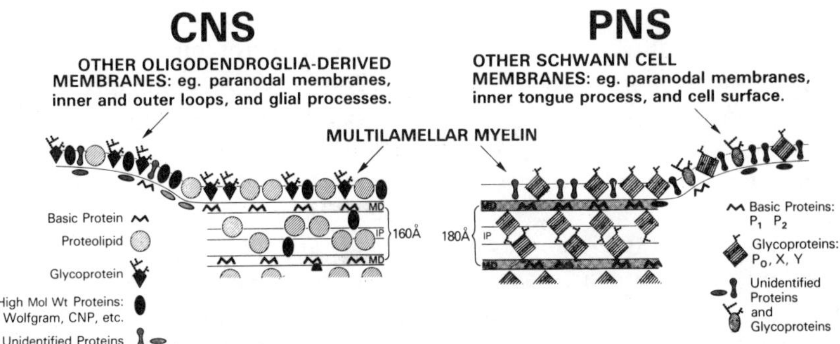

Fig. 7. Hypothetical scheme for the molecular architecture of myelin and related membranes with emphasis on glycoprotein components. Compact central and peripheral myelin are represented toward the center of the figure, while adjacent oligodendroglial and Schwann cell membranes are shown to the left and right of the figure, respectively. Much of the evidence for the localization of the major CNS myelin proteins is from the work of Braun and colleagues (reviewed in Braun, 1977). Evidence for the localization of PNS myelin proteins comes from the work of Peterson, Wood, and others, and is discussed in Section 2.3. See the text for discussion of the model and the possible significance of glycoproteins for the structure of myelin and supporting membranes. The sizes of the symbols for the various proteins are not intended to be proportional to the actual molecular sizes. The indicated dimensions of the myelin periods are those given by X-ray diffraction. (MD) major dense line; (IP) intraperiod line.

presence of basic protein in the intraperiod line (Mendell and Whitaker, 1978; Sternberger, Kies, and Webster, personal communication). Obviously, there is still uncertainty about the localization of basic proteins in the myelin lamellae, but if basic proteins are in the major dense line, it is reasonable to assume that the other major proteins of PNS and CNS myelin would be in the intraperiod line. In the case of CNS myelin, this would be proteolipid, whereas it would be the P_0 glycoprotein in PNS myelin. There is experimental evidence supporting the intraperiod localization of proteolipid (Braun, 1977), and several types of evidence indicating that the P_0 glycoprotein is at least partially in the intraperiod line of peripheral myelin were already described (Section 2.3). However, since both these proteins contain hydrophobic regions, they may span the lipid bilayer so that part of the polypeptides could also be at the major dense line. It seems likely that this difference in the intraperiod proteins of central and peripheral myelin, and particularly the fact that P_0 is glycosylated while proteolipid is not,* could be a principal factor accounting for differences in the structure of the two types

*Careful examination of rat myelin proteolipid in our laboratory indicated that it is not a glycoprotein (Poduslo, J. F., *et al.*, 1977). Similarly, protein-bound carbohydrate could not be detected in bovine myelin proteolipid (Folch-Pi and Stoffyn, 1972). However, a small amount of fucose and glucosamine was reported to be associated with the hydrophobic N-2 protein (lipophillin) purified from human myelin that is probably the human proteolipid (Gagnon *et al.*, 1971). Therefore, it is possible that proteolipid is glycosylated in some species.

of myelin. It is well known that the periodicity of peripheral myelin is slightly larger than that in central myelin, and electron-microscopic studies show that a double membrane can be seen more readily at the intraperiod line of PNS myelin than at that of CNS myelin (Peters and Vaughn, 1970). As shown in Fig. 7, the reason for this may be that the large amount of glycoprotein carbohydrate in the intraperiod line of PNS myelin gives it a hydrophillic nature and prevents complete fusion, while the more hydrophobic proteolipid interacts closely with lipids and allows greater fusion in central myelin. A similar suggestion was made by Wood and McLaughlin (1975).

Figure 7 also shows our ideas about the composition of oligodendroglia and Schwann cell membranes that are adjacent to, but distinct from, compact myelin. As we have seen (Section 3.1), subfractionation studies in the CNS have suggested that MAG, CNP, the major Wolfgram protein, and other high-molecular-weight proteins are particularly concentrated in these membranes. Glycoproteins have been reported to be present in isolated oligodendroglial plasma membranes (Poduslo, S. E., 1975). Of course, there is considerable morphological specialization among paranodal membranes, inner and outer surfaces of the sheath, glial processes, and other structures, so that there may be selective localization of biochemical components in these various oligodendroglia-derived membranes, but there was no attempt to show this in Fig. 7. When myelin is formed as an extension of the oligodendroglial membranes, the proteins that are in high concentration in these membranes would be gradually diluted by the addition of lipids, basic protein, and proteolipid, so that they are present in low concentration, if at all, in compact myelin.

Although MAG may be present in very low concentration throughout the myelin sheath, Fig. 7 does not show any MAG in compact myelin. Some indirect evidence suggesting the complete absence of MAG from compact myelin comes from study of the "floating fraction" (Section 3.4). On the other hand, the binding of ferritin-tagged Con A to the surface of fragments of isolated myelin, and to the underlying intraperiod line where the top layer was broken, suggest that MAG could be present throughout the intraperiod line (Matus *et al.*, 1973; Braun, 1977). However, other interpretations are possible, since there are other Con-A-binding glycoproteins in CNS myelin (Quarles *et al.*, 1977; Zanetta *et al.*, 1977*a*), and some redistribution of molecules may occur at the damaged surface of myelin fragments.

In the PNS, the carbohydrate of the P_0 glycoprotein may also be exposed on the surface of the Schwann cell membranes. This would be consistent with its eventual localization in the intraperiod line of compact myelin. There may also be other glycoproteins on the surface of the Schwann cell and in myelin-related membranes derived from it. Some of the minor glycoproteins recently detected in PNS myelin (see Section 2.1) may be at these sites. However, in general, there is very little known about the surface of the Schwann cell.

Finally, it must be emphasized once more that Fig. 7 is very speculative and only a hypothesis about the molecular architecture of myelin and intimately associated membranes. Nevertheless, it serves as a working model for designing experiments and considering the function of myelin-associated glycoproteins. As the technical aspects of immunohistochemical procedures at the electron-microscopic level are developed and improved, their application to intact tissue and isolated fractions should help to determine the molecular architecture of myelin and myelin-forming cells with greater certainty. For the reasons that have been described, we think it is likely that glycoproteins will prove to be important both for myelinogenesis and for demyelinating diseases. As we have seen, particular glycoproteins have now been identified in myelin and related membranes, and appear to be less heterogeneous than the complex mixtures of glycoproteins in neuronal and synaptosomal membranes (Quarles, 1975). Also, glial–axonal interactions and spiraling of the myelin sheath probably require less sophisticated recognition and contact phenomena than the complex processes involved in neuronal growth and the formation of many specific synaptic connections. For these reasons, the glycoproteins in myelin and myelin-associated membranes appear to offer great potential for relating chemical properties of glycoproteins to their structural and functional roles within the nervous system.

ACKNOWLEDGMENTS. A substantial number of coworkers have contributed to our laboratory's work on glycoproteins in myelin and related membranes. These include Dr. Roscoe O. Brady, Dr. John Everly, Dr. Mary Druse, Dr. Jean-Marie Matthieu, Dr. Andrew Zimmerman, Dr. Joseph Poduslo, Dr. Rosemary McIntyre, Dr. Laurence McIntyre, Mrs. Carol Pasnak, Dr. Norio Sakuragawa, and Mr. Gary Barbarash. The collaborative morphological studies with Dr. Henry Webster and his colleagues, including Dr. Nancy Sternberger, Dr. Yasuto Itoyama, and Dr. Bruce Trapp, have been particularly beneficial. I am grateful to all for their hard work and ideas.

References

Autilio-Gambetti, L., Gambetti, P., and Shafer, B., 1975, Glial and neuronal contribution to proteins and glycoproteins recovered in myelin fractions, *Brain Res.* **84**:336.

Braun, P. E., 1977, Molecular architecture of myelin, in: *Myelin* (P. Morell, ed.), pp. 91–115, Plenum Press, New York.

Braun, P. E., and Brostoff, S. W., 1977, Proteins of myelin, in: *Myelin* (P. Morell, ed.), pp. 201–231, Plenum Press, New York.

Brostoff, S. W., 1977, Immunological responses to myelin and components, in: *Myelin* (P. Morell, ed.), pp. 415–446, Plenum Press, New York.

Brostoff, S. W., Karkhanis, Y. D., Carlo, D. J., Reuter, W., and Eylar, E. H., 1975, Isolation and partial characterization of the major proteins of rabbit sciatic nerve myelin, *Brain Res.* **86**:449.

Brostoff, S. W., Levit, S., and Powers, J., 1977, Isolation of a disease inducing peptide from bovine PNS myelin P_2 protein, *Trans. Am. Soc. Neurochem.* **8**:205.

Carlo, D. J., Karkhanis, Y. D., Bailey, P. J., Wisniewski, H. M., and Brostoff, S. W., 1975, Allergic neuritis: Evidence for the involvement of the P_2 and P_0 proteins, *Brain Res.* **88**:580.

Druse,, M. J., Brady, R. O., and Quarles, R. H., 1974, Metabolism of a myelin-associated glycoprotein in developing rat brain, *Brain Res.* **76**:423.

Everly, J. L., Brady, R. O., and Quarles, R. H., 1973, Evidence that the major protein in rat sciatic nerve myelin is a glycoprotein, *J. Neurochem.* **21**:329.

Everly, J. L., Quarles, R. H., and Brady, R. O., 1977, Proteins and glycoproteins in myelin purified from the developing bovine and human central nervous system, *J. Neurochem.* **28**:95.

Folch-Pi, J., and Stoffyn, P. J., 1972, Proteolipids from membrane systems, *Ann. N. Y. Acad. Sci.* **195**:86.

Gagnon, J., Finch, D. D., Wood, D. D., and Moscarello, M. A., 1971, Isolation of a highly purified myelin protein, *Biochemistry* **10**:4756.

Gould, R. M., 1977, Incorporation of glycoproteins into peripheral nerve myelin, *J. Cell Biol.* **75**:326.

Greenfield, S., Brostoff, S., Eylar, E. H., and Morell, P., 1973, Protein composition of myelin of the peripheral nervous system, *J. Neurochem.* **20**:1207.

Hardwicke, P. M. D., Crawford, C. L., and Huvos, P., 1977, Skin lesions induced by sensory peripheral nerve are not due to an additional myelin protein, *J. Neurochem.* **29**:371.

Harford, J. B., Waechter, C. J., and Earl, F. L., 1977, Effect of exogenous dolichyl monophosphate on a developmental change in mannosylphosphoryl dolichol biosynthesis, *Biochem. Biophys. Res. Commun.* **76**:1036.

Herndon, R. M., Rauch, H. C., and Einstein, E. R., 1973, Immunoelectron microscopic localization of the encephalitogenic protein in myelin, *Immunol. Commun.* **2**:163.

Howe, C., and Lee, L. T., 1972, Virus–erythrocyte interactions, *Adv. Virus Res.* **17**:1.

Kitamura, K., Suzuki, M., and Uyemura, K., 1976, Purification and partial characterization of two glycoproteins in bovine peripheral nerve myelin membrane, *Biochim. Biophys. Acta* **455**:806.

Koski, C. L., and Max, S. R., 1977, A comparison of the major proteins of human motor and sensory myelin, *Neurology* **27**:358.

Marchesi, V. T., and Andrews, E. P., 1971, Glycoproteins: Isolation from cell membranes with lithium diiodosalicylate, *Science* **174**:1247.

Marchesi, V. T., Jackson, R. L., Segrest, J. P., and Kahane, I., 1973, Molecular features of the major glycoprotein of the human erythrocyte membrane, *Fed. Proc. Fed. Am. Soc. Exp. Biol.* **32**:1833.

Margolis, R., 1967, Acid mucopolysaccharides and proteins of bovine whole brain, white matter and myelin, *Biochim. Biophys. Acta* **141**:91.

Matthieu, J.-M., Quarles, R. H., Brady, R. O., and Webster, H. deF., 1973, Variation of proteins, enzyme markers, and gangliosides in myelin subfractions, *Biochim. Biophys. Acta* **329**:305.

Matthieu, J.-M., Brady, R. O., and Quarles, R. H., 1974*a*, Developmental change in a myelin-associated glycoprotein, *Dev. Biol.* **37**:146.

Matthieu, J.-M., Brady, R. O., and Quarles, R. H., 1974*b*, Anomalies of myelin associated glycoproteins in quaking mice, *J. Neurochem.* **22**:291.

Matthieu, J.-M., Quarles, R. H., Webster, H. deF., Hogan, E. L., and Brady, R. O., 1974*c*, Characterization of the fraction obtained from the CNS of Jimpy mice by a procedure for myelin isolation, *J. Neurochem.* **23**:517.

Matthieu, J.-M., Zimmerman, A. W., Webster, H. deF., Ulsamer, A. G., Brady, R. O., and

Quarles, R. H., 1974*d*, Hexachlorophene intoxication: Characterization of myelin and myelin-related fractions in rat during early postnatal development, *Exp. Neurol.* **45**:558.

Matthieu, J.-M., Everly, J. L., Brady, R. O., and Quarles, R. H., 1975*a*, [^{35}S]sulfate incorporation into myelin glycoproteins. II. Peripheral nervous tissue, *Biochim. Biophys. Acta* **392**:167.

Matthieu, J.-M., Brady, R. O., and Quarles, R. H., 1975*b*, Change in a myelin-associated glycoprotein in rat brain during development: Metabolic aspects, *Brain Res.* **86**:55.

Matthieu, J.-M., Reier, P. J., and Sawchak, J. A., 1975*c*, Proteins of rat brain myelin in neonatal hypothyroidism, *Brain Res.* **84**:443.

Matthieu, J.-M., Quarles, R. H., Poduslo, J. F., and Brady, R. O., 1975*d*, [^{35}S]sulfate incorporation into myelin glycoproteins. I. Central nervous system, *Biochim. Biophys. Acta* **392**:159.

Matthieu, J.-M., Koellreutter, B., and Joyet, M. L., 1977, Changes in CNS myelin proteins and glycoproteins after *in situ* autolysis, *Brain Res. Bull.* **2**:15.

Matthieu, J.-M., Webster, H. deF., and Devries, G. H., 1978*a*, Glial versus neuronal origin of myelin proteins studied by combined intraocular and intracranial labeling, *J. Neurochem.* **31**:93.

Matthieu, J.-M., Koellreutter, B., and Joyet, M. L., 1978*b*, Protein and glycoprotein composition of myelin and myelin subfractions from brains of Quaking mice, *J. Neurochem.* **30**:783.

Matus, A., DePetris, S., and Raff, M. C., 1973, Mobility of concanavalin A receptors in myelin and synaptic membranes, *Nature (London) New Biol.* **244**:278.

McIntyre, L. J., Quarles, R. H., and Brady, R. O., 1978*a*, Regional studies of myelin associated glycoprotein in the rat central nervous system, *Brain Res.* **149**:251.

McIntyre, L. J., Quarles, R. H., and Brady, R. O., 1978*b*, The effect of trypsin on myelin-associated glycoprotein, *Trans. Am. Soc. Neurochem.* **9**:106.

McIntyre, R. J., Quarles, R. H., Webster, H. deF., and Brady, R. O., 1978, Isolation and characterization of myelin-related membranes, *J. Neurochem.* **30**:991.

Mendell, J. R., and Whitaker, J. N., 1978, Immunocytochemical localization studies of myelin basic protein, *J. Cell Biol.* **76**:502.

Monticone, R. E., and Elam, J. S., 1975, Isolation of axonally transported glycoproteins with goldfish visual system myelin, *Brain Res.* **100**:61.

Norton, W. T., 1977, Isolation and characterization of myelin, in: *Myelin* (P. Morell, ed.), pp. 161–199, Plenum Press, New York.

Norton, W. T., and Poduslo, S. E., 1973, Myelination in rat brain: Method of myelin isolation, *J. Neurochem.* **21**:749.

Peters, A., and Vaughn, J. E., 1970, Morphology and development of the myelin sheath, in: *Myelination* (A. N. Davison and A. Peters, eds.), pp. 3–79, Charles C. Thomas, Springfield, Illinois.

Peterson, R. G., 1976, Myelin protein changes with digestion of whole sciatic nerve myelin in trypsin, *Life Sci.* **18**:845.

Peterson, R. G., and Gruener, R. W., 1978, Morphological localization of PNS myelin proteins, *Brain Res.* **152**:17.

Peterson, R. G., and Pease, D. C., 1972, Myelin imbedded in polymerized glutaraldehyde urea, *J. Ultrastruct. Res.* **41**:115.

Peterson, R. G., and Sea, C. P., 1975, Localization of the main myelin protein in PNS, *Trans. Am. Soc. Neurochem.* **6**:211.

Pleasure, D. E., Feldman, B., and Prockop, D. J., 1973, Diphtheria toxin inhibits the synthesis of myelin proteolipid and basic proteins by peripheral nerve *in vitro*, *J. Neurochem.* **20**:81.

Poduslo, J. F., and McFarlin, D. E., 1978, Immunogenicity of the major glycoprotein from central nervous system myelin, *Trans. Am. Soc. Neurochem.* **9**:105.

Poduslo, J. F., Quarles, R. H., and Brady, R. O., 1976, External labeling of galactose in surface membrane glycoproteins of the intact myelin sheath, *J. Biol. Chem.* **251**:153.

Poduslo, J. F., Everly, J. L., and Quarles, R. H., 1977, A low molecular weight glycoprotein associated with isolated myelin: Distinction from myelin proteolipid protein, *J. Neurochem.* **28**:977.

Poduslo, S. E., 1975, The isolation and characterization of a plasma membrane and a myelin fraction derived from oligodendroglia of calf brain, *J. Neurochem.* **24**:647.

Poser, C. M., 1975, Multiple sclerosis, in: *The Nervous System,* Vol. 2, *The Clinical Neurosciences* (D. B. Tower and T. N. Chase, eds.), pp. 337–345, Raven Press, New York.

Quarles, R. H., 1975, Glycoproteins in the nervous system, in: *The Nervous System*, Vol. 1, *The Basic Neurosciences* (D. B. Tower and R. O. Brady, eds.), pp. 493–501, Raven Press, New York.

Quarles, R. H., 1976, Effects of pronase and neuraminidase treatment on a myelin-associated glycoprotein in developing brain, *Biochem. J.* **156**:143.

Quarles, R. H., 1979, The biochemical and morphological heterogeneity of myelin and myelin related membrane, in: *The Biochemistry of Brain* (S. Kumar, ed.), Pergamon Press, Oxford (in press).

Quarles, R. H., and Brady, R. O., 1971, Synthesis of glycoproteins and gangliosides in developing rat brain, *J. Neurochem.* **18**:809.

Quarles, R. H., and Everly, J. L., 1977, Glycopeptide fractions prepared from purified central and peripheral rat myelin, *Biochim. Biophys. Acta* **466**:176.

Quarles, R. H., and Pasnak, C. F., 1977, A rapid procedure for selectively isolating the major glycoprotein from purified rat brain myelin, *Biochem. J.* **163**:635.

Quarles, R. H., Everly, J. L., and Brady, R. O., 1972, Demonstration of a glycoprotein which is associated with a purified myelin fraction from rat brain, *Biochim. Biophys. Res. Commun.* **47**:491.

Quarles, R. H., Everly, J. L., and Brady, R. O., 1973*a*, Evidence for the close association of a glycoprotein with myelin in rat brain, *J. Neurochem.* **21**:1171.

Quarles, R. H., Everly, J. L., and Brady, R. O., 1973*b*, Myelin-associated glycoprotein: A developmental change, *Brain Res.* **58**:506.

Quarles, R. H., Foreman, C. F., Poduslo, J. F., and McIntyre, L. J., 1977, Interactions of myelin-associated glycoproteins with immobilized lectins, *Trans. Am. Soc. Neurochem.* **8**:201.

Quarles, R. H., Sakuragawa, N., Everly, J. L., Pasnak, C. F., Webster, H. deF., and Trapp, B. D., 1978, A biochemical comparison of *Xenopus laevis* and mammalian myelin from the central and peripheral nervous system, *J. Neurobiol.* **9**:217.

Reier, P. J., Tabira, T., and Webster, H. deF., 1978, Hexachlorophene induced myelin lesions in the amphibian central nervous system: A freeze fracture study, *J. Neurol. Sci.* **35**:274.

Robertson, J. D., 1959, The ultrastructure of cell membranes and their derivatives, *Biochem. Soc. Symp.* **16**:3.

Roomi, M. W., and Eylar, E. H., 1978, Isolation of a product from the trypsin-digested glycoprotein of sciatic nerve myelin, *Biochim. Biophys. Acta* **536**:122.

Roomi, M. W., Ishaque, A., Khan, N. R., and Eylar, E. H., 1978*a*, The P_0 protein: The major glycoprotein of peripheral nerve myelin, *Biochim. Biophys. Acta* **536**:112.

Roomi, M. W., Ishaque, A., Kahn, N. R., and Eylar, E. H., 1978*b*, Glycoproteins and albumin in peripheral nerve myelin, *J. Neurochem.* **31**:375.

Savolainen, H. J., 1972, Proteins and glycoproteins of human myelin and glial cell membrane with special reference to myelin formation, *Tower Int. Technomed. J. Life Sci.* **2**:35.

Sea, C. P., and Peterson, R. G., 1975, Ultrastructure and biochemistry of myelin after isoniazid-induced nerve degeneration in rats, *Exp. Neurol.* **48**:252.

Simpson, D. L., Thorne, D. R., and Loh, H. H., 1976, Sulfated glycoproteins, glycolipids, and glycosaminoglycans from synaptic plasma and myelin membranes: Isolation and characterization of sulfated glycopeptides, *Biochemistry* **15**:5449.

Singh, H., and Spritz, N., 1974, Polypeptide components of myelin from rat peripheral nerve, *Biochim. Biophys. Acta* **351**:379.

Singh, H., Silberlicht, I., and Singh, J., 1978, A comparative study of the polypeptides of mammalian peripheral nerve myelin, *Brain Res.* **144**:303.

Sternberger, N. H., Itoyama, Y., Kies, M. W., and Webster, H. deF., 1978, Immunocytochemical method to identify basic protein in myelin forming oligodendrocytes of newborn rat CNS, *J. Neurocytol.* **7**:251.

Waechter, C. J., Kennedy, J. L., and Harford, J. P., 1976, Lipid intermediates involved in the assembly of membrane-associated glycoproteins in calf brain white matter, *Arch. Biochem. Biophys.* **174**:726.

Waehneldt, T. V., Matthieu, J.-M., and Neuhoff, V., 1977, Characterization of a myelin related fraction (SN 4) isolated from rat forebrain at two developmental stages, *Brain Res.* **138**:29.

Warren, L., Fuhrer, J. P., and Buck, C. A., 1973, Surface glycoproteins of cells before and after transformation by oncogenic viruses, *Fed. Proc. Fed. Am. Soc. Exp. Biol.* **32**:8086.

Wiggins, R. C., Benjamin, J. A., and Morell, P., 1975, Appearance of myelin proteins in rat sciatic nerve during development, *Brain Res.* **89**:99.

Wolman, M., 1957, Histochemical study of changes occurring during the degeneration of myelin, *J. Neurochem.* **1**:370.

Wolman, M., and Hestrin-Lerner, S., 1960, A histochemical contribution to the study of the molecular morphology of myelin sheath, *J. Neurochem.* **5**:114.

Wood, J. G., and Dawson, R. M. C., 1973, A major myelin glycoprotein of sciatic nerve, *J. Neurochem.* **21**:717.

Wood, J. G., and Dawson, R. M. C., 1974*a*, Some properties of a major structural glycoprotein of sciatic nerve, *J. Neurochem.* **22**:627.

Wood, J. G., and Dawson, R. M. C., 1974*b*, Lipid and protein changes in sciatic nerve during wallerian degeneration, *J. Neurochem.* **22**:631.

Wood, J. G., and Engel, E. L., 1976, Peripheral myelin glycoproteins and myelin fine structure during development of rat sciatic nerve, *J. Neurocytol.* **5**:605.

Wood, J. G., and McLaughlin, B. J., 1975, The visualization of concanavalin-A binding sites in the intraperiod line of rat sciatic nerve myelin, *J. Neurochem.* **24**:233.

Zanetta, J. P., Sarlieve, L. L., Mandel, P., Vincendon, G., and Gombos, G., 1977*a*, Fractionation of glycoproteins associated to adult rat brain myelin fractions, *J. Neurochem.* **29**:827.

Zanetta, J. P., Sarlieve, L. L., Reeber, A., Vincendon, G., and Gombos, G., 1977*b*, A protein fraction enriched in all myelin associated glycoproteins from adult rat central nervous system, *J. Neurochem.* **29**:355.

Zgorzalewicz, B., Neuhoff, V., and Waehneldt, T. V., 1974, Rat myelin proteins: Compositional changes in various regions of the nervous system during ontogenetic development, *Neurobiology* **4**:265.

Zimmerman, A. W., Quarles, R. H., Webster, H. deF., Matthieu, J.-M., and Brady, R. O., 1975, Characterization and protein analysis of myelin subfractions in rat brain: Developmental and regional comparisons, *J. Neurochem.* **25**:749.

Zimmerman, A. W., Matthieu, J.-M., Quarles, R. H., Brady, R. O., and Hsu, J. M., 1976, Hypomyelination in copper deficient rats, *Arch. Neurol.* **33**:111.

12

Axonal Transport of Complex Carbohydrates

J. S. Elam

1. The Phenomenon of Axonal Transport

The classic study of Weiss and Hiscoe (1948) provided an early indication that bulk translocation of intracellular material occurs along axons. They observed that a constricting cuff placed on a nerve caused accumulation of axoplasm proximal to the constriction. When the cuff was removed, the bolus of axoplasm migrated distally along the nerve at a rate of 1–2 mm/day. In the following years, the technique of nerve constriction was used to document the intraaxonal migration of a variety of enzymes and subcellular organelles (reviewed by Grafstein, 1969). Considerable impetus was provided to axonal-transport work by the discovery that various isotopically labeled substances could be used to trace the movement of axonal constituents (Ochs and Burger, 1958; Weiss, 1961; Miani, 1962; Droz and Leblond, 1963; Taylor and Weiss, 1965). These studies led to a general acceptance of the use of radioisotopes to label axonally transported molecules, producing a flood of publications from the mid-1960's up to the present.

Various aspects of axonal transport have been the subjects of recent reviews (Droz, 1973, 1975; Ochs, 1974, 1975; Heslop, 1975; Lasek, 1975; Lubińska, 1975; Grafstein, 1975*a,b*, 1977). A survey of this literature makes clear that there are many aspects of the transport phenomenon that are still poorly understood. However, a few basic generalizations appear to be warranted: (1) The most clear-cut evidence of axonal transport has been for large molecules including proteins, glycoproteins, and certain classes of lipids. (2) Transport occurs at a variety of rates within a given axon, ranging from a minimum of a fraction of a millimeter per day up to several hundred millimeters per day. Most rates can be catagorized broadly as rapid (100–400

J. S. Elam • Department of Biological Science, Florida State University, Tallahassee, Florida.

mm/day) or slow (0.5–10 mm/day). There is also a retrograde phase of transport (terminal to cell body) that occurs at approximately one half the rapid forward rate. (3) Rapidly transported proteins are found to be largely particulate (sedimentable at 100,000*g*), while those migrating at slower rates are approximately half particulate and half soluble. (4) The mechanism for intraaxonal displacement is unknown, but it appears to require local metabolic energy utilization and is subject to disruption by drugs (e.g., vinblastine) that depolymerize microtubules.

With these basic characteristics in mind, the scope of this review will be confined to particular aspects of complex carbohydrate transport.

2. Axonal Transport of Complex Carbohydrates

Evidence has accumulated for the intraaxonal transport of all three classes of complex carbohydrates: glycoproteins, glycosaminoglycans (mucopolysaccharides), and gangliosides. By far the greatest number of studies pertain to glycoprotein transport.

2.1. Transport of Glycoproteins

2.1.1. Methodology

Axonal transport of glycoproteins has been investigated by use of a variety of radioactive markers, including the ^{3}H or ^{14}C isotopes of fucose, glucosamine, *N*-acetylglucosamine, *N*-acetylmannosamine, and the ^{35}S isotope of inorganic sulfate. Fucose has been the most popular choice because of its specificity in labeling glycoproteins (Quarles and Brady, 1971; Margolis, R. K., and Margolis, R. U., 1972). In most instances, the precursor is presented in a single application to a group of nerve cell bodies, and radioactive glycoproteins are subsequently detected in the associated axons or nerve terminals. The preparations most frequently employed have been the visual systems of fish (McEwen *et al.*, 1971; Elam and Agranoff, 1971*b*; Forman *et al.*, 1972; Rösner *et al.*, 1973; Repérant *et al.*, 1976), birds (Marko and Cuénod, 1973; Bondy and Madsen, 1971, 1974; Rösner, 1975; Marchisio *et al.*, 1973, 1975), and mammals (Karlsson and Sjöstrand, 1971; Specht and Grafstein, 1973, 1977; Autilio-Gambetti *et al.*, 1975). In each case, the precursor is injected into the eye, and transport of glycoprotein is measured in the optic nerve, contralateral optic tract, or contralateral nerve terminal region of the brain (optic lobe or tectum for fish and birds, lateral geniculate body and superior colliculus for mammals). A typical accumulation curve for $^{35}SO_4^{2-}$-labeled molecules in the goldfish optic tectum is shown in Fig. 1.

Glycoprotein transport studies have also been conducted on various peripheral nerves following application of radioisotope to the associated cell

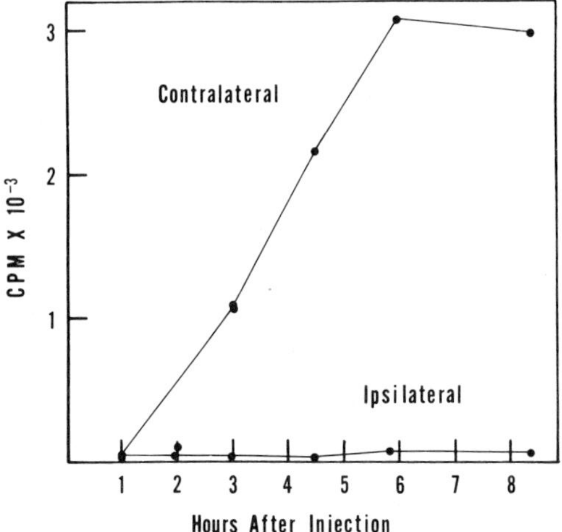

Fig. 1. Accumulation of axonally transported radioactivity in the goldfish optic tectum following intraocular injection of $^{35}SO_4^{2-}$. Radioactivity is in a chloroform-methanol-extracted residue and is approximately equally divided between sulfated glycoproteins and glycosaminoglycans. The contralateral tectum receives all fibers from the injected eye, while the ipsilateral tectum is labeled only systemically (see Elam and Agranoff, 1971*b*).

bodies (Ochs, 1972; Held and Young, 1972; Edström and Mattsson, 1972, 1973; Frizell and Sjöstrand, 1974*a,b*; Gross and Beidler, 1975; Barker *et al.*, 1975; Elam and Peterson, 1976) and in preganglionic fibers of the chick ciliary ganglion (Bennett *et al.*, 1973; DiGiamberardino *et al.*, 1973). Such studies are usually conducted *in vivo*. However, transport of glycoproteins *in vitro* has been investigated in frog sciatic nerve (Edström and Mattsson, 1972), in rabbit vagus nerve (McLean *et al.*, 1975), and in giant axons of the sea snail, *Aplysia* (Ambron *et al.*, 1974; Thompson *et al.*, 1976).

Pathways within the brain have been a less popular choice of study, perhaps because of the difficulty in performing localized isotope applications. One approach has been to label the brain generally, then determine time-dependent changes in the radioactivity of isolated synaptosomes and synaptosomal subfractions (Zatz and Barondes, 1971; Marinari *et al.*, 1972; Langley and Kennedy, 1977; Gurd, 1978). Delayed labeling of these fractions is considered to be indicative of the influx of axonally transported molecules. Recently, localized microinjections were also employed to study glycoprotein transport within the brain (Levin, 1977; Geinisman *et al.*, 1977).

When using radioactive precursors *in vivo*, it is necessary to demonstrate that labeling of axons is due to axonal transport of glycoproteins rather than to local incorporation of precursor distributed by the bloodstream. Such systemic labeling should be equal in symmetrically paired nerves (e.g., optic nerve), and any excess of radioactivity in the nerve or terminals connected to the labeled cell bodies is taken as evidence for axonal transport. In many cases, as can be seen in Fig. 1, the transported label far exceeds the level of systemic incorporation and essentially all radioactivity in the tissue can be considered to be derived from transport. When measuring transport along a nerve, it is often possible to see a crest of radioactivity that is displaced distally with time. An example from the garfish olfactory nerve is seen in Fig. 2. In this instance, the labeling of nerve segments distal to the crest provides a measure of systemic incorporation.

A more difficult issue is whether labeling along the nerve is due to the axonal transport of glycoproteins that are synthesized in the cell body rather than to the axonal transport of free precursor that is then locally incorporated along the axons. Two lines of evidence have argued against local incorporation as a major contributor to axonal glycoprotein radioactivity. The first is the widespread observation that inhibition of protein synthesis in the cell bodies during presentation of isotope greatly lowers the amount of radioactive glycoprotein later appearing in the axons or nerve terminals (Zatz and Barondes, 1971; McEwen *et al.*, 1971; Elam and Agranoff, 1971*b*; Forman *et al.*, 1972; Edström and Mattsson, 1972; Rösner *et al.*, 1973; Marchisio *et al.*, 1973; McLean *et al.*, 1975; Ambron *et al.*, 1975). This strongly implicates the cell bodies as the major site of synthesis. Second, it has been observed that in most systems the levels of free precursor isolated in the axons or nerve terminals are considerably lower ($<10\%$) than the amounts that ultimately appear in the glycoproteins (Karlsson and Sjöstrand, 1971; Bondy and Madsen, 1971; Forman *et al.*, 1972; Edström and Mattsson, 1972; Marchisio *et al.*, 1973; Bennett *et al.*, 1973; Ambron *et al.*, 1974; Frizell

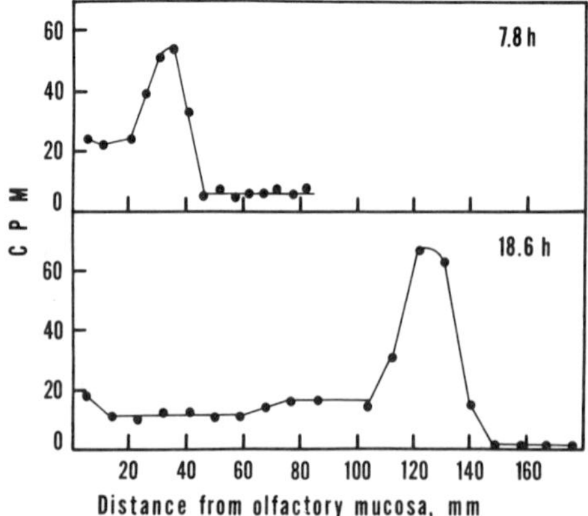

Fig. 2. Distribution of radioactivity along the garfish olfactory nerve at 7.8 and 18.6 hr after application of $^{35}SO_4^{2-}$ to the olfactory mucosa. The radioactivity shown is in a trichloroacetic acid precipitate and is approximately equally divided between sulfated glycoproteins and glycosaminoglycans (see Elam and Peterson, 1976).

and Sjöstrand, 1974*b*; McLean *et al.*, 1975). This would make a precursor–product relationship between axonally localized precursor and axonal glycoproteins seem unlikely. Axonal synthesis of glycoproteins has also been questioned on mechanistic grounds, since ribosomes and Golgi membranes appear to be confined to the cell bodies. The fact remains, however, that some studies report significant levels of precursors in nerves receiving axonally transported material (Forman *et al.*, 1971; Rösner *et al.*, 1973; Rösner, 1975; Geinisman *et al.*, 1977), and the extent to which this might contribute to a limited labeling of axons or neighboring glial cells remains difficult to assess. Specific cases of apparent axonal and terminal glycoprotein synthesis are considered in Section 3.2.

2.1.2. Rates of Glycoprotein Transport

In those instances in which rates of transport can be accurately determined, the rate of glycoprotein transport is found to equal or closely approximate the rapid-transport rate for the general population of proteins (Ochs, 1972; Forman *et al.*, 1972; Gross and Beidler, 1975; McLean *et al.*, 1975; Elam and Peterson, 1976). As in the case of proteins, glycoprotein transport rates are temperature-dependent (Elam and Agranoff, 1971*b*). However, in mammals at a given temperature, there is variation in transport rate from nerve to nerve. In the rabbit, for example, the rate of transport of [^{3}H]fucose-labeled glycoproteins is estimated to be 150, 300, and 400 mm/day in the optic, hypoglossal, and vagus nerves, repectively (Karlsson and Sjöstrand, 1971; Frizell and Sjöstrand, 1974*b*). Among cold-blooded species held at 15–25° C, the rates are generally lower, examples being goldfish optic nerve, 70–100 mm/day at 23° C (Elam and Agranoff, 1971*b*; McEwen *et al.*, 1971); frog sciatic nerve, 60–90 mm/day at 18° C (Edström and Mattsson, 1972); and the R2 giant axon of *Aplysia*, 70 mm/day at 15° C (Ambron *et al.*, 1974). The garfish olfactory nerve (Fig. 2) has the comparatively high glycoprotein transport rate of 206–208 mm/day at 23° C (Gross and Beidler, 1975; Elam and Peterson, 1976). As yet, no functional significance has been attached to the variation in glycoprotein transport rates at a given temperature.

Although the initial influx of labeled glycoproteins into the axon closely parallels the influx of amino-acid-labeled proteins, at later times there is a considerable difference in the pattern of accumulation of proteins and glycoproteins. Typically, following labeling with amino acids, a slow wave of radioactivity traverses the axon at a rate of a fraction of a millimeter to several millimeters per day. This slow-transport phase generally contains a greater amount of radioactivity than is associated with rapid transport (for reviews, see Section 1). With glycoprotein precursors, transported radioactivity usually reaches a maximum within 48 hr of cell body labeling, with no major increases at later times (McEwen *et al.*, 1971; Karlsson and

Sjöstrand, 1971; Marko and Cuénod, 1973; Bennett *et al.*, 1973). Thus, slow transport in the usual sense does not occur with glycoproteins. There have, however, been reports from several laboratories that indicate that there can be prolonged or delayed arrival of glycoproteins in the nerve terminal. In the goldfish visual system, for example, [^{3}H]fucose-labeled glycoproteins continue to accumulate in the optic tectum for up to 30 hr after a single intraocular injection, despite only a 2- 4 to 3-hr transit time in the optic nerve and tract (Forman *et al.*, 1972; Monticone and Elam, 1975). By use of protein synthesis inhibitors, Forman *et al.*, (1972) showed that this extended accumulation is due, at least in part, to an unexpectedly long period of precursor incorporation. More than half the retinal fucose incorporation occurred later than 6 hr after isotope injection. A similarly prolonged incorporation of [^{3}H]fucose was reported for retinal ganglion cells of the rabbit (Karlsson and Sjöstrand, 1971) and mouse (Specht and Grafstein, 1977), in preganglionic neurons of the chick ciliary ganglion (Bennett *et al.*, 1973), and in the R2 cell of *Aplysia* (Ambron *et al.*, 1974). In contrast, extended fucose incorporation was not observed in the pigeon retina (Marko and Cuénod, 1973) or in cells of the locus coeruleus of rat brain (Levin, 1977). Interestingly, Marchisio *et al.*, (1973) showed that retinal fucose incorporation in 7-day chick embryos reaches a peak within 4 hr after intraocular injection, while in 18-day embryos incorporation does not peak until 12 hr postinjection. This may reflect the development of a rate-limiting mechanism for precursor uptake. Retinal uptake of [^{3}H]glucosamine in goldfish (Forman *et al.*, 1971) and of $^{35}SO_4^{2-}$ in goldfish (Elam and Agranoff, 1971*b*) and rabbits (Karlsson and Linde, 1977) was found to be completed within a few hours of isotope presentation.

In addition to a prolonged period of cell body precursor incorporation, there appear to be other factors that can cause a delayed arrival of glycoproteins at certain nerve endings. Specht and Grafstein (1977) reported evidence for the arrival of a measurable portion of [^{3}H]fucose radioactivity in mouse optic nerve terminals some 6–10 days after intraocular injection. This does not appear to be due to a wave of slow transport, since radioactivity levels in the optic tract do not, as would be expected, rise prior to the arrival of the radioactivity in the nerve terminals. The authors favor the hypothesis that there is a delayed release of cell body glycoproteins that are then rapidly transported to the terminals. A delayed cell body release of rapidly transported (amino-acid-labeled) proteins was noted in the rabbit vagus nerve (McLean *et al.*, 1976) and in the garfish olfactory nerve (Gross and Beidler, 1973).

Gremo and Marchisio (1975) and Marchisio *et al.* (1975) observed a delayed arrival (6–48 hr postinjection) of [^{3}H]fucose-labeled glycoproteins in the chick optic tectum. By application of vinblastine to the retina, they were able to show that the late-arriving radioactivity was derived from the axons rather than from continued efflux from the cell bodies. The authors hypothesized that rapidly transported glycoproteins are first incorporated into axonal plasma membranes and are then more slowly transferred to the terminals through intramembranous diffusion.

A further complication in the evaluation of glycoprotein transport rates is the possible existence of discrete, multiple rates in some systems. Ambron *et al.* (1974) reported that rapidly transported glycoproteins pass along the axon of the R2 cell of *Aplysia* in a series of closely spaced discrete peaks. This pattern was more recently confirmed through radioautographic analysis (Thompson *et al.*, 1976). These studies did not determine whether this distribution results from distinct periods of cell body synthesis, periodic episodes of cell body release, or multiple rates of transport. Levin (1977) reported two distinct populations of rapidly transported glycoproteins migrating along the noradrenergic pathway from locus coeruleus to hypothalamus in rat brain. Similar discrete transport phases may occur in individual axons of many nerves, but be obscured by the large total population of axons present.

It is becoming apparent that there are subtle aspects of glycoprotein transport that may include some combination of extended cellular synthesis, delayed cellular release, axonal–terminal redistribution, and multiple transport rates.

2.2.3. Cellular and Subcellular Distribution of Axonally Transported Glycoproteins

2.1.3a. Distribution among Cell Bodies, Axons, and Terminals. Because of the extensive membrane surface of axons and synaptic terminals relative to their cell bodies of origin, it is probable that a large fraction of the glycoproteins produced in the neuronal soma will be destined for axonal transport. It might also be assumed that this fraction would increase with increasing length of the axon. As an approach to this question, a number of workers have estimated the fraction of transported glycoprotein radioactivity appearing in axons and terminals of the optic nerve relative to the total radioactivity remaining in the retina (Karlsson and Sjöstrand, 1971; Marchisio *et al.*, 1973, 1975; Sjöstrand *et al.*, 1973, Bondy and Madsen, 1974; Gremo and Marchisio, 1975; Karlsson and Linde, 1977). These values generally range from 5 to 10% of retinal radioactivity. However, because of the radioactivity in nonganglionic cells within the retina, these are certain to be underestimates of the actual fraction of neuronal glycoprotein exported. A more accurate measure was obtained from the injection of [^{3}H]fucose into the single cell body of the R2 cell of *Aplysia* (Ambron *et al.*, 1974). In this instance, 45% of the total cell glycoprotcin radioactivity is in the axon at 10 hr postinjection. It can also be noted that in retinal preparations in which both amino-acid-labeled proteins and [^{3}H]fucose-labeled glycoproteins were assessed, a greater percentage of glycoprotein than protein is found to be axonally transported (Sjöstrand *et al.*, 1973; Bondy and Madsen, 1974). This again may reflect a particular requirement for glycoproteins in axonal and synaptic structures.

Early studies on axonal transport gave rise to the generalization that rapidly transported molecules are destined mainly for the synaptic terminals, while slow transport provides material to the axon (for a review, see Grafstein, 1969). A survey of the literature on glycoprotein transport indicates that a variable and sometimes appreciable fraction of the rapidly transported material is deposited along the axon. A summary of these data appears in Table I. It is interesting to note that in the relatively long-axoned garfish olfactory nerve (15–25 cm length), a clear majority of the radioactivity remains in the axon.

2.1.3b. Subcellular Distribution within Axons and Terminals. Knowledge of the precise subcellular localization of axonally transported glycoproteins would provide additional insight into the mechanism of transport as well as the function of transported molecules after deposition. A method frequently employed as an initial approach to this problem has been a standard subcellular fractionation of the tissue containing the transport-labeled axons or synaptic endings. A summary of data of this type appears in Table II. One clear conclusion from these analyses is that axonally transported glycoproteins are largely particulate, with 80–97% of the radioactivity consistently appearing in fractions that are sedimentable at 100,000*g*. It can also be seen that a limitation on more detailed analysis has been the tendency for axonal and synaptic constituents to become broadly distributed in conventionally obtained subcellular fractions. In the case of nerve fractionation, there is considerable uncertainty about the composition of fractions labeled "nuclear," "mitochondrial," and "microsomal," since these are generally derived from tissues consisting primarily of cell bodies rather than axons. Various-sized axonal membrane fragments are probably broadly distributed in such fractions (Cancalon and Beidler, 1975).

In the fractionation of tissue containing synaptic endings, the crude mitochondrial fraction is known to contain the majority of intact synaptosomes (Gray and Whittaker, 1962) and would be expected to contain a large fraction of the axonally transported glycoproteins. However, it can be seen in Table II that a significant amount of particulate radioactivity is also recovered in crude nuclear fractions and to an even greater extent in microsomal fractions. The former probably reflects the contamination of nuclei with intact cells and large membranous structures. This has been found to be the case in goldfish optic tecta containing transported [^{3}H]proline-labeled proteins (Elam and Agranoff, 1971*a*). Purified nuclei, which are postsynaptic in the tectum, retain only a small amount of [^{3}H]proline radioactivity. This may be due to reutilization (Heacock and Agranoff, 1977). The microsomal radioactivity is most likely contributed by a combination of collapsed synaptosomes (Elam and Agranoff, 1971*a*), axonal plasma membranes

Table I. Axonal-Nerve Terminal Distribution of Axonally Transported Glycoproteins 1 Day after Injection of Radioactive Precursor

Experimental preparation	Precursor	Percentage of glycoprotein radioactivity		Tissue counted		Ref. No.[a]
		Axons	Nerve terminals	Axonal	Terminal	
Goldfish optic nerve	$[^3H]$fucose	28	72	Optic nerve and tract	Optic tectum	1
Scardinius erythrophthalamus (teleost fish) optic nerve	$[^3H]$-acetyl-glucosamine	33	67	Optic nerve and tract	Optic tectum	2
Chicken optic nerve	$[^3H]$-*N*-acetyl-mannosamine	50	50	Optic nerve	Optic lobe	3
Rabbit optic nerve	$[^3H]$fucose	12	88	Optic tract	LGB and SC[b]	4
Rabbit optic nerve	$^{35}SO_4^{2-}$[c]	17	83	Optic tract	LGB and SC[b]	5
Garfish olfactory nerve	$^{35}SO_4^{2-}$[c,d]	83	17	Olfactory nerve	Olfactory bulb	6

[a] References: (1) Forman *et al.* (1972); (2) Rösner *et al.* (1973); (3) Rösner (1975); (4) Karlsson and Sjöstrand (1971); (5) Karlsson and Linde (1977); (6) Elam and Peterson (1976).
[b] Lateral geniculate body and superior colliculus.
[c] Label is in both glycoproteins and glycosaminoglycans.
[d] Labeling measured 48 hr after injection.

Table II. Subcellular Distribution of Axonally Transported Glycoproteins

Experimental preparation	Precursor	Percentage of glycoprotein radioactivity: Crude nuclear	Crude mitochondrial	Microsomal	Soluble	Ref.
I. Nerve preparations						
a. Frog sciatic	$[^3H]$fucose	16	20	48	16	*a*
b. Frog sciatic	$[^3H]$glucosamine	12	19	55	14	*a*
c. Rabbit hypoglossal	$[^3H]$fucose	34	9	38	21	*b*
d. Goldfish optic tract	$[^3H]$glucosamine	15	34	47	5	*c*
e. *Aplysia* R2	$[^3H]$fucose		97.6 particulate		2.4	*d*
II. Nerve-ending preparations						
a. Rabbit LGB	$[^3H]$fucose	7	26	52	15	*e*
b. Rabbit SC	$[^3H]$fucose	6	51	33	9	*e*
e. Goldfish optic tectum	$^{35}SO_4^{2-}$[h]	21	37	24	18	*f*
d. Goldfish optic tectum	$[^3H]$fucose		93 particulate		7	*g*
e. Goldfish optic tectum	$[^3H]$glucosamine	19	50	28	3	*c*

[a–g] References: [a] Edström and Mattson (1972); [b] Frizell and Sjöstrand (1974*b*); [c] Forman (1971); [d] Ambron *et al.* (1974); [e] Karlsson and Sjöstrand (1971); [f] Elam and Agranoff (1971*b*); [g] McEwen *et al.* (1971).

[h] Label is in both transported glycoproteins and glycosaminoglycans.

(Cancalon and Beidler, 1975), and intraaxonal smooth endoplasmic reticulum (Droz *et al.*, 1975).

Several groups have attempted to evaluate synaptic labeling more directly by isolation of synaptosomal or subsynaptosomal fractions. Elam and Agranoff (1971*b*) isolated the synaptosomal subfraction of the crude mitochondrial fraction in $^{35}SO_4^{2-}$-labeled goldfish optic tectum. Although the synaptosomes accounted for most of the radioactivity in the crude mitochondrial fraction, the myelin and mitochondrial subfractions were also labeled. The labeling of myelin by axonally transported glycoproteins was further investigated using [^{3}H]fucose as precursor (Monticone and Elam, 1975). We found that even highly purified myelin has a specific radioactivity (dpm/μg protein) that closely approximates that of the whole tectal homogenate. Similar findings were reported for the rabbit optic system (Autilio-Gambetti *et al.*, 1975; Matthieu *et al.*, 1978), leading to speculation that there may be specific axonally transported glycoproteins that either isolate with myelin or are transferred to myelin *in vivo*. The labeling of the mitochondrial fraction is of questionable significance in light of the lack of extensive purification and the relatively meager labeling of mitochondria noted in studies cited below. Net transport of mitochondria *per se* appears to occur at slow rates (Grafstein, 1975*a*; Heslop, 1975). However, the interesting possibility that certain rapidly transported glycoproteins might be incorporated into mitochondria at the nerve terminal has not been rigorously explored.

Marko and Cuénod (1973) submitted the crude mitochondrial fraction of pigeon optic tectum to osmotic shock followed by fractionation of sucrose gradients to obtain synaptic subfractions. When the tissue contained [^{3}H]fucose-labeled transported glycoproteins, the highest specific activity was obtained in fractions rich in synaptic plasma membranes. The synaptic soluble fraction, synaptic vesicles, and mitochondrial fraction all had specific radioactivities lower than that of the whole tectal homogenate.

DiGiamberardino *et al.* (1973) examined the specific radioactivity of membranous subfractions of chick ciliary ganglion following labeling with transported [^{3}H]fucosyl glycoproteins. They found heavy labeling of fractions rich in plasma membranes, but saw equally high specific radioactivities in fractions likely to contain synaptic vesicles. Unfortunately, the extent to which the vesicles might be contaminated with plasma membrane was not determined.

Marinari *et al.* (1972) examined the labeling of rat brain synaptosomal subfractions following intraventricular infusion of [^{14}C]glucosamine. After 10 hr of labeling, the highest specific radioactivity was observed in a synaptic plasma membrane fraction, with progressively lower specific radioactivities in synaptic vesicles, the synaptic soluble fraction, and mitochondria. It should be noted that unlike the studies in which only transported molecules are labeled, the reliability of these results depends on the initial purity of the synaptosomes that were fractionated.

An alternative approach to the determination of the subcellular localization of axonally transported glycoproteins has been the use of microscopic radioautography. Bennett *et al.* (1973) examined the distribution of grains derived from transported [^{3}H]fucose in the large presynaptic terminals of the chick ciliary ganglion. At the light-microscopic level, they found that radioactivity in the synaptic bulb exceeded that in the preterminal axons, and that there was little radioactivity over postsynaptic cell bodies. Electron microscopy showed the highest grain density in the vicinity of synaptic plasma membranes and synaptic vesicles. These contained 57 and 27% of the grains, respectively, while areas containing axoplasm and mitochondria were very poorly labeled. Their radioautographic results therefore parallel those obtained by subcellular fractionation of the ciliary ganglion (DiGiamberardino *et al.*, 1973).

An electron-microscopic radioautographic analysis of [^{3}H]fucose-labeled glycoproteins undergoing transport in the R10 and R2 axons of *Aplysia* was conducted by Thompson *et al.*, (1976). They also found extensive labeling of vesicles, with an estimated specific radioactivity 4–10 times higher than that of any other labeled component. Grains were also found associated with smooth endoplasmic reticulum, multivesicular bodies, and mitochondria. The lowest grain density was over cytoplasm and axonal plasma membrane.

The latter study raises the intriguing question as to how glycoproteins still in transit might differ in subcellular localization from those that have been transferred to stationary axonal or synaptic structures. In *Aplysia*, both the distribution of sedimentable radioactivity and the concentration of vesicle-associated grains in electron microscopy seem to parallel the distribution of discrete peaks of transported radioactivity (Ambron *et al.*, 1974; Thompson *et al.*, 1976). This suggests that a vesicular localization exists during transport. Markov *et al.* (1976) examined transported [^{3}H]fucose radioactivity in the preganglionic axons of the ciliary ganglion and concluded that a majority of grains are located over profiles of smooth endoplasmic reticulum and the axolemma. On the basis of these and parallel observations for amino-acid-labeled proteins (Droz *et al.*, 1975), these authors hypothesize that glycoprotein transport occurs within the axonal smooth endoplasmic reticulum. From there, molecules would be transferred to axolemma, synaptic plasma membrane, and synaptic vesicles. As discussed in Section 2.1.2, Marchisio *et al.* (1975) proposed that some glycoproteins are displaced along the axon within the axolemma.

Subcellular fractionation analysis shows that transported glycoproteins in sciatic and hypoglossal nerves are highly particulate (see Table II); however, it is not known what proportion of the labeled molecules are migrating and what proportion has been deposited into stationary structures. Amino-acid-labeled proteins known to be migrating along avian peripheral nerve (Bray and Austin, 1969) and garfish olfactory nerve

(Cancalon and Beidler, 1975) are found to be highly particulate. The overall conclusion seems to be that glycoproteins in transit as well as those deposited in axons and terminals have a predominantly membranous and vesicular localization. A consistent picture of the precise localization within each cellular compartment awaits further study.

2.1.4. Molecular Characteristics of Transported Glycoproteins

There have been several studies on the sodium dodecyl sulfate (SDS)–polyacrylamide gel distribution of axonally transported glycoproteins. Karlsson and Sjöstrand (1971) reported that [^{3}H]fucose-labeled glycoproteins that are transported in the rabbit optic nerve yield a broad pattern of distribution on gels. Apparent molecular weights ranged from 40,000 to 100,000, with peaks of radioactivity appearing at 45,000, 60,000, 75,000, and 100,000. There was little radioactivity in the lower-molecular-weight region with the exception of a discrete peak migrating with an apparent molecular weight of 14,000. It should be noted, however, that molecular weights of glycoproteins based on SDS gel mobility can be substantially in error (Segrest *et al.*, 1971). Interestingly, the pattern for the gels of optic tract glycoproteins closely resembled the pattern for lateral geniculate body, suggesting that a similar population of glycoproteins is deposited in the axons and nerve terminals.

Studies on the frog sciatic nerve *in vitro* revealed a similarly broad distribution of transported glycoproteins on SDS gels, with most molecules labeled with either [^{3}H]glucosamine or [^{3}H]fucose having a molecular weight above 30,000 (Edström and Mattsson, 1973). As in the rabbit visual system, a discrete peak was also observed in the low-molecular-weight region (mol.wt. 18,000). In the sonic motor system of the toadfish, [^{3}H]fucose-labeled molecules were distributed as a large number of peaks between 30,000 and 200,000, again with a discrete lower-molecular-weight peak at 18,000 (Barker *et al.*, 1975). These authors note that a major fraction of the [^{3}H]fucose-labeled gel peaks are also labeled with [^{35}S]methionine, suggesting that a large proportion of the transported proteins are glycoproteins. The 13-day embryonic chick visual system shows a [^{3}H]fucose gel pattern with essentially all the radioactivity in molecules ranging from 30,000 to about 100,000 (Marchisio *et al.*, 1973). Broad peaks of label were observed with estimated molecular weights of 30,000, 40,000, 50,000, 70,000, and 80,000. Data from all these systems support the view that a variety of glycoproteins undergo transport, with their size tending to fall at the upper end of the molecular-weight range.

The apparent heterogeneity of transported glycoproteins might be due in part to the enormous number of individual neurons present in the systems studied. Utilizing the single R2 neuron of *Aplysia*, Ambron *et al.* (1974) found that most of the transported [^{3}H]fucose radioactivity was associated with five SDS gel peaks, all in excess of molecular weight 80,000. However, the level of resolution on unidimensional gels is not sufficient to conclude that these are individual glycoproteins rather than molecular-weight classes. It is interesting to note that there is a correspondence between glycoproteins that decrease with time in the cell body and those that increase with time in the axon.

Our laboratory has presented evidence that axonally transported glycoproteins give rise to glycopeptides that have a broad size distribution. In both the goldfish optic nerve and the garfish olfactory nerve, axonally transported ($^{35}SO_4^{2-}$-labeled glycoproteins yield, after protease digestion, glycopeptides that elute as a broad retarded peak on columns of Sephadex G-50 (Elam and Agranoff, 1971*b*; Elam and Peterson, 1976). This indicates a molecular-weight range of approximately 3000–10,000. No detectable amount of $^{35}SO_4$ radioactivity appeared in lower-molecular-weight glycopeptides of the dialyzable fraction. Recently, we extended our studies in the garfish olfactory nerve to include an evaluation of glycopeptides from transported glycoproteins labeled with [^{3}H]fucose and [^{3}H]glucosamine (Elam *et al.*, 1977; Elam and Peterson, 1979). Unexpectedly, both isotopes label not only a polydisperse population of conventionally sized glycopeptides but also a population of molecules that are large enough to be excluded from Sephadex G-50 and eluted as a broad peak on Sephadex G-200. The exact molecular nature of this material remains unknown, but it appears to be neither glycosaminoglycan nor undigested glycoprotein. A similar high-molecular-weight fraction was not observed in glycopeptides derived from mouse brain synaptosomes labeled with [^{14}C]glucosamine (Dutton *et al.*, 1973) or in [^{3}H]fucose labeled glycopeptides prepared from goldfish optic tectum (Elam *et al.*, 1977). Results of the garfish study also indicate that 8–10% of the [^{3}H]fucose and [^{3}H]glucosamine radioactivity is present in low-molecular-weight, dialyzable glycopeptides (Elam and Peterson, 1979). This is of interest in light of the high concentration of low-molecular-weight glycopeptides derived from synaptic plasma membranes (Gombos *et al.*, 1972; Brunngraber and Javaid, 1975).

2.1.5. Glycoprotein Transport in Developing and Regenerating Axons

It might be expected that the characteristics of glycoprotein transport under conditions of neuronal development would differ from those observed in the mature neuron. In particular, it would be anticipated that different populations of glycoproteins might be required during periods of axonal elongation, as opposed to periods of synaptogenesis and synaptic maturation. One system that has been amenable to exploration of this question is the optic pathway of the embryonic chick. Bondy and Madsen (1971) determined the ratio of contralateral to ipsilateral labeling at various embryonic ages, and concluded that small amounts of transported [^{3}H]fucose-labeled glycoprotein reach the optic lobe of 8- to 12-day embryos (contralateral/ipsilateral ratios

< 2). At 13 days, the amount of transported glycoprotein suddenly increases and continues to rise until hatching (21 days). At that time, contralateral/ipsilateral ratios of 20:1 are obtained. In a similar study, Marchisio *et al.* (1973) found no significant contralateral/ipsilateral differences at 7 embryonic days, ratios of 2:1 to 3:1 at 10 and 13 days, and a ratio of nearly 20:1 at 18 days. In subsequent studies, label was detected in optic nerve axons of 7-day embryos by radioautographic analysis (Gremo *et al.*, 1974) (see also Crossland *et al.*, 1974). Taken together, the results indicate that some glycoprotein transport occurs in growing axons at 7 days, but the amount greatly increases during and following the period of fiber connection at the optic lobe (12 days to hatching).

On the basis of the time of first arrival of transported glycoproteins in the contralateral optic lobe and the length of the optic nerve at various ages, Marchesio *et al.* (1973) concluded that the rate of glycoprotein transport increases during development, from a minimum of 60 mm/day in 10-day embryos to 110 mm/day in 18-day embryos. Despite finding age-dependent differences in the rate and amount of glycoprotein transport, these workers failed to observe any significant difference in the SDS gel molecular-size distribution of transported glycoproteins in 13-day and 18-day embryos.

Several publications have provided information about the relative amounts of retinally synthesized glycoprotein coveyed to the chick optic lobe at various embryonic ages. Bondy and Madsen (1974) reported that the proportion of retinally synthesized [^{3}H]fucose-labeled glycoprotein reaching the optic lobe at 6 hr after intraocular injection was higher in 15-day embryos and 1-day hatchlings (6–8%) than in 15-day-old chicks (4%). They again concluded that this might reflect a greater demand for axonal and synaptic glycoproteins during the period of optic nerve maturation. Marchisio *et al.* (1973), Sjöstrand *et al.*, (1973), and Gremo and Marchisio (1975) (Fig. 3) presented data that indicate that the percentage of retinally incorporated [^{3}H]fucose-labeled glycoproteins reaching the optic lobe increases between 13 and 15 embryonic days (from approximately 4 to 5.5%). It subsequently drops slightly, then increases again between day 18 and hatching (from 5 to 9%). The authors note that the two periods of increase correspond to the time of most active retinal–tectal synapse formation and the time of electrophysiological maturation of the synapses.

Another situation in which changes in the pattern of glycoprotein transport might be expected is in the regenerating neuron. Frizell and Sjöstrand (1974*a,b*) found that the amount of transported [^{3}H]fucose-labeled glycoprotein in regenerating rabbit hypoglossal nerve stands at 150–200% of the level observed in control nerves. Approximately equal elevations were observed at 1 and 4 weeks after nerve crush. No increased labeling of the hypoglossal nucleus was observed, but synthesis in nonneuronal cells may have masked any increase in the rate of synthesis in cells giving rise to the regenerating axons. A subcellular fractionation of the nerve showed a predominantly particulate distribution of radioactivity in

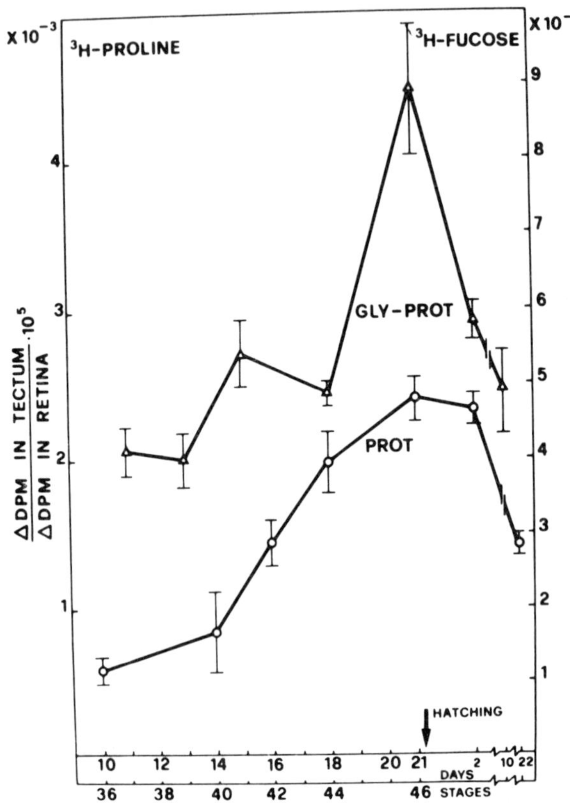

Fig. 3. Developmental changes of the relative proportion of labeled proteins and glycoproteins transported along the optic pathway and recovered in the optic tecta 6 hr after an intraocular injection of either [^{3}H]fucose or [^{3}H]proline (1 μCi). Each point represents the mean ± S.E. of more than 12 animals. From Gremo and Marchisio (1975).

both normal and regenerating nerve. Surprisingly, parallel experiments on the regenerating vagus nerve showed a 25–50% decrease in glycoprotein transport following labeling of the dorsal motor nucleus. The authors speculate that this might reflect a retrograde atrophic effect on the injured cells. The rate of glycoprotein transport did not appear to change with regeneration in either hypoglossal or vagus nerve. Recently, Frizell *et al.* (1976) documented an increased level of retrograde glycoprotein transport in regenerating hypoglossal nerve.

2.2. Transport of Glycosaminoglycans

Despite evidence documenting the presence of glycosaminoglycans in neurons (Margolis, R. U., and Margolis, R. K., 1974; Margolis, R. K., *et al.*, 1975*a*), the question of axonal transport of this class of complex carbohydrates has received relatively little attention. In 1970, we reported that axonally transported $^{35}SO_4^{2-}$ could be isolated in both sulfated glycosaminoglycans and sulfated glycopeptides in the goldfish optic system (Elam *et al.*, 1970). Label was approximately equally divided between the two classes of molecules, with glycosaminoglycan radioactivity associated with both chondroitin sulfate and heparan sulfate. Transport was shown to occur in the form of intact glycosaminoglycans rather than as free precursor, and the transport rate closely approximated that observed for proteins in the same system. Subcellular distribution of total sulfate label also closely paralleled the distribution of amino acid label in protein (Elam and Agranoff, 1971*b*).

Shortly after publication of the goldfish study, several papers provided supporting evidence for axonal transport of glycosaminoglycans. Hirosawa and Young (1971) published radioautographic evidence that some form of sulfated molecule is conveyed into axons and dendrites of mouse brain neurons. [Intradendritic migration of glycoproteins was reported by Kreutzberg *et al.* (1973).] A year later, Held and Young (1972) reported that a small portion of the transported radioactivity derived from [^{3}H]-*N*-acetylglucosamine in cat motor neurons was precipitable with cetylpyridinium chloride, suggesting the presence of transported glycosaminoglycans. Cellulose acetate electrophoretic bands corresponding to hyaluronic acid and chondroitin sulfate were recovered from the labeled fraction.

More recently, Karlsson and Linde (1977) examined $^{35}SO_4^{2-}$-labeled material undergoing rapid transport in the rabbit optic system. Preliminary characterization indicates the presence of glycosaminoglycans, possibly including chondroitin sulfate. Another recent communication has reported axonal transport of $^{35}SO_4^{2-}$-labeled glycosaminoglycans in the rat optic system (Goodrum *et al.*, 1978). The specific molecules undergoing transport were not identified, but radioactivity was reported to be divided equally between glycosaminoglycans and sulfated glycoproteins.

We examined further the question of glycosaminoglycans transport through studies in the garfish olfactory nerve (Elam and Peterson, 1976). When $^{35}SO_4^{2-}$ is utilized as precursor, the results closely parallel those obtained previously for the goldfish optic nerve. Radioactivity is rapidly transported and is divided nearly equally between glycosaminoglycans and sulfated glycopeptides. Again, both chondroitin sulfate and heparan sulfate are among the transported species. Labeling of the olfactory nerve with [^{3}H]glucosamine yields 6–7% of the transported radioactivity in the glycosaminoglycans and the remainder in various-sized glycopeptides. This distribution roughly parallels the distribution of hexosamine, indicating that the glycoproteins and glycosaminoglycans are labeled in proportion to their concentration in the nerve.

The overall conclusion from these studies is that glycosaminoglycans undergo rapid axonal transport, but probably represent a quantitatively small fraction of the total transported complex carbohydrates.

2.3. Transport of Gangliosides

As in the case of glycosaminoglycans, there have been a limited number of studies dealing with the axonal transport of gangliosides. Forman *et al.*, (1971) noted that a fraction of the [^{3}H]glucosamine-derived radioactivity undergoing rapid axonal transport in the goldfish visual system was soluble in chloroform–methanol. In a later study, Forman and Ledeen (1972) isolated gangliosides from the goldfish optic tectum 24 hr after intraocular injection of either [^{3}H]glucosamine, [^{14}C]glucosamine, or [^{3}H]-*N*-acetylmannosamine. All ganglioside species were found to be more heavily labeled in the contralateral tectum than in the ipsilateral control. Free precursor was present in the contralateral tectum, but was considered to be present in amounts too low to account for the ganglioside labeling. The specific contralateral/ipsilateral labeling ratio observed depended on the precursor used. [^{3}H]-*N*-acetylmannosamine gave a ratio of nearly 100:1, whereas [^{3}H]- and [^{14}C]glucosamine gave ratios of 5:1 and 2.8:1, respectively. This probably reflects differential efficacy of the precursors in labeling the brain through systemic routes. Transported radioactivity in individual gangliosides was approximately proportional to the resorcinol staining pattern, leading to the conclusion that those present in the tectum undergo transport in proportion to their concentration.

Rösner *et al.* (1973) also observed labeling of gangliosides in the fish optic tectum following intraocular injection of [^{3}H]-*N*-acetylmannosamine. Radioactivity was found to accumulate in both the optic tract and tectum between 6 and 24 hr after injection, with the majority ($\approx$75%) in the tectum at the latter time. As in goldfish, contralateral/ipsilateral labeling ratios of nearly 100:1 were observed. Tectal lipid labeling was severely depressed by intraocular injection of colchicine; however, levels of acid-soluble radioactivity were similarly depressed, and it was not possible to rigorously distinguish between ganglioside transport and local axonal incorporation of precursor. A time-dependent increase in the ratio of lipid to glycoprotein radioactivity in the tectum was, in fact, interpreted as possible evidence for peripheral *N*-acetylmannosamine incorporation. In parallel experiments utilizing [^{3}H]-*N*-acetylglucosamine, a 10:1 ratio of contralateral to ipsilateral radioactivity was present in the tectal lipid fraction by 18 hr post-injection. In this instance, very little acid-soluble radioactivity was detected in the tectum, and labeling was most likely due to axonal transport of gangliosides.

Rösner (1975) examined further the question of ganglioside transport through studies on the chick visual system. Following intraocular injection of [^{3}H]-*N*-acetylmannosamine, labeled gangliosides appeared in both the optic nerve and contralateral optic lobe. Labeling in the lobe was maximal by 1 day after injection, while nerve labeling increased for at least 4 days. In the nerve, acid-soluble radioactivity exceeded that incorporated into gangliosides, bringing into question the axonal transport of the latter. In the optic lobe, by contrast, soluble radioactivity remained relatively low at all times and ganglioside labeling seemed likely to be due to transport. When labeling of individual optic lobe ganglioside fractions was examined at 1 day postinjection, G_{D1a} was found to be 3-fold more heavily labeled than G_{M1}, which in turn was nearly twice as radioactive as G_{T1} and G_{D1b}. Optic lobe gangliosides labeled in the ipsilateral lobe or through intracerebral injection showed no more than 2-fold differences in the specific radioactivity of individual gangliosides, suggesting an element of specificity in the labeling due to transport.

Bondy and Madsen (1974) also provided evidence for axonal transport of [^{14}C]glucosamine and [^{3}H]-*N*-acetylmannosamine radioactivity in ganglioside fractions of the chick optic nerve. They estimated that 6 hr after injection, 6% of the retinally synthesized [^{3}H]-*N*-acetylmannosamine-labeled ganglioside and 12% of the [^{14}C]glucosamine-labeled ganglioside was in the optic lobe. Maccioni *et al.* (1977) reported additional data on the accumulation of labeled gangliosides in the chick optic lobe following intraocular injection of [^{3}H]-*N*-acetylmannosamine. They found that the accumulation was blocked by intraocular colchicine and that very low levels of precursor radioactivity were present in the lobe.

Ledeen *et al.* (1976) recently extended the evidence for ganglioside transport to the mammalian visual system. Following intraocular injection of [^{3}H]-*N*-acetylmannosamine in the rabbit, labeling was found in the optic tract (15% of total) and in nerve terminals at the lateral geniculate body and superior colliculus. The accumulation of radioactivity was found to be inhibited by intraocular injection of colchicine, and time-course experiments indicated that transport into the terminals occurred only at the rapid rate. As in the chick, an unexpectedly prolonged accumulation of ganglioside radioactivity was observed in the optic tract. Holm (1972) observed labeling of the rabbit optic system gangliosides following intraocular injection of radioactive glucosamine. However, he was not able to conclude whether the labeling was due to axonal transport of gangliosides or of free precursor.

Although more experimental evidence would be desirable, it appears quite certain that gangliosides are capable of undergoing rapid axonal transport to nerve terminals. It might be noted that in the studies on fish, the intraocular injection of protein-synthesis inhibitors was found to block the subsequent appearance of labeled gangliosides in the axon (Forman and Ledeen, 1972; Rösner *et al.*, 1973). It is not known whether this is due to an effect on cell body ganglioside synthesis, as was noted in other systems (DeVries and Barondes, 1971), or is an indication that the mechanism for conveying gangliosides into the axon requires ongoing protein synthesis. A requirement for continuing protein synthesis during cell body release of glycoproteins was noted in the R2 cell of *Aplysia* (Ambron *et al.*, 1975).

3. Turnover and Redistribution of Axonally Transported Molecules

3.1. Metabolic Turnover Rates

The presence of a population of labeled, axonally transported molecules in the nerve terminals should afford the opportunity to monitor the utilization of particular molecules in a highly localized cellular environment. This prospect becomes particularly intriguing if one includes the possibility of testing the effects of variable physiological activity. As an approach to this objective, a number of investigators have estimated the metabolic half-life of various axonally transported complex carbohydrates after their arrival at the nerve terminals. A summary of such estimates for transported glycoproteins appears in Table III. Aforementioned processes of protracted cellular synthesis or delayed cellular release or both frequently prevent the loss of radioactivity from following simple first-order kinetics. Consequently, results are expressed in terms of time for radioactivity to fall to one half its maximum value ($t_{½\ max}$). It can be seen that among warm-blooded species, turnover half-times for sugar-labeled molecules range from a minimum of 3 days for the rapid phase of [^{3}H]fucose turnover in pigeon optic lobe to 69 days for [^{3}H]fucose in the slow phase in the pigeon optic tract. Most values fall between 1 and 2 weeks, while in the goldfish, with lower body temperature, the turnover half-time is 3–4 weeks.

A more detailed analysis of glycoprotein turnover has been attempted through subcellular analysis of the transported material. Marko and Cuénod (1973) examined the slower turnover phase in [^{3}H]fucose-labeled synaptic

Table III. Turnover of Axonally Transported Glycoproteins

Precursor	Tissue	$t_{1/2max}$	Ref. No.[a]
I. [^{3}H]fucose	Chick ciliary ganglion	7–10 days	1
	Rabbit LGB and SC	15 days	2
	Rabbit optic tract	15 days	2
	Mouse LGB and SC	4–6 days[b]	3
	Pigeon optic lobe	3, 38 days	4
	Pigeon optic lobe	7, 69 days	4
	Goldfish optic tectum	20 days	5
II. [^{3}H]glucosamine	Goldfish optic tectum	20–30 days	6
III. $^{35}SO_4^{2-}$[c]	Rabbit optic tract	3 days	7
	Rabit LGB and SC	5 days	7
	Goldfish optic tectum	1, 7days	8

[a] References: (1) Bennett *et al.* (1973); (2) Karlsson and Sjöstrand (1971); (3) Specht and Grafstein (1977); (4) Marko and Cuénod (1973); (5) Forman *et al.* (1972), Monticone and Elam (1975); (6) Forman *et al.* (1971); (7) Karlsson and Linde (1977); (8) Elam and Agranoff (1971*b*).

[b] A delayed increase in labeling was noted starting at 6 days.

[c] Label is in both glycoproteins and glycosaminoglycans.

subfractions of the pigeon optic lobe. Half-times for the various membranous fractions ranged from 26 days in the synaptic-vesicle-enriched fraction to 33 days in the synaptic membrane fraction and from 58 to 63 days in the mitochondrial fraction. In a similar study of synaptic subfractions of chick ciliary ganglion, DiGiamberardino *et al.* (1973) observed roughly equal turnover rates in the soluble and various membranous fractions. A companion study utilizing quantitative radioautography (Bennett *et al.*, 1973) suggested a slightly more rapid turnover of synaptic membrane glycoproteins than of those in synaptic vesicles, mitochondria, and axoplasm.

Langley and Kennedy (1977) evaluated turnover of synaptic membrane and synaptic soluble fractions following intraventricular injection of [^{3}H]fucose in the rat. Both showed apparent half-times in excess of 20 days. The loss of radioactivity in seven glycoprotein bands on SDS acrylamide gels was also determined. Results indicated a range of individual glycoprotein half-times from 11 to 22 days, suggesting noncoordinate breakdown of synaptic glycoproteins. However, it should be recalled that synaptosomal preparations isolated after whole brain labeling may possess substantial nonsynaptosomal radioactivity.

As a whole, it would appear that transported glycoproteins, labeled with fucose or glucosamine, do not have a particularly rapid turnover time. R. K. Margolis and Gomez (1973) reported that the soluble [^{3}H]fucose-labeled glycoprotein in whole rat brain has a very rapid turnover ($t_{1/2} = 1$ day). Unusually rapid turnover of soluble, transported glycoproteins has not been reported; however, we recently found that the soluble [^{3}H]fucose-labeled molecules, transported to the goldfish optic tectum, have a turnover time that is 1/3 that of the membrane fraction (Ripellino and Elam, 1978).

Turnover of $^{35}SO_4^{2-}$-labeled molecules has been examined in the goldfish and mammalian visual system (Table III). In the former case, this label is known to be approximately equally divided between sulfated glycoproteins and glycosaminoglycans (Elam *et al.*, 1970), both of which show similar turnover kinetics (Elam, unpublished observation). It can be seen that the turnover of sulfate label is biphasic ($t_{1/2} = 1$ and 7 days) and is considerably more rapid than the turnover of fucose and glucosamine in the same system. In the rabbit, sulfate turnover is also more rapid than the corresponding fucose turnover (5 vs. 15 days in LGB and SC; 3 vs. 15 days in optic tract). Recently, Goodrum *et al.*, (1978) reported extremely rapid $^{35}SO_4^{2-}$ turnover ($t_{1/2} < 1$ day) in axonally transported molecules of the rat visual system. [^{3}H]Fucose again was found to have a relatively slower turnover. Studies on whole rat brain showed a more rapid turnover of sulfate than of fucose in particulate glycoproteins (Margolis, R. U., and Margolis, R. K., 1972; Margolis, R. K., and Gomez, 1973) and a more rapid turnover of sulfate than of glucosamine in glycosaminoglycans (Margolis, R. K., and Margolis,

R. U., 1973). It is of interest to note that Gurd (1978) has examined the turnover time of [^{3}H]leucine-labeled protein cores of lectin binding synaptic membrane glycoproteins. He reports a biphasic turnover rate, the faster phase approximating the more rapid turnover observed for $^{35}SO_4^{2-}$ and the slower phase in line with turnover times observed with [^{3}H]fucose. Whether this pattern reflects the turnover of the protein portion of molecules enriched in these glycoprotein constituents remains to be determined.

The turnover rate of transported gangliosides, like that of glycoproteins, does not appear to be extraordinarily rapid. Ledeen *et al.* (1976) reported turnover of gangliosides labeled in the rabbit optic system occurring between 2 and 3 weeks after injection of isotope. Sialic acid residues labeled with [^{3}H]-*N*-acetylmannosamine showed the same turnover kinetics as gangliosides labeled more generally with [^{14}C]serine, suggesting turnover of the molecule as a whole. In the optic lobe of the chick, labeling of transported gangliosides had not decreased by 4 days after [^{3}H]-*N*-acetylmannosamine injection, and the proportion of labeled ganglioside found in the optic lobe as opposed to the retina had not decreased by 15 days post-injection (Bondy and Madsen, 1974). Holm (1972) reported turnover half-times for [^{3}H]-*N*-acetylmannosamine-labeled gangliosides in rabbit to be 55–57 days in the optic tract and 50–80 days in the lateral geniculate.

3.2. Redistribution and Effects of Local Synthesis

Molecules that have been conveyed to nerve terminals might, in the course of their utilization and turnover, be subject to transfer from one subcellular fraction to another. One possibility that has been the subject of considerable speculation is the interchange of membrane between synaptic vesicles and synaptic plasma membrane. These two structures appear to have certain proteins and glycoproteins in common (Morgan *et al.*, 1973; Bock and Jorgensen, 1975), and the concept of reversible merging of synaptic vesicles and synaptic plasma membrane during synaptic secretion has received experimental support (Heuser and Reese, 1973; Fried and Blaustein, 1976). Although transfer of labeled glycoprotein might result in an accelerated turnover time for the donor membrane and a slower turnover of the recipient membrane, this would not be easily distinguishable experimentally from differing but independent turnover rates for the two fractions. A more convincing pattern would be a time-dependent increase in labeling of the recipient membranes with concomitant decreases in the donor membranes. Thus far, this pattern has not been observed for axonally transported label in either synaptic vesicle or synaptic plasma membrane fractions. It is possible that such exchanges reach equilibrium very quickly after arrival of the transported molecules at the nerve ending.

In studying the turnover of transported glycoproteins in ciliary ganglion, DiGiamberardino *et al.* (1973) observed a time-dependent increase in labeling of a myelin-enriched subfraction, while Bennett *et al.* (1973) observed a similarly timed increase in radioautographic grains over myelin. These observations are of interest in light of our studies on time-dependent changes in labeling of myelin in the goldfish visual system (Monticone and Elam, 1975). Myelin isolated from optic tectum at various times during the turnover of transported [^{3}H]fucose-labeled glycoproteins showed a slower turnover time than that of the whole tectal homogenate. More notably, SDS gel electrophoresis showed an increased labeling of particular-sized myelin-associated glycoproteins during the initial part of the turnover period. A similar pattern of time-delayed increases in [^{3}H]fucose-labeling of certain myelin glycoproteins has been reported for rat brain labeled through intracranial injection (Poduslo *et al.*, 1977) and more recently for rabbit optic tract labeled through intraocular injection (Matthieu *et al.*, 1978). These patterns might reflect a transfer of membranous constituents from other fractions to myelin or to axonal membranes which isolate with myelin. Alternatively, the labeling change could occur through breakdown of previously transported glycoproteins and reincorporation of [^{3}H]fucose into myelin glycoproteins, or through a metabolic alteration of existing myelin-associated glycoproteins in a manner that affects their electrophoretic mobility.

The latter possibilities raise the general question of how patterns of turnover of axonally transported glycoproteins might be affected by local synthetic processes within the nerve terminal. Reincorporation of metabolized sugars would cause underestimation of turnover rates and probable redistribution in subcellular fractions and in electron-microscopic radioautographs. Addition of nonradioactive residues to radioactive transported molecules would change their molecular properties. The extent to which such processes might be occurring within the nerve terminal is still a matter of some controversy. Several papers have reported that synaptosomes contain a variety of glycosyl transferases specific for glycoproteins, glycosaminoglycans, and gangliosides (for recent examples, see Brandt *et al.*, 1975; Den *et al.*, 1975). Synaptosomes are capable of incorporating radioactive sugars *in vitro* (Bosmann and Hemsworth, 1970; Festoff *et al.*, 1971; Dutton *et al.*, 1973), and were found to have a high-affinity glucosamine uptake system (Tan *et al.*, 1977). Barondes and associates (Barondes, 1968; Barondes and Dutton, 1969) isolated synaptosomes at various times after *in vivo* incorporation of radioactive glucosamine. They found that synaptic incorporation is very rapid and largely resistant to the effect of protein synthesis inhibitors, facts that have been interpreted as supportive of local nerve-ending glycosylation of preformed glycoproteins. Subcellular analysis indicates that at least a portion of the incorporation is into mitochondria (Dutton *et al.*,

1973). More recently, R. K. Margolis, *et al.* (1975*a*) observed a similar rapid *in vivo* incorporation of glucosamine into glycoproteins and gangliosides of the synaptosomal soluble fraction. R. K. Margolis *et al.* (1975*b*) have also observed a protracted increase in the labeling of the amino acid cores of whole rat brain glycopeptides prepared after intraperitoneal injection of radioactive amino acid. They suggest a delayed glycosylation of some proteins, possibly those that have undergone axonal transport to nerve endings in a nonglycosylated form. However, Gurd (1978) has found that protein cores of rat brain synaptic membrane glycoproteins become maximally labeled at the same time as nonglycosylated supporting membrane proteins, suggesting that any delay in glycosylation of axonally transported proteins would have to be minimal.

A more direct indication of axonal glycosylation has come from studies on the R2 cell of *Aplysia* (Ambron *et al.*, 1974). When examining axonally transported glycoproteins on acrylamide gels, the authors noted that there appeared in the axon a particular-sized glycoprotein that was never seen in the cell body. They suggested that the new molecule may have arisen as a result of metabolic alternation during or immediately after transport. Additional evidence for axonal glycosylation was obtained by intraaxonal injection of [^{3}H]-*N*-acetylgalactosamine (Ambron and Treistman, 1977). The precursor was incorporated directly into axonal glycoproteins and glycolipids, and the labeling was found to be insensitive to protein-synthesis inhibition. This suggests incorporation into molecules previously provided by axonal transport. In their study of the turnover of [^{3}H]fucose-labeled synaptic glycoproteins, Langley and Kennedy (1977) observed a delayed labeling of a particular band that may be due to axonal reutilization of transported fucose.

3.3. Transsynaptic Migration

Over and above any redistribution of axonally transported radioactivity that may occur within the axon or nerve ending, a potential exists for the spread of such radioactivity to postsynaptic cells. This could occur through the release and diffusion of radioactive monosaccharides during metabolic turnover or possibly as a result of transfer of intact complex carbohydrates from neurons to other cells. In their comprehensive radioautographic analysis of labeled glycoproteins in the ciliary ganglion, Bennett *et al.* (1973) concluded that minimal labeling occurs in postsynaptic cells. On the other hand, a growing number of studies have cited evidence for some degree of transsynaptic migration of glycoproteins (Specht and Grafstein, 1973; Graftstein and Laureno, 1973; Dräger, 1974; Wiesel *et al.*, 1974; Casagrande and Harting, 1975; Repérant *et al.*, 1977; Specht and Grafstein, 1977). This phenomenon, documented primarily in the mammalian visual system, seems to involve both diffusion of labeled material into adjacent postsynaptic tissue (as in superior colliculus to retrosplenial cortex) and the uptake by postsynaptic neurons followed by secondary axonal transport to their nerve terminals (as in lateral geniculate body to visual cortex) (Graftstein and Laureno, 1973; Specht and Grafstein, 1973, 1977). With [^{3}H]fucose as precursor, as much as 15% of the lateral geniculate radioactivity is ultimately recovered in the secondary nerve ending in the visual cortex (Specht and Grafstein, 1977). It is not yet known whether this transcellular labeling is due to the spread of free fucose or to the transfer of intact glycoproteins. An argument favoring the latter possibility is that a similar spread of transported [^{3}H]proline radioactivity in the goldfish optic tectum was found not to be diminished by inhibition of tectal protein synthesis (Neale *et al.*, 1972). Release of intact glycoproteins to the medium was noted in cultured neuroblastoma cells (Truding *et al.*, 1975).

It appears that axonally transported complex carbohydrates may suffer a variety of fates including direct metabolic breakdown, intermembranous transfer, metabolic alteration, release to the extracellular space, and, as yet unmentioned, possible retrograde transport back to the cell body. Only considerably more research will distinguish the relative quantitative importance of these processes.

4. Overview and Functional Considerations

The facts that emerge as most nearly incontestable are these: (1) A variety of high-molecular-weight glycoproteins and at least some classes of glycosaminoglycans and gangliosides are supplied to axons and synaptic terminals by the process of axonal transport. (2) Glycoproteins that are undergoing transport and that have been delivered to sites of utilization are predominantly membranous in their subcellular localization. (3) All classes of axonally transported complex carbohydrates arrive at their site of utilization within a time span of a few hours to a few days. Taken together, these facts support the view, put forth previously by others (Droz, 1973, 1975; Koenig *et al.*, 1973), that the primary function of transported complex carbohydrates is to maintain the structural and functional integrity of axonal and synaptic membranes.

The critical relationship of rapid axonal transport as a whole to synaptic maintenance can be inferred from a number of lines of evidence. Cuénod and associates demonstrated that interruption of rapid axonal transport, through application of colchicine, causes degenerative morphological and electrophysiological changes in the optic nerve terminal that mimic those produced by nerve section (Cuénod *et al.*, 1972; Perišić and Cuénod, 1972). Changes in rapid axonal transport have been correlated with other changes in synaptic membrane function, including those associated with growth and regeneration

(Section 2.1.5)(Sjöstrand *et al.*, 1973; Grafstein, 1975*b*), aging (Geinisman *et al.*, 1977), and muscular dystrophy (Komiya and Austin, 1974; McLane and McClure, 1977).

The particular roles played by various types of complex carbohydrates in synaptic and axonal maintenance are not yet clear. Glycoproteins are known to be concentrated on neuronal cell surfaces and at synaptic junctions (Brunngraber, 1970; Gombos *et al.*, 1972; Margolis, R. K., *et al.*, 1975*a*). They also appear to be constituents of synaptic vesicles (Morgan *et al.*, 1973). As a reflection of this distribution, axonal and synaptic plasma membranes and synaptic vesicles are all found to be recipients of axonally transported glycoproteins (Section 2.1.3b). The specific roles of intrasynaptic glycoproteins remain unknown; however, recent studies of Baux *et al.* (1978) suggest a possible function in neurotransmitter release. The known role of glycoproteins in mediating cell–cell interactions would suggest that particular glycoproteins, which are critical for the specific adhesion to postsynaptic cells, might need to be conveyed to the synapse. Transport of synapse-specific glycoproteins has not been demonstrated. However, Cuénod *et al.* (1973) showed that a particular glycoprotein disappears from the pigeon optic lobe during optic nerve degeneration. Also, as was noted in Section 2.1.5, the amount of glycoprotein transported in the chick optic nerve greatly increases during the period of synapse formation and functional maturation.

Radioactivity from axonally transported glycoproteins has been shown to enter postsynaptic cells (Section 3.3), a phenomenon that could conceivably be related to the trophic influence of neurons on postsynaptic cells (Guth, 1974; Grafstein, 1977). Cochicine was shown to block neurotrophic regulatory effects in skeletal muscle (Kauffman *et al.*, 1976; Sellin and McArdle, 1977), and nerve extracts can, in some cases, mimic neurotrophic effects on muscle cells in culture (Oh, 1976; Engelhardt *et al.*, 1977).

Gangliosides are also found to be associated with synaptic plasma membranes (for a review, see Ledeen *et al.*, 1976), and it is not unexpected that they are supplied by axonal transport. Like those of glycoproteins, the specific functions of gangliosides in synapses are unknown. Speculation has focused on possible roles in cell–cell adhesion, as cation binders, and as receptors (for a discussion, see Morgan *et al.*, 1976). Glycosaminoglycans are present in relatively low concentrations in synapses (Margolis, R. K., *et al.*, 1975*a*) and in nonmyelinated axons (Elam and Peterson, 1976), but their presence and transport in neurons raise the question of function. Speculation has centered on their possible roles in ordering the ionic composition and hydration at the outer surface of the neuron and as counterions in amine storage vesicles (for a related discussion, see Margolis, R. U., and Margolis, R. K., 1977). Much more needs to be known about the composition and subcellular distribution of axonally transported glycosaminoglycans and gangliosides.

It may appear paradoxical that rapidly transported complex carbohydrates generally have slow turnover rates (Section 3.1). This may be due in part to isotope-reutilization processes. However, the rate of transport has no predictive value for the rate of utilization, since in a steady state only the *amount* transported per unit time need be in balance with the amount consumed. Thus, slow turnover implies that transported complex carbohydrates are rapidly transported in small amounts. The real significance of rapid transport probably relates to the speed of chemical communication it provides between the cell body and the synapse. This permits the cell to adjust the composition of what is transported within a matter of hours, providing flexibility that may be particularly critical during development or in response to sudden changes in the postsynaptic environment. Rapid adjustment could also be made through local synaptic modification of transported molecules (Section 3.2). However, the most direct link to the genetic control point of the cell would continue to be through axonal transport.

References

Ambron, R. T., and Treistman, S. N., 1977, Glycoproteins are modified in the axon of R2, the giant neuron of *Aplysia californica*, after intra-axonal injection of ^{3}H *N*-acetylgalactosamine, *Brain Res.* **121**:287–309.

Ambron, R. T., Goldman, J. E., and Schwartz, J. H., 1974, Axonal transport of newly synthesized glycoproteins in a single identified neuron of *Aplysia californica, J. Cell Biol.* **61:** 655–675.

Ambron, R. T., Goldman, J. E., and Schwartz, J. H., 1975, Effect of inhibiting protein synthesis on axonal transport of membrane glycoproteins in an identified neuron of *Aplysia, Brain Res.* **94**:307–323.

Autilio-Gambetti, L., Gambetti, P., and Shafer, B., 1975, Glial and neuronal contribution to proteins and glycoproteins recovered in myelin fractions, *Brain Res.* **84**:336–340.

Barker, J. L., Hoffman, P. N., Gainer, H., and Lasek, R. J., 1975, Rapid transport of proteins in the sonic motor system of the toadfish, *Brain Res.* **97**:291–301.

Barondes, S. H., 1968, Incorporation of radioactive glucosamine into macromolecules at nerve endings, *J. Neurochem.* **15**:699–706.

Barondes, S. H., and Dutton, G. R., 1969, Acetoxycycloheximide effect on synthesis and metabolism of glucosamine-containing macromolecules in brain and in nerve endings, *J. Neurobiol.* **1**:99–110.

Baux, G., Simonneau, M., Tauc, L., 1978, Blocking action of Ruthenium Red on cholinergic and non-cholinergic synapses: Possible involvement of sialic acid-containing substrates in neurotransmission, *Brain Res.* **152**:633–638.

Bennett, G., DiGiamberardino, L., Koenig, H. L., and Droz, B., 1973, Axonal migration of protein and glycoprotein to nerve endings. II. Radioautographic analysis of the renewal of glycoproteins in nerve endings of chicken ciliary ganglion after intracerebral injection of ^{3}H fucose and ^{3}H glucosamine, *Brain Res.* **60**:129–146.

Bock, E., and Jorgensen, O. S., 1975, Rat brain synaptic vesicles and synaptic plasma membranes compared by crossed immuno-electrophoresis, *FEBS Lett.* **52**:37–39.

Bondy, S. C., and Madsen, C. J., 1971, Development of rapid axonal flow in the chick embryo, *J. Neurobiol.* **2**:279–286.

Bondy, S. C., and Madsen, C. J., 1974, The extent of axoplasmic transport during development, determined by migration of various radioactively-labelled materials, *J. Neurochem.* **23:** 905–910.

Bosmann, H. B., and Hemsworth, B. A., 1970, Incorporation of amino acids and monosaccharides into macromolecules by isolated synaptosomes and synaptosomal mitochondria, *J. Biol. Chem.* **245:**363–371.

Brandt, A. E., Distler, J. J., and Jourdian, G. W., 1975, Biosynthesis of chondroitin sulfate proteoglycan: Subcellular distribution of glycosyl transferases in embryonic chick brain, *J. Biol. Chem.* **250:**3996–4006.

Bray, J. J., and Austin, L., 1969, Axoplasmic transport of ^{14}C proteins at two rates in chicken sciatic nerve, *Brain Res.* **12:**230–233.

Brunngraber, E. G., 1970, Glycoproteins in neural tissue, in: *Protein Metabolism of the Nervous System* (A. Lajtha, ed.), pp. 383–407, Plenum Press, New York and London.

Brunngraber, E. G., and Javaid, J. I., 1975, Subcellular and anatomical distribution in rat brain of glycoproteins that contain mannose-rich heteropolysaccharide chains, *Biochim. Biophys. Acta* **404:**67–73.

Cancalon, P., and Beidler, L. M., 1975, Distribution along the axon and into various subcellular fractions of molecules labeled with ^{3}H leucine and rapidly transported in the garfish olfactory nerve, *Brain Res.* **89:**225–244.

Casagrande, V. A., and Harting, J. K., 1975, Transneuronal transport of tritiated fucose and proline in the visual pathways of tree shrew *Tupaia glis, Brain Res.* **96:**367–372.

Crossland, W. J., Currie, J. R., Rogers, L. A., and Cowan, W. M., 1974, Evidence for a rapid phase of axoplasmic transport at early stages in the development of the visual system of the chick and frog, *Brain Res.* **78:**483–489.

Cuénod, M., Sandri, C., and Akert, K., 1972, Enlarged synaptic vesicles in optic nerve terminals induced by intraocular injection of colchicine, *Brain Res.* **39:**285–296.

Cuénod, M., Marko, P., and Niederer, E., 1973, Disappearance of particulate tectal protein during optic nerve degeneration in the pigeon, *Brain Res.* **49:**422–426.

Den, H., Kaufman, B., McGuire, E. J., and Roseman, S., 1975, The sialic acids. XVIII. Subcellular distribution of seven glycosyltransferases in embryonic chicken brain, *J. Biol. Chem.* **250:**739–746.

DeVries, G. H., and Barondes, S. H., 1971, Incorporation of ^{14}C *N*-acetyl neuraminic acid into brain glycoproteins and gangliosides *in vivo, J. Neurochem.* **18:**101–105.

DiGiamberardino, L., Bennett, G., Koenig, H. L., and Droz, B., 1973, Axonal migration of protein and glycoprotein to nerve endings. III. Cell fraction analysis of chicken ciliary ganglion after intracerebral injection of precursors of proteins and glycoproteins, *Brain Res.* **60:** 147–159.

Dräger, U. C., 1974, Autoradiography of tritiated proline and fucose transported transneuronally from the eye to the visual cortex in pigmented and albino mice, *Brain Res.* **82:**284–292.

Droz, B., 1973, Renewal of synaptic proteins, *Brain Res.* **62:**383–394.

Droz, B., 1975, Synthetic machinery and axoplasmic transport: Maintenance of neuronal connectivity, in: *The Nervous System*, Vol. 1, *The Basic Neurosciences* (D. B. Tower, ed.), pp. 111–127, Raven Press, New York.

Droz, B., and Leblond, C. P., 1963, Axonal migration of proteins in the central nervous system and peripheral nerves as shown by radioautography, *J. Comp. Neurol.* **121:**325–345.

Droz, B., Rambourg, A., and Koenig, H. L., 1975, The smooth endoplasmic reticulum: Structure and role in the renewal of axonal membrane and synaptic vesicles by fast axonal transport, *Brain Res.* **93:**1–13.

Dutton, G. R., Haywood, P., and Barondes, S. H., 1973, ^{14}C glucosamine incorporation into specific products in the nerve ending fraction *in vivo* and *in vitro, Brain Res.* **57:**397–408.

Edström, A., and Mattsson, H., 1972, Rapid axonal transport *in vitro* in the sciatic system of the frog of fucose-, glucosamine-, and sulfate-containing material, *J. Neurochem.* **19:**1717–1729.

Edström, A., and Mattsson, 1973, Electrophoretic characterization of leucine-, glucosamine- and fucose-labelled proteins rapidly transported in frog sciatic nerve, *J. Neurochem.* **21:** 1499–1507.

Elam, J. S., and Agranoff, B. W., 1971*a*, Rapid transport of protein in the optic system of the goldfish, *J. Neurochem.* **18:**375–387.

Elam, J. S., and Agranoff, B. W., 1971*b*, Transport of proteins and sulfated mucopolysaccharides in the goldfish visual system, *J. Neurobiol.* **2:**379–390.

Elam, J. S., and Peterson, N. W., 1976, Axonal transport of sulfated glycoproteins and mucopolysaccharides in the garfish olfactory nerve, *J. Neurochem.* **26:**845–850.

Elam, J. S., and Peterson, N. W., 1979. Axonal transport of glycoproteins in the garfish olfactory nerve: Isolation of high molecular weight glycopeptides labeled with ^{3}H fucose and ^{3}H glucosamine (submitted).

Elam, J. S., Goldberg, J. M., Radin, N. S., and Agranoff, B. W., 1970, Rapid axonal transport of sulfated mucopolysaccharide proteins, *Science* **170:**458–460.

Elam, J. S., Peterson, N. W., and Ripellino, J. A., 1977, High molecular weight glycopeptides from axonally transported glycoproteins, *Trans. Am. Soc. Neurochem.* **8:**254.

Engelhardt, J. K., Ishikawa, K., Mori, J., and Shimabukuro, Y., 1977, Neurotrophic effects on the electrical properties of cultured muscle produced by conditioned medium from spinal cord explaints, *Brain Res.* **128:**243–248.

Festoff, B. W., Appel, S. H., and Day, E., 1971, Incorporation of ^{14}C glucosamine into synaptosomes *in vitro, J. Neurochem.* **18:**1871–1886.

Forman, D. S., 1971, A symmetrical double-label method for studying the rapid axonal transport of radioactivity from labelled D-glucosamine in the gold-fish visual system, *Acta Neuropathol. (Berlin) Suppl.* **V:**171–178.

Forman, D. S., and Ledeen, R. W., 1972, Axonal transport of gangliosides in the goldfish optic nerve, *Science* **177:**630–633.

Forman, D. S., McEwen, B. S., and Grafstein, B., 1971, Rapid transport of radioactivity in goldfish optic nerve following injections of labelled glucosamine, *Brain Res.* **28:**119–130.

Forman, D. S., Grafstein, B., and McEwen, B. S., 1972, Rapid axonal transport of ^{3}H fucosyl glycoproteins in the goldfish optic system, *Brain Res.* **48:**327–342.

Fried, R. C., and Blaustein, M. P., 1976, Synaptic vesicle recycling in synaptosomes *in vitro, Nature (London)* **261:**255–256.

Frizell, M., and Sjöstrand, J., 1974*a*, Transport of proteins, glycoproteins and cholinergic enzymes in regenerating hypoglossal neurons, *J. Neurochem.* **22:**845–850.

Frizell, M., and Sjöstrand, J., 1974*b*, The axonal transport of ^{3}H fucose labelled glycoproteins in normal and regenerating peripheral nerves, *Brain Res.* **78:**109–123.

Frizell, M., McLean, W. G., and Sjöstrand, J., 1976, Retrograde axonal transport of rapidly migrating labelled proteins and glycoproteins in regenerating peripheral nerves, *J. Neurochem.* **27:**191–196.

Geinisman, Y., Bondareff, W., and Telser, A., 1977, Transport of ^{3}H fucose labelled glycoproteins in the septo-hippocampal pathway of young adult and senescent rats, *Brain Res.* **125:** 182–186.

Gombos, G., Morgan, I. G., Waehneldt, T. V., Vincendon, G., and Breckenridge, W. C., 1972, Glycoproteins of the synaptosomal plasma membrane, *Adv. Exp. Med. Biol.* **25:**101–113.

Goodrum, J. F., Toews, A. D., and Morell, P., 1978, Rapid axonal transport of ^{3}H fucose and ^{35}S sulfate labelled macromolecules in rat visual system, *Soc. Neurosci. Abs.* **4:**33.

Grafstein, B., 1969, Axonal transport: Communication between soma and synapse, *Adv. Biochem. Psychopharmacol.* **1:**11–25.

Grafstein, B., 1975*a*, Principles of anterograde axonal transport in relation to studies of neuronal connectivity, in: *The Use of Axonal Transport for Studies of Neuronal Connectivity* (W. M. Cowan and M. Cuénod, eds.), pp. 49–67, Elsevier, Amsterdam.

Grafstein, B., 1975*b*, The eyes have it: Axonal transport and regeneration in the optic nerve, in:

The Nervous System, Vol. 1, *The Basic Neurosciences* (D. B. Tower, ed.), pp. 147–151, Raven Press, New York.

Grafstein, B., 1977, Axonal transport: The intracellular traffic of the neurone, in: *Handbook of the Nervous System,* Vol. I, *Cellular Biology of Neurones* (E. R. Kandel, ed.), American Physiological Society, Washington, D.C.

Grafstein, B., and Laureno, R., 1973, Transport of radioactivity from eye to visual cortex in the mouse, *Exp. Neurol.* **39:**44–57.

Gray, E. G., and Whittaker, V. P., 1962, The isolation of nerve endings from brawn: An electron microscopic study of cell fragments derived from homogenization and centrifugation, *J. Anat.* **96:**79–86.

Gremo, F., and Marchisio, P. C., 1975, Dynamic properties of axonal transport of proteins and glycoproteins: A study based on the effects of metaphase blocking drugs in the developing optic pathway of chick embryos, *Cell Tissue Res.* **161:**303–316.

Gremo, F., Sjöstrand, J., and Marchisio, P. C., 1974, Radioautographic analysis of ^{3}H-fucose labelled glycoproteins transported along the optic pathway of chick embryos, *Cell Tissue Res.* **153:**465–476.

Gross, G. W., and Beidler, L. M., 1973, Fast axonal transport in the C-fibers of the garfish olfactory nerve, *J. Neurobiol.* **4:**413–428.

Gross, G. W., and Beidler, L. M., 1975, A quantitative analysis of isotope concentration and rapid transport velocities in the C-fibers of the garfish olfactory nerve, *J. Neurobiol.* **6:** 213–232.

Gurd, J. W., 1978, Biosynthesis of synaptic membranes. Incorporation of ^{3}H leucine into proteins and glycoproteins of rat brain synaptic membrane, *Brain Res.* **147:**201–204.

Guth, L., 1974, Axonal regeneration and functional plasticity in the central nervous system, *Exp. Neurol.* **45:**606–654.

Heacock, A. M., and Agranoff, B. W., 1977, Reutilization of precursor following axonal transport of ^{3}H proline-labelled protein, *Brain Res.* **122:**243–254.

Held, I., and Young, I. J., 1972, Transport of radioactivity derived from labelled *N*-acetylglucosamine in mammalian motor axons, *J. Neurobiol.* **3:**153–161.

Heslop, J. P., 1975, Axonal flow and fast transport in neurons, *Adv. Comp. Physiol. Biochem.* **6:**75–163.

Heuser, J. E., and Reese, T. S., 1973, Evidence for recycling of synaptic vesicles membrane during transmitter release at the frog neuromuscular junction, *J. Cell Biol.* **57:**315–344.

Hirosawa, K., and Young, R. W., 1971, Autoradiographic analysis of sulfate metabolism in the cerebellum of the mouse, *Brain Res.* **30:**295–309.

Holm, M., 1972, Gangliosides of the optic pathway: Biosynthesis and biodegradation studied *in vivo, J. Neurochem.* **19:**623–629.

Karlsson, J.-O., and Linde, A., 1977, Axonal transport of ^{35}S sulphate in retinal ganglion cells of the rabbit, *J. Neurochem.* **28:**293–297.

Karlsson, J.-O., and Sjöstrand, J., 1971, Rapid intracellular transport of fucose-containing glycoproteins in retinal ganglion cells, *J. Neurochem.* **18:**2209–2216.

Kauffman, F. C., Albuquerque, E. X., Warnick, J. E., and Max, S. R., 1976, Effect of vinblastine on neural regulation of metabolism in rat skeletal muscle, *Exp. Neurol.* **50:**60–66.

Koenig, H. L., DiGiamberardino, L., and Bennett, G., 1973, Renewal of proteins and glycoproteins of synaptic constituents by means of axonal transport, *Brain Res.* **62:**413–417.

Komiya, Y., and Austin, L., 1974, Axoplasmic flow of proteins in the sciatic nerve of normal and dystrophic mice, *Exp. Neurol.* **43:**1–12.

Kreutzberg, G. W., Schubert, P., Toth, L., and Rieske, E., 1973, Intradendritic transport to postsynaptic sites, *Brain Res.* **62:**399–404.

Langley, O. K., and Kennedy, P., 1977, The metabolism of nerve terminal glycoproteins in the rat brain, *Brain Res.* **130:**109–120.

Lasek, R. J., 1975, Axonal transport and the use of intracellular markers in neuroanatomical investigations, *Fed. Proc. Fed. Am. Soc. Exp. Biol.* **34**:1603–1609.

Ledeen, R. W., Skrivanek, J. A., Tirri, L. J., Margolis, R. K., and Margolis, R. U., 1976, Gangliosides of the neuron: Localization and origin, *Adv. Exp. Med. Biol.* **71**:83–103.

Levin, B. E., 1977, Axonal transport of ^{3}H fucosyl glycoproteins in noradrenergic neurons in the rat brain, *Brain Res.* **130**:424–432.

Lubińska, L., 1975, On axonal flow, *Int. Rev. Neurobiol.* **17**:241–296.

Maccioni, H. J., Landa, C., Arce, A., and Caputto, R., 1977, The biosynthesis of brain gangliosides—Evidence for a "transient pool" and an "end product pool" of gangliosides, *Adv. Exp. Med. Biol.* **83**:267–281.

Marchisio, P. C., Sjöstrand, J., Aglietta, M., and Karlsson, J.-O., 1973, The development of axonal transport of proteins and glycoproteins in the optic pathway of chick embryos, *Brain Res.* **63**:273–284.

Marchisio, P. C., Gremo, F., and Sjöstrand, J., 1975, Axonal transport in embryonic neurons: The possibility of a proximo–distal axolemmal transfer of glycoproteins, *Brain Res.* **85**: 281–285.

Margolis, R. K., and Gomez, Z., 1973, Rapid turnover of fucose in the water-soluble glycoproteins of brain, *Biochim. Biophys. Acta* **313**:226–228.

Margolis, R. K., and Margolis, R. U., 1972, Disposition of fucose in brain, *J. Neurochem.* **19**: 1023–1030.

Margolis, R. K., and Margolis, R. U., 1973, The turnover of hexosamine and sialic acid in glycoproteins and mucopolysaccharides of brain, *Biochim. Biophys. Acta* **304**:413–420.

Margolis, R. K., Margolis, R. U., Preti, C., and Lai, D., 1975*a*, Distribution and metabolism of glycoproteins and glycosaminoglycans in subcellular fractions in brain, *Biochemistry* **14**: 4797–4803.

Margolis, R. K., Preti, C., Chang, L., and Margolis, R. U., 1975*b*, Metabolism of the protein moiety of brain glycoproteins, *J. Neurochem.* **25**:707–709.

Margolis, R. U., and Margolis, R. K., 1972, Sulfate turnover in mucopolysaccharides and glycoproteins of brain, *Biochim. Biophys. Acta* **264**:426–431.

Margolis, R. U., and Margolis, R. K., 1974, Distribution and metabolism of mucopolysaccharides and glycoproteins in neuronal perikarya, astrocytes, and oligodendroglia, *Biochemistry* **13**:2849–2852.

Margolis, R. U., and Margolis, R. K., 1977, Metabolism and function of glycoproteins and glycosaminoglycans in nervous tissue, *Int. J. Biochem.* **8**:85–91.

Marinari, U. F., Morgan, I. G., Mack, G., and Gombos, G., 1972, Synthesis of synaptic glycoproteins. II. Delayed labelling of the glycoproteins of synaptic vesicles and synaptosomal plasma membranes, *Neurobiology* **2**:176–182.

Marko, P., and Cuénod, M., 1973, Contribution of the nerve cell body to renewal of axonal and synaptic glycoproteins in the pigeon visual system, *Brain Res.* **62**:419–423.

Markov, D., Rambourg, A., and Droz, B., 1976, Smooth endoplasmic reticulum and fast axonal transport of glycoproteins, an electron microscope radioautographic study of thick sections after heavy metals impregnation, *J. Microsc. Biol. Cell.* **25**:57–60.

Matthieu, J.-M., Webster, H. DeF., DeVries, G. H., Corthay, S., and Koellreutter, B., 1978, Glial versus neuronal origin of myelin proteins and glycoproteins studied by combined intraocular and intracranial labelling, *J. Neurochem.* **31**:93–102.

McEwen, B. S., Forman, D. S., and Grafstein, B., 1971, Components of fast and slow axonal transport in the goldfish optic nerve, *J. Neurobiol.* **2**:361–377.

McLane, J. A., and McClure, W. O., 1977, Rapid axoplasmic transport in dystrophic mice, *J. Neurochem.* **29**:865–872.

McLean, W. G., Frizell, M., and Sjöstrand, J., 1975, Axonal transport of labelled proteins in sensory fibres of rabbit vagus nerve *in vitro, J. Neurochem.* **25**:695–698.

McLean, W. G., Frizell, M., and Sjöstrand, J., 1976, Labelled proteins in rabbit vagus nerve between the fast and slow phases of axonal transport, *J. Neurochem.* **26**:77–82.

Miani, N., 1962, Evidence of a proximo–distal movement along the axon of a phospholipid synthesized in the nerve-cell body, *Nature (London)* **193**:887–888.

Monticone, R. E., and Elam, J. S., 1975, Isolation of axonally transported glycoproteins with goldfish visual system myelin, *Brain Res.* **100**:61–71.

Morgan, I. G., Zanetta, J.-P., Breckenridge, W. C., Vincedon, G., and Gombos, G., 1973, The chemical structure of synaptic membranes, *Brain Res.* **62**:405–411.

Morgan, I. G., Tettamanti, G., and Gombos, G., 1976, Biochemical evidence on the role of gangliosides in nerve-endings, *Adv. Exp. Med. Biol.* **71**:137–150.

Neale, J. H., Neale, E. A., and Agranoff, B. W., 1972, Radioautography of the optic tectum of the goldfish after intraocular injection of ^{3}H proline, *Science* **176**:407–410.

Ochs, S., 1972, Fast transport of materials in mammalian nerve fibers, *Science* **176**:252–260.

Ochs, S., 1974, Systems of material transport in nerve fibers (axoplasmic transport) related to nerve function and trophic control, in: *Trophic Function of the Neuron* (D. Drachman, ed.), *Ann. N. Y. Acad. Sci.* **228**:202–223.

Ochs, S., 1975, Axoplasmic transport, in: *The Nervous System,* Vol. 1, *The Basic Neurosciences* (D. B. Tower, ed.), pp. 137–146, Raven Press, New York.

Ochs, S., and Burger, E., 1958, Movement of substance proximo–distally in nerve axons as studied with spinal cord injection of radioactive phosphorus, *Am. J. Physiol.* **194**:499–506.

Oh, T. H., 1976, Neurotrophic effects of sciatic nerve extracts on muscle development in culture, *Exp. Neurol.* **50**:376–386.

Perišić, M., and Cuénod, M., 1972, Synaptic transmission depressed by colchicine blockade of axoplasmic flow, *Science* **175**:1140–1142.

Poduslo, J. F., Everly, J. L., and Quarles, R. H., 1977, A low molecular weight glycoprotein associated with isolated myelin: Distinction from myelin proteolipid protein, *J. Neurochem.* **28**:977–986.

Quarles, R. H., and Brady, R. O., 1971, Synthesis of glycoproteins and gangliosides in developing rat brain, *J. Neurochem.* **18**:1809–1820.

Repérant, J., Lemire, M., Miceli, D., and Peyrichoux, J., 1976, A radioautographic study of the visual system in fresh water teleosts following intraocular injection of tritiated fucose and proline, *Brain Res.* **118**:123–131.

Repérant, J., Miceli, D., and Raffin, J., 1977, Transneuronal transport of tritiated fucose and proline in the avian visual system, *Brain Res.* **121**:343–347.

Ripellino, J. A., and Elam, J. S., 1978, Differentiated turnover of axonally transported glycoproteins, *Soc. Neurosci. Abs.* **4**:36.

Rösner, H., 1975, Incorporation of scialic acid into gangliosides and glycoproteins of the optic pathway following an intraocular injection of *N*-^{3}H acetylmannosamine in the chicken, *Brain Res.* **97**:107–116.

Rösner, H., Wiegandt, H., and Rahmann, H., 1973, Sialic acid incorporation into gangliosides and glycoproteins of the fish brain, *J. Neurochem.* **21**:655–665.

Segrest, J. P., Jackson, R. L., Andrews, E. P., Marchesi, V. T., 1971, Human erythrocyte membrane glycoprotein: A re-evaluation of the molecular weight as determined by SDS polyacrylamide gel electrophoresis, *Biochem. Biophys. Res. Commun.* **44**:390–395.

Sellin, L. C., and McArdle, J. J., 1977, Colchicine blocks neurotrophic regulation of the resting membrane potential in reinnervating skeletal muscle, *Exp. Neurol.* **55**:483–492.

Sjöstrand, J., Karlsson, J.-O., and Marchisio, P. C., 1973, Axonal transport in growing and mature retinal ganglion cells, *Brain Res.* **62**:395–397.

Specht, S., and Grafstein, B., 1973, Accumulation of radioactive protein in mouse cerebral cortex after injection of ^{3}H fucose into the eye, *Exp. Neurology* **41**:705–722.

Specht, S. C., and Grafstein, B., 1977, Axonal transport and transneural transfer in mouse visual system following injection of ^{3}H fucose into the eye, *Exp. Neurology* **54**:352–368.

Tan, C. H., Peterson, N. A., and Raghupathy, E., 1977, Characteristics of D-glucosamine uptake by rat brain synaptosomes, *J. Neurochem.* **29**:261–265.

Taylor, A. C., and Weiss, P., 1965, Demonstration of axonal flow by the movement of tritium-labelled protein in mature optic nerve fibers, *Proc. Natl. Acad. Sci. U.S.A.* **54**:1521–1527.

Thompson, E. B., Schwartz, J. H., and Kandel, E. R., 1976, A radioautographic analysis in the light and electron microscope of identified *Aplysia* neurons and their processes after intrasomatic injection of L-^{3}H fucose, *Brain Res.* **112**:251–281.

Truding, R., Shelanski, M. L., and Morell, P., 1975, Glycoproteins released into the culture medium of differentiating murine neuroblastoma cells, *J. Biol. Chem.* **250**:9348–9345.

Weiss, P., 1961, The concept of perpetual neuronal growth and proximo–distal substance convection, in: *Regional Neurochemistry* (S. S. Kety and J. Elkes, eds.), pp. 220–242, Pergamon, Oxford.

Weiss, P., and Hiscoe, H. L., 1948, Experiments on the mechanism of nerve growth, *J. Exp. Zool.* **107**:315–396.

Wiesel, T. N., Hubel, D. H., and Lam, D. M. K., 1974, Autoradiographic demonstration of ocular-dominance columns in the monkey striate cortex by means of transneuronal transport, *Brain Res.* **79**:273–279.

Zatz, M., and Barondes, S. H., 1971, Rapid transport of fucosyl glycoproteins to nerve endings in mouse brain, *J. Neurochem.* **18**:1125–1133.

13

Regional Aspects of Neuronal Glycoprotein and Glycolipid Synthesis

Richard T. Ambron and James H. Schwartz

1. Introduction

The neuron is a highly differentiated cell consisting of distinct regions—cell body, axon, terminal—that differ in structure and function. It is likely that characteristic differences in macromolecular composition underlie regional differentiation. Because synthesis of macromolecules occurs primarily, if not exclusively, in the cell body, there are constraints on how regional biochemical differences might be established. We can think of three mechanisms: (1) differential export of macromolecules from the cell body and subsequent selective axonal transport to terminals; (2) selective import from enveloping glial cells; and (3) specific modifications of macromolecules in the axon or at terminals. There is much evidence to support the first mechanism (Ambron *et al.*, 1974*a*; Dahlström, 1971; Ochs, 1972). As for the second, transfer of materials between glial cells and neurons has frequently been postulated (Singer, 1968), but only recently was it convincingly demonstrated (Lasek *et al.*, 1977). Finally, there is some evidence for modification of macromolecules by glycosylation at synapses in mouse brain (Dutton *et al.*, 1973) and in *Aplysia* axons (Ambron and Treistman, 1977) (also see below).

Our approach to the problem of determining how regional specialization is established has been to examine the synthesis and fate within *single* identified neurons of two classes of membrane macromolecules, glycoproteins and glycolipids. We focused on these membrane components because it is likely

Richard T. Ambron • Department of Anatomy, Division of Neurobiology and Behavior, College of Physicians and Surgeons, Columbia University, New York, New York. **James H. Schwartz** • Departments of Physiology and Neurology, Division of Neurobiology and Behavior, College of Physicians and Surgeons, Columbia University, New York, New York.

that they are involved specifically in a variety of important cellular activities, can be readily labeled, and are rapidly moved toward synapses by axonal transport (Bennett *et al.*, 1973; Edström and Mattsson, 1973; Forman *et al.*, 1972; Grafstein, 1977) (see also Chapter 12).

Since glycoproteins are components of organelles, it was also our strategy to use individual labeled glycoproteins as markers characteristic of specific organelles. Thus, by examining the distribution of a given glycoprotein within a neuron, we would be able to determine the distribution of the organelle in which it resides. In this way, we also hoped to study how membrane glycoproteins synthesized in the cell body are inserted into surface membranes of the various regions of the cell. Our studies were carried out on identified single neurons to avoid the difficulties encountered in trying to resolve the great number of glycoprotein species present in samples of complex nervous tissue.

1.1. Advantages of Using Identified Neurons of *Aplysia*

Within the CNS of the marine mollusk *Aplysia californica* are many neurons whose cell bodies range from 100 to 400 μm in diameter; some are as large as 800 μm. The majority are located near the ganglion surface and may be easily seen with a dissecting microscope. About 70 neurons have been characterized electrophysiologically and pharmacologically in the abdominal ganglion (Frazier *et al.*, 1967; Koester and Kandel, 1977; Strumwasser, 1969) and 4 in the cerebral ganglion (Weinreich *et al.*, 1973; Weinreich and Yu, 1977). Since they can be identified as individual neurons, it is possible to study the same neuron in different specimens, permitting a multidisciplinary approach to related biochemical and physiological processes in a single neuron. The neurophysiology of identified neurons has been studied in great detail (Kandel *et al.*, 1967*b*), and several behavioral reflexes and long-term alterations in synaptic activity have been elucidated on a cellular basis (for a review, see Koester and Kandel, 1977).

We have concentrated particularly on R2 and L10, two cholinergic neurons of the abdominal ganglion, and on a pair of serotonergic cells, the giant cerebral neurons (GCNs) of the cerebral ganglion (Fig. 1). These neurons are readily studied in the isolated CNS. R2 and the GCNs are especially well suited for the study of axonal transport, since the distribution of radioactive materials along their axons can be measured in millimeter segments of nerve obtained by sequential sectioning. The major axon of R2 courses unbranched within the right connective to the circumesophageal ganglia, a distance of 2–4 cm (Fig. 1) (Coggeshall, 1967). In contrast, the axon of each GCN bifurcates close to the cell body into two axons of about equal cross-sectional area. One of these enters the buccal connective, where it runs a length of between 0.5 and 1 cm to make contact with identified cells in the buccal ganglion; the other enters the longer (1.5–2.5 cm) lateral lip nerve to innervate muscle (Fig. 1). Thus, in the GCN it is possible to examine how materials are distributed

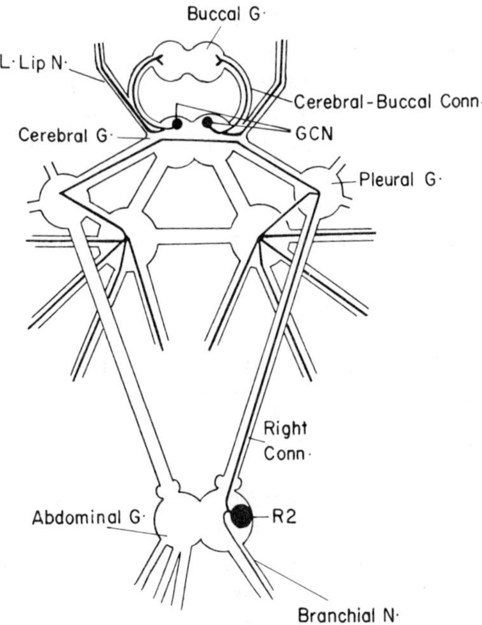

Fig. 1. Diagram of the CNS of *Aplysia* showing the axon distribution of R2 and the pair of giant metacerebral neurons (GCN). (G) Ganglion; (N) nerve; (Conn.) connective.

between two branches of an axon. The electrophysiology of the pair of GCNs and their role in controlling feeding behavior in *Aplysia* is currently under study (Cohen *et al.*, 1978; Weiss *et al.*, 1978).

L10 is an interneuron that is useful for studying synapses. It makes synaptic connections with 40 or more follower cells in the abdominal ganglion (Kandel *et al.*, 1967*a*), mediating cholinergic inhibition to some cells, excitation to others, and dual excitation–inhibition to yet another group of cells (Wachtel and Kandel, 1971). It also makes electrotonic connections with a fourth class of cells (Waziri, 1969).

1.2. Advantages of Direct Pressure Injection

Microinjection procedures with large animal cells are a tradition in developmental biology and have been used in neurobiology for almost a decade to define the geometry of neurons and to trace their connections. Procion dyes (Stretton and Kravitz, 1968), radioactive substances (e.g., see Eisenstadt *et al.*, 1973; Globus *et al.*, 1968), and cobalt salts (Pitman *et al.*, 1972) have been injected iontophoretically into nerve cells for subsequent detection by histochemical techniques or by radioautography (for a review, see Kater and Nicholson, 1975).

Unlike iontophoresis, pressure permits injection of relatively large amounts of material without restricting the chemical nature of the substance to be injected. Thus, for our studies, we injected the uncharged sugars [^{3}H]-L-fucose and [^{3}H]-*N*-acetyl-D-galactosamine (Fig. 2). In addition, much larger amounts of radioactivity from these sugar precursors are incorporated into single neurons after pressure injection than after prolonged incubations in solutions containing even high concentrations of the same substance (Ambron *et al.*, 1974*b*; Ambron and Treistman, 1977). Injection resulted in a localized and highly concentrated distribution of label within the injected neuron (Fig. 3).

2. Distribution of [^{3}H]Glycoproteins and [^{3}H]Glycolipids in R2 after Intrasomatic Injection of [^{3}H]Fucose or [^{3}H]-*N*-Acetylgalactosamine

When we investigated the distribution of labeled macromolecules in the cell body of R2 after intracellular injection, most of the [^{3}H]glycoproteins and all the [^{3}H]glycolipids were found to be associated with membranes (Ambron *et al.*, 1974*b*) [Table I(B)]. The glycoproteins were integral components, since they were not removed from the membrane by repeated washing with hypotonic alkaline buffer, nor could they be extracted by treatment with Triton X-100, deoxycholate, or other mild detergents (Ambron *et al.*, 1974*b*). Only by using hot sodium dodecyl sulfate (SDS) in the presence of β-mercaptoethanol

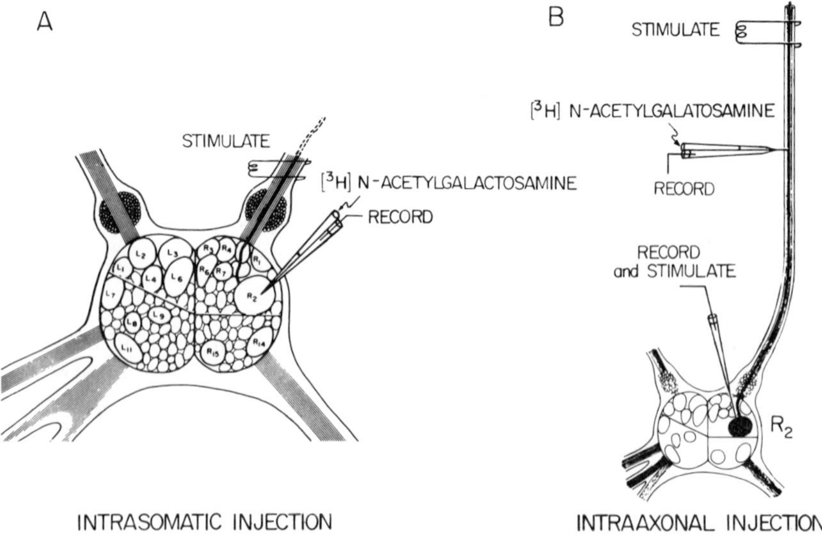

Fig. 2. (A) Diagram of the dorsal surface of the abdominal ganglion of *Aplysia*. The cell body of R2 is impaled by a double-barreled electrode for recording and injection. (B) Intraaxonal injection of [^{3}H]-*N*-acetylgalactosamine into the axon of R2 within the right connective.

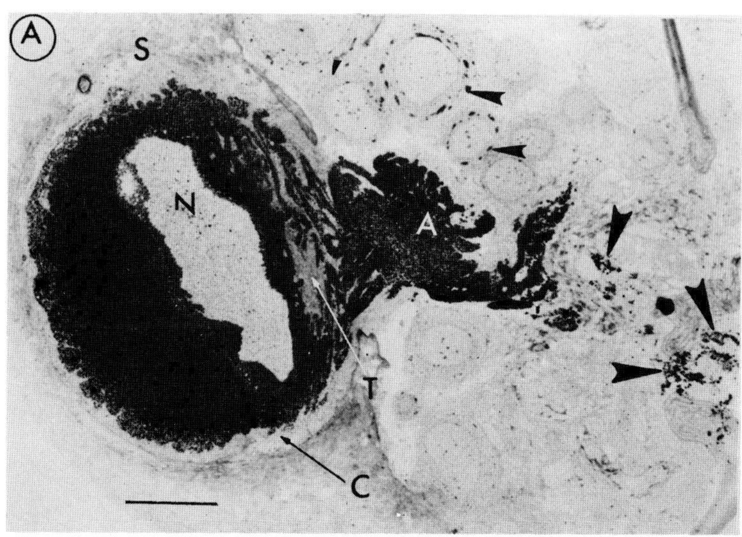

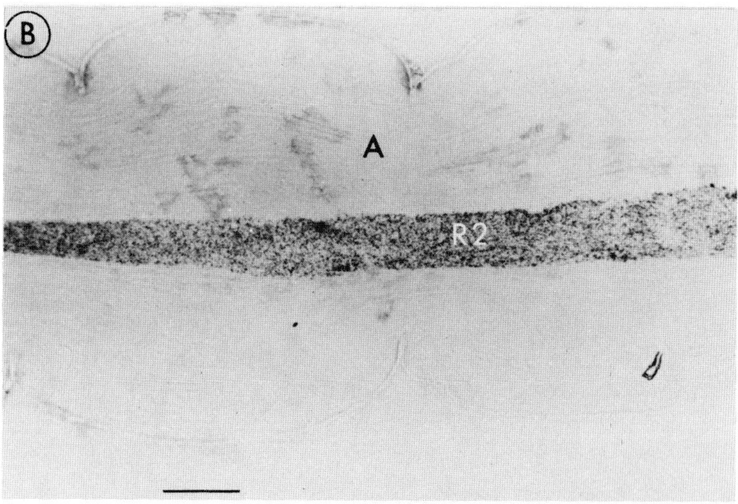

Fig. 3. Light-microscopic radioautograph of 2-μm transverse section of an abdominal ganglion stained with methylene blue and basic fuchsin. The cell body of R2 was injected with [^{3}H]fucose; after 10 hr of maintenance in culture, the tissue was prepared for microscopy. The cytoplasm is heavily labeled, as are the main axon (A) and regions of the neuropil containing branches of the axon of R2 (large arrowheads). Relatively unlabeled are the nucleus (N), glial cells and connective tissue capsule (C) that surrounds R2, glial indentations into the cell body at the trophospongium (T), and the connective tissue sheath (S). Neurons that surround R2 are not labeled. Small arrowheads indicate dense fuchsinophilic granules in the cytoplasm of several of the neurons. Calibration bar: 100 μm. From Thompson *et al.* (1976). (B) Unstained light-microscopic radioautograph of a 2-μm longitudinal section of the right connective containing the region of the axon of R2. Essentially all the label is restricted to the axon of R2 (R2). No grains appear over adjacent axons (A). Calibration bar: 50 μm. Prepared in collaboration with Dr. Ludmiela Shkolnik.

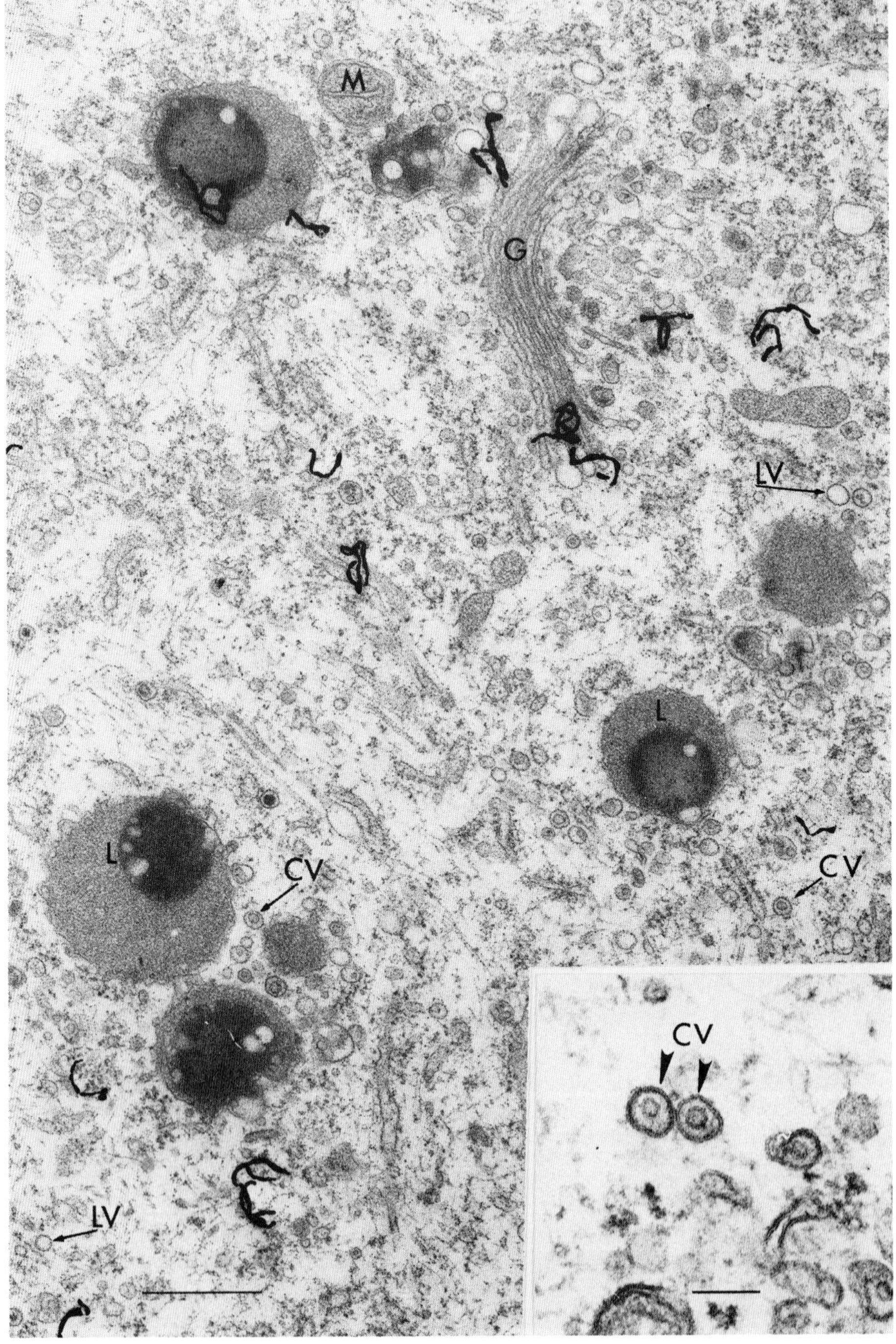

Fig. 5. Electron-microscopic radioautography of cytoplasm of R2 22 hr after intrasomatic injection of [^{3}H]-*N*-acetylgalactosamine. A morphometric analysis (Williams, 1969) of a series of similar radioautographs is given in Table II. (M) Mitochondrion; (G) Golgi cisternae; (LV) lucent vesicles; (L) lipochondrion; (CV) compound vesicles. Calibration bar: 0.5 μm. *Inset:* Higher magnification showing compound vesicles (CV). Calibration bar: 0.1 μm. Prepared in collaboration with Dr. L. Shkolnik.

Table I. Distribution of [^{3}H]Glycoprotein and [^{3}H]Glycolipid in R2 22 hr after Intrasomatic Injection of [^{3}H]-*N*-acetylgalactosamine[a]

	Cell body	Axon
	(% of total radioactivity)[b]	
(A) Soluble glycoprotein	9.0 ± 1.8	6.0 ± 1.5
(B) Membranes		
Glycoprotein	39.0 ± 2.7	41.3 ± 4.2
Glycolipid	14.7 ± 3.2	12.1 ± 1.4
(C) Total macromolecules	62.3 ± 2.9	59.5 ± 4.1

[a]Since incorporated radioactivity was shown to be restricted to the injected neuron, it was unnecessary to remove the cell body from the ganglion or the axon from the right connective. Tissue was homogenized at 0°C in ground glass tissue grinders containing 50 mM tris-HCl (pH 7.6). The homogenate was fractionated at 105,000*g* for 45 min. Soluble glycoproteins were precipitated from the resulting supernatant with 5% trichloroacetic acid (TCA) containing 0.5% phosphotungstic acid. Glycolipid was extracted from the pellet with 20 volumes of chloroform methanol (2:1 and 1:2). Glycoprotein was extracted from the residue with hot SDS-β-mercaptoethanol. Total macromolecules is the sum of the radioactivity in glycolipid and glycoprotein (Ambron *et al.*, 1974*b*).
[b]Values are means ±S.E. ($N = 6$).

were we able to release them quantitatively. Our original analysis of fucose-labeled membrane glycoproteins in R2 by polyacrylamide gel electrophoresis (PAGE) resolved five major components. These ranged in molecular weight from 90,000 to 180,000 daltons (Fig. 4), although the largest glycoprotein, designated Component I, was shown to migrate anomalously (Ambron *et al.*, 1974*b*). Comparison of mobilities in a variety of gel electrophoretic systems in SDS suggests that some of the glycoproteins are labeled by both precursor sugars.

Chemical evidence that these labeled macromolecules were indeed glycoproteins and not glycosaminoglycans or other uncharacterized polysaccharides was that they could be digested by pronase, but were unaffected by *Proteus vulgaris* chondroitanase ABC and beef testis hyaluronidase. In addition, they were not precipitated by cetylpyridinium bromide.

After injection of either [^{3}H]fucose or [^{3}H]-*N*-acetylgalactosamine, approximately 25% of the particulate radioactivity was found in glycolipids (Table I). Analysis of the [^{3}H]-*N*-acetylgalactosamine-labeled glycolipids in R2 by polysilicic acid thin-layer chromatography revealed three major and two minor components (Sherbany *et al.*, 1976). The two most polar species predominated at all times examined. Glycolipids labeled with [^{3}H]fucose have not yet been characterized.

The foregoing biochemical studies established that the newly synthesized glycoproteins are primarily associated with membranes, and this was confirmed by radioautography in the electron microscope. Cell bodies of *Aplysia* neurons contain a variety of organelles; representative examples of these structures in the cytoplasm of R2 are shown in Fig. 5. Thompson *et al.* (1976) used the quantitative morphometric procedure of Williams (1969) to localize radioactivity to the intracytoplasmic membranes of L10 after injection with

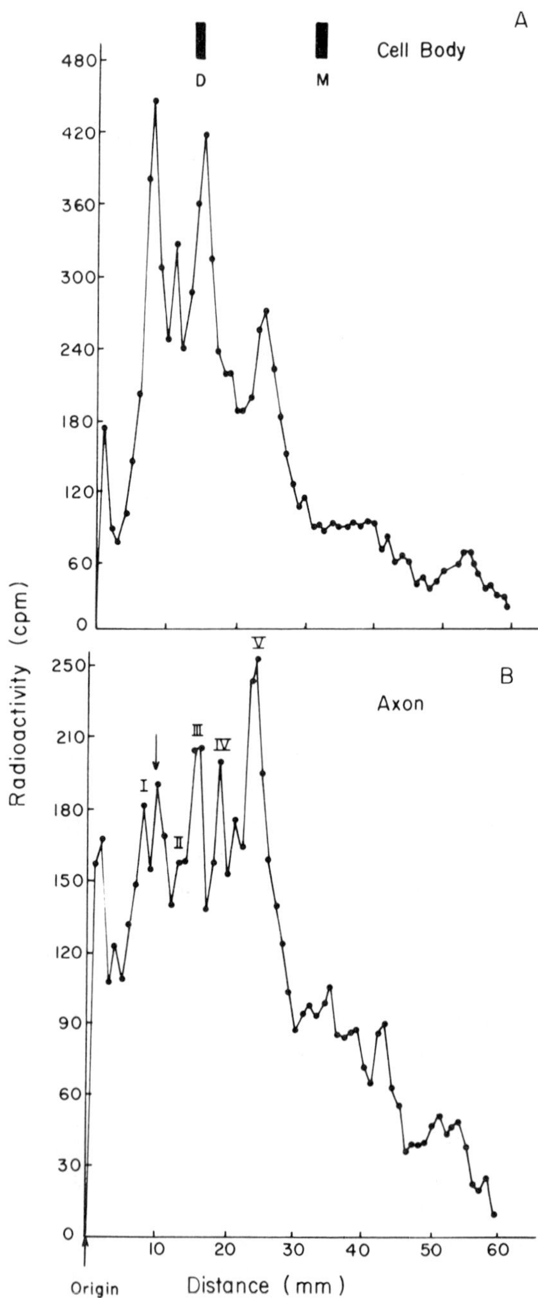

Fig. 4. SDS-PAGE of particulate [^{3}H]glycoproteins in cell body (A) and axon (B) of R2 24 hr after intrasomatic injection of [^{3}H]fucose. After electrophoresis, the gels were frozen and cut into sequential millimeter segments for counting by scintillation. Dimer (D) and monomer (M) of dansylated bovine serum albumin (BSA). From Ambron *et al.* (1974*a*).

[^{3}H]fucose. Similar results were subsequently obtained in R2. In the cell body of both neurons shortly after injection, more than half the silver grains appeared over the Golgi apparatus. It is likely that in *Aplysia* neurons, as in other tissues (Kraemer, 1971), Golgi membranes are the site of fucose incorporation. The other organelles that became labeled were vesicles, multivesicular bodies, large terminal lysosomes, smooth endoplasmic reticulum, mitochondria, and peroxisomes. In contrast to the results obtained with [^{3}H]fucose, the distribution of incorporated radioactivity from [^{3}H]-*N*-acetylgalactosamine in cell bodies of R2 and GCN was somewhat different; whereas both precursors were incorporated into the Golgi apparatus, a substantial proportion of grains from *N*-acetylgalactosamine were also associated with membranes of the rough endoplasmic reticulum, even at long times after injection [Table II(A)]. Although we cannot rule out the possibility that the grains over the rough endoplasmic reticulum reflect localization of indigenous structural glycoproteins, it is more likely that radioactivity associated with this structure represents glycosylation of polypeptides in the process of being synthesized. Thus, they would be destined for eventual incorporation into other organelles. Recent evidence suggests that sugars are added to nascent glycoproteins during their elaboration on the membrane of the rough endoplasmic reticulum (Czichi and Lennarz, 1977; Katz *et al.*, 1977). Since *N*-acetylgalactosamine is a linkage sugar in certain types of glycoproteins (Kornfeld and Kornfeld, 1976), it would not be surprising to find that this sugar is added at an early stage in synthesis.

In addition, [^{3}H]-*N*-acetylgalactosamine failed to label mitochondria and terminal lysosomes within neurons in isolated ganglia maintained in culture for as long as 24 hr after injection [Table II(A)]. The external membrane of the cell body was not labeled at short intervals after injection, but contained significant amounts of radioactivity after periods longer than 15 hr. Glial cells, either on the surface of the injected neurons or interdigitating into the region of the axon–hillock (the so-called "trophospongium" characteristic of large invertebrate neurons), were not labeled, nor was the nucleus [Table II(A)]. This restricted distribution of incorporated radioactivity was experimentally crucial, since it permitted specific biochemical and morphological studies that would otherwise be impossible.

3. Association of Individual Glycoprotein Components with Subcellular Organelles in the Cell Body

Since a relatively small number of [^{3}H]glycoprotein components became labeled after intrasomatic injection, we thought it might be possible to assign individual glycoprotein species to their *resident* organelles. The association of incorporated radioactivity with precursor organelles (Golgi and the reticula), on the one hand, and degradative organelles (multivesicular bodies and termi-

Table II. Grain Distribution over the Cytoplasm of R2 15 hr after Intrasomatic Injection of [^{3}H]-*N*-acetylgalactosamine

Additions:	(A) None			(B) Anisomycin (18 μM)		
	Silver grains (%)	Effective area circles (%)	Relative specific activity[a]	Silver grains (%)	Effective area circles (%)	Relative specific activity[a]
(1) Cytoplasmic components						
Golgi membranes	9.2	1.0	*9.2*	5.8	1.0	*5.8*
Rough endoplasmic reticulum	12.4	4.5	*2.8*	0.6	0.7	0.9
Smooth endoplasmic reticulum	31.8	9.9	*3.2*	9.5	16.7	0.6
Lucent vesicles	2.5	1.1	*2.3*	2.9	1.5	*1.9*
Compound vesicles	3.2	1.0	*3.2*	22.9	3.6	*6.4*
Multivesicular bodies	3.6	0.8	*4.4*	4.0	1.0	*4.0*
Lipochondria	4.5	6.9	0.7	4.8	4.4	1.1
Peroxisomes[b]	0	0	—	10.0	1.5	6.7
Cytoplasm	21.3	58.2	0.4	34.2	61.4	0.6
Mitochondria	2.9	2.2	1.3	2.1	1.9	1.1
Nucleus	0.5	1.0	0.5	0	0.4	0
(2) Outside of cell	2.9	10.9	0.3	3.2	6.0	0.5
(3) External membrane	4.9	2.2	*2.2*	0	0	—
(4) Total grains examined	381	—	—	652	—	—
(5) Total area circles	1309	—	—	2197	—	—

[a] Relative specific activity: Silver grains (%) divided by Effective area circles (%). Italicized entries were labeled more than expected from a random distribution of grains; $p < 0.05$. The cytoplasm and outside in both cells were labeled less than expected from a random distribution of grains; $p < 0.05$.

[b] In normal cells, peroxysomes are quite rare (0–0.5% of effective area circles). The high values obtained in anisomycin-treated cells suggest that the drug might have caused an increase in formation of this organelle, but we do not have a sufficient number of experiments to be sure.

nal lysosomes), on the other, suggested that we might not be able to show residence of any given single glycoprotein component in only one organelle, since during its life cycle a component would be expected to pass through organelles of origin (during synthesis) and to be destined for an organelle of turnover or destruction. Despite this reservation, we approached this problem in R2 using a variety of subcellular fractionation methods.

3.1. External Membrane

The cell body of R2 is so large that it can be readily dissected by hand, and the nucleus and cytoplasm separated from the external membrane (Lasek and Dower, 1971). Although the isolated membrane is still covered by glial cells and connective tissue, these were shown by radioautography not to contain any radioactivity [Table II(A)]. We tested the efficacy of the dissection method by determining the distribution of choline acetyltransferase; 95% of this soluble enzyme was found in the isolated cytoplasm. When the external membrane was examined 15 hr after an injection, only one of the five [^{3}H]glycoproteins, Component III, was greatly enriched relative to the cytoplasm (Fig. 6).

In keeping with current ideas of asymmetry in the disposition of carbohydrates at cell surfaces (Rothman and Lenard, 1977), we found that [^{3}H]glycopeptides could be released from the external membrane of R2 by treating the neuron *in situ* with trypsin. To ensure that the trypsin did not form holes in the membrane, the somatic resting potential was monitored with intracellular recording electrodes during the digestion. These results indicated that sugar moieties of Component III extend outward from the surface of R2. To test whether Component III is accessible to selective labeling from the outside, we used galactose-oxidase and sodium borotritide (Steck and Dawson, 1974). Since it is thought that the oxidase is unable to penetrate the membrane, it has been widely used in a convenient procedure for labeling macromolecules of the cell surface. After treatment of R2 with borotritide *in situ*, several radioactive proteins were labeled. When these were resolved by gel electrophoresis, one was found to have the same mobility as Component III. From these studies, we concluded that Component III is the most rapidly labeled glycoprotein of the cell surface of R2. Glycoprotein III is also exported into the axon of R2, and it will be of great interest to know whether or not it is enriched in the axolemma.

3.2. Vesicles

Two types of vesicles are plentiful in L10 and R2. The first, characteristic of these cholinergic neurons, is a relatively electron-lucent structure having a mean diameter of 90 nm (see Fig. 5). It resembles the cholinergic vesicle isolated from the electric organ of *Torpedo* (Israel *et al.*, 1970). The second type,

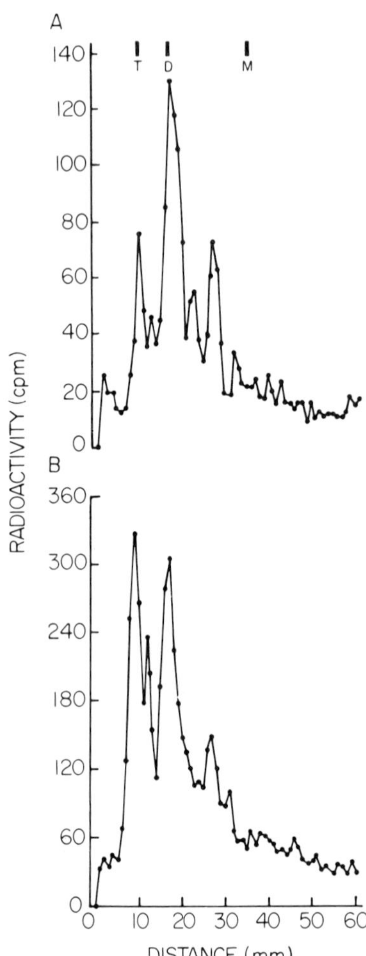

Fig. 6. SDS-PAGE of [^{3}H]glycoproteins from external membrane (A) and cytoplasm (B). R2 was injected with [^{3}H]fucose, and after 15 hr, the cell body was removed from the ganglion. The nucleus, cytoplasm, and external membranes were separated by manual dissection at 4°C (Lasek and Dower, 1971). Membranes of the cytoplasm were isolated by centrifugation. The membrane pellet and the external membrane were separately extracted with hot SDS–β-mercaptoethanol and electrophoresed (Ambron *et al.*, 1974*b*). In this electrophoretic system, Component III migrates with the dimer (D) of BSA. (M) Monomer and (T) trimer of BSA.

called the compound vesicle (Thompson *et al.*, 1976), contains a small vesicle (about 40-nm mean diameter) enclosed within a larger vesicle (82-nm mean diameter) (Fig. 5, inset). Vesicles with this appearance are widely distributed in *Aplysia* neurons. Despite a similar morphology, however, evidence suggests that they function differently in different cells. For example, in the serotonergic GCN, compound vesicles contain serotonin (Goldman *et al.*, 1976). Yet, when [^{3}H]serotonin is injected intrasomatically into the cholinergic R2, it is not taken up into compound vesicles. Membranes of both types of vesicles were labeled in radioautographs obtained from R2 and GCN after injection of [^{3}H]-*N*-acetylgalactosamine [Table II(A)].

To determine which [^{3}H]glycoprotein components might be associated with vesicles, we isolated by hand dissection the cytoplasm from an R2 that had been injected with [^{3}H]-*N*-acetylgalactosamine. The cytoplasm was added to a homogenate of carrier nervous tissue, and large organelles were removed by differential centrifugation. The supernatant was then filtered on a column of glass beads having a nominal mean pore diameter of 200 nm. Radioactivity was eluted from the column in two peaks (Fig. 7). The totally included material consisted mainly of soluble precursors and soluble [^{3}H]glycoprotein. The partially included radioactivity emerged from the column in the same volume as that previously shown to elute [^{3}H]serotonin-containing vesicles obtained from axons of the GCN after injection with [^{3}H]serotonin (see the Fig. 7 caption for details). Radioactive glycoproteins in this fraction could be sedimented by high-speed centrifugation, indicating that they were associated with membranes. Electron microscopy of the resulting pellet confirmed the presence of compound vesicles and other vesicle profiles of larger diameter; no mitochondria, terminal lysosomes, or other large organelles were present. SDS-PAGE of the [^{3}H]-*N*-acetylgalactosamine-labeled membranes from this fraction showed them to be enriched in glycoprotein Component I.

It was fortunate that Component I, and not one of the other glycoproteins, was found in this fraction, since it enabled us to take advantage of the selective effects of anisomycin. When ribosomal protein synthesis in R2 was blocked by *prolonged* treatment with the antibiotic, injected sugar precursors continued to be incorporated into Component I, but not into the other glycoproteins (Ambron *et al.*, 1975). Under these conditions, 85–90% of the radioactivity incorporated into membranes was characterized biochemically

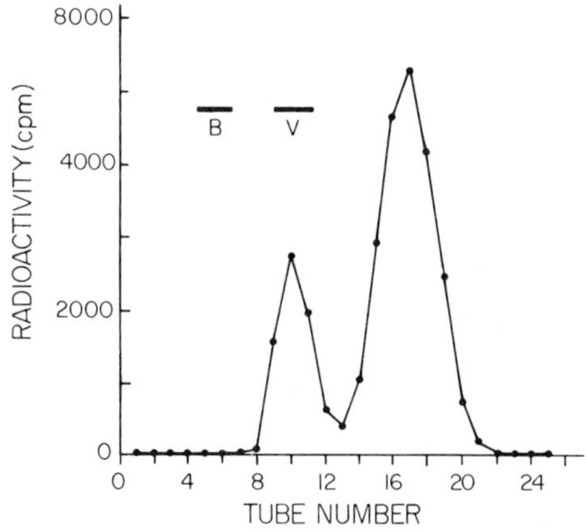

Fig. 7. Glass bead column filtration of cytoplasm obtained from an R2 injected with [^{3}H]-*N*-acetylgalactosamine. At 22 hr after injection, the cell body was isolated and hand-dissected. Cytoplasm, from which large organelles had been removed by differential centrifugation, was added to a column (1 × 55 cm) of glass beads (nominal pore size, 200 nm) previously treated with a 1% solution of polyethylene glycol to reduce tissue adsorption. The column was standardized by filtration of a suspension of polystyrene beads, 300 nm nominal diameter (B). To determine the elution characteristics of vesicles, GCN was injected with [^{3}H]serotonin. After 15 hr, the lip nerve and the cerebrobuccal connective were homogenized and centrifuged at 15,000*g*. The resulting supernatant was filtered on the column. (V) Fractions containing membranes labeled with serotonin.

as Component I (Fig. 8). Thus, in anisomycin-treated cells, the distribution of radioactivity, as determined either by subcellular fractionation or by radioautography, reflects the distribution of this [^{3}H]glycoprotein. Subcellular fractionation by filtration on glass beads of the cytoplasm from an R2 exposed to anisomycin was again consistent with the idea that Component I is a constituent of vesicles; all the labeled membranes emerged from the column in the same fraction as did serotonin-labeled vesicles. When the Component-I-labeled membrane fraction from the column was further purified by centrifugation through a discontinuous sucrose gradient, membranes could be seen at all the gradient interfaces (0.4, 0.7, and 1.2 M). Radioactivity from Component I, however, was present only above the 0.4 M sucrose interface, indicating that it is contained in a small structure of low density. Isolation of the organelles at this interface by high-speed centrifugation yielded a pellet that, in the electron microscope, appeared highly enriched in vesicles.

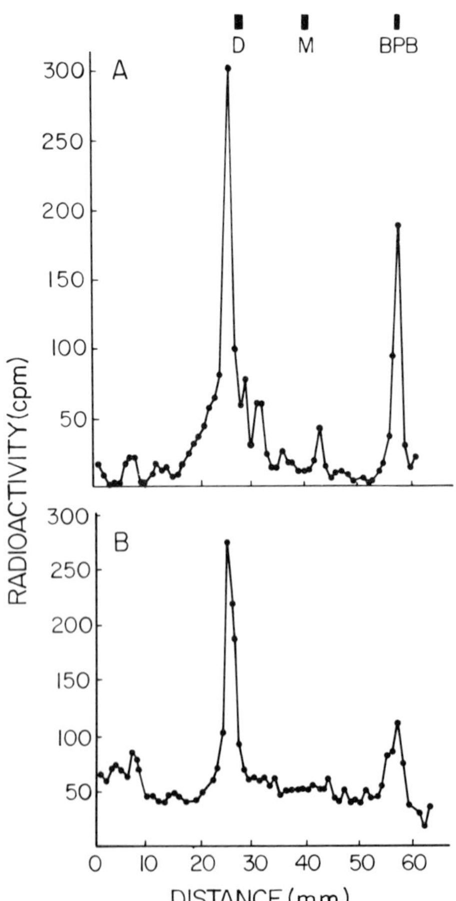

Fig. 8. Discontinuous gel electrophoresis in SDS of the particulate fraction from an anisomycin-treated R2 injected with [^{3}H]-*N*-acetylgalactosasmine (A) or [^{3}H]fucose. The isolated CNS was exposed to 18 μM anisomycin for 4 hr before and for the 15 hr after injection. The membrane fraction from the cell body of R2 was obtained as described in the Table I footnote. Electrophoresis was performed as described by Laemmli (1970). (BPB) Bromophenol blue. [^{3}H]glycolipid migrated with the tracking dye.

Although these fractionation studies strongly suggested that Component I is a constituent of vesicles, the resolution achieved was not sufficient to distinguish which type of vesicle. To obtain this information, we analyzed the distribution of grains on radioautographs from an R2 treated with anisomycin. We found that fewer organelles were labeled than in normal cells (see below), but that a much greater percentage of silver grains were found over compound vesicles [compare Table II(A) and (B)]. Furthermore, comparison of the relative specific activities of organelles in normal and anisomycin-treated cells showed an increase only in compound vesicles (Table II). An increase in relative specific activity would be expected for that organelle that contains [^{3}H]Component I. These data, then, provide evidence that Component I is a constituent of compound vesicles.

It is not yet clear why Component I is selectively labeled in cells after prolonged exposure to anisomycin. One explanation is that the polypeptide precursor pool for this glycoprotein is much larger than that for all the others. Thus, more would be available for glycosylation after cessation of protein synthesis. We think the explanation is more complex, however. Formation of a glycoprotein occurs in a series of sequential, vectorial processes thought to take place in distinct membranous compartments (Palade, 1975). Sugars can be added to the nascent polypeptide chain within cisternae of the rough endoplasmic reticulum (Czichi and Lennarz, 1977; Katz *et al.*, 1977). It is only at this site that anisomycin would have an *immediate* effect. Glycosylation can also occur on membranes of the smooth endoplasmic reticulum and in the Golgi region (Fleischer and Zambrano, 1974; Schachter *et al.*, 1970). There is also some evidence that glycosylation can take place in areas of the cell other than the perikaryon, e.g., in the axon (Ambron and Treistman, 1977) and at terminals (Dutton *et al.*, 1973). Presumably, anisomycin would not *directly* affect these sites, but would reduce sugar incorporation only by halting synthesis of precursor polypeptides.

As already described above, a morphometric analysis of treated cells showed essentially no labeling of the rough endoplasmic reticulum and great reduction in the relative specific activity of the smooth endoplasmic reticulum and Golgi membranes when compared to normal cells (Table II). The proportion of silver grains associated with compound vesicles, however, was greatly increased and its relative specific activity doubled. This is consistent with the idea that glycosylation can occur in these structures.

We have been able to isolate and purify Component I from untreated, injected neurons using the chaotropic agent lithium diiodosalicylate (Marchesi and Andrews, 1971). Preliminary characterization of the molecule indicated that it has several labeled oligosaccharide sidechains that differ in size and composition. On treatment with anisomycin, there was a dramatic change in the labeling pattern; most of the radioactivity was now associated with a single short chain. These data, when considered together with those from radioautographs discussed above, suggest that chain type is determined in some way by the specific site in the cell at which glycosylation takes place. If our ideas are correct, then the short chain of Component I would be characteristic of specific glycosylation reactions occurring in the Golgi region or on the membrane of the compound vesicle.

4. Selective Export of Newly Synthesized Glycoproteins into Axons

Glycoproteins and glycolipids synthesized after injection did not remain in the cell body, but after an initial short delay were exported into axons where they were rapidly translocated toward terminals. By 10 hr after injection of R2, about 45% of the macromolecules labeled with [^{3}H]fucose were exported from the cell body into the axons (Ambron *et al.*, 1974*a*). The most rapidly moving [^{3}H]glycoproteins and [^{3}H]glycolipids were transported along axons of GCN or R2 at a rate of about 70 mm/day at 15° C, a rate that is similar to that reported for fast axonal transport in other poikilotherms (Elam and Agranoff, 1971; Forman *et al.*, 1972).

As expected from the association of label with organelles in the cell body already described, more than 90% of the macromolecular radioactivity in axons of injected R2 (Table I) or GCN was also found to be particulate. Approximately 20% of this labeled, exported material could be characterized as glycolipid, which was also shown to move by rapid axonal transport (Sherbany *et al.*, 1976). Previous studies have indicated that materials moving along axons of other animals by fast transport are mainly particulate and consist primarily of membranous organelles (Grafstein, 1977).

The pattern of [^{3}H]glycolipids exported into the axon of R2 was the same as that seen in the cell body. In contrast, gel electrophorograms of membrane [^{3}H]glycoproteins from axons of R2 (see Fig. 4B) and GCN were strikingly different from those previously obtained from cell bodies. All the somatic [^{3}H]glycoproteins were present in axons, but their relative proportions varied. Axon and cell body patterns were complementary and reciprocal; the lower-molecular-weight components (80,000–120,000 daltons) predominated in axons, whereas higher-molecular-weight components (120,000–180,000 daltons) were characteristic of cell bodies. It is likely that these differences in composition resulted from preferential export of the lower-molecular-weight components into the axon, since selective degradation of the components, at least in the cell body, has been ruled out (Ambron *et al.*, 1974*a*). Within the axon tree of the same neuron, however, the [^{3}H]glycoproteins are transported in apparently the same proportions. No differences in patterns were observed between gel electropherograms of ^{3}H components from axons in the lip nerve and cerebrobuccal connective of the GCN. These results are in accord with the observations of Barker *et al.* (1976) in the frog, but disagree with those of

Anderson and McClure (1973), who found differences in proteins being transported in the two branches of dorsal root ganglion cells of the cat.

Radioautography demonstrated that all the exported [^{3}H]glycoproteins were restricted to the axons of the injected neuron (see Fig. 3B). Processes of the injected cells were examined for two additional purposes: to trace the axon distribution of the identified neuron using a nondestructive intracellular marker at the level of the light microscope, and, with the electron microscope, to study the structure of identified synapses. This was the aim of Thompson *et al.* (1976), who determined the three-dimensional configuration of the intraganglionic axons of R2. Presumptive synapses of L10 were also identified and their fine structure described. Use of [^{3}H]sugar precursors as intracellular markers has continued, and synapses of neurons undergoing behavioral modification are now being analyzed (Bailey *et al.*, 1979; Thompson and Bailey, 1979).

The second objective of radioautography was to find out which membranous organelles contain the [^{3}H]glycoprotein components that are being rapidly transported along axons. Quantitative radioautographic analyses of L10 (Thompson *et al.*, 1976) and GCN have shown that a major proportion of the exported label is associated with vesicles. In each of these neurons, morphometry revealed a striking increase in the relative specific activities of axonal vesicles compared to those in the cell body. These results strongly suggest that newly synthesized vesicles are preferentially exported into the axon.

It is important to note that newly synthesized glycoproteins must be components of completed organelles before they can be exported from the cell body into the axon. *Brief** exposure to anisomycin, which inhibits peptide bond formation by ribosomes, almost completely blocked the appearance of [^{3}H]glycoproteins in the axon, yet glycosylation of protein in the cell body was inhibited by only 50% (Ambron *et al.*, 1975). PAGE showed that all the normal glycoprotein components were available in substantial amounts in somatic membranes of cell bodies inhibited under these conditions. Since treatment with anisomycin did not interfere with the translocation of materials already exported into the axon, we concluded that synthesis and insertion into somatic membranes is not a sufficient condition for export; completion of organelles is required, and this, in turn, requires synthesis of new protein (Ambron *et al.*, 1975).

4.1. Selective Axonal Transport

The terminal region is differentiated from the rest of the axon, and must therefore have some special macromolecular components. Differences could arise if there were selective transport of macromolecules to the synapse. In single *Aplysia* axons, we found that soluble and particulate [^{3}H]glycoproteins

* Less than 1 hr. Exposure for 4 hr or longer was required to inhibit glycosylation of all glycoproteins except Component I as described above (Ambron *et al.*, 1975).

were distributed differently along the same axon, suggesting that mechanisms exist within the axon that can distinguish between exported glycoproteins (Ambron *et al.*, 1974*a*). Further indication for selective translocation was obtained by examining the distribution of [^{3}H]glycoproteins in segments of the axon of R2 at various distances from the cell body. We found that the proportion of labeled Component V appeared to increase in the distal regions of the axon of R2, although all the [^{3}H]glycoproteins were present in each of the regions examined (Fig. 9). The implication of this observation is that Component V, being the predominant [^{3}H]glycoprotein in the axon (see Fig. 4B)(Dahlströn, 1971), is associated with *axonal* vesicles.

We have shown that Component I is associated with vesicles in the cell body, but that this glycoprotein contains a relatively small proportion of the macromolecular radioactivity in the axon. Thus, vesicles in the two regions contain different labeled glycoproteins. One way in which this could occur is that the vesicles that predominate in the cell body, and that contain Component I, are not destined for transport to synapses, but are to remain primarily associated with the somatic region of the neuron. Vesicles rich in Component V, on the other hand, would be rapidly labeled and selectively exported for transport toward synapses. Alternatively, Component I may be a precursor of Component V. The processing of materials by selective proteolysis is known to occur in axons (Loh and Gainer, 1975). We are currently carrying out experiments to distinguish between these two possibilities.

5. Modification of Macromolecules in the Axon of R2

Macromolecules synthesized in the cell body can be modified within the axon specifically by the addition of sugars. Gel patterns of fucose-labeled glycoproteins in the axon showed that a new component, which is never seen in the cell body, appeared in the axon (see Fig. 4B, arrow). The appearance of this component suggested that one of the [^{3}H]glycoproteins synthesized in the cell body from injected [^{3}H]fucose might be modified in the axon. Because of the large size of the axon of R2 within the right connective (40–60 μm in diameter), it is possible to inject labeled sugar precursors directly into the axon (see Fig. 2B)(Treistman and Schwartz, 1974). When [^{3}H]fucose was injected, no incorporation occurred (Ambron *et al.*, 1974*a*; Treistman and Schwartz, 1974). Injected [^{3}H]-*N*-acetylgalactosamine, however, was incorporated into both glycoprotein and glycolipids, most of which were associated with membranes (Ambron and Treistman, 1977). Examination of the [^{3}H]glycoproteins by discontinuous gel electrophoresis in SDS (Laemmli, 1970) resolved five major components (Ambron and Treistman, 1977). The highest-molecular-weight component had a mobility similar to that of the new glycoprotein found in the axon after cell body injection (Fig. 4B, arrow). The other major labeled components in the axon had molecular weights lower than

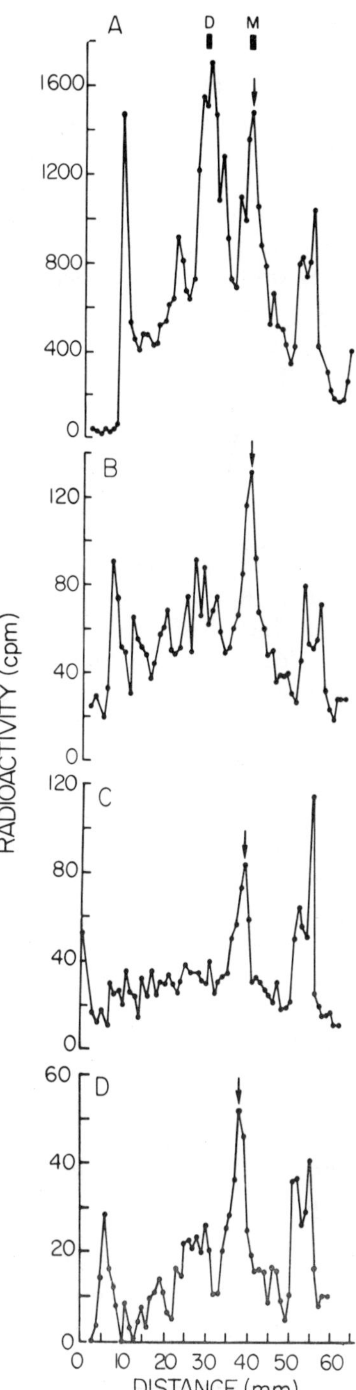

Fig. 9. Enrichment of [^{3}H]glycoprotein V along the axon of R2 after intrasomatic injection of [^{3}H]-*N*-acetylgalactosamine. At 22 hr after injection, the abdominal ganglion attached right connective were frozen and sequentially cut into millimeter segments (Ambron *et al.*, 1974*a*). Each segment was homogenized separately in 10% TCA containing 1% phosphotungstic acid, and the precipitated labeled macromolecules were collected on a glass fiber pad. Lipid was removed from the pad by washing with ethanol–ether (1:2, vol./vol.). [^{3}H]Glycoproteins were extracted from the pads with SDS–β-mercaptoethanol. To obtain enough radioactivity for a gel pattern, extracts from three consecutive segments of the axon were combined. Electrophoresis was performed as described by Laemmli (1970). (A) Cell body; (B) combined axon segments 7, 8, 9 mm from cell body; (C) axon segments 13, 14, and 15 mm from cell body; (D) axon segments 36, 37, and 38 mm from cell body. (↓) [^{3}H]Component V. The two radioactive peaks seen at the bottom of the gel are soluble glycoproteins that have been precipitated by the TCA. They are not present in gels from washed membrane fractions.

those of the major glycoproteins synthesized by the cell body. The pattern of axonal [^{3}H]glycoproteins also differed from the pattern of labeled glycoproteins derived from surrounding glial cells and connective tissue (Ambron and Treistman, 1977). These tissues can be selectively labeled by *incubating* the connectives in the presence of [^{3}H]-*N*-acetylgalactosamine, since a negligible amount of sugar is incorporated into axoplasm under these conditions. Analysis of radioautographs of the injected axon showed that most of the grains were associated with vesicles. This is consistent with the idea that these structures are the sites of axonal glycosylation.

6. Afterword

Our studies on the synthesis of glycoproteins and glycolipids in selected, identified *Aplysia* neurons was not undertaken for the purpose of demonstrating that invertebrate cells make use of biochemical pathways and mechanisms common to neurons of higher animals. Nor was the characterization of components from the molluscan neurons carried out to find new and exotic substances. Rather, our strategy was to use identified neurons to obtain information relating biochemical processes to neuronal function. Function and electrophysiological properties are most obviously anchored in cell structure, and the use of identified neurons permits us to localize the particular processes under study, not only to a single nerve cell, but also—and as critically—to anatomical regions of a single neuron. For example, in the studies described, our goal is to examine the origin and fate of membranes within the various regions of single neurons.

An awareness that cell biology is the essential bridge between biochemistry and physiology is by no means new with us. Cell biologists have approached important physiological problems with experimentally advantageous tissues for many years. The use by Palade (1975) of the pancreas to study secretion is a paradigm. The nervous system differs from other tissues in one crucial respect, however: interesting aspects of nerve cell function are likely to involve biochemical *differences* between neurons and not their universal properties. Therefore, brain cannot be directly approached as a *tissue* in order to obtain biochemical information relevant to the unique functioning of individual cells. Furthermore, the totality of a nerve cell's function is constituted of biochemical processes that take place in different regions of the same neuron.

ACKNOWLEDGMENTS. This work was supported by research grants NS 12066 from the National Institute of Neurological Disorders and Stroke and BNS 77-16505 from the National Science Foundtion (to J.H.S.) and a Senior Investigatorship (to R.T.A.) from the New York Heart Association. We thank Dr. Ludmiela Shkolnik for her electron micrographs.

References

Ambron, R. T., and Treistman, S. N., 1977, Glycoproteins are modified in the axon of R2, the giant neuron of *Aplysia californica*, after intra-axonal injection of ^{3}H-*N*-acetylgalactosamine, *Brain Res.* **121**:287–309.

Ambron, R. T., Goldman, J. E., and Schwartz, J. H., 1974*a*, Axonal transport of newly synthesized glycoproteins in a single identified neuron of *Aplysia californica, J. Cell Biol.* **61**: 665–675.

Ambron, R. T., Goldman, J. E., Thompson, E. B., and Schwartz, J. H., 1974*b*, Synthesis of glycoproteins in a single identified neuron of *Aplysia californica, J. Cell Biol.* **61**:649–664.

Ambron, R. T., Goldman, J. E., and Schwartz, J. H., 1975, Effect of inhibiting protein synthesis on axonal transport of membrane glycoproteins in an identified neuron of *Aplysia, Brain Res.* **94**:307–323.

Anderson, L. E., and McClure, W. O., 1973, Differential transport of protein in axons: Comparison between sciatic nerve and dorsal columns of cats, *Proc. Natl. Acad. Sci. U.S.A.* **70**: 1521–1525.

Bailey, C. H., Thompson, E. B., Chen, M., and Hawkins, R., 1978, Insights into the morphological basis of presynaptic facilitation in the gill withdrawal reflex of *Aplysia:* Analysis of the fine structure of a modulatory synapse, *Abstr. Soc. Neurosci.*, Eighth Annual Meeting, St. Louis, p. 187.

Barker, J. L., Neale, J. H., and Gainer, H., 1976, Rapidly transported proteins in sensory, motor, and sympathetic nerves of the isolated frog nervous system, *Brain Res.* **105**:497–515.

Bennett, G., DiGiamberardino, L., Koenig, H. L., and Droz, B., 1973, Axonal migration of protein and glycoprotein to nerve endings. II. Radioautographic analysis of the renewal of glycoproteins in nerve endings of chicken ciliary ganglion after intracerebral injection of [^{3}H] fucose and [^{3}H]glucosamine, *Brain Res.* **60**:129–146.

Coggeshall, R. E., 1967, A light and electron microscope study of the abdominal ganglion of *Aplysia californica, J. Neurophys.* **30**:1263–1287.

Cohen, J. L., Weiss, K. R., and Kupfermann, I., 1978, Motor control of buccal muscles in *Aplysia, J. Neurophysiol.* **41**:157–180.

Czichi, U., and Lennarz, W. J., 1977, Localization of the enzyme system for glycosylation of proteins via lipid-linked pathway in rough endoplasmic reticulum, *J. Biol. Chem.* **252**: 7901–7904.

Dahlström, A., 1971, Axoplasmic transport (with particular respect to adrenergic neurons), *Philos. Trans. R. Soc. London Ser. B* **261**:325–358.

Dutton, G. R., Hayward, P., and Barondes, S. H., 1973, ^{14}C-glucosamine incorporation into specific products in the nerve ending fraction *in vivo* and *in vitro, Brain Res.* **57**:377–408.

Edström, A., and Mattsson, H., 1973, Electrophoretic characterization of leucine-, glucosamine-, and fucose-labeled proteins rapidly transported in frog sciatic nerve, *J. Neurochem.* **21**: 1499–1507.

Eisenstadt, M., Goldman, J. E., Kandel, E. R., Koike, H., Koester, J., and Schwartz, J. H., 1973, Intrasomatic injection of radioactive precursors for studying transmitter synthesis in identified neurons of *Aplysia californica, Proc. Natl. Acad. Sci. U.S.A.* **70**:3371–3375.

Elam, J. S., and Agranoff, B. W., 1971, Rapid transport of protein in the optic system of the goldfish, *J. Neurochem.* **18**:375–387.

Fleischer, B., and Zambrano, F., 1974, Golgi apparatus of kidney: Preparation and role in sulfatide formation, *J. Biol. Chem.* **249**:5995–6003.

Forman, D. S., Grafstein, B., and McEwen, B. S., 1972, Rapid axonal transport of ^{3}H-fucosyl glycoproteins in the goldfish optic system, *Brain Res.* **48**:327–342.

Frazier, W. T., Kandel, E. R., Kupfermann, I., Waziri, R., and Coggeshall, R. E., 1967, Morphological and functional properties of identified neurons in the abdominal ganglion of *Aplysia californica, J. Neurophysiol.* **30**:1288–1351.

Globus, A., Lux, H. D., and Schubert, P., 1968, Somadendritic spread of intracellularly injected glycine in cat spinal motor neurons, *Brain Res.* **11**:440–445.

Goldman, J. E., Kim, K. S., and Schwartz, J. H., 1976, Axonal transport of [^{3}H]serotonin in an identified neuron of *Aplysia californica, J. Cell Biol.* **70**:304–318.

Grafstein, B., 1977, Axonal transport: The intracellular traffic of the neuron, in: *The Handbook of the Nervous System,* Vol. I, *Cellular Biology of Neurones* (E. R. Kandel, ed.) pp. 236–262, American Physiological Society, Washington, D.C.

Israel, M., Gautron, J., and Lesbats, B., 1970, Subcellular fractionation of the electric organ of *Torpedo marmorata, J. Neurochem.* **17**:1441–1450.

Kandel, E. R., Frazier, W. T., and Coggeshall, R. E., 1967*a*, Opposite synaptic actions mediated by different branches of an identifiable interneuron in *Aplysia, Science* **155**:346–349.

Kandel, E. R., Frazier, W. T., Waziri, R., and Coggeshall, R. E., 1967*b*, Direct and common connections among identified neurons in *Aplysia, J. Neurophysiol.* **30**:1352–1376.

Kater, S. B., and Nicholson, C. (eds.), 1975, *Intracellular Staining in Neurobiology*, Springer-Verlag, New York.

Katz, F. N., Rothman, J. E., Lingappa, V. R., Blobel, G., and Lodish, H. F., 1977, Membrane assembly *in vitro*—Synthesis, glycosylation, and asymmetric insertion of a transmembrane protein, *Proc. Natl. Acad. Sci. U.S.A.* **74**:3278–3282.

Koester, J., and Kandel, E. R., 1977, Further identification of neurons in the abdominal ganglion of *Aplysia* using behavioral criteria, *Brain Res.* **121**:1–20.

Kornfeld, R., and Kornfeld, S., 1976, Comparative aspects of glycoprotein structure, *Annu. Rev. Biochem.* **45**:217–238.

Kraemer, P. M., 1971, Glycoproteins, in: *Biomembranes,* Vol. I (L. A. Manson, ed.), pp. 67–190, Plenum Press, New York.

Laemmli, V. K., 1970, Cleavage of structural proteins during the assembly of the head of bacteriophage T4, *Nature (London)* **227**:680–685.

Lasek, R., and Dower, W. J., 1971, Analysis of nuclear DNA in individual nuclei of giant neurons, *Science* **172**:278–280.

Lasek, R. J., Gainer, H., and Barker, J., 1977, Cell-to-cell transfer of glial proteins to the squid giant axon, *J. Cell Biol.* **74**:501–523.

Loh, Y. P., and Gainer, H., 1975, Low molecular weight specific protein in identified molluscan neurons. II. Processing, turnover, and transport, *Brain Res.* **92**:193–206.

Marchesi, V. T., and Andrews, G. P., 1971, Glycoproteins: Isolation from cell membranes with lithium diiodosalicylate, *Science* **174**:1247–1248.

Ochs, S., 1972, Fast transport of materials in mammalian nerve fibers, *Science* **176**:252–260.

Palade, G., 1975, Intracellular aspects of the process of protein synthesis, *Science* **189**:347–358.

Pitman, R. M., Tweedle, C. D., and Cohen, M. S., 1972, Branching of central neurons: Intracellular cobalt injection for light and electron microscopy, *Science* **176**:412–414.

Rothman, J. G., and Lenard, J., 1977, Membrane asymmetry, *Science* **195**:743–758.

Schachter, H., Jabbal, I., Hudgin, R. L., Pinterio, L., McGuire, E. J., and Roseman, S., 1970, Intracellular localization of liver sugar nucleotide glycoprotein glycosyltransferases in a Golgi-rich fraction, *J. Biol. Chem.* **245**:1090–1100.

Sherbany, A. A., Ambron, R. T., and Schwartz, J. H., 1976, Synthesis and axonal transport of membrane glycolipids in R2, a single identified neuron of *Aplysia californica, Abstr. Soc. Neurosci.*, Sixth Annual Meeting, Toronto, p. 49.

Singer, M., 1968, Penetration of labeled amino acids into the peripheral nerve fiber from surrounding body fluids, in: *Ciba Found. Symp.: Growth of the Nervous System* (G. E. W. Wolstenholme and M. O'Connor, eds.), pp. 200–215, J. & A. Churchill, London.

Steck, T. L., and Dawson, G., 1974, Topographical distribution of complex carbohydrates on the erythrocyte membrane, *J. Biol. Chem.* **249**:2135–2142.

Stretton, A. O. W., and Dravitz, E. A., 1968, Neuronal geometry: Determination with a technique of intracellular dye injection, *Science* **162**:132–134.

Strumwasser, F., 1969, Types of information stored in single neurons, in: *Invertebrate Nervous Systems* C. A. G. Wiersma, ed.), p. 291, University of Chicago Press.

Thompson, E. B., Schwartz, J. H., and Kandel, E. R., 1976, A radioautographic analysis in the light and electron microscope of identified *Aplysia* neurons and their processes after intrasomatic injection of ^{3}H-fucose, *Brain Res*, **112**:251–281.

Thompson, E. B., and Bailey, C. H., 1979, Two different and compatible intraneuronal labels for ultrastructural study of synaptically related cells, *Brain Res.* (in press).

Treistman, S. N., and Schwartz, J. H., 1974, Injection of radioactive materials into an identified axon of *Aplysia, Brain Res.* **68**:358–364.

Wachtel, H., and Kandel, E. R., 1971, Conversion of synaptic excitation to inhibition of a dual chemical synapse, *J. Neurophysiol.* **34**:56–68.

Waziri, R., 1969, Electrical transmission mediated by an identified cholinergic neuron of *Aplysia, Life Sci.* **8**:469–476.

Weinreich, D., and Yu, Y. T., 1977, The characterization of histidine decarboxylase and its distribution in nerves, ganglia and in single neuronal cell bodies from the CNS of *Aplysia californica, J. Neurochem.* **28**:361–367.

Weinreich, D., McCaman, M. W., McCaman, R. E., and Vaughn, J. G., 1973, Chemical, enzymatic, and ultrastructural characterization of 5-hydroxy-tryptamine-containing neurons from the ganglia of *Aplysia californica* and *Tritonia diomedia, J. Neurochem.* **20**: 969–976.

Weiss, K. R., Cohen, J. L., and Kupfermann, I., 1978, Modulatory control of buccal musculature by a serotonergic neuron (metacerebral cell) in *Aplysia, J. Neurophysiol.* **41**:181–203.

Williams, M. A., 1969, The assessment of electron microscopic autoradiographs, *Adv. Opt. Electron. Microsc.* **3**:219–272.

14

Complex Carbohydrates of Cultured Neuronal and Glial Cell Lines

Glyn Dawson

1. Introduction

The normal development of the CNS requires that neurons derived from the germinal epithelium of the neural tube segregate into "nuclei," columns, and laminae and produce axons that ignore numerous incompatible neurons and eventually locate "target" cells at considerable distances from their cell body. These axons are then enveloped by a myelin sheath (derived from the plasma membrane of oligodendroglial cells of neuroectodermal origin) and surrounded by cells, collectively known as glial cells or astrocytes, that both facilitate the targeting of axons and provide structural and metabolic support (Rakic, 1974). It is generally believed that complex carbohydrates are involved in determining the specificity of these interactions (Pfenninger and Rees, 1976), as well as in a number of other functions ranging from facilitation of ion transport and regulation of neurotransmitter release and uptake to memory consolidation, but direct evidence is hard to obtain. This chapter will attempt to assess the evidence both for the existence of neurospecific complex carbohydrates in cell lines of neurotumor origin and for the existence of a specific functional role for them.

A remarkable feature of nervous tissue is the high concentration, relative to other tissues, of a group of lipid-linked complex carbohydrates, the glycolipids. Gray matter (and by inference the neuron) is enriched in sialoglycosphingolipids (Svennerholm, 1964), whereas myelin is enriched in galactosylceramides, galactosyldiglycerides, and their sulfated and sialylated derivatives (O'Brien and Sampson, 1965). The glycosaminoglycan heparan sulfate is also

Glyn Dawson • Departments of Pediatrics and Biochemistry, Joseph P. Kennedy, Jr., Mental Retardation Research Center, University of Chicago, Chicago, Illinois.

considerably enriched in nervous tissue, especially in glial cells, and abnormalities in its metabolism rapidly lead to severe neurological dysfunction (Matalon and Dorfman, 1976; Dawson and Lenn, 1976), whereas abnormal metabolism of other glycosaminoglycans such as dermatan sulfate and keratan sulfate leads to visceral and skeletal abnormalities only. The evidence for specific glycoproteins in nervous tissue is less clear, although progress is being made in terms of coding for neuronal recognition (Pfenninger and Rees, 1976), neural cell aggregation (Hausmann and Moscona, 1975), adhesion (Stallcup, 1977; Santala *et al.*, 1977), and receptors (Michaelis, 1975).

Since the CNS consists of three major unique cell types, namely, neurons and two types of glial cells, astrocytes and oligodendrocytes, the initial step has been to study the complex carbohydrate composition of individual cell types. Neuroanatomists have detected many morphological subtypes of these three cells, and have started to use fluorescent conjugated specific lectin probes to look for variations in cell-surface complex carbohydrates (Wood *et al.*, 1974; Pfenninger and Rees, 1976). Thus, Pfenninger and Maylié-Pfenninger (1975) used ferritin–lectin conjugates to show a different pattern of surface complex carbohydrates in growth cones from spinal cord, superior cervical ganglia, and dorsal root ganglia, and it is possible that such an approach will be as useful as that of fluorescent histochemistry in understanding the biochemistry and function of neurotransmitters. The complex carbohydrate composition of partially pure bulk isolated neurons, axons, astrocytes, and oligodendroglia has been reported (Abe and Norton, 1974; Deshmukh *et al.*, 1974; DeVries *et al.*, 1976; Fluharty *et al.*, 1975; Hamberger and Svennerholm, 1971; Margolis, R. U., and Margolis, R. K., 1974; Norton and Poduslo, 1971; Poduslo, 1975; Poduslo and Norton, 1975; Radin *et al.*, 1972; Robert *et al.*, 1975; Sarlieve *et al.*, 1976; Simpson *et al.*, 1976), and some cell-type specificity has been noted. Analytical studies have also been carried out on hand-dissected neurons (Tamai *et al.*, 1975), but there has been limited success in attempts to isolate viable individual neuronal cells that can be cultured in large amounts and used to study complex carbohydrates.

Limitations in attempting to study what is essentially a nondividing cell, the mature neuron, have stimulated the search for tumors of neuronal origin that can be adapted to tissue culture, cloned for individual traits such as the presence of specific enzymes and receptors, and induced to undergo "differentiation" under carefully controlled conditions. A mouse C1300 adrenal gland (sympathetic nervous system) neuroblastoma arose spontaneously in an A/J strain mouse in 1940 and was passaged in host mice until 1969, when it was adapted to tissue culture by Augusti-Tocco and Sato (1969), Klebe and Ruddle (1969), Seeds *et al.* (1970), Schubert *et al.* (1969), and Amano *et al.* (1972) and given the designations NB2a, NB41A, N18, and others. The morphological appearance of these cell lines is shown in Fig. 1. Subsequently, Schubert *et al.* (1976) adapted several chemically induced neuronal tumor cell lines of rat CNS origin to tissue culture (designated B103, B65, B50, and so on) and showed them to express certain neuron-specific properties. The C1300 neuronal cell lines have been the most extensively studied, and are characterized by the production of processes up to 3000 μm long that morphologically resemble axons; by the ability to convert [^{3}H]tyrosine to dopa, dopamine (3,4-dihydroxyphenylamine), and norepinephrine (Rosenberg, 1973; Haffke and Seeds, 1975); and by the synthesis of relatively large amounts of sialoglycosphingolipids (gangliosides) (Dawson *et al.*, 1971; Yogeeswaran *et al.*, 1973; Stoolmiller *et al.*, 1973; Dawson and Stoolmiller, 1976). Such cells often contain choline acetylase and acetylcholinesterase, show dense core vesicles on electron microscopy, and are electrically excitable (Rosenberg, 1973; Nelson, 1973; Schubert *et al.*, 1973) (Table I).

Glial cells are more readily cultured than neurons, and a line of normal (NN) astroblasts has been established (Shein *et al.*, 1970). However, relatively few studies have been carried out on the NN cell line thus far, and the bulk-isolated glial cell fractions are usually both contaminated by membranes and nonviable. Benda *et al.* (1968) cloned a rat glioma to produce a series of cell lines of which C-6 (Fig. 1) has been the most intensely studied. C-6 cells produce uniformly well-differentiated astrocytomas (grades I and II), whereas those from other clones such as C-2_1 produce more pleotrophic (grade III) tumors. Major biochemical characteristics of rat C-6 include production of S-100 protein (Benda *et al.*, 1968) and nerve growth factor (Monard *et al.*, 1973), inducibility of cytoplasmic glycerol phosphate dehydrogenase by cortisol (de Vellis and Brooker, 1973), and elevation of cyclic AMP (cAMP) in response to catecholamines (Gilman and Nirenberg, 1971), as well as the presence of certain neurotransmitter-degrading enzymes (Schrier and Thompson, 1974). Biochemical properties are summarized in Table I. C-6 is generally regarded as an immature glioblastoma, in the absence of clear biochemical diagnostic criteria for designating it as an astrocytoma. It does not appear to exhibit the biochemical or morphological properties associated with oligodendrocytes (or their Schwann cell counterparts in the peripheral nervous system), the cells that produce and maintain the myelin sheath, although such properties could be inducible.

Several oligodendroglioma cell lines have been established from cloned cell lines derived from a methylcholanthrene-induced mouse G-26 glioma (Zimmerman, 1955; Sundarraj *et al.*, 1975) (Fig. 1). Similarly, a cell line with Schwann-cell characteristics (RN-2) has been derived from an ethylnitrosourea-induced rat tumor (Pfeiffer and Wechsler, 1972). Cell lines established from the G26 tumor are considered to be immature glial cells of neuroectodermal origin that exhibit oligodendroglial-like properties such as synthesis of S-100 protein, 2′,3′-cyclic nucleotide 3′-phosphohydrolase, a neurotransmitter-sensitive adenylate cyclase, and sulfogalactosylceramide (sulfatide); such cell lines have been designated G26-15, -19, -20, -24, -28, and -29 (Sundarraj *et al.*, 1975; Dawson *et al.*, 1977) (Table I).

Thus, provided one is cognizant that all these clonal neurotumor lines are

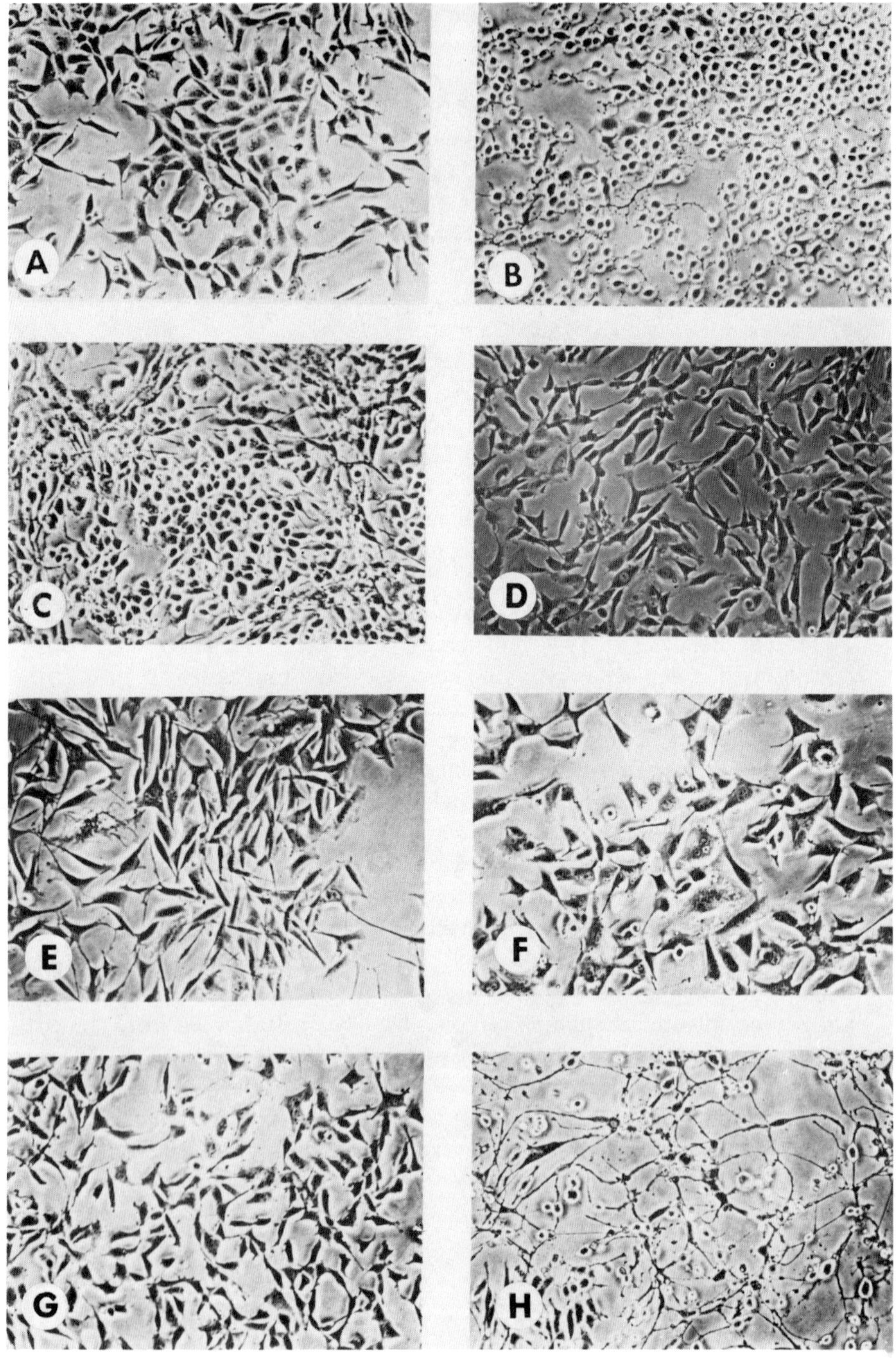
A
B
C
D
E
F
G
H

Table I. Some Biochemical Properties of Neuroblastoma, Astrocytoma, and Oligodendroglioma Cell Lines

Property	Neuroblastoma (NB2a, N18, N4TG1, etc.)	Astrocytoma (glioma RGC-6)	Oligodendroglioma (G26-20)[a]
S-100	−	+	+
14-3-2 protein	+	−	NT
NS-1 antigen	−	+	+
Electrical activity	+	−	NT
Gangliosides	++	−	+
Sulfatide	−	−	+
Endorphin receptors	+[b]	−	−
Inducible α-glycerol-phosphate dehydrogenase	−	+	NT
Inducible acetyl-cholinesterase	+	−	NT
Inducible catechol-*O*-methyltransferase	+	+	NT
2′,3′-Cyclic nucleotide 3′-phosphohydrolase	−	+	+
Neurite formation in response to Bt_2cAMP	++	+	+
Norepinephrine-sensitive adenylate cyclase	+	+	+
Nerve growth factor	−	+	NT
Norepinephrine-sensitive (inducible) lactate dehydrogenase	−	+	−

[a](NT) Not tested. [b]For cell line N4TG1 only.

transformed cells, with some properties different from those of the parental cell; that there is no really reliable set of neuronal, astrocytic, or oligodendrocytic biochemical markers with which to classify such cells; and that there is variability of expression of differentiated characteristics in these generally immature neuroblasts or glioblasts, the cell lines discussed above fulfill most of the requirements for studying the regulation of complex carbohydrate metabolism in specific neural cell types.

Fig. 1. Morphological appearance of neurotumor cell lines under phase-contrast microscopy. All cells were grown as monolayers in Falcon plastic dishes using modified Eagle's medium supplemented with 10% fetal calf serum unless otherwise stated. (A) C1300 mouse neuroblastoma clone NB2a; (B) NB2a following treatment with 10^{-3} M dibutyryl cyclic AMP (Bt_2cAMP) for 3 hr; (C) opiate-receptor-positive C1300 mouse neuroblastoma clone N4TG1 grown on gelatin; (D) rat glioma cell line RGC-6; (E) C57Bl/6 mouse oligodendroglioma clone G26-19; (F) mouse oligodendroglioma clone G26-24; (G) mouse oligodendroglioma clone G26-20; (H) mouse oligodendroglioma clone G26-20 following 24-hr treatment with Bt_2cAMP.

2. Glycolipid Metabolism in Neurotumor Cell Lines

2.1. Neuroblastomas

2.1.1. Neuroblastoma-Specific Glycosphingolipids

Because of the association of *N*-acetylgalactosamine-containing sialoglycosphingolipids with gray matter, it was anticipated that neuroblastoma cell lines would contain these complex gangliosides. This has turned out to be the case, although such cells appear unable to synthesize long-chain trisialogangliosides unless grown as *in vivo* tumors in AJ/6 mice (Dawson and Stoolmiller, 1976). Dawson *et al.* (1971) characterized the major gangliosides in neuroblastoma clone NB41A as G_{M2}, G_{M1}, G_{D1a}, and G_{M3}. Two other clones, NB2a and N18, showed the same qualitative profile (Table II). This was confirmed by Yogeeswaran *et al.* (1973), who extended the analyses to an 8-azoguanine-resistant NB2a hybrid (NA) and hybrid clones derived from fusion with thymidine-kinase-deficient mouse L-cells. The NA clone contained 70% less sialoglycolipid than the parental NB2a clone, much of which could be attributable to a reduction in G_{M3}, but the presence of an additional ganglioside, G_{D1b}, was reported. The hybrid cells contained more sialoglycolipid than either the NA or LMTK$^-$ parents, namely, G_{M2}, G_{M3}, G_{M1} and G_{D1a} in order of declining concentration. We examined clones derived from fusion of N18 and LMTK$^-$ (NL-1F) (Minna, 1973) and found only trace amounts of gangliosides more complex than G_{M2} (Table II). When neuroblastoma cells

Table II. Glycosphingolipid Composition of Representative Neuroblastoma (NB2a), Astrocytic Glioma (RGC-6), and Oligodendroglioma (G26-20) Cell Lines[a]

Glycosphingolipid[b]	Neuroblastoma NB2a	Astrocytic glioma RGC-6	Oligodendroglioma G26-20
	(μmol/10^9 cells)		
GlcCer	0.50	0.53	0.90
GalCer	0.15	0.05	0.35
LacCer	0.25	0.35	0.20
Sulfo-GalCer	<0.01	<0.01	0.35
Asialo-G_{M2}	0.25	<0.05	<0.10
Tetrahexosylceramide	1.20	<0.05	0.10
G_{M3}	0.60	2.05	1.75
G_{M2}	1.40	<0.10	<0.10
G_{D3}	0.10	0.15	0.50
G_{M1}	0.70	<0.10	0.10
G_{D1}	0.90	<0.10	<0.10

[a] All cell lines were grown in modified Eagle's media supplemented with 10% fetal calf serum. Quantitative variability results from differences in cell density, passage number, and culture conditions; different clonal lines also show quantitative differences.

are cultured under a wide variety of conditions such as high serum (10–20%), low serum (1–5%), serum-free media, heat-inactivated serum, monolayer culture, suspension culture, high and low density, and passage number, one observes many quantitative differences but the same qualitative ganglioside composition. One possible explanation for such variability is the degree of "differentiation" or neurite extension exhibited by the cell, and, as will be discussed later, we have observed glycosphingolipid (GSL) differences between differentiated and undifferentiated cells. A second is the susceptibility of glycosyltransferases to stimulation by such neurologically active agents as norepinephrine, 5-hydroxytryptamine (serotonin), N^6,O^2-dibutyryl adenosine 3′,5′ cyclic monophosphate (Bt_2cAMP), corticosteroids, and less obvious factors such as metal (e.g., manganese) ion concentration and serum or cell-produced "factors."

The major neutral glycosphingolipids in neuroblastoma cell strains were glucosylceramide (GlcCer), lactosylceramide (LacCer), and a tetrahexosylceramide [GalNAc$(Gal)_2$GlcCer], with variable amounts of asialo-G_{M2} (Dawson *et al.*, 1971), trihexosylceramide (Yogeeswaran *et al.*, 1973), and traces of galactosylceramide (GalCer) (Table II). Once again, significant quantitative variation was found within a given clone, in contrast to the quantitative stability of other cultured cells such as human skin fibroblasts (Dawson *et al.*, 1972*a*). The existence of asialo-G_{M2} in such cells is of interest because of the possible role of this glycosphingolipid as an intermediate in ganglioside biosynthesis, suggesting the sequence LacCer $\longrightarrow$ asialo-G_{M2} $\longrightarrow$ G_{M2} rather than LacCer $\longrightarrow$ G_{M3} $\longrightarrow$ G_{M2}, as shown in Fig. 2. *In vitro* glycosyltransferase assays have failed to confirm its role as a major metabolic intermediary. Thus, conversion of LacCer to G_{M3} *in vitro* is always at least five times as efficient as conversion of LacCer to asialo-G_{M2} (Dawson *et al.*, 1972*b*; Kemp and Stoolmiller, 1976), although this could be attributed to differences in enzyme or nucleotide sugar stability. The existence in neuroblastoma cells of all the theoretical intermediates in the synthesis of G_{M1} from GlcCer, in detectable quantities, offers a distinct advantage over whole brain, where even G_{M3} is barely detectable. Kemp and Stoolmiller (1976) presented metabolic evidence, based on short-term isotope labeling studies, for the sequence $G_{M3} \longrightarrow G_{M2} \longrightarrow G_{M1}$.

2.1.2. Neuroblastoma-Specific Glycosphingolipid Metabolic Pathways

The key regulatory step in glycosphingolipid biosynthesis appears to be at the level of LacCer, where competition among four transferases and four nucleotide sugars, namely, UDP-Gal, CMP-*N*-acetylneuraminic acid (CMP-NeuAc), UDP-*N*-acetylgalactosamine (UDP-GalNac), and UDP-*N*-acetylglucosamine (UDP-GlcNac), at the multienzyme complex level (Fig. 2) determines the complex glycosphingolipid composition of a particular tissue. Recent studies, such as the observed enhancement of G_{M3} synthesis in cells

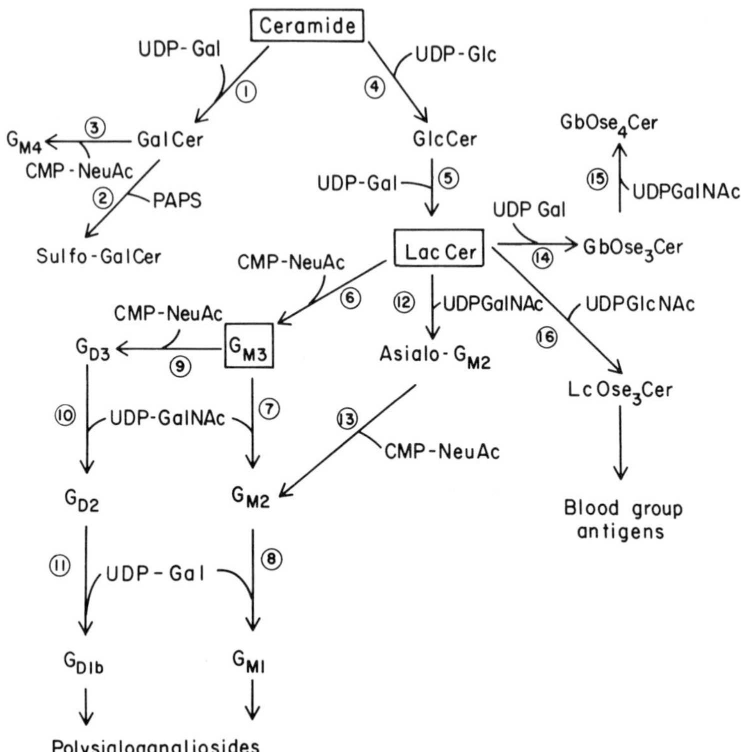

Fig. 2. Proposed pathways for the biosynthesis of glycosphingolipids. Nomenclature is according to IUPAC-IUB recommendations [*Lipids* **12**:455–468 (1977)]. Transferase activities 1–3 are characteristic of oligodendroglial cells; transferase activities 4–11 are characteristic of neuronal cells; transferase activities 4–6 and 9 are associated with astrocytes. *N*-Acetylgalactosaminyltransferase activity 7 is absent from astrocytes and many nonneuronal cell lines; transferases 12 and 13 have a doubtful role in ganglioside synthesis. Galactosyltransferase activity 14 and *N*-acetylgalactosaminyltransferase activity 15 are associated mainly with nonneuronal tissue; *N*-acetylglucosaminyltransferase activity 16 is associated primarily with secretory tissue and bone marrow. Since all the 16 glycosyltransferases are membrane-bound and have not yet been purified, it is not possible to make definitive statements with regard to substrate or cell-type specificity. For example, it is not clear whether *N*-acetylgalactosaminyltransferase activities 7, 10, 12, and 15 are a single enzyme or four discrete enzymes.

treated with short-chain fatty acids such as butyrate (Fishman *et al.*, 1976) and those on synchronized human KB cells (Chatterjee *et al.*, 1975), are promising new approaches to studying regulation of GSL synthesis in cells.

The relative amounts of the four derived glycosphingolipids, $GbOse_3Cer$, G_{M3}, asialo-G_{M2}, and $LcOse_3Cer$, vary considerably from tissue to tissue, and it is of interest to note that all four are virtually undetectable in normal nervous tissue, whereas all but GlcNAc-Gal-GlcCer ($LcOse_3Cer$, a precursor of blood group glycosphingolipids) are detectable in various neuroblastoma cell lines. A potentially useful neuroblastoma (i.e., neuronal) enzyme marker is the transferase responsible for the conversion of G_{M3} to G_{M2}, the first step in the synthesis of higher gangliosides. Neuroblastoma cell lines such as NB2a incorporate more [^{3}H]-$GlcNH_2$ into complex gangliosides than do other neurotumor cell lines (Dawson, 1979). Incorporation is not enhanced by coculture in the presence of mouse oligodendroglioma cells—in fact it is somewhat inhibited—but addition of Bt_2cAMP did cause a 2-fold stimulation (Dawson and Kernes, 1978). All neuroblastoma cell lines thus far studied show high UDP-GalNac : G_{M3}*N*-acetylgalactosaminyltransferase (EC 2.4.1-) activity under *in vitro* conditions, but this activity is not detectable in extracts of rodent glioma cell lines such as RGC-6 (Duffard *et al.*, 1977; Stoolmiller *et al.*, 1978) (Table III). However, despite the failure to detect this enzyme activity in many nonneuronal cell strains, a significant level of G_{M2} synthesis has been observed in nonneural cells such as the mouse epithelial line 3T3, and the supression of hexosaminytransferase activity following viral transformation has been well documented (Brady and Fishman, 1974).

Other glycosyltransferase activities have been measured in a search for neuronal marker enzymes, but the results are less clear. Thus, UDP-Gal : G_{M2} galactosyltransferase activity is relatively high in neuroblastoma cell strains (Table III), but some activity is detectable *in vitro* in most cells, presumably from nonspecific galactosyltransferases. In NB2a and NB41A cell lines, we find significant levels of galactosyltransferase activity but virtually no sulfotransferase activity, a characteristic restricted to oligodendrogliomas. These latter *in vitro* findings were confirmed by the failure of NB2a, NB41A, or N18 cells to incorporate carrier-free H_2[^{35}S]O_4 into sulfatide (Dawson *et al.*, 1977; Dawson, 1979) and by the presence of detectable GalCer, but not sulfo-GalCer, in the cells (Dawson, 1979). However, Sarlieve and Mandel (1975) and Sarlieve *et al.* (1976) came to precisely the opposite conclusion, namely, that adrenergic clone NIE-115 expressed sulfotransferase but not galactosyltransferase activity at both high and low density.

Glycosphingolipid catabolism in neuroblastoma cell strains has been fol-

Table III. Glycosyltransferase Activity in Mouse Neurotumor Cell Strains Compared to 17-Day-Old Chick Brain

Enzyme	Chick brain	G26-24	NB41A
	(cpm/mg protein/hr)		
UDP-Gal: Cer galactosyltransferase	5414	5087	2687
UDP-Gal: 1,2-dipalmitin galactosyltransferase	1536	439	936
UDP-Gal: G_{M2} galactosyltransferase	7764	1130	4918
PAPS[a]: GalCer sulfotransferase	2026	670	<10

[a](PAPS) 3-Phosphoadenosine 5′-phosphosulfate.

lowed by pulse–chase labeling studies that indicate an average turnover half-time of approximately 36 hr (Stoolmiller *et al.*, 1973), and by measurement of lysosomal hydrolase activity. Dawson *et al.* (1974) showed that NB41A neuroblastoma cells take up and metabolize a wide variety of glycosphingolipids and that catabolism can be blocked by specific inhibitors such as *N*-hexyl-glucosylsphingosine (a β-glucosidase inhibitor that effectively creates a "Gaucher" neuron). Direct measurement of lysosomal hydrolase activity (Table IV) indicated considerable variability in enzyme levels among different cell lines. Of particular interest is the fact that G_{M1}-β-galactosidase specific activity is higher in neuroblastoma cell lines than in other cultured cells, whereas GalCer-β-galactosidase activity is much lower in neuroblastoma cell lines than in mouse oligodendroglioma cell lines. No clear-cut differences in either arylsulfatase A or synthetic substrate β-galactosidase activity were noted, although both enzymes have been claimed to be cortisol-inducible in neuroblastoma cell line M1 (Farooqui *et al.*, 1977).

2.1.3. Evidence for a Functional Role for Glycosphingolipids

There have been several attempts to establish functional relationships between neuronal-specific glycosphingolipid glycosyltransferase activities and neurotransmitters. Thus, Stoolmiller *et al.* (1974) found that incuba-

Table IV. Lysosomal Hydrolase Levels in Cultured Mouse Cell Strains[a]

	Mouse oligodendroglioma				Mouse neuroblastoma			Mouse
	G26-15	G26-19	G26-20	G26-24	NB2a	N18	NB41A	LMTK⁻
Enzyme	(nmol 4MU liberated/mg protein/hr)							
N-Acetyl-β-D-hexosaminidase	35	130	80	43	136	138	150	74
β-D-Galactosidase	110	140	40	35	106	30	80	42
Arylsulfatase A	28	35	20	18	8	5	7	20
α-L-Fucosidase	15	160	50	37	104	22	55	54
α-D-Mannosidase	8	2	4	4	4	3	5	3
α-D-Galactosidase	40	50	24	14	32	8	20	39
β-D-Glucoronidase	20	5	12	15	23	15	20	6
	(cpm/mg protein/hr)							
GalCer-β-D-galactosidase	246	341	134	95	10	5	48	105
G_{M1}-β-D-galactosidase	1300	1300	625	ND	2400	ND	2250	900

[a] Lysosomal hydrolase assays were carried out on 600*g* supernatant solutions derived from sonifer-disrupted confluent monolayers of 10^7 cells grown in modified Eagle's medium supplemented with 10% fetal calf serum as described previously (Rushton and Dawson, 1975; Dawson and Tsay, 1977). (ND) Not determined.

tion of neuroblastoma clone NB41A with 10^{-6} M norepinephrine produced a 6-fold increase in the $(G_{M1} + G_{M2})/G_{M3}$ ratio and a stimulation of UDP-GalNac : G_{M3}: *N*-acetylgalactosaminyltransferase. As yet, there is no evidence to link norepinephrine and related neurotransmitters with gangliosides in any functional sense, although Van Heyningen (1974) implicated polysialoglycosphingolipids as the physiological receptors for 5-hydroxytryptamine (serotonin). Preliminary studies with neuroblastoma cells (NB2a and NB41A) in this laboratory indicate that incubation with 5×10^{-4} M serotonin for up to 24 hr produces a more than 50% inhibition of [^{3}H]GlcNH$_2$ incorporation into $G_{D1a} + G_{D3'}$ but a 2-fold stimulation of incorporation into G_{M2} and G_{M1}. If a ganglioside such as G_{D3} is part of the physiological receptor for serotonin, our data would therefore indicate autoregulation of receptor synthesis, a phenomenon widely noted for neurotransmitters (Raff, 1976).

Moskal *et al.* (1974) reported that adrenergic C1300 neuroblastoma cell line NIE-115, when cultured in the presence of 10^{-3} M Bt$_2$cAMP for 24 hr, expressed enhanced galactosyltransferase and sialotransferase activity. We confirmed this observation in NB2a cells (Dawson and Kernes, 1978). A concomitant elevation of glutamate decarboxylase activity was observed in NIE-SB1 cells grown in 10^{-3} M Bt$_2$cAMP for 24 hr, suggesting a possible relationship between ganglioside and inhibitory neurotransmitter (γ-aminobutyric acid) synthesis. The studies also showed that different neuroblastoma cell clones have widely varying glycosphingolipid glycosyltransferase activities. Treatment with cholera toxin (1 μg/ml) will elicit an adenylate cyclase response in neuroblastoma cells (Dawson, 1979), which must be mediated through G_{M1} ganglioside at the cell surface (Cuatrecasas, 1973). Mullin *et al.* (1977) speculated that sialoglycosphingolipids could have an important role in mediating the membrane potential changes associated with neuronal electrical activity. Neuroblastoma cell strain NB2a was labeled for 24 hr with either [^{14}C]acetate or [$^{-3}$H]-GlcNH$_2$ and then incubated with Schwartz–Mann cholera toxin (1 μg/ml) for 3 hr. The result was a profound morphological change in the cells (extensive somal shrinkage and neurite extension) (cf. Figs. 1 A and B) and a stimulation of incorporation of label into all glycosphingolipids (Dawson, 1979). However, when neuroblastoma clone N4TG1 was cultured under conditions that either suppressed or stimulated neurite formation, there was no difference in total incorporation of [^{3}H]-GlcNH$_2$ into glycosphingolipids, and G_{D1a} synthesis was actually enhanced in "amorphous" cells (Dawson, 1979).

2.2. Astrocytomas

Rat glioma RGC-6 has a simple glycosphingolipid composition consisting mainly of G_{M3} ganglioside, with lesser amounts of GlcCer, LacCer, and G_{D3} and trace amounts of tetrahexosyl- and trihexosylceramides (see Table

II). Labeling studies with [^{3}H]-GlcNH$_2$ (Dawson, 1979) confirmed the virtual inability of RGC-6 cells to synthesize hexosamine-containing sialoglycosphingolipids. This lack of UDP-GalNAc : G$_{M3}$-*N*-acetylgalactosaminyltransferase activity (see Table III) is shared by all other astrocytoma cell lines, as well as by many uncharacterized solid gliomas of human and rodent origin (Stoolmiller *et al.*, 1979). Although the inability of astrocytoma cells to synthesize gangliosides such as G$_{M1}$ and G$_{D1a}$ may indeed reflect a difference between neurons and astrocytes (Dawson *et al.*, 1971; Stoolmiller *et al.*, 1973; Duffard *et al.*, 1977), one must bear in mind that these are transformed cells and that transformation of mouse 3T3 or hamster AL/N cells with DNA viruses, RNA viruses, or chemical carcinogens causes the specific suppression of the transferase required for conversion of G$_{M3}$ to G$_{M2}$ (Brady and Fishman, 1974). Further, bulk-isolated astrocytes appear to have the same ganglioside content as bulk-isolated neurons, although the former may be contaminated with dendrites and membrane fragments, whereas the neurons are shorn of their processes (Norton and Poduslo, 1971; Hamberger and Svennerholm, 1971). Glioma cell line RGC-6 contains virtually undetectable levels of GalCer and sulfo-GalCer and does not incorporate ^{35}S into any glycolipid (Dawson, 1979), although Sarlieve *et al.* (1976) reported high levels of sulfotransferase and detectable galactosyltransferase activity in C-6 (but not in NN astroblast) cell extracts.

2.3. Oligodendrogliomas and Schwannomas

2.3.1. Oligodendroglioma-Specific Glycosphingolipids

We have examined many cell strains and lines for the ability to incorporate carrier-free H$_2$[^{35}S]O$_4$ into sulfo-GalCer (sulfatide) (Dawson, 1979). Positive results were obtained with only three clones (G26-19, -20, and -24) (Dawson *et al.*, 1977), derived from a methylcholanthrene-induced mouse G26 oligodendrocyte tumor (Zimmerman, 1955; Sundarraj *et al.*, 1975). Confluent monolayers of G26-15, -19, -20, and -24 cells were found to contain varying amounts of GlcCer, GalCer, LacCer, and G$_{M3}$ in addition to sulfatide (see Table II). Maximum sulfatide synthesis occurred during log phase growth before the cells reached confluency. Small amounts of trihexosyl- and tetrahexosylceramides and a ganglioside tentatively identified as G$_{D3}$ were also detected (Table II). The hexosamine-containing sialoglycosphingolipid content is thus similar to that of astrocytoma cells but quite distinct from that of neuroblastoma cells.

2.3.2. Oligodendroglioma-Specific Glycosphingolipid Metabolic Pathways

The unique ability of these neurotumor cells to synthesize sulfatide, an important and probably functional component of the myelin sheath (Karlsson, 1977), offered an opportunity to study the regulation of its synthesis. When G26-24 or G26-20 cells were cocultured in the presence of partially "differentiated" neuroblastoma (NB2a) cells, we observed no enhancement of sulfatide synthesis (Dawson, 1979), suggesting that myelination may not necessarily be stimulated by neuronal trophic factors. However, we found that the PAPS : GalCer : sulfotransferase enzyme is inducible by agents, such as hydrocortisone and triiodothyronine, believed to be involved in initiation of the process of myelination. Hydrocortisone and other corticosteroids were shown to initate precocious myelination in developing brain (Balázs and Richter, 1973), and de Vellis and Brooker (1973) reported induction of α-glycerol-phosphate dehydrogenase in RGC-6 cells by levels of hydrocortisone in the range 10^{-7} to 10^{-6} M. Culture of G26-19 and G26-20 cells in the presence of hydrocortisone (cortisol) for 30 hr gave a maximum 6-fold stimulation of $H_2[^{35}S]O_4$ incorporation into sulfatide over the next 24-hr period (Dawson and Kernes, 1978). The optimum concentration for sulfotransferase induction was 5×10^{-6} M, and the specific effect on the enzyme was confirmed by *in vitro* PAP[^{35}S] : GalCer : sulfotransferase assay, which showed a 3- to 4-fold enhancement under the aforementioned conditions of growth in the presence of hydrocortisone. No stimulation of ^{35}S incorporation into glycosaminoglycan was observed under these conditions, indicating that increased synthesis or availability of PAPS was not responsible for the stimulation. A similar induction of PAPS : GalCer : sulfotransferase activity was observed with cortisone and dexamethasone, but estradiol, testosterone, and androsterone had essentially no effect. It was not possible to induce sulfotransferase activity in the sulfatide-negative clone G26-15 or in any of the neuroblastoma cell lines tested.

Apart from a report by Coles *et al.* (1970) that testosterone will induce di-GalCer synthesis in female C57BL/6 mouse kidney, there have been no previous reports of induction of glycosphingolipid synthesis by steroids, although effects on lysosomal hydrolase levels have been noted. Thus, Nishimura and Shimoda (1975) reported that cortisol treatment of young mice stimulated lysosomal hydrolase levels but did not alter the level of glycosphingolipids in thymus. We have not observed any stimulation of lysosomal hydrolase activity in any of the steroid-treated oligodendroglioma or neuroblastoma cell lines studied, in contrast to a recent report by Farooqui *et al.* (1977) that claimed a 2- to 4-fold stimulation of arylsulfatases A and B and β-galactosidase activity in rat glioblastoma (C-6) cells, neuroblastoma clonal line M1, and normal hamster (NN) astroblasts. Suzuki *et al.* (1969) and Carubelli and Griffin (1970) also found that hydrocortisone stimulated β-glucuronidase, neuraminidase, and deoxyribonuclease activity to some extent in HeLa cells. Farooqui *et al.* (1977) attributed the higher V_{max} of the arylsulfatases and β-galactosidase in hydrocortisone-treated cells to enzyme induction, although no data were presented.

3. Glycoprotein Metabolism in Neurotumor Cell Lines

3.1. Neuroblastomas

In common with other eukaryotic cells, neuroblastoma cell lines contain a large number of acidic and neutral glycoproteins with the general structure [NeuNAc-Gal-GlcNAc-Man]$_{2-3}$-Man-GlcNAc[Fuc]-GlcNAc-Asn and [Man]$_{5-8}$GlcNAc-GlcNAc-Asn. These can be labeled with radioactive GlcNH$_{2'}$ fucose, or mannose, and their composition varies with stage of growth cycle and cell density. This discussion of neuroblastoma glycoproteins will not concern itself with soluble glycoproteins such as the cytoplasmic and lysosomal enzymes, but will concentrate on the evidence for specific cell-surface glycoproteins in these cell lines. Particular attention has been focused on neuroblastoma plasma membrane glycoproteins in the hope that they might give some insight into the functional role of glycoproteins at the neuronal cell surface.

3.1.1. Glycoproteins Associated with Neurite Formation (Differentiation)

Neuroblastoma cells with elongated processes, often called "neurites," take on the appearance of differentiated neurons and are stained by the neuron-specific Bodian procedure, whereas those with rounded neuroblast morphology do not (Schubert *et al.*, 1973). Glick *et al.* (1973) showed differences in the chromatographic elution profile of trypsin-released glycopeptides (labeled with [^{3}H]fucose, [^{14}C]fucose or [^{3}H]-GlcNH$_2$) from the surface of neuroblastoma cells that either formed neurites (N18 and NS20) or failed to form neurites (N1 and N-1A-103), when challenged with serum-free media or Bt$_2$cAMP. Similarly, Brown (1971) demonstrated the appearance of a new class of trypsin-releasable surface glycopeptides when "undifferentiated" C1 cells were induced to extend neurites by treatment with bromodeoxyuridine (BrdU). These differences have been confirmed to some extent by studying the binding of concanavalin A–ferritin conjugates to "differentiated" and "undifferentiated" cells. "Differentiation" in neuroblastoma cells may therefore be equated with the appearance at the cell surface of new specific glycoproteins, some of which could have a role in cell recognition and cell-contact phenomena, although it is possible that one is simply observing the unmasking or modification of preexisting glycoproteins. Further evidence for neuronal glycoprotein specificity was obtained by Akeson and Hershman (1974), who prepared antisera to N18 cells with neurites, absorbed the sera with nonneurite cells, and found that the resulting antisera cross-reacted with brain tissue. It is possible that this potentially neuron-specific antigen (or antigens) is related to the 78,000-dalton glycoprotein that is preferentially exposed to lactoperoxidase-catalyzed iodination in morphologically differentiated neuroblastoma cells (Truding *et al.*, 1974).

Garvican and Brown (1977) reported that differentiated NB41A cells showed increased incorporation of fucose into glycoproteins of molecular weight 60,000 and 70,000, compared to undifferentiated cells. However, Bt_2cAMP-induced differentiation stimulated both glycosylation and low-molecular-weight-protein synthesis, whereas BrdU stimulated only glycosylation of membrane proteins. This supports the hypothesis of Siman-Tov and Sachs (1975) that the mechanisms of Bt_2cAMP- and of BrdU-induced differentiation are different. However, Truding *et al.* (1975) observed no such distinctions when NB2aE cells were labeled in spinner culture for 12–60 hr with either $[^3H]$-$GlcNH_2$ or $[^3H]$fucose, the media removed and replenished, and the proteins released over a 2-hr period separated by sodium dodecyl sulfate–polyacrylamide gel electrophoresis. Three major $[^3H]$-$GlcNH_2$/Fuc-labeled regions corresponding to molecular weights of 55,000, 66,000, and 87,000 daltons were detected, but none of these components was observed in plasma membrane preparations. In the presence of either 1×10^{-3} M Bt_2cAMP or 10^{-5} M BrdU for 60 hr, the released ^3H-labeled glycoprotein was almost exclusively in the 66,000-dalton class. However, when $[^3H]$- or $[^{14}C]$leucine was used as a precursor, neither Bt_2cAMP nor BrdU had any effect on the secretion profile of labeled protein. These results are in basic agreement with a previous report that 55,000-, 77,000-, and 95,000-dalton fractions were released by $[^3H]$leucine-labeled neuroblastoma cells in serum-free media (Schubert *et al.*, 1973).

Truding *et al.* (1975) interpreted their $[^3H]$-Fuc/$GlcNH_2$-labeling data to mean that the 66,000- to 70,000-molecular-weight fraction was preferentially glycosylated and released into the culture media under conditions of morphological differentiation, implying the activation of a glycosyltransferase complex. However, the absence of changes when leucine is used as precursor could also be explained if glycoprotein represents only a small fraction of the total protein secreted. In addition, Truding *et al.* (1975) presented some evidence for identifying the 55,000-dalton glycoprotein released by "differentiated" neuroblastoma cells as tubulin. Tubulin, which contains a small amount (1–2%) of carbohydrate (Margolis, R. K. *et al.*, 1972), had previously been found in neuroblastoma cells (Olmsted *et al.*, 1970), and is almost certainly involved in neurite formation (Schubert and Jacob, 1970).

3.1.2. Glycoproteins and Neuronal Cell Adhesion

Cloned neural cell lines, especially neuronal tumor lines B103, B65, and B50, derived from ethylnitrosourea-treated rat embryos (Schubert *et al.*, 1974, 1976), will adhere preferentially to monolayers of cells obtained by dissociation of chick or rat embryonic neural tissue (Stallcup, 1977; Santala *et al.*, 1977). B103 cells were found to bind to cells derived from any part of the embryonic nervous system but not to rat or chick embryonic liver or fibroblasts or to Chinese hamster ovary cells, indicating a high degree of neural specificity. Similar results were obtained with [^{3}H]leucine or [^{125}I]lactoperoxidase (protein)-labeled plasma-membrane-enriched fractions from B103, B65, and B50 neuronal tumor cell lines, and it was calculated that approximately one cell is required to bind one cell equivalent of plasma membrane vesicles. The plasma membranes of neuroblastoma cells apear to be rich in fucosylated complex carbohydrates (Glick *et al.*, 1973; Truding *et al.*, 1974; Mathews *et al.*, 1976), and in NB2a cells, approximately one quarter of the fucosyl-glycoproteins have a high turnover rate (Hudson and Johnson, 1977). Since the binding of neuronal tumor cell lines to one another or to embryonic chick tectum was abolished by mild trypsinization or treatment with either glutaraldehyde or formaldehyde, it was concluded that specific proteins, possibly neuronal-specific glycoproteins, were involved. Further, the ability of B103 membrane vesicles to bind cells was abolished by either lowering the temperature to 4° or fixing the cells, suggesting that energy and perhaps a transglycosylation reaction (Roseman, 1970; Umbreit and Roseman, 1975) are necessary for adhesion.

To explain differences between plasma membrane–cell and cell–cell binding, Santala *et al.* (1977) proposed that neuronal cells contain four major adhesion "determinants," A, a, B, and b, that can all be labeled with [^{3}H]-GlcNH_2 and are therefore in all probability either glycoprotein or glycolipid in nature. Membranes derived from such cells express only a or b, and adhesion or binding requires A–a or B–b binding (Fig. 3). On the basis of their experiments, these authors proposed that B103 cells expressed A, a, and b, B65 cells only B and b, and B50 cells, A, a, B, and b glycoproteins. This explains why B65 membranes (containing only b) bind B65 and B50 cells (but not B103 cells, which lack B), whereas B103 membranes (containing a *and* b) bind B103, B65, and B50 cells. The authors were careful not to exclude the possiblity that plasma membrane preparations could contain nonfunctional A or B (glycoprotein or glycolipid) determinants. All cells were postulated to contain a universal adhesion determinant C, to explain the low level of binding to nonneuronal cells such as RGC-6, Chinese hamster ovary (CHO), and embryonic liver cells.

The complexity of the adhesion process between different neural cell lines was also noted by Stallcup (1977), using B50 neuronal tumor cell lines and a variety of cell monolayers. Cultured B50 cells adhered rapidly to other nerve cell lines (e.g., B65, B104, XKC, SW16), glial cell lines (e.g., B9, B49, B90, C68), muscle cell lines (L6), and unclassified types (e.g., B82, B108) derived by Schubert *et al.* (1974). However, by pretreating the B50 cells with trypsin, coating them with an anti-nerve antiserum, or lowering the temperature from 20° to 0° C, he was able to distinguish at least three distinct subclasses of binding. Stallcup (1977) also proposed that binding involves pairs of interlocking or complementary surface components on the cell lines and designated these

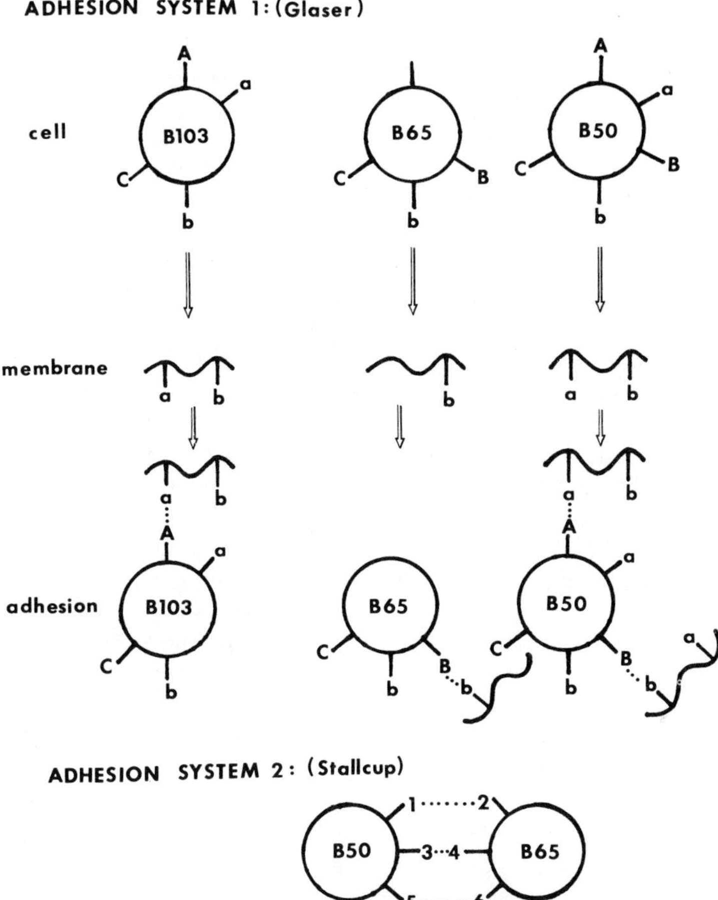

Fig. 3. Proposed models for the role of complex carbohydrates in adhesion between neurotumor cell lines. System 1 (Santala *et al.*, 1977) requires the variable expression of pairs of determinants A–a, and B–b and the constant expression of C on different cell lines. Membranes derived from such cells express only a, b, etc., but can bind to cells if their corresponding adhesion site, A, B, etc., is expressed. System 2 (Stallcup, 1977) requires a system of complementary complex carbohydrates, 1–2, 3–4, 5–6, for adhesion, as described in the text.

pairs as 1–2, 3–4, and 5–6 (Fig. 3). These pairs are not analogous to the A–a and B–b pairs of Santala *et al.* (1977) in that adhesion system 1–2 was both trypsin-and antibody-sensitive (with a further temperature-sensitive subdivision) and 3–4 was trypsin- and antibody-insensitive but very temperature-sensitive, whereas system 5–6 was both trypsin- and temperature-sensitive, but antibody-insensitive. System 3–4 adhesions were interrupted by incubation with almond *N*-acetyl-β-hexosaminidase, suggesting involvement of a terminal HexNAc residue on a glycoprotein or glycolipid, whereas β-galactosidase and α-mannosidase had no effect. Component 1 was found in brain, spleen, and liver, so its eventual purification and characterization should be possible.

These cell-adhesion studies on neurotumor cell lines may eventually be able to delineate the role of cell-membrane glycoproteins in neural cell interaction. However, it must be remembered that there is considerable evidence for altered complex carbohydrate composition and metabolism in transformed cells, so that extrapolation of the data presented above to the normal developing nervous system may not be valid. Thus far, no one has described the existence in neural cell line culture media of the soluble glycoprotein aggregating factors described by Hausmann and Moscona (1975) for embryonic chick retinal cells. A direct role for carbohydrate residues in cell adhesion has proved difficult to demonstrate in "normal" cells, and Hausmann and Moscona (1975) found no evidence for the involvement of carbohydrate in the adhesion process. Studies in this laboratory showed the embryonic chick retinal aggregating glycoprotein to contain mannose, galactose, GlcNAc, and sialic acid, and removal of at least 95% of the sialic acid residues with *Vibrio cholerae* neuraminidase did not diminish its aggregating properties. In contrast, Balsamo and Lilien (1974) reported that *N*-acetylhexosaminidase destroyed the ability of their retina aggregating factor (a *sialo*glycoprotein!) to bind cells, and that α-mannosidase destroyed the ability of their cerebral lobe "factor" to bind retinal cells.

Because of the interest in B50, B65, and B103 surface complex carbohydrates, we examined the ganglioside composition of their derived tumors relative to that of other rodent neurotumors (Fig. 4). G_{M3} was the major ganglioside in all three, in common with rat gliomas A6 and others, but in marked contrast to C1300 neuroblastoma. However, the presence of complex gangliosides was also noted, and these could have a role in adhesion.

3.2. Astrocytomas (Glioblastomas)

There have been few studies on glycoprotein metabolism in glioblastoma cell lines, and the glial-specific S-100 protein synthesized by rat astrocytoma RGC-6 (C-6) (Benda *et al.*, 1968) and CHB cells (Pfeiffer, 1973) at high density is nonglycosylated. However, it seems probable that cell-surface glycoproteins (as well as other complex carbohydrates) are involved in glial–glial and glial–neuronal cell interactions in developing nervous tissue.

3.2.1. Glycoproteins Associated with Glial Cell Differentiation

Santala and Glaser (1977) suggested that density-dependent affinity changes could be of importance in differentiation of the nervous system if one assumes that occupancy of an adhesive site is the first step in induction of glial-specific proteins such as S-100 protein (Benda *et al.*, 1968), glycoprotein

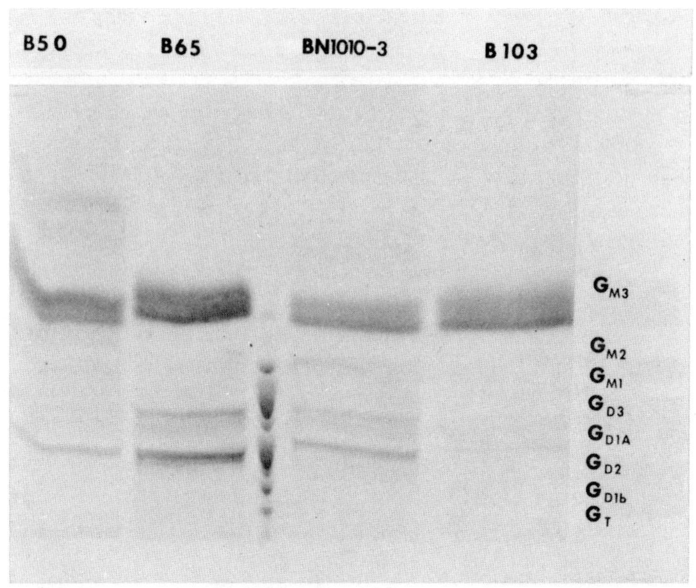

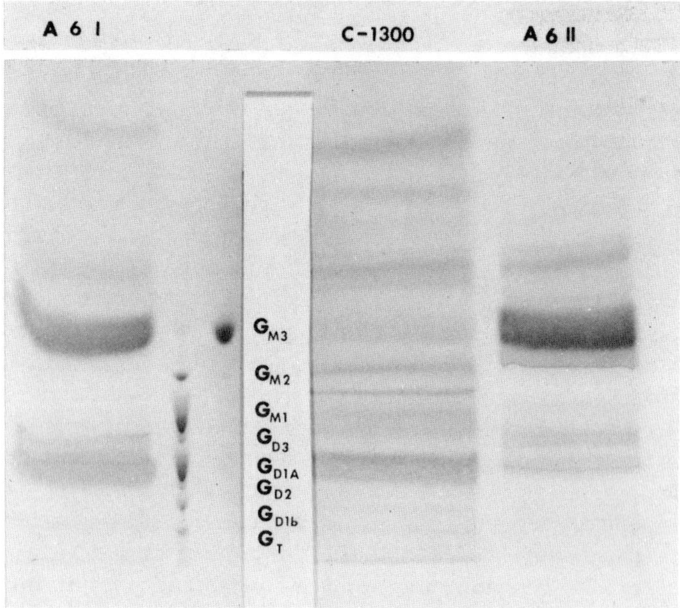

Fig. 4. Thin-layer chromatographic separations of gangliosides derived from rat (B50, B65, BN1010-3, and B103) neuronal-like, rat glioma (A61, A611), and mouse C1300 neuroblastoma cell lines grown as solid tumors in host animals. Silica gel-G plates were developed two times in chroloform–methanol and 0.02% $CaCl_2$ and sprayed with resorcinol. It can be clearly seen that G_{M3} is the major ganglioside in all the rat tumors.

enzymes such as α-glycerophosphate dehydrogenase (de Vellis and Brooker, 1973), ornithine decarboxylase (Bachrach, 1976), and catechol-*O*-methyltransferase (Silberstein *et al.*, 1972), or changes in a catecholamine-sensitive adenylate cyclase. Another interesting mechanism for enzyme regulation in nervous tissue was proposed by Stefanovic *et al.* (1975) on the basis of their observation that treatment of hamster astroblast (NN) cell lines and mouse neuroblastoma cell lines S21, N18, and N115 for brief periods with *Clostridium perfringens* neuraminidase caused an immediate increase in activity of two sialoglycoprotein membrane-associated enzymes, acetyl- and butyrylcholinesterase. Removal of sialic acid residues caused an increase in V_{max} for these glycoprotein enzymes, either by relieving a total steric block or by allosteric activation of inactive enzyme molecules. Siman-Tov and Sachs (1972) had previously suggested that most acetylcholinesterase activity is cell-membrane-associated and that membrane differences could explain the variation in acetylcholinesterase activity observed in different neurotumor cell lines. It is therefore possible that these esterases, and probably other membrane-associated glycoprotein enzymes, are activated by direct removal of sialic acid residues.

3.2.2. Glycoproteins and Glial Cell Adhesion

Santala and Glaser (1977) showed that C-6 cells adhere poorly to other C-6 cells or to neuroblastoma cell lines and attributed this to a failure of C-6 cells to express certain cell-surface proteins, possibly glycoproteins. A more detailed investigation showed that cell–cell adhesion characteristics varied considerably with cell density. Thus, cell suspensions prepared from monolayer C-6 cultures with a density less than 10^5 cells/cm^2 showed maximal affinity for plasma membranes derived from either astrocytoma C-6 or neuroblastoma B103 and B65 cells, whereas cells from monolayers of density greater than 10^6 cells/cm^2 had the least affinity. The complexity of the adhesion system is evidenced by the fact that the binding affinity of plasma membranes to cells was independent of the density of the cells that the membranes were derived from. To determine the basis for this loss of adhesiveness with increasing cell density, adhesion studies were carried out with the plasma membrane concentration reduced by 75%. Under these conditions, there was no reduction in the percentage of membrane bound to the C-6 monolayer. Thus, the change in adhesiveness must result from a decrease in affinity rather than a reduction in the number of binding sites, since if the cells at higher density have fewer binding sites, lowering the plasma membrane concentration would have resulted in an increased percentage of membranes being bound. Since it is generally believed that glycoproteins (as well as other complex carbohydrates) are involved in cell adhesion (Roseman, 1970; Barondes, 1970), it seems likely that astrocyte cell-surface glycoprotein composition must be density-dependent, and that perhaps the decrease in adhesion results from completion (e.g., sialylation) of the oligosaccharide units.

3.3. Oligodendrogliomas and Other Glioblastoma Cell Lines: Glycoproteins Associated with Oligodendroglial Cell Differentiation

Oligodendroglial cell lines designated G26-15 through -29 (Sundarraj *et al.*, 1975) were found to cross-react with the glial-specific NS-1 antigen. The NS-1 antigen occurs in higher concentration in white matter than in gray matter, is reduced in concentration in mutant mice with myelination defects, is not found in C1300 neuroblastoma or in nonneural tissue, and is therefore probably a glial-specific glycoprotein (Schachner, 1974). Treatment of G26-19 cells with 1 mM Bt_2cAMP or 0.1 mM norepinephrine resulted in a 10% decrease in NS-1 concentration (Sundarraj *et al.*, 1975).

Preliminary studies in this laboratory indicate increased synthesis of certain glycoproteins in G26-15, -19, -20, and -24 cells as a result of culture for 2–4 days in the presence of corticosteroids such as cortisol (hydrocortisone) (Fig. 5). Although the degree of stimulation was much less than the 6-fold elevation of PAPS : GalCer sulfotransferase activity observed in G26-20 cells under similar conditions of hydrocortisone treatment (see Section 2.3.2), the glycoproteins involved appear to be exposed on the cell surface, since they were also labeled (Fig. 5) following neuraminidase, galactose oxidase (GaO), and $NaB[^3H]_4$ treatment (Steck and Dawson, 1974).

3.4. Uncharacterized Gliomas

Other glioma cell lines that may prove useful in elucidating the role of complex carbohydrates in nervous tissue include BN1010-1 and BN1010-3, derived from an ethylnitrosourea-induced CNS tumor in a CDF rat (Wechsler *et al.*, 1969; Pfeiffer, 1973), which secrete the tripeptide thyrotropin-releasing factor during log phase of growth under slightly acid (pH 6.8) culture conditions (Grimm-Jorgensen *et al.*, 1976). Precursors of other biologically active peptides such as ACTH are known to be glycoproteins (Eipper *et al.*, 1976), and the existence of biologically active glycopeptides is a distinct possibility.

Thus far, there have been no reported studies on glycoprotein metabolism in human glioma cell lines despite the ready availability of glial tumors from neurosurgical sources. Indeed, the "human" glioblastoma CHB, which had been the subject of several studies (Lightbody *et al.*, 1970; Pfeiffer *et al.*, 1972), was subsequently found to have a normal rat karyotype (Stoolmiller *et al.*, 1973). Serendipitously, since they did not intend to study glycoprotein metabolism, Clark and Perkins (1971) reported that cloned tumor astrocyte line 1181N1, originally derived from a human cerebral glioblastoma multiforme, contains both a catecholamine-sensitive adenylate cyclase and a histamine-sensitive adenylate cyclase. Maximum sensitivity of cyclase induction was observed during log phase, which is consistent with the hypothesis of Santala and Glaser (1977) (based on studies on C-6 cells) that surface glycoproteins are active (adhesive) as receptors during rapid cell division. Thus, the nature of the cyclase-linked neurotransmitter complex carbohydrate receptors in 118N1 cells is worthy of further study.

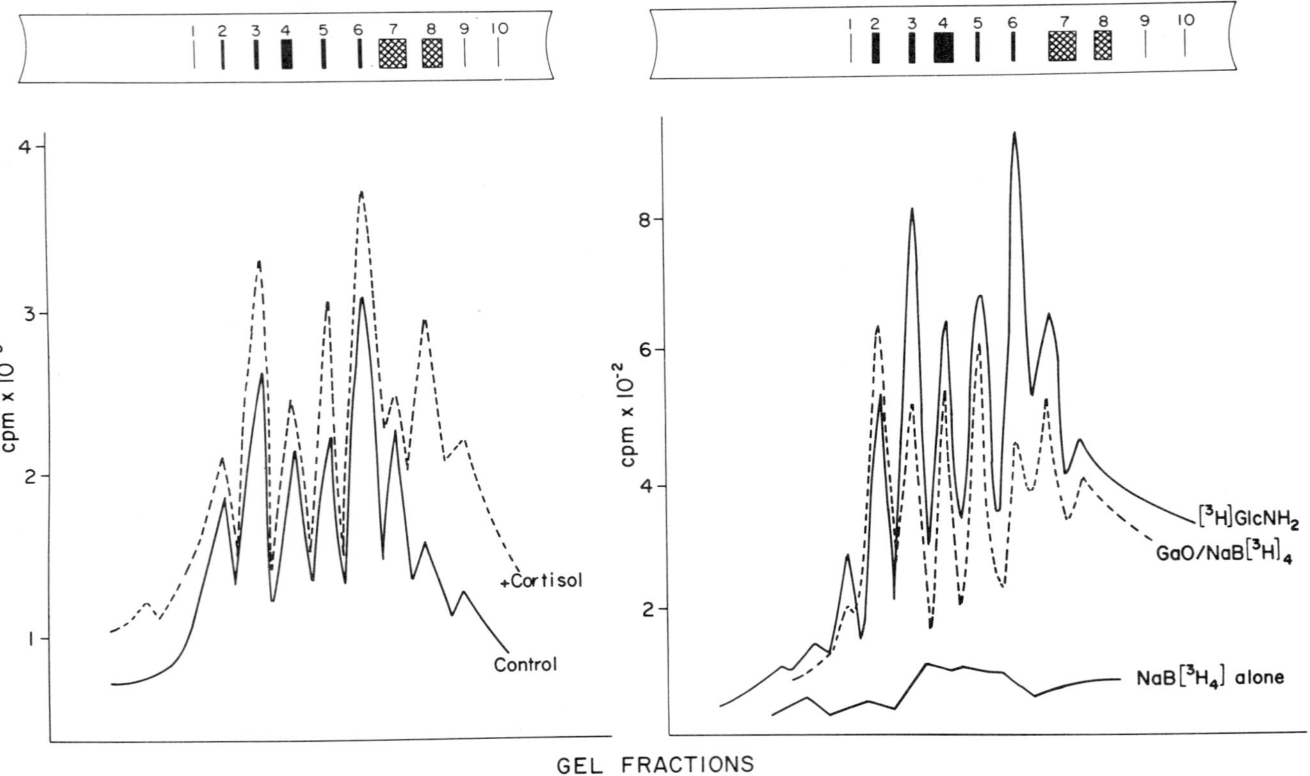

Fig. 5. Effect of hydrocortisone on the incorporation of [^{3}H]-GlcNH$_2$ into the major membrane glycoproteins of mouse oligodendroglioma clones G26-19 and G26-20.

4. Collagen Synthesis in Neurotumor Cell Lines

Neither glial (RGC-6) nor neuroblastoma (NB41A) cells are capable of synthesizing significant amounts of collagen (as judged by the conversion of [^{14}C]proline into [^{14}C]hydroxyproline). The percentage of radioactive amino acids attributable to [^{14}C]hydroxyproline was less than 0.5%, in contrast to connective tissue cells under comparable conditions, where the percentage conversion approaches 35% (Dehm and Prockop, 1971). However, Church *et al.* (1973) reported that rat peripheral neurinoma clonal line (RN-2) can synthesize substantial amounts of procollagen and collagen, although the degree of glycosylation was not reported. Following labeling in serum-free medium with [^{3}H]-Pro for 2–3 days, 14% of the medium proline was hydroxylated compared with 20% in 3T6 fibroblasts and less than 1% in nonfibroblastic cell lines. Since cells of neural (neuroectodermal) origin, such as RN2, appear to synthesize collagen, it was concluded that Schwann cells themselves may be responsible for a large part of the collagen seen in peripheral neurinomas.

5. Glycosaminoglycan Synthesis in Neurotumor Cell Lines

The structure and composition of glycosaminoglycans in nervous tissue are discussed in Chapter 3 and have been reviewed previously (e.g., Stoolmiller *et al.*, 1973; Dawson and Lenn, 1976).

5.1. Neuroblastomas

Mouse C1300 neuroblastoma cells cultured as a monolayer in modified Eagle's medium supplemented with 10% fetal calf serum were found to secrete very little glycosaminoglycan (GAG) into the media, and analysis of intracellular GAGs revealed small amounts of predominantly sulfated GAGs. Studies with carrier-free $H_2[^{35}S]O_4$ (which labels sulfated GAGs) and [^{14}C]acetate (which labels hyaluronic acid also) followed by ion-exchange and Sephadex G-50 chromatography showed that heparan sulfates constituted approximately 60% of the total GAGs, and that 30–50% had *N*-sulfate groups. Hyaluronic acid was a minor component in both cells and media (approximately 10% of total GAGs), and the remaining 30% was not conclusively identified (Stoolmiller *et al.*, 1972). The material was tentatively identified as sulfated glycoprotein, although the presence of small amounts of dermatan and chondroitin sulfates could not be ruled out. Similar results were obtained with other neuroblastoma cell strains, suggesting that neurons synthesize little GAG, but that heparan sulfates, which are associated primarily with the plasma membrane (Kraemer, 1971), predominate. This was confirmed by labeling studies with [^{3}H]-GlcNH$_2$ and $H_2[^{35}S]O_4$ in which the $^{3}H/^{35}S$ ratio in secreted GAGs was 0.6 in neuroblastoma compared with 4.0 in oligodendroglioma cell lines and the amount of GAG synthesized was much less in the oligodendrogliomas (Dawson, 1979).

5.2. Astrocytomas

Rat astrocytoma cell line RGC-6 secretes approximately 20 times as much GAG as neuroblastoma cell lines (Stoolmiller *et al.*, 1973). Dorfman and Ho (1970) isolated the GAGs from 750 ml of conditioned media and harvested cells; they found both to contain hyaluronic acid, chondroitin 4-sulfate, and heparan sulfates in the approximate ratio 10:3:2; this was confirmed by labeling studies with $H_2[^{35}S]O_4$ and [^{14}C]acetate. It was calculated that 25% of the GAG in the media was derived from the fetal calf serum (hyaluronic acid and chrondroitin 4-sulfate in the ratio 2:1), but that all the heparan sulfate was derived by *de novo* synthesis (Stoolmiller *et al.*, 1973).

Galligani *et al.* (1975) showed that addition of β-D-xylosides to RGC-6 cells and NB41A neuroblastoma cells can bypass the need for both core protein and the first enzyme in chondroitin sulfate chain initiation (xylosyltransferase) and produce a 6-fold stimulation in the synthesis and secretion of long-chain chondroitin 4-sulfate xylosides. No stimulation of heparan sulfate synthesis was observed under these conditions. Similar results were observed both in differentiating chondrocytes (Levitt and Dorfman, 1972), where BrdU irreversibly blocks both differentiation and GAG synthesis, and in cultured chick neural retinas (Morris and Dorfman, 1976). The addition of β-xylosides can also overcome the inhibitory effect of BrdU on GAG synthesis in differentiating cells, suggesting that chondroitin sulfate synthesis is most likely regulated at the gene level by the availability of core protein or xylosyltransferase or both. The implication that GAGs may have a functional role in retina histogenesis suggests a wider role in nervous system differentiation (Morris and Dorfman, 1976). The fact that free β-xylosides do not act as chain initiators for heparan or dermatan sulfate may result from the increased structural complexity of these GAGs (Dawson and Lenn, 1976), or it may indicate that their synthesis is regulated differently.

5.3. Oligodendrogliomas and Schwannomas

Mouse oligodendroglioma cell lines G26-19, -20, and -24 were quantitatively quite distinct from other neurotumor cell lines such as C-6 and NB41A in that they secreted large amounts of hyaluronic acid, together with lesser amounts of heparan sulfate and chondroitin sulfates (Dawson, unpublished results. Rat Schwannoma cell lines (RN-2) showed a similar qualitative GAG pattern, but the amount per milligram of protein was intermediate between that in neuroblastomas and astrocytomas (Stoolmiller *et al.*, 1973). Culturing G26-19, -20, and -24 cell strains in the presence of 5×10^{-6} M hydrocortisone, conditions that produce a 6-fold enhancement of sulfo-GalCer synthesis (see Section 2.3.2), had little effect on sulfated GAG synthesis.

In conclusion, it has been shown that all neurotumor cell strains are capable of synthesizing hyaluronic acid, chondroitin 4-sulfate, and heparan sulfate, but that the level of GAG synthesis and the relative amounts vary greatly when expressed in terms of milligrams of protein or cell number, being highest in cells of glial origin and least in those of neuronal origin.

6. Relevance of Studies on Neurotumor Cell Lines to Whole Brain and Relationship to Studies on Individual Cell Types

6.1. Neurons

Neuroblastoma cell lines appear to be at least a partially valid model for studying the function of complex carbohydrates in neurons in that they synthesize appreciable quantities of gangliosides, a group of sialoglycosphingolipids found in highest concentration in axons and dendrites rather than neuronal soma (Ledeen *et al.*, 1976), and some of the glycoprotein receptors, antigens, and neurotransmitter-synthesizing enzymes associated with neurons. However, most neuroblastoma cell lines do not synthesize the most complex gangliosides (G_{DIb}, G_{TIa}, G_{TIb}, G_Q) under *in vitro* tissue culture conditions, although they can at least partially regain this ability when grown subcutaneously in host mice (Dawson and Stoolmiller, 1976). Dimpfel *et al.* (1977) showed a direct relationship between the presence of long-chain gangliosides (G_{DIb} and G_{TI}) and the ability to bind ^{125}I-labeled tetanus toxin. Significant binding was seen with homogenates of embryonic or adult CNS and with primary cultures of embryonic rat CNS, but not with neuroblastoma cell line NB2a, neuroblastoma–glioma hybrid line (108CC15), a human embryonic oligodendroglioma cell line, or C-6 glioma cell line. However, it would appear to be imperative to test a number of neuroblastoma cell lines such as the adrenergic NIE-115 or the endorphin-receptor-positive N4TG1 before a definitive negative statement about complex ganglioside synthesis can be made. One problem in attempting to evaluate the usefulness of neuroblastoma cell lines is the difficulty in culturing pure neuronal populations. Stern (1972) cultured dissociated CNS and spinal cord and showed that feeding G_{M2} ganglioside to such cells could produce abnormal cells (e.g., spinal cord neurons) that morphologically resembled those in Tay–Sachs brain. Schengrund and Repman (1977) cultured dissociated rat embryonic cerebra for up to 21 days and showed that the ganglioside composition (G_{M3} up to complex trisialogangliosides) resembled that of whole embryonic brain. However, after 35 days in culture, G_{M3} became the major ganglioside and G_{DIb} and G_T were no longer detectable. One interpretation was that the simplification of the ganglioside pattern reflected an increasing proportion of glial cells in the culture, although the authors noted that neuroblastoma cell lines continuously maintained in tissue culture also lacked G_{D1b} and G_T. To complicate the picture further, neurons can be isolated from immature nervous tissue by proteolytic digestion and sucrose density gradient ultracentrifugation. Analyses of such isolated neuronal perikarya failed to reveal any ganglioside enrichment (Norton and Poduslo, 1971; Hamberger and Svennerholm, 1971; Abe and Norton, 1974). However, this apparent discrepancy was attributed to the fact that such cells are shorn of their axons and dendrites. This contention is supported by the fact that isolated neurons have been found to exhibit high levels of UDP-Glc:Cer glucosyltransferase activity (Radin *et al.*, 1972) and other ganglioside-synthesizing enzymes compared with bulk-isolated glial cells. Other studies showed bulk-isolated neurons to exhibit low levels of sulfotransferase activity (Benjamins *et al.*, 1974), but appreciable levels of UDP-Gal: galactosyltransferase activity (Köhlschutter and Hershkowitz, 1973), which was reduced to 20% of normal in "Jimpy" neurologically mutant mice. These observations are in agreement with our findings in neuroblastoma cell lines NB2a and NB41A.

Neuroblastoma cell lines synthesize smaller amounts of GAGs in comparison to other neurotumor cell lines, with sulfated species predominating. This is in qualitative agreement with GAG studies by R. U. Margolis and R. K. Margolis (1974) on isolated neurons, although they found the actual levels of GAG to be similar in neurons and astrocytes, but considerably lower in oligodendroglial cells. These data, together with the cell-adhesion studies of Stallcup (1977) and Santala *et al.* (1977), provide further evidence for the potential usefulness of neuroblastoma cell lines in studying neuronal function and organization.

6.2. Astrocytes

Since short-term cultures of astrocytes can be obtained from normal brain biopsies, a number of analytical and metabolic studies have been carried out. Typical findings are a low level of complex sialoglycosphingolipids, GalCer, and no detectable sulfatide (Embree *et al.*, 1973; Noble *et al.*, 1976), which is in agreement with our studies on astrocytoma and glioma cell lines. A major difference between cultured glial cells, glioblastomas, and bulk-isolated glial cells is the report that isolated astrocytes and neurons contain equal levels of complex sialoglycosphingolipids (gangliosides) (Norton and Poduslo, 1971; Hamberger and Svennerholm, 1971; Abe and Norton, 1974). However, this ability to synthesize gangliosides has not been confirmed by *in vitro* biosynthetic studies, since Radin *et al.* (1972) and Deshmukh *et al.* (1974) observed that the first enzyme in ganglioside synthesis, UDP-Glc-Cer:glucosyltransferase is much more active in neurons than in glial cells. One explanation is that the glial cell fraction is contaminated by fractured axons and dendrites, which are enriched in gangliosides. Lim and Mitsunobu (1974) reported that embryonic epithelioid brain cells (flat cells) could be transformed into glial-like cells by a protein (or proteins) they termed the "glial maturation factor." The flat cells incorporated [^{3}H]-$GlcNH_2$ into G_{M1}, G_{D1a}, G_{M2}, and G_{M3} at a level much lower than that observed in neuroblastoma cell lines, and following "transformation" there was a 50% reduction of incorporation into G_{M1}, G_{D1a}, and G_{M2}, but a 3-fold stimulation of incorporation of label into G_{M3} (Lim and Dawson, 1978). These data are somewhat reminiscent of those reported for viral transformation of cells (Brady and Fishman, 1974), and may indicate that ganglioside changes are an integral part of nervous system development and differentiation.

There have been few studies on glycoprotein metabolism in isolated or cultured glial cells, although Gielen and Hinze (1974) reported levels of CMP-NeuNAc synthetase and sialotransferase activity toward glycosphingolipid and glycoprotein substrates to be comparable in isolated rat brain neuronal and glial fractions, in general agreement with observations on neurotumor cell lines. Fluharty *et al.* (1975) found accumulation of heparan sulfate in astrocytes cultured from a patient with the Sanfilippo syndrome (see Chapter 16), again confirming the importance of heparan sulfate in nervous tissue. Thus, the consensus that emerges from studies on both isolated glial cells and glioma cell lines is that we have thus far failed to identify any glial-specific complex carbohydrates.

6.3. Oligodendrocytes

Isolated oligodendroglial cells have a lipid composition consistent with their being a major constituent of white matter (Poduslo, 1975), and this has been confirmed by *in vitro* studies. Thus, Benjamins *et al.* (1974) found a higher level of PAP[^{35}S]: GalCer sulfotransferase activity in isolated oligodendroglial cells than in neurons or glia, and Deshmukh *et al.* (1974) reported that synthesis of monogalactosyldiglyceride and GalCer was higher in oligodendroglia than in either glial- or neuronal-enriched fractions. Radin *et al.* (1972) also reported that GalCer-metabolizing enzymes predominated in oligodendroglial-enriched fractions rather than in neurons or glia. Oligodendroglia were isolated intact from white matter and shown to incorporate ^{35}S label into sulfo-GalCer and other glycosphingolipids when maintained for up to 2 days in culture (Poduslo and McKhann, 1977; Pleasure *et al.*, 1977; Szuchet *et al.*, 1978; Szuchet *et al.*, unpublished). These data are in good general agreement with our studies on mouse oligodendroglial cell lines and suggest that such cells will be useful for studying their differentiation into myelin-producing cells.

Silberberg *et al.* (1972) studied newborn rat cerebellar slices under culture conditions where the oligodendroglia divide more than once prior to elaborating myelin. Expression of both UDP-Gal: Cer galactosyltransferase activity and 2′,3′-cyclic nucleotide 3′-phosphohydrolase activity is irreversibly blocked by addition of BrdU (20 μg/ml) at any time up to 5 days in culture. After 10 days, both activities become refractory to BrdU, suggesting concurrent genetic determination (Latovitzki and Silberberg, 1977).

In this laboratory (Dawson and Kernes, unpublished), we have shown that exposure of G26-20 and -24 mouse oligodendroglioma cells to BrdU (10 μg/ml) for 12–24 hr will block both sulfatide synthesis and the induction of GalCer sulfotransferase activity by hydrocortisone. Synthesis of other glycosphingolipids such as LacCer and G_{M3}, together with both sulfated GAGs (heparan sulfate and chondroitin 6-sulfate) and hyaluronic acid, was essentially unaffected by BrdU treatment. Such observations indicate that both neurotumor cell lines and dissociated brain cultures can be useful for studying the mechanisms of CNS differentiation.

7. Conclusions

The use of neurotumor cell lines for the study of complex carbohydrates of the nervous system offers a number of advantages over studies on whole brain or isolated normal brain cell types that in general outweigh the disadvantages. The major disadvantages are the fact that transformed cells *per se* exhibit deranged complex carbohydrate metabolism and that brain functionality is derived from the harmonious interaction of cell types arranged in a highly specific layered and regional manner. Although the first critique can never be wholly answered, it can be circumvented by studying only clones of cells that express specific neuronal differentiated characteristics, such as the mouse neuroblastoma N4TG1 (developed by Nirenberg and his associates), which expresses opiate peptide receptors (Miller *et al.*, 1978); the noradrenergic rat pheochromocytoma, which responds to nerve growth factor (Greene and Tischler, 1976); or the mouse oligodendroglioma, which synthesizes sulfatide (Dawson *et al.*, 1977) and expresses certain other biochemical characteristics associated with myelin-producing cells. One can therefore take advantage of the fact that large quantities of monoclonal and therefore genetically identical cells can be obtained and the mechanism of a particular receptor function or differentiation process studied and dissected. It is in this area that the neurotumor cell lines will be of greatest value, and all the preliminary evidence indicates that complex carbohydrates will play a major role. Similarly, the problem of the specific three-dimensional architecture of the brain, while not being answered by such studies, can at least be approached by using cloned cell lines in the manner outlined by Stallcup (1977), Santala *et al.* (1977), and others. One can envision such studies eventually relating to the lectin–antibody fluorescent labeling studies being carried out on whole brain sections in the laboratories of Cotman, Pfenninger, Wood, and others. These studies, while elegant in certain respects, are not specific or sensitive enough at this time to satisfy the neurochemist, but eventually the two approaches will find common ground and facilitate our understanding of the precise role of complex carbohydrates in the nervous system.

References

Abe, T., and Norton, W. T., 1974, The characterization of sphingolipids from neurons and astroglia of immature rat brain, *J. Neurochem.* **23:**1025.

Akeson, R., and Hershman, H. R., 1974, Neural antigens of morphologically differentiated neuroblastoma cells, *Proc. Natl. Acad. Sci. U.S.A.* **71:**187.

Amano, T., Richelson, E., and Nirenberg, M., 1972, Neurotransmitter synthesis by neuroblastoma clones, *Proc. Natl. Acad. Sci. U.S.A.* **69:**258.

Augusti-Tocco, G., and Sato, G., 1969, Establishment of functional clonal lines of neurons from mouse neuroblastoma, *Proc. Natl. Acad. Sci. U.S.A.* **64:**311.

Bachrach, U., 1976, Induction of ornithine decarboxylase in glioma and neuroblastoma cells, *FEBS Lett.* **68:**63.

Balázs, R., and Richter, D., 1973, Effect of hormones on the biochemical maturation of the brain, in: *Biochemistry of the Developing Brain* (E. Himwich, ed.), Vol. 1, pp. 254–305, Marcel-Dekker, New York.

Balsamo, J., and Lilien, J., 1974, Functional identification of three components which mediate tissue-type embryonic cell adhesion, *Nature (London)* **251:**522.

Barondes, S. H., 1970, Brain glycomacromolecules and interneuronal recognition, in: *Aspects of Molecular Neurobiology* (F. O. Schmitt, ed.), pp. 747–768, Rockefeller University Press, New York.

Benda, P., Lightbody, L., Sato, G., Levine, L., and Sweet, W., 1968, Differentiated rat glial cell strain in tissue culture, *Science* **161:**370.

Benjamins, J. A., Guarnieri, M., Miller, K., Sonneborn, M., and McKhann, G. M., 1974, Sulfatide synthesis in isolated oligodendroglial and neuronal cells, *J. Neurochem.* **23:**751.

Brady, R. O., and Fishman, P. H., 1974, Biosynthesis of glycolipids in virus-transformed cells, *Biochim. Biophys. Acta* **355:**121.

Brown, J. C., 1971, Surface glycoprotein characteristic of the differentiated state of neuroblastoma C-1300, *Exp. Cell. Res.* **69:**440.

Carubelli, R., and Griffin, M. J., 1970, Neuraminidase activity in HeLa cells: Effect of hydrocortisone, *Science* **170:**1110.

Chatterjee, S., Velicher, L. L., and Sweeley, C. C., 1975, Glycosphingolipid glycosyl hydrolases and glycosidases of synchronized human KB cells, *J. Biol. Chem.* **250:**61.

Church, R. C., Tanzer, M. C., and Pfeiffer, S. E., 1973, Procollagen and collagen synthesis by a clonal line of rat Schwann cells, *Proc. Natl. Acad. Sci. U.S.A.* **70:**1943.

Clark, R. B., and Perkins, J. H., 1971, Regulation of adenosine 3′:5′-cyclic monophosphate concentration in cultured human astrocytoma cells by catecholamines and histamine, *Proc. Natl. Acad. Sci. U.S.A.* **68:**2757.

Coles, L., Hay, J. B., and Gray, G. M., 1970, Factors affecting the glycosphingolipid composition of murine tissues, *J. Lipid Res.* **11:**158.

Cuatrecasas, P., 1973, Interaction of *Vibrio cholerae* endotoxin with cell membranes, *Biochemistry* **12:**3547.

Dawson, G., 1979, Regulation of glycosphingolipid metabolism in neurotumor cell strains, *J. Biol. Chem.* **254:**155.

Dawson, G., and Kernes, S., 1978, Induction of sulfogalactosylceramide (sulfatide) synthesis by hydrocortisone (cortisol) in mouse G-26 oligodendroglioma cell strains, *J. Neurochem.* **31:**1091.

Dawson, G., and Lenn, N. J., 1976, Polysaccharide metabolism disorders, in: *Handbook of Clinical Neurology* (P. J. Vinken, G. W. Bruyn, and H. L. Klawans, eds.), Vol. 27, Part I, pp. 143–168, American Elsevier, New York.

Dawson, G., and Stoolmiller, A. C., 1976, Comparison of ganglioside composition of established mouse neuroblastoma cell strains grown *in vivo* and in tissue culture, *J. Neurochem.* **26:** 225.

Dawson, G., and Tsay, G. C., 1977, Substrate specificity of human α-L-fucosidase, *Arch. Biochem. Biophys.* **184:**12.

Dawson, G., Kemp, S. F., Stoolmiller, A. C., and Dorfman, A., 1971, Biosynthesis of glycosphingolipids by mouse neuroblastoma (NB41A), rat glia (RGC-6) and human glia (CHB-4) in cell culture, *Biochem. Biophys. Res. Commun.* **44:**687.

Dawson, G., Matalon, R., and Dorfman, A., 1972*a*, Glycosphingolipids in cultured human skin fibroblasts. I. Characterization and metabolism in normal fibroblasts, *J. Biol. Chem.* **247:**18.

Dawson, G., Stoolmiller, A. C., and Kemp, S. F., 1972*b*, Biosynthesis of glycosphingolipids in cloned cell strains of neurological origin, *Trans. Am. Soc. Neurochem.* **3:**68.

Dawson, G., Stoolmiller, A. C., and Radin, N. S., 1974, Inhibition of β-glucosidase by *N*-(*n*-hexyl)-*O*-glucosylsphingosine in cell strains of neurological origin, *J. Biol. Chem.* **249:**4638.

Dawson, G., Sundarraj, N., and Pfeiffer, S. E., 1977, Synthesis of myelin glycosphingolipids [galactosylceramide and galactosyl (3-*O*-sulfate) ceramide (sulfatide)] by cloned cell lines derived from mouse neurotumors, *J. Biol. Chem.* **252:**2777.

Dehm, P., and Prockop, D. J., 1971, Synthesis and extrusion of collagen by freshly isolated cells from chick embryo tendon, *Biochim. Biophys. Acta* **240:**358.

Deshmukh, D. S., Flynn, T., and Pieringer, R. A., 1974, The biosynthesis and concentration of galactosyl diglyceride in glial and neuronal enriched fractions of actively myelinating rat brain, *J. Neurochem.* **22:**479.

de Vellis, J., and Brooker, G., 1973, Induction of enzymes by glucocorticoids and catecholamines in a rat glial cell line, in: *Tissue Culture of the Nervous System* (G. Sato, ed.), pp. 231–243, Plenum Press, New York.

DeVries, P., Hadfield, M. G., and Cornbrooks, C., 1976, The isolation and lipid composition of myelin-free axons from rat CNS, *J. Neurochem.* **26:**725.

Dimpfel, W., Huang, R. T. C., and Habermann, E., 1977, Gangliosides in nervous tissue cultures and binding of ^{125}I-labeled tetanus toxin, a neuronal marker, *J. Neurochem.* **29:**329.

Dorfman, A., and Ho, P.-L., 1970, Synthesis of acid muco-polysaccharides by glial tumor cells in tissue culture, *Proc. Natl. Acad. Sci. U.S.A.* **66:**495.

Duffard, R. O., Fishman, P. H., Bradley, R. M., Lauter, C. J., Brady, R. O., and Trams, E. G., 1977, Ganglioside composition and biosynthesis in cultured cells derived from CNS, *J. Neurochem.* **28:**1161.

Eipper, B. A., Mains, R. E., and Guenzi, D., 1976, High molecular weight forms of adrenocorticotropic hormone and glycoproteins, *J. Biol. Chem.* **25:**4121.

Embree, L. J., Hess, H. H., and Shein, H. M., 1973, Microchemical studies of lipids, proteins and nucleic acids in polyoma virus–transformed hamster astroglia, *J. Neuropathol. Exp. Neurol.* **32:**542.

Farooqui, A. A., Elkouby, A., and Mandel, P., 1977, Effect of hydrocortisone and thyroxine on arylsulphatases A and B of cultured cells of neuronal and glial origin, *J. Neurochem.* **29:**365.

Fishman, P. H., Bradley, R. M., and Henneberry, R. C., 1976, Butyrate-induced glycolipid biosynthesis in HeLa cells: Properties of the induced sialyltransferase, *Arch. Biochem. Biophys.* **172:**618.

Fluharty, A. L., Davis, M. L., Trammell, J. L., Stevens, R. L., and Kihara, H., 1975, Mucopolysaccharides synthesized by cultured glial cells derived from a patient with Sanfilippo A syndrome, *J. Neurochem.* **25:**4.

Galligani, L., Hopwood, J., Schwartz, N. B., and Dorfman, A., 1975, Stimulation of synthesis of free chondroitin sulfate chains by β-D-xylosides in cultured cells, *J. Biol. Chem.* **250:** 5400.

Garvican, J. H., and Brown, G. L., 1977, A comparative analysis of the protein components of plasma membranes isolated from differentiated and undifferentiated mouse neuroblastoma cells in tissue culture, *Eur. J. Biochem.* **76:**251.

Gielen, W., and Hinze, D. H., 1974, Acetylneuraminat-Cytidylyltransferase und Sialyltransferase in isolierten neuronal- und Gliazellen des Rattengehirns, *Z. Physiol. Chem.* **355:**895.

Gilman, A. G., and Nirenberg, M. W., 1971, Effect of catecholamines on the adenosine 3′:5′-cyclic monophosphate concentrations of clonal satellite cells of neurons, *Proc. Natl. Acad. Sci. U.S.A.* **68:**2165.

Glick, M. C., Kimhi, Y., and Littauer, V. Z., 1973, Glycopeptides from surface membrane of neuroblastoma cells, *Proc. Natl. Acad. Sci. U.S.A.* **70:**1682.

Greene, L. A., and Tischler, A. S., 1976, Establishment of a noradrenergic clonal line of rat adrenal pheochromacytoma cells which respond to nerve growth factor, *Proc. Natl. Acad. Sci. U.S.A.* **73:**2424.

Grimm-Jorgensen, Y., Pfeiffer, S. E., and McKelvy, J. F., 1976, Metabolism of thyrotropin releasing factor in two clonal cell lines of nervous system origin, *Biochem. Biophys. Res. Commun.* **70:**167.

Haffke, S. C., and Seeds, N. W., 1975, Neuroblastoma: The *E. coli* of neurobiology?, *Life Sci.* **16:**1649.

Hamberger, A., and Svennerholm, L., 1971, Composition of gangliosides and phospholipids of neuronal and glial cell enriched fractions, *J. Neurochem.* **18:**1821.

Hausmann, R. E., and Moscona, A. A., 1975, Purification and characterization of the retina-specific cell-aggregating factor, *Proc. Natl. Acad. Sci. U.S.A.* **72:**916.

Hudson, J. E., and Johnson, T. C., 1977, Rapidly metabolized glycoproteins in a neuroblastoma cell line, *Biochim. Biophys. Acta* **497:**567.

Karlsson, K. A., 1977, Aspects on structure and function of sphingolipids in cell surface membranes, in: *Structure of Biological Membranes* (S. Abrahamsson and I. Pascher, eds.), pp. 245–274. Plenum Press, New York.

Kemp, S. F., and Stoolmiller, A. C., 1976, Biosynthesis of glycosphingolipids in cultured mouse neuroblastoma cells, *J. Biol. Chem.* **251:**7626.

Klebe, R. J., and Ruddle, F. H., 1969, Neuroblastoma: Cell culture analysis of a differentiating stem cell system, *J. Cell Biol.* **43:**69a.

Kõhlschutter, A, and Herschkowitz, N. N., 1973, Sulfatide synthesis in neurons: A defect in mice with a hereditary myelination disorder, *Brain Res.* **50:**379.

Kraemer, P. M., 1971, Heparan sulfates of cultured cells. I. Membrane associated and cell-sap species in Chinese hamster cells, *Biochemistry* **10:**1437.

Latovitzki, N., and Silberberg, D. H., 1977, UDP-galactose: ceramide galactosyltransferase and 2′,3′-cyclic nucleotide 3′-phosphohydrolase activities in cultured newborn rat cerebellum: Association with myelination and concurrent susceptibility to 5-bromodeoxyuridine, *J. Neurochem.* **29:**611.

Ledeen, R. W., Skrivanek, J. A., Tirri, L. J., Margolis, R. K., and Margolis, R. U., 1976, Gangliosides of the neuron: Localization and origin, in: *Ganglioside Function: Biochemical and Pharmacological Implications* (G. Porcellati, B. Ceccarelli, and G. Tettamanti, eds.), pp. 83–103, Plenum Press, New York.

Levitt, D., and Dorfman, A., 1972, The irreversible inhibition of differentiation of limb bud mesenchyme by bromodeoxyuridine, *Proc. Natl. Acad. Sci. U.S.A.* **69:**1253.

Lightbody, J., Pfeiffer, S. E., Kornblith, P. L., and Herschman, H., 1970, Biochemically differentiated clonal human glial cells in tissue culture, *J. Neurobiol.* **1:**411.

Lim, R., and Dawson, G., 1978, Regulation of complex carbohydrate metabolism by a glial maturation factor (unpublished).

Lim, R., and Mitsunobu, K., 1974, Brain cells in culture: Morphological transformation by a protein, *Science* **185:**63.

Margolis, R. K., Margolis, R. U., and Shelanski, M. L., 1972, The carbohydrate composition of brain microtubule protein, *Biochem. Biophys. Res. Commun.* **47:**432.

Margolis, R. U., and Margolis, R. K., 1974, Distribution and metabolism of mucopolysaccharides and glycoproteins in neuronal perikarya, astrocytes and oligodendroglia, *Biochemistry* **13:**2849.

Matalon, R., and Dorfman, A., 1976, The mucopolysaccharidoses (a review), *Proc. Natl. Acad. Sci. U.S.A.* **73**(2):630.

Mathews, R. A., Johnson, T. C., and Hudson, J. E., 1976, Synthesis and turnover of plasma-membrane proteins and glycoproteins in a neuroblastoma cell line, *Biochem. J.* **154:**57.

Michaelis, E. K., 1975, Glutamate-binding glycoprotein in rat brain, *Biochem. Biophys. Res. Commun.* **65:**1004.

Miller, R. J., Dawson, G., Kernes, S. M., and Wainer, B., 1978, Enkephalin receptors in cultured cells: Role of membrane glycolipids, *Proc. 7th Int. Congr. Pharmacol.,* Paris.

Minna, J., 1973, Genetic analysis of the mammalian nervous system using somatic cell culture techniques, in: *Tissue Culture of the Nervous System* (G. Sato, ed.), pp. 161–185, Plenum Press, New York.

Monard, D., Solomon, F., Rentsch, M., and Gysin, R., 1973, Glia-induced morphological differentiation in neuroblastoma cells, *Proc. Natl. Acad. Sci. U.S.A.* **70:**1894.

Morris, J. E., and Dorfman, A., 1976, Inhibition by 5-bromo-2′-deoxyuridine of differentiation-dependent changes in glycosaminoglycans of the retina, *Biochem. Biophys. Res. Commun.* **69:**1065.

Moskal, J. R., Gardner, D. A., and Basu, K., 1974, Changes in glycolipid glycosyltransferases and glutamate decarboxylase and their relationship to differentiation in neuroblastoma cells, *Biochem. Biophys. Res. Commun.* **61:**751.

Mullin, B. R., Pacuszka, T., Lee, G., Kohn, L. D., Brady, R. O., and Fishman, P. H., 1977, Thyroid gangliosides with high affinity for thyrotropin: Potential role in thyroid regulation, *Science* **199:**77.

Nelson, P. G., 1973, Electrophysiological studies of normal and neoplastic cells in tissue culture, in: *Tissue Culture of the Nervous System* (G. Sato, ed.), pp. 135–158, Plenum Press, New York.

Nishimura, K., and Shimoda, R., 1975, Effects of hydrocortisone on mouse thymus β-galactosidase and glycolipid composition, *Jpn. J. Exp. Med.* **45**(3):241.

Noble, E. F., Syapin, D. J., Vigran, R., and Rosenberg, A., 1976, Neuraminidase-releasable surface sialic acid of cultured astroblasts exposed to ethanol, *J. Neurochem.* **27:**217.

Norton, W. T., and Poduslo, S. E., 1971, Chemical composition of neuronal perikarya and astroglia, *J. Lipid Res.* **12:**84.

O'Brien, J. S., and Sampson, E. L., 1965, Lipid composition of the normal human brain: Gray matter, white matter, and myelin, *J. Lipid Res.* **6:**537.

Olmsted, J. B., Carlson, K., Klebe, R., Ruddle, F., and Rosenbaum, J., 1970, Isolation of microtubule protein from cultured mouse neuroblastoma cells, *Proc. Natl. Acad. Sci. U.S.A.* **65:**129.

Pfeiffer, S. E., 1973, Clonal lines of glial cells, in: *Tissue Culture of the Nervous System* (G. Sato, ed.), pp. 203–230, Plenum Press, New York.

Pfeiffer, S. E., and Wechsler, W., 1972, Biochemically differentiated neoplastic clone of Schwann cells, *Proc. Natl. Acad. Sci. U.S.A.* **69:**2885.

Pfeiffer, S. E., Kornblith, P. L., Cares, H. L., Seals, J., and Levine, L., 1972, S-100 protein in human acoustic neurinomas, *Brain Res.* **41:**187.

Pfenninger, K. H., and Maylié-Pfenninger, M. F., 1975, Distribution and fate of lectin binding sites on the surface of growing neuronal processes, *J. Cell Biol.* **67:**332a.

Pfenninger, K. H., and Rees, R. P., 1976, From the growth cone to the synapse: Properties of membranes involved in synapse formation, in: *Neuronal Recognition* (S. M. Barondes, ed.) pp. 131–178, Plenum Press, New York.

Pleasure, D., Abramsky, D., Silberberg, D., Quinn, B., Parris, J., and Saida, T., 1977, Lipid synthesis by an oligodendroglial fraction in suspension culture, *Brain Res.* **134**:377.

Poduslo, S. E., 1975, The isolation and characterization of a plasma membrane and a myelin fraction derived from oligodendroglia of calf brain, *J. Neurochem.* **24**:647.

Poduslo, S. E., and McKhann, G. M., 1977, Synthesis of cerebrosides by intact oligodendroglia maintained in culture, *Neurosci. Lett.* **5**:159.

Poduslo, S. E., and Norton, W. T., 1975, Methods for the isolation and characterization of neurons and glia, in: *Methods in Enzymology* (S. P. Colowick and N. O. Kaplan, eds.), Vol. 35, Part B, pp. 561–579, Academic Press, New York.

Radin, N. S., Brenkert, A., Arora, R. C., Sellinger, O. Z., and Flangas, A. L., 1972, Glial and neuronal localization of cerebroside-metabolizing enzymes, *Brain Res.* **39**:163.

Raff, M., 1976, Self-regulation of membrane receptors, *Nature* (*London*) *New Biol.* **259**:265.

Rakic, P., 1974, Mode of cell migration to the superficial layers of fetal monkey neocortex, *J. Comp. Neurol.* **145**:61.

Robert, J., Freysz, L., Sensenbrenner, M., Mandel, P., and Rebel, G., 1975, Gangliosides of glial cells: Comparison of normal astroblasts with isolated neurons and glia, *FEBS Lett.* **50**:144.

Roseman, S., 1970, The synthesis of complex carbohydrates by multiglycosyltransferase systems and their potential function in intercellular adhesion, *Chem. Phys. Lipids* **5**:270.

Rosenberg, R. N., 1973, Regulation of neuronal enzymes in cell culture, in: *Tissue Culture of the Nervous System* (G. Sato, ed.), pp. 107–132, Plenum Press, New York.

Rushton, A. R., and Dawson, G., 1975, Glycosphingolipid β-galactosidases of cultured mammalian cells, *Biochim. Biophys. Acta* **388**:92.

Santala, R., and Glaser, L., 1977, The effect of cell density on the expression of cell adhesive properties in a cloned rat astrocytoma (C-6), *Biochem. Biophys. Res. Commun.* **79**:285.

Santala, R., Gottlieb, D. I., Littman, D., and Glaser, L., 1977, Selective cell adhesion of neuronal cell lines, *J. Biol. Chem.* **252**:7625.

Sarlieve, L. L., and Mandel, P., 1975, Presence d'activité sulfotransferasique dans un clone de cellules neuronales en culture, *C. R. Acad. Sci.* 169:5.

Sarlieve, L. L., Neskovic, N. M., Preysz, L., Mandel, P., and Rebel, G., 1976, Ceramide galactosyltransferase and cerebroside sulphotransferase in chicken brain cellular fractions and glial and neuronal cells in cultures, *Life Sci.* **18**:251.

Schachner, M., 1974, NS-1 (nervous system antigen-1), a glial cell–specific antigenic compound of the surface membrane, *Proc. Natl. Acad. Sci. U.S.A.* **71**:1795.

Schengrund, C.-L., and Repman, M. A., 1977, Cell culture of sixteen-day old rat embryo cerebra and associated changes in ganglioside pattern, *J. Neurochem.* **29**:923.

Schrier, B. K., and Thompson, E. J., 1974, On the role of glial cells in the mammalian nervous system, *J. Biol. Chem.* **249**:1769.

Schubert, D., and Jacob, F., 1970, 5-Bromodeoxyuridine induced differentiation of a mouse neuroblastoma, *Proc. Natl. Acad. Sci. U.S.A.* **67**:247.

Schubert, D., Humphreys, S., Baroni, C., and Cohn, M., 1969, *In vitro* differentiation of a mouse neuroblastoma, *Proc. Natl. Acad. Sci. U.S.A.* **64**:316.

Schubert, D., Harris, A. J., Heinemann, S., Kidokoro, Y., Patrick, J., and Steinbach, J. H., 1973, Differentiation and interaction of clonal cell lines of nerve and muscle, in: *Tissue Culture of the Nervous System* (G. Sato, ed.), pp. 55–84, Plenum Press, New York.

Schubert, D., Heinemann, S., Carlisle, W., Tarikas, H., Kimes, B., Patrick, J., Steinback, J. H., Culp, W., and Brandt, B. L., 1974, Clonal cell lines from the rat central nervous system, *Nature* (*London*) **249**:224.

Schubert, D., Tarikas, H., and La Corbiere, M., 1976, Neurotransmitter regulation of adenosine 3′,5′-monophosphate in clonal nerve, glia and muscle cell lines, *Science* **192**:471.

Seeds, N. W., Gilman, A. G., Amano, T., and Nirenberg, M., 1970, Regulation of axon formation by clonal lines of a neural tumor, *Proc. Natl. Acad. Sci. U.S.A.* **66**:160.

Shein, H. M., Britva, A., Hess, H. H., and Selkoe, D. J., 1970, Isolation of hamster brain astroglia by *in vitro* cultivation and subcutaneous growth and content of cerebroside ganglioside, RNA and DNA, *Brain Res.* **19**:497.

Silberberg, D. H., Benjamins, J., Herschkowitz, N., and McKhann, G. M., 1972, Incorporation of radioactive sulfate into sulfatide during myelination in cultures of rat cerebellum, *J. Neurochem.* **19**:11.

Silberstein, S. D., Shein, H. M., and Berv, K. R., 1972, Catechol-*O*-methyl transferase and monoamine oxidase activity in cultured rodent astrocytoma cells, *Brain Res.* **41**:245.

Siman-Tov, R., and Sachs, L., 1972, Enzyme regulation in neuroblastoma cells: Selection of clones with low acetylcholinesterase activity and the independent control of acetyl cholinesterase and choline-*O*-acetyltransferase, *Eur. J. Biochem.* **30**:123.

Siman-Tov, R., and Sachs, L., 1975, Induction mechanisms for cholinesterase in mouse neuroblastoma cells, *Dev. Biol.* **45**:382.

Simpson, D. L., Thorne, D. R., and Loh, H. H., 1976, Sulfated glycoproteins, glycolipids, and glycosaminoglycans from synaptic plasma and myelin membranes: Isolation and characterization of sulfated glycopeptides, *Biochemistry* **15**:5449.

Stallcup, W. B., 1977, Specificity of adhesion between cloned neural cell lines, *Brain Res.* **126**: 475.

Steck, T. L., and Dawson, G., 1974, Topographical distribution of complex carbohydrates in the erythrocyte membrane, *J. Biol. Chem.* **249**:2135.

Stefanovic, A., Mandel, P., and Rosenberg, A., 1975, Activation of acetyl- and butylcholinesterase by enzymatic removal of sialic acid from intact neuroblastoma and astroblastoma cells in culture, *Biochemistry* **14**:5257.

Stern, J., 1972, The induction of ganglioside storage in nervous system cultures, *Lab. Invest.* **26**:509.

Stoolmiller, A. C., Dawson, G., and Dorfman, A., 1973, The metabolism of glycosphingolipids and glycosaminoglycans, in: *Tissue Culture of the Nervous System* (G. Sato, ed.), pp. 247–280, Plenum Press, New York.

Stoolmiller, A. C., Dawson, G., and Kemp, S. F., 1974, Regulation of ganglioside synthesis in mouse neuroblastoma cells (NB41A and NB2A) in tissue culture, *Fed. Proc.* **33**:1226.

Stoolmiller, A. C., Dawson, G., Kemp, S. F., and Schachner, M., 1979, Synthesis of glycosphingolipids in mouse glial tumors, *J. Neurochem.* **32**:637.

Sundarraj, N., Šchachner, M., and Pfeiffer, S. E., 1975, Biochemically differentiated mouse glial lines carrying a nervous system specific cell surface antigen (NS-1), *Proc. Natl. Acad. Sci. U.S.A.* **72**:1927.

Suzuki, H., Ohsawa, M., and Endo, H., 1969, Effect of hydrocortisone on HeLa cells with special referance to the RNA metabolism, *Endocrinol. Jpn.* **16**:121.

Svennerholm, L., 1964, The gangliosides, *J. Lipid Res.* **5**:145.

Szuchet, S., Arnasorn, B. G. W., and Polak, P. E., 1978, A new method for oligodendrocyte isolation, *Biophys. J.* **21**:51a.

Szuchet, S., Polak, P. E., and Dawson, G., unpublished (1978), Isolation and biochemical characterization of oligodendroglia isolated from ovine brain.

Tamai, Y., Araki, S., Komai, Y., and Satake, M., 1975, Studies on neuronal lipid, *J. Neurosci. Res.* **1**:161.

Truding, R., Shelanski, M. L., Daniels, M. P., and Morell, P., 1974, Comparison of surface membranes isolated from cultured murine neuroblastoma cells in the differentiated or undifferentiated state, *J. Biol. Chem.* **249**:3973.

Truding, R., Shelanski, M. L., and Morell, O., 1975, Glycoproteins released into the culture medium of differentiating murine neuroblastoma cells, *J. Biol. Chem.* **250**:9348.

Umbreit, J., and Roseman, S., 1975, A requirement for reversible binding between aggregating embryonic cells before stable adhesion, *J. Biol. Chem.* **250**:9360.

Van Heyningen, W. E., 1974, Gangliosides as membrane receptors for tetanus toxin, cholera toxin and serotonin, *Nature* (*London*) **249**:415.

Wechsler, W., Kleihues, P., Matsumoto, S., Zulch, K. J., Ivankovic, S., Preussman, R., and Druckney, H., 1969, Pathology of neurogenic tumors, *Ann. N. Y. Acad. Sci.* **159**:360.

Wood, J. G., McLaughlin, B. J., and Barber, R. P., 1974, The visualization of concanavalin A binding sites in Purkinje cell somata and dendrites of rat cerebellum, *J. Cell Biol.* **63**:541.

Yogeeswaran, G., Murray, R. K., Pearson, M. L., Sanwal, B. D., McMorris, F. A., and Ruddle, F. H., 1973, Glycosphingolipids of clonal lines of mouse neuroblastoma and neuroblastoma × L cell hybrids, *J. Biol. Chem.* **248**:1231.

Zimmerman, H. M., 1955, The nature of gliomas as revealed by animal experimentation, *Am. J. Pathol.* **31**:1.

15

Complex Carbohydrates of Secretory Organelles

Giuliana Giannattasio, Antonia Zanini, and Jacopo Meldolesi

1. Introduction

Over the last few years, considerable attention has been devoted to the role played by complex carbohydrates in the structure and function of secretory granules and vesicles. These intracellular storage organelles, which are delimited by a single membrane, are endowed with the ability to discharge their content into the extracellular space by exocytosis. Since each secretory system is characterized by the specificity of its secretion products, the various granules and vesicles constitute an extremely heterogenous family of organelles. In some of them, complex carbohydrates represent secretion products of known physiological significance. This is the case, for instance, with the gonadotropin and thyrotropin granules of the anterior pituitary, the B granules of follicular thyroid cells, and the immunoglobulin vesicles of plasmocytes. Other secretory granules (e.g., those of the exocrine pancreas, parotid gland, and liver) contain a mixture of many physiologically important secretion products, some of which are complex carbohydrates. Finally, in a variety of other systems, the occurrence of complex carbohydrates (often in small amounts) in the segregated content of secretory organelles has also been reported, even if no clear information is yet available on whether these components have any function after their discharge. However, complex carbohydrates do not reside only in the content of secretory organelles, since in all systems so far investigated they have also been found in the limiting membrane, where at least some of them are oriented according to a specific geometry.

Giuliana Giannattasio, Antonia Zanini, and Jacopo Meldolesi • Department of Pharmacology and CNR Center of Cytopharmacology, University of Milan, Milan, Italy.

A discussion of granule and vesicle complex carbohydrates that have known functions outside the producing cells (e.g., as hormones, enzymes, or antibodies) is beyond the scope of this chapter. Rather, we will concentrate on the components of the segregated matrix as well as on the membrane components, in view of the possible general significance of these molecules in some of the fundamental processes that take place in secretory cells (including neurons), such as the intracellular storage of secretion products, the biogenesis of the membranes participating in the secretory process, and the specific recognition and fusion–fission of membranes in exocytosis.

2. Complex Carbohydrates of Secretory Granules and Vesicles

2.1. Membranes

Although it is likely that glycolipids are widely distributed in membranes of secretory granules and vesicles, so far they have been studied in only a few secretory systems. In chromaffin granule membranes, isolated from ox adrenal glands, Geissler *et al.* (1977) recently showed that the concentration of gangliosides is approximately 3-fold higher than in the corresponding microsomal membranes. Two hematosides, containing *N*-glycolyl- and *N*-acetylneuraminic acid, respectively, and several unidentified minor components were found in the granule membranes. In contrast, gangliosides were not detected in either the synaptic vesicles of rat brain (Morgan *et al.*, 1973) or in zymogen granules of the exocrine pancreas (Meldolesi *et al.*, 1971). The latter organelles contain small amounts of cerebrosides [monohexo- and dihexoceramide (Meldolesi *et al.*, 1971)].

With respect to the membrane glycosaminoglycans (GAGs), the available information is limited. Geissler *et al.* (1977) found that approximately 80% of the GAGs present in intact chromaffin granules is located in their matrix (see Table I and Section 2.2), while the rest is in the membrane. In both these pools, chondroitin sulfate is the major component; heparan sulfate is present in smaller amount and is more concentrated in the membranes. A significant concentration of GAGs was also found in the limiting membrane of the zymogen granules of the exocrine pancreas, where the major component is heparan sulfate (Kronquist *et al.*, 1977; Reggio and Palade, 1978).

As far as the glycoprotein composition of the granule membranes is concerned, a large degree of variability has been found in the various secretory systems. As discussed in detail elsewhere (Meldolesi *et al.*, 1978), two groups can be identified. The first includes the granules and vesicles whose membrane protein complement is entirely different from that of the other cellular membranes. Thus, in chromaffin granule membranes, the major glycoprotein is the enzyme dopamine β-hydroxylase (DBH) (Hört-

Table I. Distribution of Complex Carbohydrates in Bovine Chromaffin Granules[a]

Complex carbohydrates	Soluble	Membrane	Total[b]
Glycosaminoglycans (hexosamine)	80%	20%	3.4
Chondroitin sulfate/heparan sulfate ratio	20:1	8:1	16:1
Glycoproteins			
Hexosamine	67%	33%	5.8
Glucosamine/galactosamine ratio	1:2	9:1	1:1
Sialic acid	82%	18%	5.4
Neutral sugars	61%	39%	9.3
Gangliosides (sialic acid)	4%	96%	1.6
Protein	77%	23%	—

[a]Data from Geissler *et al.* (1977).
[b]Whole granule concentration expressed as micromoles per 100 mg protein.

nagl *et al.*, 1972; Hörtnagl, 1976), which is also present in the content, but is not found elsewhere (see also Sections 2.2 and 3) (for a recent review, see Winkler, 1976). The same situation has been described in the large noradrenaline-storing vesicles of sympathetic terminals (for a recent review, see Lagercrantz, 1976). Specific glycoproteins are also present in the membranes of secretory granules of the exocrine pancreas and parotid gland. In the first organelle, one major component (called GP_2, mol.wt. 74,000) and a few minor components have been described (Ronzio *et al.*, 1978), while in the latter organelle the composition is more elaborate (Castle *et al.*, 1975; Wallach *et al.*, 1975).

The membranes of the secretory organelles of the second group are characterized by a less-pronounced molecular specificity. In particular, only some of their proteins and glycoproteins are unique, while the others are shared with different cellular membranes. This group includes the specific and azurophylic granules of polymorphonuclear leukocytes (Baggiolini *et al.*, 1977) as well as the synaptic vesicles of rat brain (Morgan *et al.*, 1973) and the neurosecretory granules of the neurohypophysis (Vilhardt *et al.*, 1975). Most membrane glycoproteins of the latter two organelles are also found in the synaptic plasmalemma of their homologous systems. Recently, the existence of sulfated glycoproteins in the plasmalemma of brain synaptosomes was reported (Simpson *et al.*, 1976), but synaptic vesicles have not yet been analyzed.

Studies carried out in other secretory systems, such as the liver and plasmocytes, failed to reveal the existence of specific secretory organelles. At least in the liver, the available evidence suggests that exocytosis is carried out by organelles whose membrancs are indistinguishable from those of the Golgi complex. These secretory granules, identified morphologically by their content of lipoprotein particles, contain several membrane glycoproteins, some of which are enzymes or receptors, such as 5′-nucleotidase and the insulin receptor (Bergeron *et al.*, 1973; Farquhar *et al.*, 1974; Little and Widnell, 1975).

As will be discussed in Section 3, it is generally agreed that the orientation of molecular components (in particular that of complex carbohydrates) in the transverse plane of membranes is of critical importance in a number of cellular functions. In fact, it is now known that all biological membranes are asymmetrical because although most of their components can move freely in the lateral plane of the membrane, major restraints exist to their reorientation across the lipid bilayer. No information on the transverse distribution of glycolipids in the membranes of secretory granules has been reported. However, more is known about GAGs and, especially, glycoproteins. In general, it can be said that in all systems so far investigated, the bulk of these components has been found to be located in the internal leaflet, with the sugar residues freely exposed at the intraluminal surface. Such an orientation was clearly demonstrated for 5′-nucleotidase (Farquhar *et al.*, 1974; Little and Widnell, 1975) and the insulin receptor (Bergeron *et al.*, 1973) in the lipoprotein-rich granules of the liver, for DBH in chromaffin granules (König *et al.*, 1976), and for GP_2 in pancreatic zymogen granules (Ronzio *et al.*, 1978). However, recent evidence obtained in the latter two systems (Meyer and Burger, 1976; Ronzio *et al.*, 1978), as well as in the specific and azurophylic granules of polymorphonuclear leukocytes (Baggiolini *et al.*, 1977), suggests that glycoproteins can also be exposed at the cytoplasmic surface of granule membranes. This localization has been questioned on the basis of results showing that (1) the transverse distribution of sugars in granule membranes is extremely asymmetrical (Eagles *et al.*, 1975) and (2) the contribution of sialic acid residues to the net surface charge of intact chromaffin and pancreatic β granules is very little (Matthews *et al.*, 1972; Eagles *et al.*, 1976; Howell and Tyhurst, 1977). However, it should be acknowledged that these results are not incompatible with the cytoplasmic orientation of a few carbohydrate chains. In this respect, it is interesting that in pancreatic zymogen granules, the glycoproteins exposed at the external surface were reported to be not only much less in amount but also molecularly different from those facing the content of the organelles (Ronzio *et al.*, 1978).

2.2. Internal Matrix

Complex carbohydrates have been detected in the internal matrix of a variety of secretory granules and vesicles. Knowledge in this field is still incomplete, however, because most of the studies have been conducted exclusively by cytochemical and radioautographic techniques. The identification and quantitative estimation of the various complex carbohydrates were achieved in only a few cases. There are at least two reasons for this limitation. On one hand, quantity is often a problem, because in many secretory cells the number of granules is not large and the concentration of complex carbohydrates in the granule matrix is very low; on the other hand, satisfactory techniques for the isolation of secretory granules and vesicles and for the separation of the limiting membranes from the contents are available for only a limited number of cell systems.

Among the secretory organelles investigated so far, the chromaffin granules of the adrenal medulla are those that contain the highest levels of complex carbohydrates (for a recent review, see Winkler, 1976). In fact, as demonstrated originally by A. D. Smith and Winkler (1967) and Wallace *et al.* (1973), the two major proteins of the matrix, chromogranin A and DBH [the latter is also found in the granule membrane (see Sections 2.1 and 3)] are both glycoproteins; moreover, chromaffin granules also contain GAGs (Fillion *et al.*, 1971; Margolis, R. U., and Margolis, R. K., 1973; Pletscher *et al.*, 1974; Geissler *et al.*, 1977). Recently, a complete analysis of the complex carbohydrates of chromaffin granules was reported by Geissler *et al.* (1977). As can be seen in Table II, chromogranin A was found to contain mainly galactose, *N*-acetylgalactosamine (GalNAc), and sialic acid; since its oligosaccharide chains bear bound sulfate residues, chromogranin A is a sulfated glycoprotein (Margolis, R. U., and Margolis, R. K., 1973). DBH, on the other hand, contains primarily mannose, galactose, GalNAc, and fucose. As already mentioned in Section 2.1, the main GAG of the soluble content of chromaffin granules is chondroitin sulfate; heparan sulfate is present in small amount, while the other GAGs are absent.

Due to difficulties in their isolation, large adrenergic vesicles have been investigated less extensively. In general, their composition appears analogous to that of chromaffin granules. However, chromogranin A is present only in trace amounts; moreover, a large proportion of DBH is membrane-bound, while the rest of the enzyme (8–37%) is the dominant protein of the vesicle matrix (Lagercrantz, 1976; Nelson *et al.*, 1978). The data on GAGs are quite conflicting. The presence of heparan sulfate, dermatan sulfate, a

Table II. Carbohydrate Composition of Soluble Proteins from Chromaffin Granules[a]

Carbohydrate	Total soluble proteins	Soluble DBH	Chromogranin A
Fucose	0.4	1.32	0.2
Mannose	1.6	9.55	0.7
Galactose	6.6	3.35	4.9
GlcNAc[b]	1.4	5.78	Trace
GalNAc[b]	7.8	1.79	6.1
Sialic acid	6.1	1.88	6.2

[a]Data from Geissler *et al.* (1977), expressed as micromoles sugar per 100 mg protein.
[b](GlcNAc) *N*-acetylglucosamine; (GalNAc) *N*-acetylgalactosamine.

dermatan sulfate–chondroitin sulfate hybrid, chondroitin sulfate, and hyaluronic acid was reported in synaptic vesicles isolated from brain (Åborg *et al.*, 1972; Pycock *et al.*, 1975) and from splenic nerve terminals (Åborg *et al.*, 1972; Blaschke *et al.*, 1976). Even vesiculin, a molecule originally described as a low-molecular-weight protein localized within the content of synaptic vesicles from the electric organ of *Torpedo marmorata*, was recently recognized to be a GAG (Stadler and Whittaker, 1978). However, it was shown that purified synaptosomes and synaptosomal subfractions contain very little sulfated GAG (Margolis R. K., *et al.*, 1975*a*; Simpson *et al.*, 1976), although hyaluronic acid may be present in synaptic vesicles (Vos *et al.*, 1969; Margolis, R. K., *et al.*, 1975*a*). Thus, the earlier results may be largely due to contamination by other subcellular fractions rich in sulfated GAGs and glycoproteins, such as the microsomes (Margolis, R. U., and Margolis, R. K., 1977; Kiang *et al.*, 1978) (see also Chapter 3). The concentration of GAGs is also very low in the amine-storing granules of blood platelets (Pletscher *et al.*, 1974; Anderson *et al.*, 1974), whereas their concentration is much higher in the α granules that do not contain amines (Anderson *et al.*, 1974).

A different situation exists in another type of amine-storing granules, the basophil granules of mast cells. In these organelles, heparin accounts for approximately 30% of the content (Uvnäs, 1974; Lagunoff and Pritzl, 1976). Other components are basic proteins (33–65%), histamine, and in some species, serotonin. By histochemical staining methods, sulfated GAGs, together with basic proteins, were also detected in the granules of human polymorphonuclear leukocytes of the three series (Dunn and Spicer, 1969). The major component was identified as chondroitin sulfate (Olsson, 1969; Jaques, 1975). Moreover, the granules of guinea pig neutrophils (especially the secondary granules) also contain glycoproteins (Noseworthy *et al.*, 1975).

Complex carbohydrates have been found in the content of other protein-storing organelles. The occurrence of ^{35}S-labeled macromolecules (tentatively identified as GAGs) was first demonstrated by electron-microscopic radioautography in granules of the exocrine pancreas and parotid gland (Berg and Young, 1971; Berg and Austin, 1976). Recently, these studies were extended using biochemical techniques. In the pancreas, it was demonstrated that the intragranular concentration of sulfated macromolecules is low ($\approx$0.5 nmol/mg protein) (Kronquist *et al.*, 1977) and that the major component is heparan sulfate (Kronquist *et al.*, 1977; Reggio and Palade, 1978). Moreover, the distribution of these molecules in the content of zymogen granules is probably uneven, with preferential concentration in the outer layer (Reggio and Palade, 1976). Evidence suggesting the discharge of the sulfated macromolecules from pancreatic acinar cells along with the other secretion products was also reported (Tartakoff *et al.*, 1974; Berg, 1978).

Ample experimental evidence supports the existence of glycoproteins in granules of the hypothalamo-neurohypophyseal system (Tasso, 1973; Hvas and Thorn, 1973; Jones and Swann, 1975). In particular, recent experiments by Haddad *et al.*, (1977) confirmed that these glycoproteins, while located within the granules, migrate along the axon from their site of synthesis (in the paraventricular and supraoptic nuclei of the hypothalamus) to the neurohypophysis, where they are released concomitantly with the peptide hormones and neurophysins. Moreover, Tasso *et al.* (1977) suggested that glycoproteins are more concentrated in vasopressin- than in oxytocin-containing granules. Large glycoproteins are also present in the secretory granules of the intermediate pituitary cells (Hopkins, 1975).

In studies by Pelletier (1971, 1974), complex carbohydrates were shown by cytochemistry and radioautography to be present in corticotropin and growth hormone granules of the anterior pituitary. More recently, detailed investigations were carried out by our group on prolactin granules in the rat (Giannattasio and Zanini, 1976) and in the cow (Zanini *et al.*, 1979*b*), taking advantage of the fact that these organelles can be isolated as a pure fraction (Zanini and Giannattasio, 1973) and are very stable and therefore retain their organization even after solubilization of the limiting membrane by treatment with mild detergents (Giannattasio *et al.*, 1975). Thus, in this system, a satisfactory separation of the granule content from the limiting membranes can be achieved (Fig. 1). In both animal species, the prolactin granules were found to contain GAGs (accounted for by heparan sulfate, as a major component, and chondroitin sulfate; in the cow, hyaluronic acid was also detected), as well as sulfated glycoproteins. The biochemical analysis of the complex carbohydrates of cow prolactin granules is reported in Table III. Two points should be emphasized: (1) the much higher concentration of glycoproteins with respect to GAGs and (2) the clear difference in GAG composition of prolactin as compared with chromaffin granules (see Table I). The latter observation suggests a specificity of the intragranular GAG mixture in different secretory cell systems.

The presence of glycoproteins in a rat pituitary large granule fraction (prolactin + growth hormone granules) was also reported by Slaby and Farquhar (1976).

The secretory systems in which intragranular complex carbohydrates have been detected also include the islets of Langerhans, adrenal cortex, interstitial cells of the testis, ovarian follicles, corpus luteum, and parathyroid gland (Nakagami, *et al.*, 1971; Lalli and Clermont, 1975; Haddad *et al.*, 1977; Morrissey *et al.*, 1978). In contrast, no conclusive negative results have ever been reported. It therefore seems reasonable to conclude that the presence of complex carbohydrates might be a general feature of the segregated content of all secretory granules. The possible functional significance of these molecules will be considered in Section 3.2.

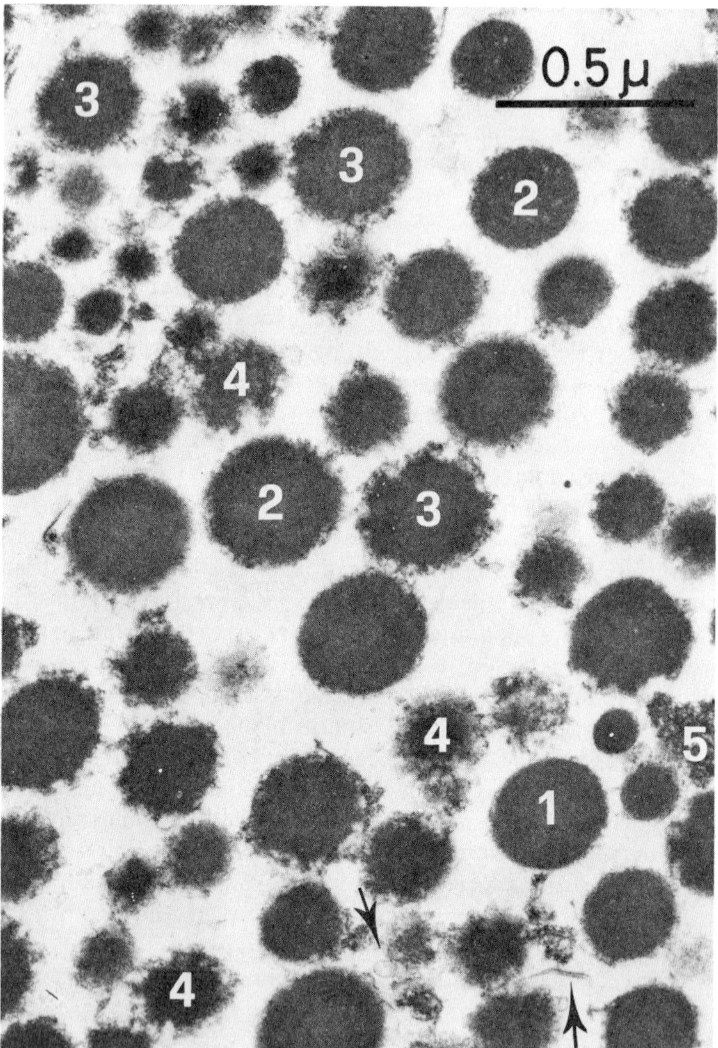

Fig. 1. Membraneless prolactin granules isolated from the pituitary of the cow. A fraction of intact prolactin granules, prepared according to Zanini and Giannattasio (1973) and containing 0.5 mg protein/ml, was treated with 0.2% lubrol PX and then loaded onto a 1.2 M sucrose cushion. Centrifugation at 30,000*g* max for 60 min yielded a pellet that contained most of the prolactin originally present in the granule fraction. This pellet contains recognizable prolactin granules devoid of their limiting membranes. Some of these membraneless granules are very well preserved (1), others have a well-preserved core surrounded by an irregular (2) or moth-eaten (3) outline, and some appear partially (4) or completely (5) transformed into masses of twisted filaments. Small remnants of the limiting membranes are indicated by arrows.

Table III. Complex Carbohydrates in the Matrix of Bovine Prolactin Granules[a]

Complex carbohydrate composition	Monosaccharide composition of matrix glycoproteins: Carbohydrate	μmol/100 mg protein
Glycosaminoglycans	Fucose	0.072
—as μmol hexosamine/100 mg protein:....................0.086		
Hyaluronic 13%	Mannose	0.373
Heparan sulfate 52%		
Chondroitin sulfate 35%	Galactose	0.449
Glycoproteins		
—as μmol of hexosamine/100 mg protein:....................0.631	GlcNAc[b]	0.522
	GalNAc[b]	0.109
	Sialic acid	0.314

[a]Data fron Zanini *et al.* (1979*b*).
[b](GlcNAc) *N*-acetylglucosamine; (GalNAc) *N*-acetylgalactosamine.

3. Role of Complex Carbohydrates in the Assembly and Function of Secretory Organelles

There is now ample evidence, obtained in a wide variety of cell systems, that implicates the Golgi complex in both the biosynthesis of complex carbohydrates and the assembly and maturation of secretory granules and vesicles (for recent reviews, see Palade, 1975; Holtzman, 1977; Winkler, 1977; Morré, 1977; Meldolesi *et al.*, 1978). Recent results suggest that these two phenomena are not independent. Rather, the possibility exists that the biosynthesis of complex carbohydrates is among the processes by which secretory proteins, synthesized by membrane-bound polysomes and immediately segregated within the endoplasmic reticulum cisternae as a mixture of soluble molecules not yet available for discharge, are assembled in discrete membrane-bounded organelles capable of discharging their content by exocytosis. In the remainder of this review, the possible role of complex carbohydrates in the biogenesis and function of secretory organelles will be considered.

In view of our present knowledge of the secretory process in eukaryotic cells, the fact that complex carbohydrates of secretory organelles are located primarily within the content as well as at the inner surface of the limiting membrane can be readily explained. In fact, it is now known that most of the enzymes responsible for protein glycosylation are located at the intraluminal surface of endomembranes. Thus, the only available substrates are those segregated within, or exposed at the inner surface of, the cavities they delimit. The glycosylation of many glycoproteins begins already in the endoplasmic reticulum, where the core sugars are attached by dolichol-phosphate-dependent mechanism onto growing chains and to newly synthesized peptides. Eventually, the oligosaccharide chains are completed along a unidirectional assembly line that includes the Golgi complex, where a number of transferases specific for the distal and terminal sugars are located (Hirano *et al.*, 1972; Cook, 1977). The same localization was reported for enzymes responsible for GAG synthesis and sulfation (Horwitz and Dorfman, 1968; Freilich *et al.*, 1975). In contrast, no sugar transferases have been found in the membranes of typical secretory granules.

The fate of the glycosylated macromolecules (i.e., whether they will be discharged to the extracellular space or will remain firmly attached to membranes in the cell) seems to depend primarily on their peptide backbone, rather than on their oligosaccharide side chains. In general, we can assume that the peptides glycosylated while segregated within the Golgi cisternae will be assembled into the granule content; in contrast, the glycosylated peptides attached to the inner surface of Golgi membranes (e.g., by means of a hydrophobic sequence or by a specific interaction with integral membrane components) will have a better chance to remain membrane-bound even after transfer to the secretory granules. In the latter case, glycosylation will introduce into the molecules a large hydrophilic moiety that will act as a further restraint to their transmembrane reorientation (Bretscher, 1973). Analogously, the same process might operate to select secretory protidoglycans destined to be discharged to the extracellular space from those that remain attached to the surface of the cells (see below). However, it must be admitted that this simple scheme is inadequate to explain the data concerning DBH. The dual localization of this glycoprotein enzyme at the intraluminal surface of the membrane and in the segregated content of chromaffin granules (see Sections 2.1 and 2.2) is not an artifact, since the two pools were recently found to turn over at greatly different rates (Geissler *et al.*, 1977; Gagnon *et al.*, 1976), as is to be expected for true membrane and secretory proteins (Meldolesi, 1974). Yet, no molecular, immunological, or kinetic differences in the soluble and membrane-bound enzymes have been detected so far (Winkler, 1976; Aunis *et al.*, 1977). The reasons that apparently identical DBH molecules have different localization and fate in the chromaffin granules remain unclear.

Another open question concerns the presence of complex carbohydrates (primarily glycoproteins) at the cytoplasmic surface of secretory organelles (see Section 2.1). Whether these superficial glycoproteins derive from the specific translocation of intraluminally oriented molecules, or are the products of soluble (or cytoplasmically oriented) sugar transferases, is still a matter of speculation. In this respect, however, two series of experiments should be mentioned. On one hand, kinetic data obtained in the liver (Autuori *et al.*, 1975; Svensson *et al.*, 1976) have been interpreted as indicating that a sialoglycoprotein, originally synthesized within the Golgi complex, might eventually be transferred to the soluble cytoplasm and finally attached to the cytoplasmic surface of endoplasmic reticulum membranes. If confirmed, these data would support the Golgi origin of the complex carbohydrates facing the cytoplasm. On the other hand, the alternative process is supported by the results of studies on brain. In this system, there is evidence indicating that soluble and membrane glycoproteins and gangliosides can be glycosylated by local transferases that are not segregated within Golgi cisternae (Den *et al.*, 1975; Margolis, R. K., *et al.*, 1975*a,b*; Ledeen *et al.*, 1976).

3.1. Membrane Recognition and Biogenesis

The possible functions of complex carbohydrates in the membranes of secretory organelles will now be considered. A role in the process of membrane recognition (as a preliminary step in exocytosis) has been advocated for the glycoproteins oriented toward the cytoplasm (Meyer and Burger, 1976; Baggiolini *et al.*, 1977). At present, this possibility is entirely speculative, since it is based only on the analogy with other recognition processes in which glycoproteins are directly involved (such as the antigen–antibody and some hormone–receptor interactions). Moreover, it should be emphasized that the ability of specific recognition and fusion is not confined to granule and plasma membranes in exocytosis, but is shared by other endomembranes [such as those of the Golgi complex and endoplasmic reticulum (see the reviews by Palade, 1975; Meldolesi, 1974; Meldolesi *et al.*, 1978)], in which the existence of complex carbohydrates exposed at the cytoplasmic surface has not been clearly demonstrated.

On the other hand, the glycoproteins and GAGs attached to the inner leaflet of the granule membranes are often considered as molecules in transit from their site of synthesis (in the Golgi complex) to their final subcellular location (in the plasmalemma). Thus, the role of these complex carbohydrates is believed to be in the biogenesis and turnover of the plasmalemma (for recent reviews, see Bennett and Leblond, 1977; Cook, 1977). This conclusion is based primarily on the common pattern of transverse asymmetry of the two membranes [on exocytosis, the inner leaflet of the granule membranes becomes continuous with the outer plasmalemma leaflet, where membrane saccharides are exclusively located (Nicolson and Singer, 1974)], as well as on kinetic experiments in which labeled sugars, especially fucose, were used as precursor (reviewed by Bennett and Leblond, 1977). The available evidence in favor of this process now appears soundly based. However, a number of important details remain to be clarified. In this respect, the following considerations should be taken into account.

1. Some membrane complex carbohydrates play important roles in the secretory organelles themselves. For example, this may be the role of

membrane-bound DBH, an enzyme that catalyzes the last step in the biosynthesis of noradrenaline.

2. Most probably the transport of complex carbohydrates to the cell surface is not carried out exclusively by secretory granules and vesicles. Other organelles, such as the "pinocytotic" vesicles [the movement of which is probably bidirectional (Steinman *et al.*, 1976)], are also involved. An elegant cytochemical demonstration of the role of nonsecretory vesicles in providing the plasmalemma with its complement of extracellularly exposed complex carbohydrates was recently given by Pfenninger and Maylié-Pfenninger (1977) in cultured growing neurites.

Among typical secretory organelles, the situation is also probably quite variable. Those of the first group, characterized by a peculiar membrane composition (see Section 2.1) seem to play a minor role in the biogenesis of the plasmalemma, as suggested by the fact that DBH and GP_2 [the major glycoprotein components of chromaffin and zymogen granule membranes, respectively, which are both oriented inward (see Section 2.1)] are not normal constituents of the plasmalemma and appear at the cell surface only temporarily, at the moment of exocytosis (Winkler *et al.*, 1974; Ronzio *et al.*, 1978).

3. The Golgi-to-plasmalemma transport most likely concerns individual macromolecules rather than entire membrane patches. Actually, experimental evidence that has been reviewed elsewhere (Meldolesi *et al.*, 1978) suggests that the process is specific, i.e., that of the mixture of complex carbohydrates localized at the inner leaflet of the granule and vesicle membranes only some are transferred to the plasmalemma, while the others are not.

4. The Golgi-to-plasmalemma transport could be bidirectional, at least in some systems. In fact, the partial molecular overlapping among Golgi, granule, and plasma membranes (see Section 2.1) (also see Meldolesi *et al.*, 1978) can be interpreted as an indication of the existence of a pool of specific, right-side-oriented macromolecules, which shuttle between the endo- and exomembrane compartments of the cell.

3.2. Storage of Secretion Products

It is now recognized that the secretion products stored within secretory organelles are not freely soluble but are organized in supramolecular complexes (for reviews, see Ceccarelli *et al.*, 1974). Thus, in some granules (such as the prolactin granules of the pituitary and the specific granules of eosinophilic leukocytes and mast cells), the internal matrix is solid and remains structured even after removal of the limiting membrane (see Fig. 1) (Giannattasio *et al.*, 1975; Jansson, 1970; Murer and Baggiolini, personal communication); moreover, other granules are either osmotically inactive or much less active than expected from their chemical composition. This intriguing phenomenon, which is of great importance for the economy of secretory cells, implies the existence of extensive binding among the segregated molecules of the granule content. The mechanisms of this process are still poorly understood. In the case of biogenic amines, evidence has been reported indicating that they are stored in association with nucleotides and that Ca^{2+} ions might act as bridges between the amine–nucleotide complexes (Pletscher *et al.*, 1974). The critical role of Ca^{2+} in promoting the aggregation of negatively charged secretion products has also been proposed for protein-storing granules, such as those of the parotid gland (Wallach and Schramm, 1971) and of B cells in pancreatic islets (Howell *et al.*, 1978). However, recent results cast some doubt on this hypothesis, since it was shown that most of the Ca^{2+} can be withdrawn from the parotid granules (by *in vitro* exposure to a Ca^{2+} ionophore) without major consequences for their internal organization (Flashner and Schramm, 1977).

In another series of reports, complex carbohydrates have been implicated in storage through the possible binding of the positively charged residues of secretion products to their sulfate and carboxyl groups. However, the evidence in favor of this hypothesis is still inconclusive. Thus, in chromaffin granules, the amount of GAGs is insufficient to account for the binding of catecholamines, which are found in high concentration in these organelles (DaPrada *et al.*, 1972; Margolis, R. U., and Margolis, R. K., 1973), and chromogranin A, which comprises 40–50% of the soluble proteins of the granules, has a very low amine-binding capacity (Smith, W. J., and Kirshner, 1967). It has therefore been calculated that GAGs plus glycoproteins could account for the binding of only a minor part of the granule catecholamines (Margolis, R. U., and Margolis, R. K., 1973). The situation with adrenergic vesicles (which contain less GAGs and chromogranin A than chromaffin granules) is even less favorable. Moreover, no significant role is recognized for GAGs in serotonin storage within platelet granules (Pletscher *et al.*, 1974; Anderson *et al.*, 1974). In contrast, GAGs are present in mast cell granules in sufficient amount to play an important role in histamine-binding, either directly (Lagunoff, 1974) or through their interaction with basic proteins (Uvnäs, 1974).

In the case of protein-storing granules, the available evidence is still too fragmentary to permit definite conclusions. However, it is already clear that at least in prolactin granules of the pituitary and in pancreatic zymogen granules, the amount of complex carbohydrates is much too small to account for the direct binding of all segregated secretory proteins (Kronquist *et al.*, 1977; Reggio and Dagorn, 1978; Reggio and Palade, 1978; Zanini *et al.*, 1979*b*).

In our opinion the considerations discussed above should not necessarily be interpreted as an indication that complex carbohydrates have no connection with the storage of secretion products. It should be emphasized that the latter is a complex, probably multifactorial phenomenon (Reggio and Dagorn, 1978; Giannattasio *et al.*, 1979) which, in many systems, occurs at a relatively slow rate (see Palade, 1975; Winkler, 1977). In this respect, it is tempting to speculate that the small amount of complex carbohydrates present in the segregated content of immature secretory organelles might provide the initial "condensation cores" around which the interaction of secretion products could eventually develop through direct intermolecular binding or Ca^{2+} bridging or both. This hypothetical interpretation appears consistent, on one hand, with the relative specificity of the complex carbohydrates present in the various secretory organelles and, on the other, with the observation that in a variety of systems glycosylated macromolecules are distributed unevenly in the granule content (Tasso, 1973; Reggio and Palade, 1976; Simson, 1977; Zanini *et al.*, 1979*a*).

Alternative functions for the complex carbohydrates of secretory organelles can also be proposed. The increasing evidence that at least in some systems the content GAGs and glycoproteins are released concomitantly with the other secretion products (Berg and Young, 1971; Margolis, R. K., *et al.*, 1973; Berg and Austin, 1976; Geissler *et al.*, 1977; Haddad *et al.*, 1977; Berg, 1978; Morrissey *et al.*, 1978) raises the possibility that these compounds might display a functional activity after their release. So far, however, such activity has been demonstrated only for mast cell heparin. The final possibility is that intragranular complex carbohydrates have no function whatsoever, but are simply copacked with secretion products by chance, as a consequence of the relative lack of specificity of the mechanisms that control the intracellular distribution of the molecules traveling through the Golgi complex. Admittedly this pessimistic hypothesis is quite unlikely. However, it could be considered as a good indication of the uncertainties existing at present in the field.

ACKNOWLEDGMENTS. We wish to thank the following friends and colleagues for sending preprints and reprints of their recent articles: G. Bennet, N. B. Berg, M. Da Prada, R. K. Margolis, R. U. Margolis, G. E. Palade, G. Pelletier, H. A. Reggio, R. A. Ronzio, J. A. V. Simson, and H. Winkler.

References

Åborg, C. H., Fillion, G., Nosal, R., and Uvnäs, B., 1972, A sulphomucopolysaccharide–protein complex in the adrenergic vesicle (granule) fraction from nerves and tissues, *Acta Physiol. Scand.* **86**:427.

Anderson, P., Slorach, S. A., and Uvnäs, B., 1974, 5-HT storage in rat and rabbit blood platelets: The separation of ATP-containing and sulphomucopolysaccharide-containing granules, *Acta Physiol. Scand.* **90**:522.

Aunis, D., Bouclier, M., Pescheloche, M., and Mandel, P., 1977, Properties of membrane-bound dopamine β-hydroxylase in chromaffin granules from bovine adrenal medulla, *J. Neurochem.* **29**:439.

Autuori, F., Svensson, H., and Dallner, G., 1975, Biogenesis of microsomal membrane glycoprotein in rat liver. II. Purification of soluble glycoproteins and their incorporation into microsomal membranes, *J. Cell Biol.* **67**:700.

Baggiolini, M., Bretz, U., and Dewald, B., 1977, Biochemical and structural properties of the vacuolar apparatus of polymorphonuclear leukocytes, in: *Movement, Metabolism and Bactericidal Mechanisms of Phagocytes* (F. Rossi, P. Patriarca, and D. Romeo, eds.), pp. 89–102, Piccin Medical Books, Padua.

Bennett, G., and Leblond, C. P., 1977, Biosynthesis of the glycoproteins present in plasma membrane, lysosomes and secretory materials, as visualized by radioautography, *Histochem. J.* **9**:393.

Berg, N. B., 1978, Sulfate metabolism in the exocrine pancreas. II. The production of sulfated macromolecules by the mouse exocrine pancreas, *J. Cell Sci.* **31**:199.

Berg, N. B., and Austin, B. P., 1976, Intracellular transport of sulfated macromolecules in parotid acinar cells, *Cell Tissue Res.* **165**:215.

Berg, N. B., and Young, R. W., 1971, Sulfate metabolism in pancreatic acinar cells, *J. Cell Biol.* **50**:469.

Bergeron, J. J. M., Evans, W. H., and Geschwind, I. I., 1973, Insulin binding to rat liver Golgi fractions, *J. Cell Biol.* **59**:771.

Blaschke, E., Bergqvist, U., and Uvnäs, B., 1976, Identification of the mucopolysaccharides in catecholamine-containing subcellular particle fractions from various rat, cat and ox tissues, *Acta Physiol. Scand.* **97**:110.

Bretscher, M. S., 1973, Membrane structure: Some general principles, *Science* **181**:622.

Castle, J. D., Jamieson, J. D., and Palade, G. E., 1975, Secretion granules of the rabbit parotid gland. Isolation, subfractionation and characterization of the membrane and content subfractions, *J. Cell Biol.* **64**:182.

Ceccarelli, B., Clementi, F., and Meldolesi, J., (eds.), 1974, *Cytopharmacology of Secretion*, Raven Press, New York.

Cook, G. M. W., 1977, Biosynthesis of plasma membrane proteins, in: *The Synthesis, Assembly and Turnover of Cell Surface Components*, (G. Poste and G. L. Nicolson, eds.), pp. 85–136, Elsevier/North-Holland, Amsterdam.

DaPrada, M., VonBerlepsch, K., and Pletscher, A., 1972, Storage of biogenic amines in blood platelets and adrenal medulla: Lack of evidence for direct involvement of glycosaminoglycans, *Arch. Pharmacol.* (*New Ser.*) **275**:315.

Den, H., Kaufman, B., McGuire, E. J., and Roseman, S., 1975, The sialic acids. XVIII. Subcellular distribution of seven glycosyltransferases in embryonic chicken brain, *J. Biol. Chem.* **250**:739.

Dunn, W. B., and Spicer, S. S., 1969, Histochemical demonstration of sulfated mucosubstances and cationic proteins in human granulocytes and platelets, *J. Histochem. Cytochem.* **17**:668.

Eagles, P. A. M., Johnson, L. N., and VanHorn, C., 1975, The distribution of concanavalin A receptor sites on the membrane of chromaffin granules, *J. Cell Sci.* **19**:33.

Eagles, P. A. M., Johnson, L. N., and VanHorn, C., 1976, The distribution of anionic sites on the surface of the chromaffin granule membrane, *J. Ultrastruct. Res.* **55**:87.

Farquhar, M. G., Bergeron, J. J. M., and Palade, G. E., 1974, Cytochemistry of Golgi fractions prepared from rat liver, *J. Cell Biol.* **60**:8.

Fillion, G., Nosal, R., and Uvnäs, B., 1971, The presence of a sulphomucopolysaccharide–protein complex in adrenal medullary cell granules, *Acta Physiol. Scand.* **84**:286.

Flashner, Y., and Schramm, M., 1977, Retention of amylase in the secretory granules of parotid gland after extensive release of Ca^{++} by ionophore A-23187, *J. Cell Biol.* **74**:789.

Freilich, L. S., Lewis, R. C., Reppucci, A. C., Jr., and Silbert, J. E., 1975, Glycosaminoglycan synthesizing activity of an isolated Golgi preparation from cultured mast cells, *Biochem. Biophys. Res. Commun.* **63**:663.

Gagnon, C., Schatz, R., Otten, U., and Thoenen, H., 1976, Synthesis, subcellular distribution and turnover of dopamine β-hydroxylase in organ cultures of sympathetic ganglia and adrenal medulla, *J. Neurochem.* **27**:1083.

Geissler, D., Martinek, A., Margolis, R. U., Margolis, R. K., Skrivanek, J. A., Ledeen, R., König, P., and Winkler, H., 1977, Composition and biogenesis of complex carbohydrates of ox adrenal chromaffin granules, *Neuroscience* **2**:685.

Giannattasio, G., and Zanini, A., 1976, Presence of sulfated proteoglycans in prolactin secretory granules isolated from the rat pituitary gland, *Biochim. Biophys. Acta* **439**:349.

Giannattasio, G., Zanini, A., and Meldolesi, J., 1975, Molecular organization of rat prolactin granules. I. *In vitro* stability of intact and "membraneless" granules, *J. Cell Biol.* **64**:246.

Giannattasio, G., Zanini, A., Rosa, P., Margolis, R. K., Margolis, R. U., and Meldolesi, J., 1979, Molecular organization of prolactin granules. III. Intracellular transport of glycosaminoglycans and glycoproteins of bovine prolactin granule matrix (in prep.).

Haddad, A., Pelletier, G., Marchi, F., and Brasileiro, I. L. G., 1977, Light microscope radioautographic study of glycoprotein secretion in the hypothalamic–neurohypophysial system of the rat after L-fucose-^{3}H injection, *Cell Tissue Res.* **177**:67.

Hirano, H., Parkhouse, B., Nicolson, G. L., Lennox, E. S., and Singer, S. J., 1972, Distribution of saccharide residues on membrane fragments from a myeloma-cell homogenate: Its implications for membrane biogenesis, *Proc. Natl. Acad. Sci. U.S.A.* **69**:2945.

Holtzman, E., 1977, The origin and fate of secretory packages, especially synaptic vesicles, *Neuroscience* **2**:327.

Hopkins, C. R., 1975, Synthesis and secretion of a large glycoprotein in the pars intermedia, *J. Endocrinol.* **65**:225.

Hörtnagl, H., 1976, Membranes of the adrenal medulla: A comparison of membranes of chromaffin granules with those of the endoplasmic reticulum, *Neuroscience* **1**:9.

Hortnagl, H., Winkler, H., and Lochs, H., 1972, Membrane proteins of chromaffin granules: Dopamine β-hydroxylase, a major constituent, *Biochem. J.* **129**:187.

Horwitz, A. L., and Dorfman, A., 1968, Subcellular sites for synthesis of chondromucoprotein of cartilage, *J. Cell Biol.* **38**:358.

Howell, S. L., and Tyhurst, M., 1977, Distribution of anionic sites on surface of B cell granule and plasma membranes: A study using cationic ferritin, *J. Cell Sci.* **27**:289.

Howell, S. L., Tyhurst, M., Duvefelt, H., Andersson, A., and Hellerström, C., 1978, Role of zinc and calcium in the formation and storage of insulin in the pancreatic B-cell, *Cell Tissue Res.* **188**:107.

Hvas, S., and Thorn, N. A., 1973, Hexosamine and heparin in homogenate and subcellular fractions from bovine neurohypophysis, *Acta Endocrinol.* **74**:209.

Jansson, S. E., 1970, Uptake of 5-hydroxytryptamine by mast cell granules *in vitro*, *Acta Physiol. Scand.* **82**:35.

Jacques, L. B., 1975, The mast cells in the light of new knowledge of heparin and sulfated mucopolysaccharides, *Gen. Pharmacol.* **6**:235.

Jones, C. W., and Swann, R. W., 1975, A glycoprotein in the neurosecretory granules of the neurohypophysis, *J. Physiol.* (*London*) **245**:45P.

Kiang, W.-L., Crockett, C. P., Margolis, R. K., and Margolis, R. U., 1978, Glycosaminoglycans and glycoproteins associated with microsomal subfractions of brain and liver, *Biochemistry* **17**:3841.

Kònig, P., Hörtnagl, H., Kostron, H., Sapinsky, H., and Winkler, H., 1976, The arrangement of dopamine β-hydroxylase (EC 1.14.2.1) and chromomembrin B in the membrane of chromaffin granules, *J. Neurochem.* **27**:1539.

Kronquist, K. E., Elmahdy, A., and Ronzio, R. A., 1977, Synthesis and subcellular distribution of heparan sulfate in the rat exocrine pancreas, *Arch. Biochem. Biophys.* **182**:188.

Lagercrantz, H., 1976, On the composition and function of large dense-cored vesicles in sympathetic nerves, *Neuroscience* **1**:81.

Lagunoff, D., 1974, Analysis of dye binding sites in mast cell granules, *Biochemistry* **13**:3982.

Lagunoff, D., and Pritzl, P., 1976, Characterization of rat mast cell granule proteins, *Arch. Biochem. Biophys.* **173**:554.

Lalli, M. F., and Clermont, Y., 1975, Leydig cells and their role in the synthesis and secretion of glycoproteins, *Anat. Rec.* **181**:403.

Ledeen, R. W., Skrivanek, J. A., Tirri, L. J., Margolis, R. K., and Margolis, R. U., 1976, Gangliosides of the neuron: Localization and origin, in: *Ganglioside Function: Biochemical and Pharmacological Implications* (G. Porcellati, B. Ceccarelli, and G. Tettamanti, eds.), pp. 83–103, Plenum Press, New York.

Little, J. S., and Widnell, C. C., 1975, Evidence for the translocation of 5′-nucleotidase across hepatic membranes *in vitro*, *Proc. Natl. Acad. Sci. U.S.A.* **72**:4013.

Margolis, R. K., Jaanus, S. D., and Margolis, R. U., 1973, Stimulation of acetylcholine of sulfated mucopolysaccharide release from the perfused cat adrenal gland, *Mol. Pharmacol.* **9**:590.

Margolis, R. K., Margolis, R. U., Preti, C., and Lai, D., 1975*a*, Distribution and metabolism of glycoproteins and glycosaminoglycans in subcellular fractions of brain, *Biochemistry* **14**:4797.

Margolis, R. K., Preti, C., Chang, L., and Margolis, R. U., 1975*b*, Metabolism of the protein moiety of brain glycoproteins, *J. Neurochem.* **25**:707.

Margolis, R. U., and Margolis, R. K., 1973, Isolation of chondroitin sulfate and glycopeptides from chromaffin granules of adrenal medulla, *Biochem. Pharmacol.* **22**:2195.

Margolis, R. U., and Margolis, R. K., 1977, Metabolism and function of glycoproteins and glycosaminoglycans in nervous tissue, *Int. J. Biochem.* **8**:85.

Matthews, E. K., Evans, R. J., and Dean, P. M., 1972, The ionogenic nature of the secretory-granule membrane: Electrokinetic properties of isolated chromaffin granules, *Biochem. J.* **130**:825.

Meldolesi, J., 1974, Secretory mechanisms in pancreatic acinar cells: Role of the cytoplasmic membranes, in: *Cytopharmacology of Secretion* (B. Ceccarelli, F. Clementi, and J. Meldolesi, eds.), pp. 71–85, Raven Press, New York.

Meldolesi, J., Jamieson, J. D., and Palade, G. E., 1971, Composition of cellular membranes in the pancreas of the guinea pig. II. Lipids, *J. Cell Biol.* **49**:130.

Meldolesi, J., Borgese, N., De Camilli, P., and Ceccarelli, B., 1978, Cytoplasmic membranes and the secretory process, in: *Cell Membrane Fusion* (G. Poste and G. L. Nicolson, eds.), pp. 509–627, Elsevier/North-Holland, Amsterdam.

Meyer, D. I., and Burger, M. M., 1976, The chromaffin granule surface localization of carbohydrate on the cytoplasmic surface of an intracellular organelle, *Biochim. Biophys. Acta* **443**:428.

Morgan, I. G., Zanetta, J. G., Breckenridge, W. G., Vincendon, G., and Gombos, G., 1973, The chemical structure of synaptic membranes, *Brain Res.* **62**:405.

Morré, D. J., 1978, The Golgi apparatus and membrane biogenesis, in: *Membrane Fusion* (G. Poste and G. L. Nicolson, eds.), pp. 1–83, Elsevier/North-Holland, Amsterdam.

Morrissey, J. J., Hamilton, J. W., and Cohn, D. V., 1978, The secretion of parathormone and glycosylated proteins by parathyroid cells in culture, *Biochem. Biophys. Res. Commun.* **82**: 1279.

Nakagami, K., Warshawsky, H., and Leblond, C. P., 1971, The elaboration of protein and carbohydrate by rat parathyroid cells as revealed by electron microscope radioautography, *J. Cell Biol.* **51**:596.

Nelson, D. L., Harden, T. K., and Molinoff, B., 1978, Adrenergic storage vesicles in rat heart: Quantitation of membrane-bound and membrane-enclosed dopamine β-hydroxylase, *J. Pharmacol. Exp. Ther.* **205**:357.

Nicolson, G. L., and Singer, S. J., 1974, The distribution and asymmetry of mammalian cell surface saccharides utilizing ferritin conjugated plant agglutinins as specific saccharide stains, *Proc. Natl. Acad. Sci. U.S.A.* **68:**942.

Noseworthy, J., Smith, G. H., Himmelhoch, S. R., and Evans, W. H., 1975, Protein and glycoprotein electrophoretic patterns of enriched fractions of primary and secondary granules from guinea pig polymorphonuclear leukocytes, *J. Cell Biol.* **65:**577.

Olsson, I., 1969, Intracellular transport of glycosaminoglycans (mucopolysaccharides) in human leukocytes, *Exp. Cell Res.* **54:**318.

Palade, G. E., 1975, Intracellular aspects of the process of protein secretion, *Science* **189:**347.

Pelletier, G., 1971, Detéction des glycoproteines dans les cellules corticotropes de l'hypophyse du rat, *J. Microsc.* **11:**327.

Pelletier, G., 1974, Autoradiographic studies of synthesis and intracellular migration of glycoproteins in the rat anterior pituitary gland, *J. Cell Biol.* **62:**185.

Pfenninger, K. H., and Maylié-Pfenninger, M. F., 1977, Localized appearance of new lectin receptors on the surface of growing neurites, *J. Cell Biol.* **75:**54a.

Pletscher, A., Da Prada, M., Berneis, K. H., Steffen, H., Lütold, B., and Weder, H. G., 1974, Molecular organization of amine storage organelles of blood platelets and adrenal medulla, in: *Cytopharmacology of Secretion* (B. Ceccarelli, F. Clementi, and J. Meldolesi, eds.), pp. 257–264, Raven Press, New York.

Pycock, C., Blaschke, E., Bergqvist, U., and Uvnäs, B., 1975, On the possible involvement of sulphomucopolysaccharides in the storage of catecholamines within the central nervous system, *Acta Physiol. Scand.* **95:**373.

Reggio, H. A., and Dagorn, J. C., 1978, Ionic interactions between bovine chymotrypsinogen A and chondroitin sulfate A. B. C., *J. Cell. Biol.* **78:**951.

Reggio, H., and Palade, G. E., 1976, Sulfated compounds in the secretion and zymogen granule content of the guinea pig pancreas, *J. Cell Biol.* **70:**360a.

Reggio, H. A., and Palade, G. E., 1978, Sulfated compounds in the zymogen granules of the guinea pig pancreas, *J. Cell Biol.* **77:**288.

Ronzio, R. A., Kronquist, K. E., Lewis, D. S., MacDonald, R. J., Mohrlok, S. H., and O'Donnell, J. J., Jr., 1978, Glycoprotein synthesis in the adult rat pancreas. IV. Subcellular distribution of membrane glycoproteins, *Biochim. Biophys. Acta* **508:**65.

Simpson, D. L., Thorne, D. R., and Loh, H. H., 1976, Sulfated glycoproteins, glycolipids, and glycosaminoglycans from synaptic plasma and myelin membranes: Isolation and characterization of sulfated glycopeptides, *Biochemistry* **15:**5449.

Simson, J. A. V., 1977, The influence of fixation on the carbohydrate cytochemistry of rat salivary gland secretory granules, *Histochem. J.* **9:**645.

Slaby, F., and Farquhar, M. G., 1976, The major polypeptides of somatotrophic and mammotrophic granules from the rat anterior pituitary, *J. Cell Biol.* **70:**92a.

Smith, A. D., and Winkler, H., 1967, Purification and properties of an acidic protein from chromaffin granules of bovine adrenal medulla, *Biochem. J.* **103:**483.

Smith, W. J., and Kirshner, N., 1967, A specific soluble protein from the catecholamine storage vesicles of bovine adrenal medulla. I. Purification and chemical characterization, *Mol. Pharmacol.* **3:**52.

Stadler, H., and Whittaker, V. P., 1978, Identification of vesiculin as a glycosaminoglycan, *Brain Res.* **153:**408.

Steinman, R. M., Scott, E. B., and Cohn, Z. A., 1976, Membrane flow during pinocytosis: A stereological analysis, *J. Cell Biol.* **68:**665.

Svensson, H., Elhammer, Å., Autuori, F., and Dallner, G., 1976, Biogenesis of microsomal membrane glycoproteins in rat liver. IV. Characteristics of a cytoplasmic lipoprotein having properties of a membrane precursor, *Biochim. Biophys. Acta* **455:**383.

Tartakoff, A., Greene, L. J., and Palade, G. E., 1974, Studies on the guinea pig pancreas: Fractionation and partial characterization of exocrine proteins, *J. Biol. Chem.* **249:**7420.

Tasso, F., 1973, Localisation cytochimique ultrastructurale de glycoprotéines dans les granules neurosécrétoires de la post-hypophyse du rat, *J. Microsc.* **18:**115.

Tasso, F., Rua, S., and Picard, D., 1977, Cytochemical duality of neurosecretory material in the hypothalamo–posthypophysial system of the rat as related to hormonal content, *Cell Tissue Res.* **180:**11.

Uvnäs, B., 1974, The molecular basis for the storage and release of histamine in rat mast cell granules. *Life Sci.* **14:**2355.

Vilhardt, H., Baker, R. V., and Hope, D. B., 1975, Isolation and protein composition of membranes of neurosecretory vesicles and plasma membranes from the neural lobe of the bovine pituitary gland, *Biochem. J.* **148:**57.

Vos, J., Kuriyama, K., and Roberts, E., 1969, Distribution of acid mucopolysaccharides in subcellular fractions of mouse brain, *Brain Res.* **12:**172.

Wallace, E. F., Krantz, M. J., and Lovenberg, W., 1973, Dopamine β-hydroxylase: A tetrameric glycoprotein, *Proc. Natl. Acad. Sci. U.S.A.* **70:**2253.

Wallach, D., and Schramm, M., 1971, Calcium and the exportable protein in rat parotid gland: Parallel subcellular distribution and concomitant secretion, *Eur. J. Biochem.* **21:**433.

Wallach, D., Kirshner, N., and Schramm, M., 1975, Non-parallel transport of membrane proteins and content proteins during assembly of the secretory granule in rat parotid gland, *Biochim. Biophys. Acta* **375:**87.

Winkler, H., 1976, The composition of adrenal chromaffin granules: An assessment of controversial results, *Neuroscience* **1:**65.

Winkler, H., 1977, The biogenesis of adrenal chromaffin granules, *Neuroscience* **2:**657.

Winkler, H., Schneider, F. H., Rufener, C., Nakane, P. K., and Hörtnagl, H., 1974, Membranes of adrenal medulla: Their role in exocytosis, in: *Cytopharmacology of Secretion* (B. Ceccarelli, F. Clementi, and J. Meldolesi, eds.), pp. 127–139, Raven Press, New York.

Zanini, A., and Giannattasio, G., 1973, Isolation of prolactin granules from rat anterior pituitary glands, *Endocrinology* **92:**349.

Zanini, A., Giannattasio, G., and Meldolesi, J., 1979*a*, Intracellular events in prolactin secretion, in: *Synthesis and Release of Adenohypophyseal Hormones* (M. Jutisz and K. McKerns, eds.), Plenum Press, New York (in press).

Zanini, A., Giannattasio, G., Margolis, R. K., Margolis, R. U., Nussdorfer, G., and Meldolesi, J., 1979*b*, Molecular organization of prolactin granules. II. Characterization of glycosaminoglycans and glycoproteins of bovine prolactin granule matrix (in prep.).

16

Glycoprotein Storage Diseases and the Mucopolysaccharidoses

Glyn Dawson

1. Introduction

Complex carbohydrates are degraded under acidic conditions, in organelles known as lysosomes, by the sequential action of a number of exo- and endoglycosidases. Numerous human neurological and neurovisceral storage diseases, known collectively as the lysosomal storage diseases, have been found to result from the inherited deficiency of any one of these glycosidases. Two groups of these disorders, namely, the inborn errors of glycoprotein catabolism (the "oligosaccharidoses") and the inborn errors of glycosaminoglycan catabolism (the "mucopolysaccharidoses"), will be discussed in this chapter. All are characterized at the light-microscopic or electron-microscopic level by the presence of numerous storage vacuoles (engorged lysosomes) containing either amorphous, granular, or electron-dense multilamellar material (multilamellar cytosomes, zebra bodies, Lafora bodies). Biochemically, they have been characterized by the isolation of a specific storage product or products and by a specific enzyme deficiency. The fundamental basis for the neurological degeneration is unknown, although recent studies (Purpura and Suzuki, 1976) indicate that secondary neuronal membrane changes (meganeurite and aberrant synapse formation) may be responsible.

2. Glycoproteins of Nervous Tissue

Numerous glycoproteins have been isolated from nervous tissue (Margolis, R.K., and Margolis, R. U., 1970; Brunngraber *et al.*, 1973) (for a

Glyn Dawson • Departments of Pediatrics and Biochemistry, Joseph P. Kennedy, Jr., Mental Retardation Research Center, University of Chicago, Chicago, Illinois.

detailed review, see Chapter 3), but none has been characterized in detail. However, from the results of carbohydrate analyses and by inference from the detailed structural information available on analogous plasma glycoproteins such as the immunoglobulins, it appears that two major types of asparaginylglucosamine-linked oligosaccharide units are represented (Spiro, 1969; Kornfeld and Kornfeld, 1976).

2.1. Acidic (Sialo)glycoproteins

These have the following basic structure:

```
NeuAc(8-2α)NeuAc ┐                                                  |
|                ┘                                                 Asn
|   (α,2-6)                                                         |
|                                                        (β,1-1)    |
Gal(β,1-4)GlcNAc(β,1-2)Man(α,1-3/6) ┐ Man(β,1-4)GlcNAc(β1-4)GlcNAc
                                    ┘ 2–3                           |
                                                         (α,1-6)    |
                                                                   Fuc
```

In immature rat brain plasma membrane glycoproteins, it was observed that 20% of the *N*-acetylneuraminic acid (NeuAc) residues are 8-*O*-substituted with an additional NeuAc residues (Finne *et al.*, 1977).

2.2 Neutral (Manno)glycoproteins

These have the following basic structure:

[Man(α,1-2)Man(α,1-2)Man(α,1-3/6]$_{2-3}$-
Man(β,1-4)GlcNAc(β,1-4)GlcNAc(β,1-1)Asn

Both types of glycoproteins show considerable heterogeneity.

2.3. Other Types

There is good evidence for the existence of sulfated glycoproteins, in which galactose and glucosamine residues are sulfated at C-6 (Margolis, R. K., and Margolis, R. U., 1970; Brunngraber *et al.*, 1973; Simpson *et al.*, 1976), and complex galactose and *N*-acetylglucosamine (GlcNAc)-rich glycoproteins, structurally related to mucin blood group glycoproteins or keratan sulfate (Kornfeld and Kornfeld, 1976), in which the linkage to protein is of the *O*-glycoside *N*-acetylgalactosamine (GalNAc)-serine/threonine type, but these have not been characterized structurally. Collagen is also an intrinsic component of peripheral nervous tissue, since it is synthesized by cells of neuroectodermal origin (Church *et al.*, 1973) and is presumably glycosylated by Glc(α,1-2)Gal-hydroxylysine linkages (Spiro, 1969).

3. Inborn Errors of Glycoprotein Catabolism

Glycoproteins are catabolized within lysosomes by the sequential action of exoglycosidases (neuraminidase I, neuraminidase II, β-galactosidase, *N*-acetyl-β-glucosaminidase B, (*N*-acetyl-β-hexosaminidase B) α-mannosidase, β-mannosidase, and α-fucosidase) and endoglycosidases (endo-β-*N*-acetylglucosaminidase and *N*-aspartyl-β-glucosaminidase), as shown in Fig. 1. It is interesting to note that all these enzymes are themselves glycoproteins. Endo-β-*N*-acetylglucosaminidase activity can be demonstrated in nervous tissue using mannoglycopeptide substrates prepared from ovalbumin (Chien and Dawson, 1978). The substrate specificity is presumably similar to that of the H (mannose-rich glycopeptides) and D (mannose core derived from sialoglycoproteins) activities described previously (Koide and Muramatasu, 1974; Trimble and Maley, 1977; Chien *et al.*, 1977). Their existence suggests that glycoprotein catabolism could be initiated in this manner, as shown in Fig. 1. This idea is further supported by the discovery that patients with inborn errors of glycoprotein catabolism store mainly oligosaccharide rather than glycopeptide material, and that GlcNAc is the terminal reducing sugar residue in this storage material.

3.1. Neuraminidase Deficiency

The neuraminidases of nervous tissue have been reported to be associated both with lysosomes and with the plasma membrane, especially synaptosomal membrane (Schengrund and Rosenberg, 1970; Gielen, 1974). Such membrane-associated neuraminidases appear to act preferentially on sialoglycosphingolipids (gangliosides) rather than glycoproteins, but this remains to be substantiated. One can arbitrarily classify the neuraminidases as neuraminidase I [hydrolyzing NeuAc(α,2-8)NeuAc linkages found in glycoproteins (Finne *et al.*, 1977) and gangliosides (Merat and Dickerson, 1973)] and neuraminidase II [hydrolyzes NeuAc(α,2-3)Gal- and NeuAc(α,2-6)Gal or -GalNAc linkages and is probably not a single enzyme]. Recent studies have implied that the basic defect in a group of inherited lysosomapathies known as the mucolipidoses is a deficiency of lysosomal glycoprotein neuraminidase, probably neuraminidase II. The most compelling evidence for this has been the isolation of the following three major sialooligosaccharides:

NeuAc(α,2-6)Gal(β,1-4)GlcNAc(β,1-2)Man(α,1-3)Man(β,1-4)GlcNAc

NeuAc(α,2-6)Gal(β,1-4)GlcNAc(β,1-2)Man(α1-3)↘
Man(β,1-4)GlcNAc
Gal(β,1-4)GlcNAc(β,1-2)Man(α,1-6)↗

NeuAc(α,2-6)Gal(β,1-4)GlcNAc(β,1-2)Man(α,1-3)↘
Man(β,1-4)GlcNAc
NeuAc(α,2-6)Gal(β,1-4)GlcNAc(β,1-2)Man(α,1-6)↗

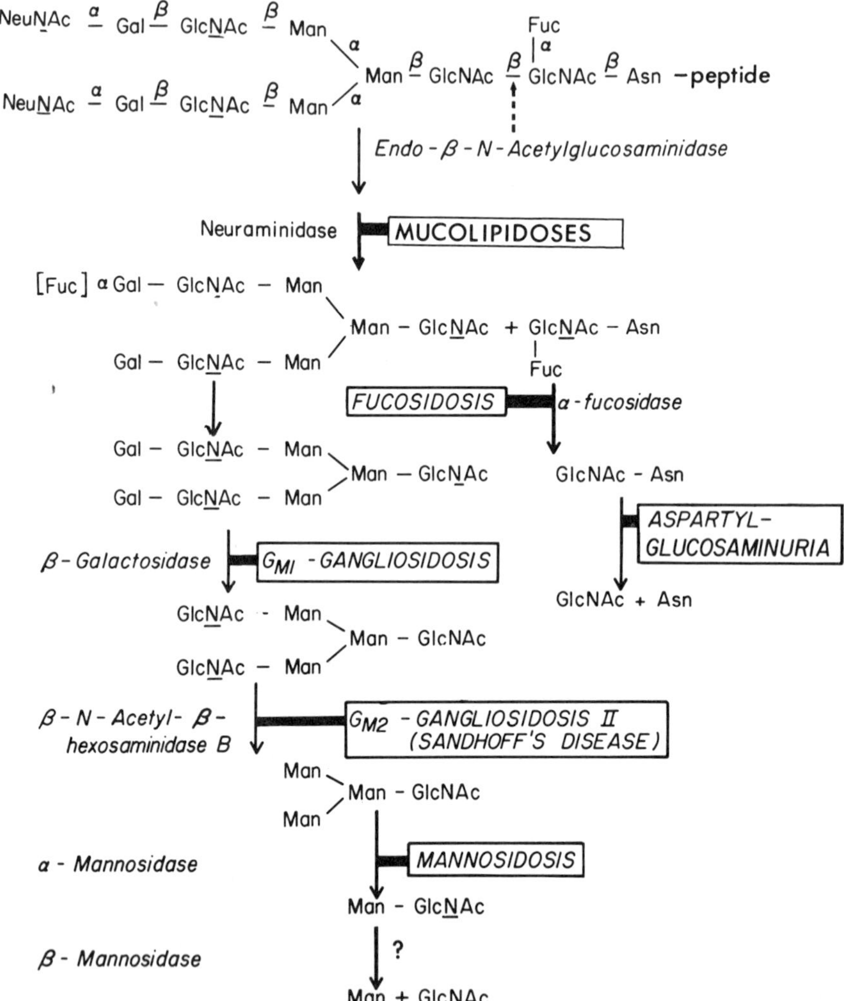

Fig. 1. Proposed pathway for catabolism of typical brain glycoprotein showing location of enzyme defects leading to oligosaccharide storage diseases. Initial steps in hydrolysis involve exoglycosidases: neuraminidase(s), α-fucosidase, β-galactosidase, *N*-acetyl-β-glucosaminidase (Hex "B"), (*N*-acetyl-β-hexosaminidase B) α-mannosidase, β-mannosidase, *N*-acetyl-β-D-glucosaminidase. Endo-β-*N*-acetylglucosaminidase normally acts at the mannoglycopeptide stage, but in storage lysosomes can act slowly to yield sialo-oligosaccharides. *N*-Aspartyl-β-glucosaminidase will normally degrade asparaginylglycopeptides.

from the urine of patients with "I-cell disease" (mucolipidosis II) and related "mucolipidoses" (Strecker *et al.*, 1977*a*). Similar oligosaccharide storage material has not yet been isolated from nervous tissue of these patients, and previous studies (Gilbert *et al.*, 1973; Martin *et al.*, 1975) showed the NeuAc(α,2-8)NeuAc- and NeuAc(α,2-3)Gal-linked sialoglycosphingolipid levels to be normal in mucolipidosis brain. A deficiency of glycoprotein neuraminidase activity was reported in fibroblasts cultured from patients with neurovisceral storage diseases classified as mucolipidosis I and II (Spranger and Wiedemann, 1970; Cantz *et al.*, 1977; Strecker *et al.*, 1976; Thomas *et al.*, 1976). Substrates used have included sialolactose [NeuAc(α,2-3)Gal-Glc], fetuin [NeuAc(α,2-3 or 6)Gal-linkage], and a chromogenic substrate, 2-(3′methoxyphenyl)NeuAc (MPN) (Kelly and Graetz, 1977). However, fibroblasts from mucolipidosis II patients are also markedly deficient in many other lysosomal hydrolases, especially β-galactosidase, α-fucosidase, *N*-acetyl-β-hexosaminidase, α-mannosidase, and arylsulfatase A (Leroy *et al.*, 1972; Gilbert *et al.*, 1973; Dawson *et al.*, 1972). In none of these cases have we been able to demonstrate gross sialoglycolipid storage or a deficiency of G_{M3} neuraminidase activity

This generalized deficiency of lysosomal hydrolases in "I-cell" cultured fibroblasts and a concomitant excess in the culture media (and also the patient's sera and urine) was attributed to a defect in the oligosaccharide unit of the secreted hydrolases that prevents their normal high-affinity uptake and pinocytosis (Hickman *et al.*, 1974). Since neuraminidase is the first exoglycosidase to act on a glycoprotein oligosaccharide unit (as shown in Fig. 1), a generalized deficiency of *all* lysosomal hydrolases would be expected to produce a sialooligosaccharide storage material and could be responsible for the impaired uptake of the other lysosomal hydrolases, since these enzymes are themselves sialoglycoproteins. However, the available data point to the involvement of mannose or phosphate groups rather than sialic acid residues in this high-affinity uptake phenomenon (Hieber *et al.*, 1976; Kaplan *et al.*, 1977).

Other phenotypes in which a neuraminidase deficiency appears to be the primary defect include an 8-month-old female with coarse facies, hepatosplenomegaly, and dysostosis multiplex but normal mental development who secreted sialic-acid-rich macromolecules and manifested a specific deficiency of neuraminidase when tested with fetuin and MPN (Kelly and Graetz, 1977), the cherry red spot–myoclonus syndrome (O'Brien, 1977), and two patients with a form of mucolipidosis I, involving late onset of neurological symptoms. Durand *et al.* (1977) termed this latter disorder "sialidosis" to distinguish it from mucolipidosis II and III, in which other enzymes are also defective. Sialidosis may be a specific lysosomal NeuNAc(α2-6)Gal-neuraminidase deficiency, since NeuNAc(α2-3)Gal linkages were minor components of the stored oligosaccharides and sialoglycolipids did not accumulate.

3.2. β-Galactosidase Deficiency (G_{M1}-Gangliosidoses)

An almost total deficiency of G_{M1}-ganglioside-β-D-galactosidase and synthetic substrate (4-methylumbelliferyl or *p*-nitrophenyl-β-D-galactoside) β-galactosidase activity was reported in the neurovisceral storage disease G_{M1}-(generalized) gangliosidosis (Okada and O'Brien, 1968). The enzyme has been purified (Alhadeff *et al.*, 1975) and appears to be coded for by a gene on chromosome 12 (Rushton and Dawson, 1977). Several variants of the disorder have been reported with varying degrees of neurological, skeletal, and hepatosplenic involvement, and partial deficiencies of synthetic substrate β-galactosidase activity have been observed in a number of storage diseases including the mucolipidoses and the Hunter–Hurler syndrome (King *et al.*, 1973).

3.2.1. G_{M1}-Gangliosidosis Type I (Generalized)

G_{M1}-gangliosidosis (Type I) is a neurovisceral storage disease resulting in death between ages 3 and 5. Such patients show massive storage of G_{M1} ganglioside in nervous tissue (Ledeen *et al.*, 1965), especially in neurons, where the accumulation results in production of membranocytoplasmic bodies and meganeurite formation (Purpura and Suzuki, 1976). There is little extraneural storage of G_{M1} ganglioside, but oligosaccharide storage occurs in liver and other tissues including the brain (Wolfe *et al.*, 1974; Tsay and Dawson, 1976) (Table I). Wolfe *et al.*, (1974) characterized the glycoprotein-derived oligosaccharide storage material as

$$[\text{Gal}(\beta,1\text{-}4)\text{GlcNAc}(\beta,1\text{-}2)\text{Man}]_{2\text{ or }3}(\beta,1\text{-}3\text{ and }6)\text{Man}(\beta,1\text{-}4)\text{GlcNAc}$$

as shown in Fig. 1.

A glycopeptide

$$[\text{Gal}(\beta,1\text{-}4)\text{GlcNAc}(\beta,1\text{-}3)]_3\ \text{Gal}(\beta,1\text{-}3)[\text{Gal}(\beta,1\text{-}6)]\ \text{GalNAc-O-Ser/Thr}$$

was isolated from the liver of a patient (Feldges *et al.*, 1973) with a slightly atypical form of the disease by Tsay *et al.* (1975). Urine from such patients contains a complex mixture of galactoglycopeptides.

3.2.2. G_{M1}-Gangliosidosis Type II

This is a milder, more slowly progressive form of the disease in which massive accumulation of G_{M1} ganglioside has been observed in nervous tissue, but storage of glycoprotein-derived oligosaccharide material appears minimal (Table I), and there is little organomegaly (O'Brien, 1972*a*). Studies on β-galactosidase isoenzymes (Pinsky *et al.*, 1974) and cell hybridization experiments (Galjaard *et al.*, 1975) have thus far failed to resolve the differences among Types I, II, III, and IV, so it is not altogether clear whether the

Table I. Carbohydrate Composition of Low-Molecular-Weight Oligosaccharide Fraction Isolated from Nervous Tissue of Patients with Inborn Errors of Glycoprotein Catabolism[a]

		G_{M1}-gangliosidosis		G_{M2}-gangliosidosis			
Sugar	Mannosidosis[b]	Type I (infantile)	Type II (juvenile)	Type I	Type II	Fucosidosis	AGU[b]
	(Relative moles of monosaccharide per amino acid)						
Fucose	0	0	0	0	0	2.5	0
Mannose	4.6	3.5	4.0	0	3.6	3.3	<0.1
Galactose	0	2.1	2.0	0	0	2.4	<0.1
GlcNAc	1.0	3.0	3.0	0	3.0	3.0	1.0
NeuNAc	0	0	0	0	0	0	<0.1
Asn	0	0	0	0	0	0	1.0
Total (μmol/g fresh wt)	2.8	1.0	0.5	—	2.1	19.9	ND

[a] For proposed structures, see Fig. 1.
[b] Data from Öckerman (1969). (AGU) Aspartylglucosaminuria.

glycoprotein and glycolipid hydrolyzing activities can be resolved. In general, the patients with the Type II neurological form of G_{M1}-gangliosidosis have a higher (5–10% of normal) residual level of β-galactosidase activity than the severe neurovisceral cases.

3.2.3. *Other β-Galactosidase Deficiency States*

O'Brien *et al.* (1976) described a case of spondyloepiphyseal dysplasia with corneal clouding, normal intelligence, and a deficiency of "acid" β-galactosidase activity. The G_{M1}-β-galactosidase activity was 7% of normal and the asialofetuin (glycoprotein) β-galactosidase was 1% of normal. Orii *et al.* (1972), Wenger *et al.* (1974), and Lowden *et al.* (1974) have previously reported similar cases in which a slow progressive neurodegenerative disorder, characterized by some skeletal abnormalities, ataxia, and slurred speech, but near-"normal" intelligence, is associated with a profound (<4% of normal) deficiency of synthetic β-galactosidase activity in leukocytes and fibroblasts. Some of these cases may show corneal clouding (associated with mucopolysaccharidoses) and a cherry red spot in the macula (associated with neuronal sphingolipidoses) (Orii *et al.*, 1972; Suzuki *et al.*, 1977), and the degree of neurological degeneration can be minimal or quite severe. Koster *et al.* (1976) reported a total deficiency of β-galactosidase in fibroblasts from a 29-year-old male with skeletal abnormalities and normal intelligence. Fusion of these cells with fibroblasts from either Type I or Type II patients gave β-galactosidase complementation. Similar results were obtained by Galjaard *et al.* (1975), who showed that Type I × Type IV hybrids had normal β-galactosidase activity, whereas Type IV × Type IV, Type I × Type II, and Type II × Type II hybrids were still deficient. These findings suggest either that the mutation in Type IV patients involves a different gene locus or that the low β-galactosidase levels are an *in vitro* artifact and Type IV patients have a different metabolic disorder.

3.2.4. *Mucolipidoses and Mucopolysaccharidoses*

Reports of partial β-galactosidase activities in tissues from such patients can be attributable to the effect of lysosomal storage of polyanionic glycosaminoglycans such as heparan sulfate and dermatan sulfate (Kint *et al.*, 1973; Rushton and Dawson, 1977). Normal activity generally can be restored *in vitro* by the addition of chloride ion to the extract. Fibroblasts cultured from such patients do not normally show a β-galactosidase abnormality, with the exception of I-cell disease (mucolipidosis II), in which there is a profound deficiency.

3.3. *N*-Acetyl-*β*-Glucosaminidase Deficiency (G_{M2}-Gangliosidoses) (Sandhoff's Disease)

Sandhoff's disease is a neurovisceral storage disease resulting from the low activity of the *N*-acetyl-*β*-hexosaminidase A and B isoenzymes (Sandhoff *et al.*, 1971). The B isoenzyme of *N*-acetyl-*β*-D-hexosaminidase is a β_6 hexamer coded for by a gene on human chromosome 5, whereas the A isoenzyme is an $\alpha_3\beta_3$ hexamer coded for by genes on human chromosomes 5 and 15 (Lalley *et al.*, 1974). Their combined deficiency results in the storage of G_{M2} ganglioside (GalNAc(*β*,1-4)[NeuNAc]Gal-GlcCer), asialo-G_{M2} ganglioside, and glycoprotein-derived oligosaccharides (Ng Ying Kin and Wolfe, 1974) such as [GalNAc(*β*,1-2)Man(*α*,1-3/6]$_2$Man(*β*,1-4)GlcNAc (Fig. 1) in brain (Table I). Massive storage of this oligosaccharide, together with globoside [GalNAc(*β*,1-3)Gal-Gal-GlcCer] and asialo-G_{M2}, is also observed in liver and other nonneural tissue of such patients (Snyder *et al.*, 1972). GlcNAc(*β*,1-2)Man linkages in glycoproteins are catabolyzed only by the B isoenzyme of *N*-acetylhexosaminidase, since an abnormality in the A isoenzyme alone [known as Tay–Sachs disease or G_{M2}-gangliosidosis (Okada and O'Brien, 1969)] results in the massive neural accumulation of G_{M2} ganglioside (Svennerholm, 1964), but no oligosaccharide accumulation. The B isoenzyme, which by inference is responsible for glycoprotein catabolism, is apparently less important than the A isoenzyme in nervous tissue, since the two disorders are neuropathologically similar. Further, the low level of oligosaccharide storage in both G_{M1}- and G_{M2}-gangliosidosis brain (Tsay and Dawson, 1976) indicates either a slow rate of turnover of brain asialoglycoproteins or some alternative neuronal metabolic pathway involving endoglycosidases.

3.4. *α*-Mannosidase Deficiency (Mannosidosis)

A deficiency of lysosomal *α*-mannosidases A and B isoenzymes results in a rapidly fatal neurovisceral storage disease known as mannosidosis. Both isoenzymes are coded for by a gene on chromosome 19 (Champion and Shows, 1977), and the nonlysosomal isoenzyme (C) is unaffected (Carroll *et al.*, 1972). Mannose-rich oligosaccharides have been isolated (but not characterized)(Öckerman, 1969) from central nervous tissue (Table I) and probably correspond in structure to the oligosaccharides isolated from the urine of such patients (Nordén *et al.*, 1973, 1974), for example:

(A) Man(*α*,1-3)Man(*β*,1-4)GlcNAc

(B) Man(*α*,1-2)Man(*α*,1-2)Man(*α*,1-3)Man(*β*,1-4)GlcNAc

(C) Man(*α*,1-2)Man(*α*,1-2)Man(*α*,1-6)↘
Man(*α*,1-2)Man(*α*,1-3)↗ Man(*β*,1-4)GlcNAc

Type A is common to both acidic and neutral glycoproteins (Fig. 1), whereas Type C is more typical of the neutral type. Type B is an unusual linear oligosaccharide with no known counterpart in glycoproteins, apart from the $(Glc)_{2-3}$(Man) $(GlcNAc)_2$-dolichol pyrophosphate intermediate in glycoprotein synthesis (see Chapter 4). An inherited deficiency of β-mannosidase has not yet been described. The accumulation of glycoprotein-derived oligosaccharide units in neuronal lysosomes appears to be the primary cause of the neurological dysfunction associated with the disease. The disorder also occurs in Angus cattle, and Jolly *et al.* (1976) reported "natural" enzyme replacement therapy in a chimeric male mannosidosis calf that received 77% of its lymphocytes from its normal twin sister. α-Mannosidase activity was low in all tissues (except lymphocytes, in which it was almost normal), but storage of oligosaccharide in brain (and other tissue) was slightly less than normal. Although the calf succumbed to mannosidosis, the observations offer some hope for eventual enzyme replacement therapy of such inborn errors of glycoprotein catabolism.

3.5. α-Fucosidase Deficiency (Fucosidosis)

A deficiency of lysosomal α-fucosidase results in a rapidly debilitating neurovisceral storage disease termed fucosidosis (Durand *et al.*, 1969). Studies with synthetic α-fucosides indicate that the enzyme is coded for by a gene on human chromosome 1 (Hirschhorn *et al.*, 1976), although by using natural substrates we obtained evidence for the existence of at least two α-fucosidases in human tissue, one hydrolyzing Fuc(α,1-2)Gal and Fuc(α,1-6)GlcNAc linkages and the other hydrolyzing Fuc(α,1-3 and 4)GlcNAc linkages. This could explain some of the observed clinical heterogeneity in fucosidosis (Borrone *et al.*, 1974). Tsay *et al.* (1976) isolated a decasaccharide with the apparent structure [Fuc(α,1-2)Gal(β,1-4)GlcNAc(α,1-2)Man]$_2$(α,1-3/6)-Man(β,1-4)GlcNAc, a Fuc(α,1-6)GlcNAc disaccharide (Table I), and two minor glycosphingolipid storage compounds, H-antigen [Fuc(α,1-2)Gal-GlcNAc-Gal-Glc-Cer] and Fuco-G_{M1} (Fuc(α,1-2)GalNAc[NeuAc]-Gal-GalCer) from the brain of one patient with fucosidosis. Cultured skin fibroblasts from patients with fucosidosis stored only the glycopeptide (Fuc(α,1-6)GlcNAc-Asn) from the most common type of glycoprotein oligosaccharide unit (Tsay and Dawson, 1976). The storage of glycopeptides rather than oligosaccharides in fibroblast cultured patients with inborn errors of glycoprotein catabolism may be attributable to the low level of endoglycosidase activity in these cells. The Fuc(α,1-6)GlcNAc disaccharide is a common component of glycoproteins (Fig. 1), but thus far we have been unable to detect in normal nervous tissue the glycoprotein precursor of the fucodecasaccharide stored in fucosidosis brain. Recently, Strecker *et al.* (1977*b*) reported the isolation and characterization of ten different fucooligosaccharides and glycopeptides from the urine of patients with fucosidosis. Such complexity may reflect impaired catabolism of blood group glycoproteins, but is consistent with the complex pattern of excreted oligosaccharides and glycoproteins in other inborn errors of glycoprotein catabolism including mannosidosis, G_{M1}-gangliosidosis, and aspartylglucosaminuria.

3.6. *N*-Aspartyl-β-Glucosaminidase Deficiency (Aspartylglucosaminuria)

A deficiency of *N*-aspartyl-β-glucosaminidase activity has been reported in patients with aspartylglucosaminuria (AGU) (Jenner and Pollitt, 1967; Autio, 1972; Palo *et al.*, 1973). The storage of *N*-aspartyl-GlcNAc in brain (Table I) and other tissues was originally reported in two patients from England (Jenner and Pollitt, 1967), but almost all subsequent cases have originated from Finland (Autio, 1972; Palo *et al.*, 1972). The glycopeptide is an integral part of the linkage region of both acidic (sialo-) and neutral (manno)glycopeptides, and studies on the purified enzyme indicate that it will hydrolyze only asparaginylglycopeptides. Catabolism of the oligosaccharide unit as shown in Fig. 1 apparently proceeds normally in AGU brain, although studies in several laboratories indicated the presence of at least 20 minor glycopeptides in AGU urine in addition to the major GlcNAc-Asn glycopeptide (Palo *et al.*, 1973; Akasaki *et al.*, 1976; Sugahara *et al.*, 1976). These minor abnormal glycopeptides can be acidic, e.g., NeuAc-Gal-GlcNAc-Gal-GlcNAc-Asn (Sugahara *et al.*, 1975), or neutral, e.g., Man(α,1-6)Man(β,1-4)GlcNAc(β,1-4)GlcNAc-Asn (Akasaki *et al.*, 1976). Electron-microscopic studies of brain from AGU patients have revealed lysosomal storage bodies similar to those seen in mannosidosis.

3.7. Other Genetic Disorders That Involve Glycoprotein Storage

3.7.1. Gaucher's Disease (β-Glucosidase Deficiency)

Although the major storage material in Gaucher's disease is a glycosphingolipid (glucosylceramide), ancillary storage of a thermostable glycoprotein, believed to be an essential component of active β-glucosidase, was reported (Ho *et al.*, 1973). The presence of glycoprotein "co-factors" or "activators" may be necessary for the activity of other lysosomal hydrolases such as *N*-acetyl-β-hexosaminidase B, α-galactosidase, and β-galactosidase (Li *et al.*, 1973), and it is possible that an abnormality in the activator for G_{M2} ganglioside hydrolysis is the basic defect in the AB variant (Sandhoff *et al.*, 1971) of Tay–Sachs disease.

3.7.2. Inborn Errors of Collagen Metabolism

Several inborn errors of collagen-chain assembly and metabolism have been described (McKusick and Martin, 1975), but none primarily involves the oligosaccharide unit. Nervous system involvement has been described only in Moya-Moya [a variant of Ehrlers–Danlos syndrome resulting from defective amino acid hydroxylation (Richman *et al.*, 1977)], and even here the neurological problems are caused by cerebral artery deterioration.

3.7.3. Metachromatic Leukodystrophy

Metachromatic leukodystrophy is a progressive demyelinating disorder, with typical onset at age 3–7 heralded by gait disturbance and ataxia (Moser, 1972). The adult form with onset at age 30–40 presents as a presenile dementia. Tissues show a profound deficiency of arylsulfatase A activity (Table II) and sulfogalactosylceramide is stored in central and peripheral nervous system myelin and to a lesser extent in visceral tissue (Moser, 1972). There have been some reports (Brunngraber *et al.*, 1976) of sulfated glycopeptide storage.

3.8. Effect of Oligosaccharide Storage on CNS Function

The intralysosomal accumulation of oligosaccharide material in primary oligosaccharidoses, such as mannosidosis and fucosidosis, results in generalized ballooning of the cytoplasm of nerve cells throughout the cerebral cortex, brainstem, and spinal cord, and to a lesser extent in the basal ganglia. Electron-microscopic examination reveals a somewhat amorphous or granu-

Table II. Inborn Errors of Metabolism Associated with Specific Sulfatase Deficiencies

Sulfatase	Disease	Storage material in nervous tissue
1. Iduronate-2-sulfate	Hunter's disease	Heparan sulfate Dermatan sulfate
2. Galactosyl-3-sulfate Ascorbate-2-sulfate (arylsulfatase A)	Metachromatic leukodystrophy	Sulfogalactosylceramide (sulfatide)
3. Galactosyl-6-sulfate GalNAc-6-sulfate	Morquio's disease	None[a] Chondroitin 6-sulfate[a] Keratan sulfate[a]
4. GalNAc-4-sulfate (arylsulfatase B)	Maroteaux–Lamy syndrome	Dermatan sulfate
5. GlcNAc-6-sulfate	Glucosamine sulfate sulfatase deficiency	Heparan sulfate Keratan sulfate[a]
6. $GlcNSO_3H$ $GlcNSO_3H$-6-sulfate	Sanfilippo A syndrome	Heparan sulfate
7. Steroid (arylsulfatase C)	Steroid sulfatase	None[a] Δ-5-3-β-Hydroxysteroid sulfates[a]

[a] Material stored in extraneural tissue.

lar storage material with few of the membranocytoplasmic bodies associated with the gangliosidoses. Neurological degeneration progresses more slowly than in the gangliosidoses, but presumably results from the eventual distortion and disruption of the neuronal membrane.

4. Glycosaminoglycans of Nervous Tissue

As discussed in Chapter 3, glycosaminoglycans are minor constituents of nervous tissue (0.02% of fresh weight in human brain). Hyaluronic acid, a repeating polymer of [GlcUA(β,1-3)GlcNAc(β,1-4)] disaccharide units, is found largely in the extracellular matrix and appears to have an important role in CNS development and differentiation (Toole, 1976). Heparan sulfate is a heterogeneous repeating polymer of

$$\begin{bmatrix} \text{IdUA}(\longrightarrow 2\text{SO}_3\text{H})(\alpha,1\text{-}4)\text{GlcNAc}(6\text{SO}_3\text{H})(\alpha,1\text{-}4) \\ \textit{or} \qquad\qquad\qquad\qquad\qquad\qquad \textit{or} \\ \text{GlcUA} \text{ ------- } (\beta,1\text{-}4)\text{GlcNSO}_3\text{H} \text{ ------- } (\beta,1\text{-}4) \end{bmatrix}$$

disaccharide units (Fig. 2) attached to a protein core, and is generally associated with the plasma membrane of cells (Kraemer, 1971). Chondroitin 4/6-sulfate contains [GlcUA(β,1-3)GalNAc-(4 or 6SO_3H)(β,1-4)] repeating disaccharide units attached to protein by a Gal(β,1-3)Gal(β,1-4)Xyl-Ser linkage (Rodén, 1970). The ratio hyaluronic acid/heparan sulfate/chondroitin sulfates in whole mammalian brain is approximately 4 : 1 : 5, although this may differ according to cell type and brain region (Margolis, R. U., and Margolis, R. K., 1974). Dermatan sulfate may be a minor constituent of brain parenchyma and contains [IdUA(β,1-3)GalNAc-(4 or 6SO_3H)(β,1-4)] repeating disaccharide units, while keratan sulfate contains [Gal(6SO_3H)(β,1-4)-GlcNAc(-6SO_3H)(β,1-3)] repeating disaccharide units and occurs as skelatal and corneal forms, but is not found in nervous tissue. The high-molecular-weight (20,000–10^6) polydisperse oligosaccharide chains of heparan, chondroitin, and dermatan sulfates are covalently attached to protein through a Gal(β,1-3)Gal(β,1-4)Xyl-Ser linkage region, keratan sulfate contains both GalNAc-Ser/Thr and GlcNAc-Asn linkages, and hyaluronic acid is probably not associated covalently with protein (Rodén, 1970).

5. Inborn Errors of Glycosaminoglycan Catabolism (the Mucopolysaccharidoses)

Glycosaminoglycans (mucopolysaccharides) are normally degraded by a combination of exo- and endoglycosidases, and the deficiency of one of these

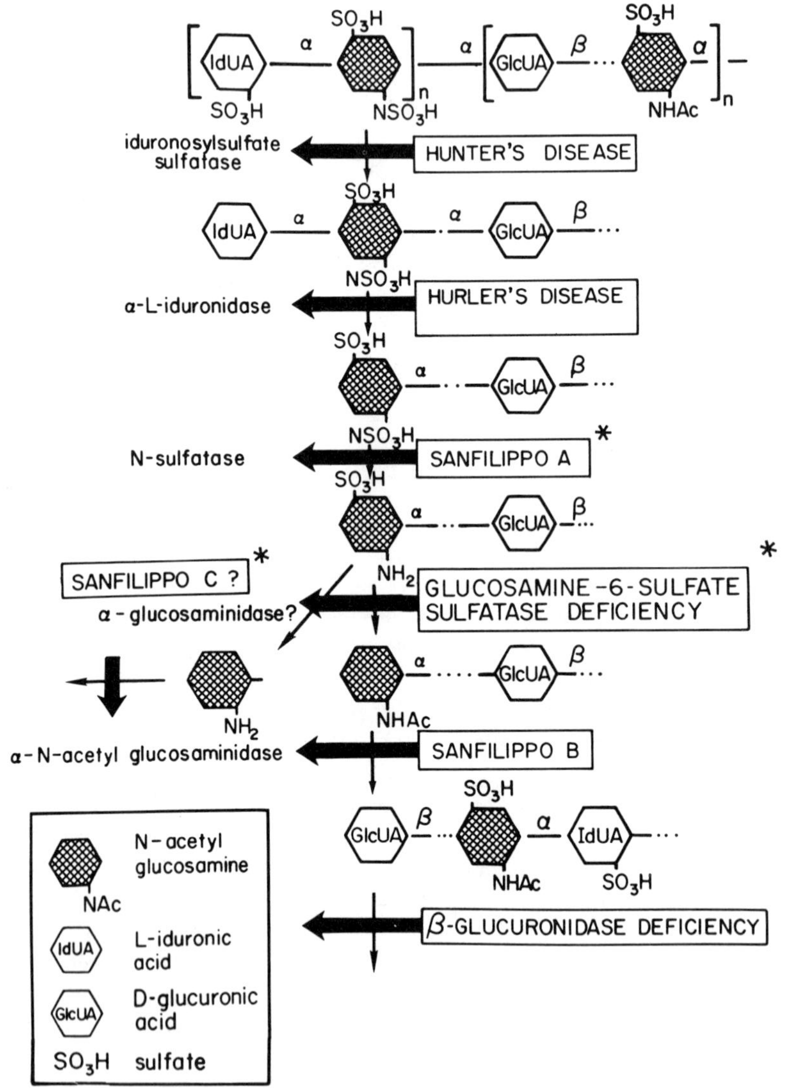

Fig. 2. Schematic pathway for catabolism of heparan sulfate disaccharide units by exoglycosidases showing location of enzyme defects leading to mucopolysaccharide storage diseases.

enzymes typically results in major visceral and skeletal abnormalities that may be accompanied by severe neurological degeneration. The mucopolysaccharidoses were classified by McKusick (1972), and the first evidence that they involved catabolic enzyme deficiencies came from the cross-correction studies of Fratantoni *et al.* (1969) and by analogy to the sphingolipidoses.

5.1. α-L-Iduronidase Deficiency (Hurler–Scheie Syndrome): Mucopolysaccharidosis I and V

Hurler's disease is a neurovisceral storage disease that results from an autosomal recessively inherited deficiency of the lysosomal hydrolase α-L-iduronidase (Matalon and Dorfman, 1972; Dorfman *et al.*, 1972), as shown in Figs. 2 and 3 and Table III. Iduronic acid is a major component of both heparan sulfate and dermatan sulfate, and both these glycosaminoglycans accumulate in tissues (including brain) and urine of patients with Hurler's disease (Dorfman and Matalon, 1972; Matalon and Dorfman, 1976). Electron-microscopic examination of cerebral neurons reveals characteristic lysosomal storage bodies, the so-called "zebra bodies," that appear to result from the primary accumulation of glycosaminoglycans and the secondary accumulation of membrane-associated glycosphingolipids such as G_{M3} and G_{D3} (Dorfman and Matalon, 1972).

A deficiency of α-L-iduronidase activity toward the synthetic substrate phenyl α-L-iduronide was also demonstrated in patients with the clinically milder Scheie syndrome (Bach *et al.*, 1972; Matalon and Dorfman, 1976). Scheie fibroblasts do not "correct" the abnormal sulfate-incorporation patterns observed in skin fibroblasts cultured from patients with Hurler's disease when cocultured with Hurler fibroblasts (Barton and Neufeld, 1971), whereas fibroblasts from patients with other mucopolysaccharidoses, such as Hunter's disease, will cross-correct (Fratantoni *et al.*, 1969). Thus, the two disorders are probably allelic mutations. Preliminary observations indicate that patients with the Scheie syndrome excrete much more dermatan sulfate than heparan sulfate and that cell-free extracts of Scheie fibroblasts release iduronic acid from desulfated heparin preparations with the structure [IdUA(α,1-4)-GlcNAc(α,1-4)]$_n$, whereas Hurler fibroblast extracts do not (Matalon and Deanching, 1977). Extracts of both cells are unable to catabolize dermatan sulfate preparations. Since storage of heparan sulfate is generally associated with neurological dysfunction and storage of dermatan sulfate with organomegaly, these findings (which need to be confirmed) would explain the lesser degree of neurological involvement in the Scheie syndrome.

5.2. Iduronate Sulfate Sulfatase Deficiency (Hunter's Disease): Mucopolysaccharidosis II

Hunter's disease is an X-linked, recessively inherited neurovisceral storage disease (Dorfman and Matalon, 1972) that results from a deficiency of the

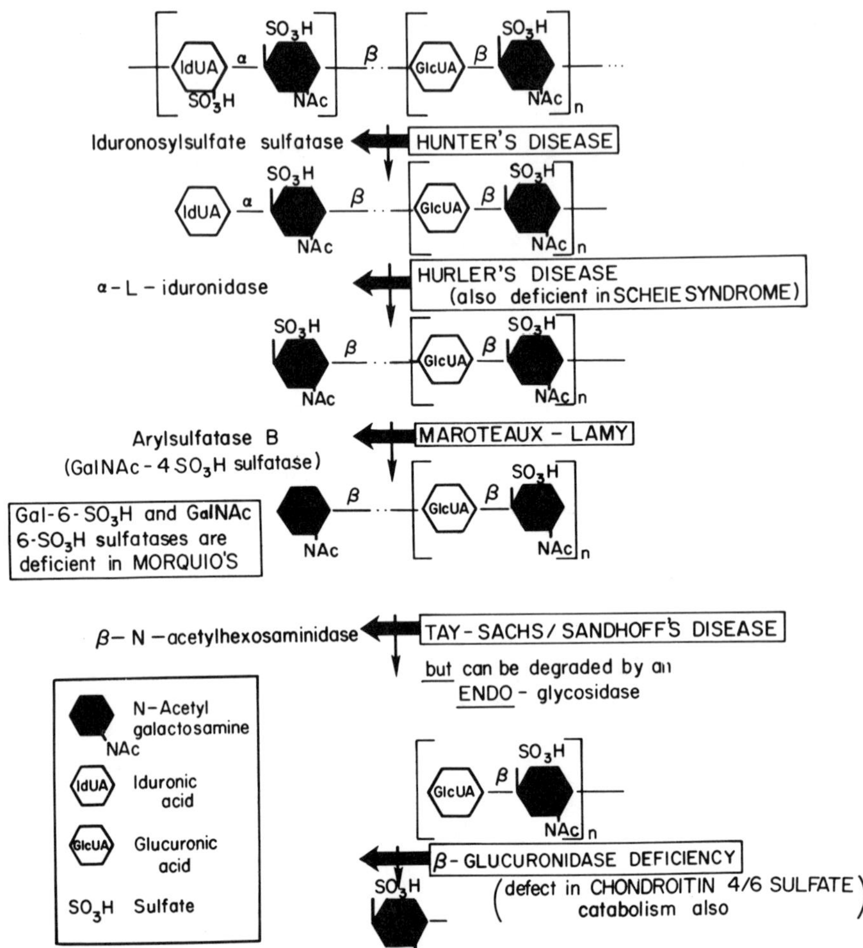

Fig. 3. Schematic pathway for catabolism of dermatan sulfate disaccharide units by exoglycosidases showing location of enzyme defects leading to mucopolysaccharide storage diseases.

enzyme iduronate sulfate sulfatase (Bach *et al.*, 1973; Coppa *et al.*, 1973; Sjöberg *et al.*, 1973) (Table II), as shown in Figs. 2 and 3. The disorder can be diagnosed *in vitro* by a failure to hydrolyze *O*-(α-idopyranosyluronic acid-2-sulfate)(1-4)-2,5-anhydro-D-[^{3}H]mannitol-6-sulfate to a [^{3}H]monosulfate disaccharide that is then hydrolyzed by α-L-iduronase *in vitro* to a [^{3}H]anhydromannitol-6-sulfate (Hall *et al.*, 1978). Since the ratio of excess heparan sulfate to dermatan sulfate stored and excreted by patients with Hunter's disease is somewhat less than that observed in patients with Hurler's disease (Dorfman and Matalon, 1972), this may be an explanation for the milder neurological symptoms associated with Hunter's disease. However, both disorders are clinically heterogeneous, and microscopic examination of nervous tissue reveals the characteristic lysosomal "zebra bodies" in both Hurler and Hunter patients (Dorfman and Matalon, 1972). Because the assay for Hunter's disease is technically difficult, the most common method of diagnosis is that of a male with the Hurler–Hunter syndrome who has a normal level of α-L-iduronidase, or where cultured fibroblasts will "cross-correct" the abnormal [^{35}S]glycosaminoglycan accumulation in Hurler, but not Hunter, fibroblasts (Fratantoni *et al.*, 1969).

5.3. Sulfamidase Deficiency (Sanfilippo A Syndrome): Mucopolysaccharidosis IIIa

The enzyme sulfamidase will hydrolyze the *N*-sulfate group in *N*-sulfatylglucosamine or *N*-sulfatylglucosamine-6-sulfate residues of heparin and heparan sulfate and is deficient in patients with Sanfilippo A disease (Table II) (Bach *et al.*, 1973; Matalon and Dorfman, 1974), as shown in Fig. 2. This is a severe progressive mental retardation syndrome with bony abnormalities, but lack of corneal clouding and overt hepatosplenomegaly, which is further characterized biochemically by the storage and excretion of heparan sulfate (Dorfman and Matalon, 1972). Very little dermatan sulfate is stored or excreted in comparison with the Hunter–Hurler syndrome, and it is significant that both corneal clouding and visceromegaly are much less pronounced. Enzymatic diagnosis (sulfamidase deficiency) is based on the inability of tissue extracts to

Table III. Inborn Errors of Glycosaminoglycan Catabolism Associated with Specific Glycosidase Deficiencies

Enzyme	Disease	Storage material in nervous tissue
α-L-Iduronidase	Hurler's disease Scheie syndrome	Heparan sulfate Dermatan sulfate
β-Glucuronidase	β-D-Glucuronidase deficiency	Chondroitin sulfate
N-Acetyl-α-glucosaminidase	Sanfilippo B syndrome	Heparan sulfate
β-Galactosidase	G_{M1}-gangliosidosis variant	G_{M1} ganglioside Keratan sulfate linkage region [a] Gal-Gal-Xyl-Ser?
β-Galactosidase	Mucopolysaccharidosis IV B	Keratan sulfate [a]
β-Galactosidase	Mucolipidosis	Unidentified
N-Acetyl-β-hexosaminidase	Sandhoff's disease	G_{M2} ganglioside; no glycosaminoglycans
β-Xylosidase	Unknown	Xyl-Ser?

[a] Material stored in extraneural tissue.

release [^{35}S]sulfate from *N*-[^{35}S]sulfate-labeled heparin-derived oligosaccharides, a normal level of *N*-acetyl-α-D-glucosaminidase activity [deficient in Sanfilippo B (O'Brien, 1972*b*)], and cross-correction studies (Kresse *et al.*, 1971).

5.4. *N*-Acetyl-α-D-Glucosaminidase Deficiency (Sanfilippo B Syndrome): Mucopolysaccharidosis IIIb

Heparan sulfate contains a mixture of α-linked GlcNAc and $GlcNSO_3H$ residues in the hexosamine–uronic acid disaccharide repeating unit. The α-linked GlcNAc residues are hydrolyzed by an *N*-acetyl-α-glucosaminidase that is deficient in Sanfilippo B patients (O'Brien, 1972*b*; von Figura and Kresse, 1972) (Table III). The A and B forms of this neurodegenerative syndrome are clinically indistinguishable and were initially separated on the basis of cultured skin fibroblast cross-correction studies (Kresse *et al.*, 1971) and the fact that only about half the Sanfilippo patients showed a deficiency of *N*-acetyl-α-glucosaminidase activity (O'Brien, 1972*b*; von Figura and Kresse, 1972). Kresse (1975) reported that fibroblasts from one patient with the Sanfilippo syndrome would cross-correct the abnormal sulfate incorporation in either Type A or B fibroblasts. This disorder was designated Sanfilippo C, and there was speculation that it could result from a deficiency of an α-glucosaminidase enzyme. It is also possible that this patient could have an *N*-acetylglucosamine-6-sulfate sulfatase deficiency as recently described by Di-Ferrante *et al.* (1978).

The basis of the severe mental retardation in Sanfilippo patients appears to be the accumulation of heparan sulfate, believed to be a functional constituent of neuronal and glial cell membranes, and the secondary accumulation of short-chain gangliosides (G_{M3}, G_{M2}, and G_{D3}) and neutral glycosphingolipids (Kint *et al.*, 1973). This results in abnormal ballooning of neurons (containing intracytoplasmic granular metachromatic material), but does not cause demyelination. The early and rather severe involvement of the cortical neurons in the second and fourth layers in Sanfilippo patients has been correlated with their rapid mental deterioration, and the relative intactness of the third and fifth layer is consistent with their better retention of motor function (Dekaban and Patton, 1971).

5.5. *N*-Acetylgalactosaminyl-6-Sulfate and Galactose-6-Sulfate Sulfatase Deficiency (Morquio's Disease): Mucopolysaccharidosis IV

Morquio's disease is an autosomal recessively inherited mucopolysaccharidosis involving severe skeletal deformities, spondylepiphyseal dysplasia, and often corneal clouding, but minimal visceromegaly or neurological degeneration. However, abnormal neuronal cytoplasmic globules have been reported in the brain of patients with Morquio's syndrome, which may represent glycosaminoglycan storage material (Gilles and Deuel, 1971; Koto *et al.*,

1978). Such patients store and excrete large quantities of chondroitin sulfate and keratan sulfate. These two glycosaminoglycans appear to share a common polypeptide chain, perhaps existing as heterogeneous dimers, and the major point of structural similarity is that they contain either GalNAc, GlcNAc, or galactose that is sulfated at the C-6 position. To test the possibility that Morquio's disease could result from a specific sulfatase deficiency, Matalon *et al* (1974*a*) prepared a GalNAc(6-[^{35}S]O_3H)[GlcUA-GalNAc(6-[^{35}S]O_3H)] heptasaccharide from embryonic chick epiphyseal chondroitin sulftate and showed that homogenates of fibroblasts cultured from patients with Morquio's disease liberated only 10–15% as much sulfate as did similar preparations from controls and patients with Hurler–Hunter or Sanfilippo syndromes. Further, analysis of the partially degraded keratan sulfate secreted by patients with Morquio's disease (Dawson, unpublished) revealed a galactose/GlcNAc ratio of 16 : 12 and showed Gal-6-SO_3H to be the terminal nonreducing sugar in the abnormal stored material. More recent studies (Horwitz and Dorfman, 1978) suggest that the basic defect is the inability to hydrolyze both 6-sulfated galactose and *N*-acetylgalactosamine residues in chondroitin sulfate and keratan sulfate (Table II).

5.6. *N*-Acetylglucosamine-6-Sulfate Sulfatase Deficiency (Mucopolysaccharidosis VIII?)

DiFerrante *et al.* (1978) described a 5-year-old boy with retarded growth and mental development (I.Q. 50–60), mild osteochondrodystrophy, hepatomegaly, and excessive and coarse hair, but clear corneas. Clinically, the patient showed features of both the Sanfilippo and Morquio syndromes. Biochemical studies indicated excessive excretion of heparan sulfate and keratan sulfate and accumulation of [^{35}S]glycosaminoglycans by cultured fibroblasts. Fibroblast homogenates from this patient were unable to desulfate *N*-acetylglucosamine-6-sulfate (GlcNAc 6-SO_3H) and its [^{3}H]alditol (Table II), but readily desulfated [^{3}H]galactitol-6-SO_3H, [^{3}H]-*N*-acetylgalactosaminitol-6-SO_3H, and tetrasaccharides containing terminal GalNAc 6-SO_3H residues, in complete contrast to fibroblasts cultured from three patients with Morquio's disease. Sulfamidase, *N*-acetyl-α-glucosaminidase, α-iduronidase, and related enzymes were normal. A second case has now been reported (Matalon *et al.*, unpublished), which confirms its classification as Mucopolysaccharidosis VIII, a deficiency of *N*-acetylglucosamine-6-sulfate sulfatase, with clinical and biochemical features resembling the Morquio and Sanfilippo phenotypes.

5.7. *N*-Acetylgalactosamine-4-Sulfate Sulfatase (Arylsulfatase B) Deficiency (Maroteaux–Lamy Syndrome): Mucopolysaccharidosis VI

The Maroteaux–Lamy syndrome is characterized by hepatosplenomegaly, skeletal abnormalities, and gross corneal opacities, but only mild, progressive mental retardation. The storage material is primarily dermatan sulfate, and a total deficiency of arylsulfatase B activity (Table II) can be demonstrated in tissues and cultured skin fibroblasts (Stumpf *et al.*, 1973). Biochemical studies with natural substrates indicate that the disorder results from a specific *N*-acetylgalactosamine-4-sulfate sulfatase deficiency (Matalon *et al.*, 1974*b*). Clinically, the disorder resembles a milder form of the Hurler–Hunter syndrome, from which it can be distinguished both enzymatically and by the absence of heparan sulfate excretion and storage. The existence of this phenotype is further evidence that abnormal heparan sulfate catabolism rapidly leads to severe neurological dysfunction.

5.8. β-Glucuronidase Deficiency: Mucopolysaccharidosis VII

A 2-year old patient with some mental retardation, coarse facial features, hepatosplenomegaly, and skeletal abnormalities, who excreted primarily chondroitin sulfate, was found to have a specific deficiency of lysosomal β-glucuronidase activity (Sly *et al.*, 1973; Hall *et al.*, 1973) (Table III). Glucuronic acid is a constituent of some molecules of heparan sulfate (Fig. 2) and dermatan sulfate (Fig. 3), both of which showed slight accumulation. However, the content of hyaluronic acid was normal despite its high GlcUA content, indicating that it is degraded primarily by an endoglycosidase, hyaluronidase.

5.9. Multiple Sulfatase Deficiency

Biochemical studies on the human mucopolysaccharidoses and sphingolipidoses have revealed the existence of at least six specific lysosomal sulfatases and one nonlysosomal sulfatase, as shown in Table II. Almost all these sulfatases appear to be deficient (10% of normal) in an autosomal recessively inherited disorder known as multiple sulfatase deficiency (MSD). MSD presents as a primary demyelinating metachromatic leukodystrophy and increasing ichthyosis and hepatosplenomegaly (Austin, 1973; Murphy *et al.*, 1971; Couchot *et al.*, 1974). The accumulating glycosaminoglycans in liver and urine appear to be mainly heparan sulfate, together with dermatan and chondroitin sulfates, but there is some disagreement in the literature. At the electron-microscopic level, the disorder can be distinguished from classic cases of metachromatic leukodystrophy and diffuse cerebral sclerosis (Couchot *et al.*, 1974), indicating that glycosaminoglycans and possibly other sulfated complex carbohydrates are stored in nervous tissue of MSD patients.

5.10. Steroid Sulfatase Deficiency

Steroid sulfatase (arylsulfatase C) is a nonlysosomal enzyme with a pH optimum of 7–8 (microsomal), and has been shown to be specifically deficient in an X-linked syndrome characterized by ichthyosis and problems during pregnancy and childbirth (Table II). No neurological symptoms have been reported and no specific glycosaminoglycan abnormalities have been found, but patients secrete large amounts of Δ-5-3-β-hydroxy-steroid sulfates in urine (Shapiro *et al.*, 1977).

5.11. Other Glycosidase Deficiencies

Three mucopolysaccharidoses result from specific glycosidase deficiencies, as shown in Table III: α-L-iduronidase deficiency (Hurler's disease), β-D-glucuronidase deficiency, and *N*-acetyl-α-D-glucosaminidase deficiency (Sanfilippo B syndrome). Other mucopolysaccharidoses could result from deficiencies as discussed below.

β-Galactosidase deficiency: A deficiency of β-galactosidase would be expected to result in accumulation of material related structurally to keratan sulfate, and this has been observed in certain patients with variant forms of G_{M1}-(generalized) gangliosidosis (Table III). The structure of the keratan-sulfate-related material accumulating in liver (Tsay *et al.*, 1975) was found to be

$$\mathrm{Gal}(\beta,1\text{-}4)[\mathrm{GlcNAc}(\beta,1\text{-}3)\mathrm{Gal}]_2\mathrm{GlcNAc}(\beta,1\text{-}4)[\mathrm{Gal}(\beta,1\text{-}6)]\mathrm{GalNAc\text{-}Ser}$$

which corresponds to the desulfated linkage region of keratan sulfate (Kieras, 1974). The major cerebral storage material was G_{M1} ganglioside (Ledeen *et al.*, 1965). In visceral tissue from typical cases of G_{M1}-gangliosidosis, a galactooligosaccharide derived from impaired glycoprotein catabolism is the major storage product (see Section 3.2.).

A Morquio-like syndrome with β-galactosidase deficiency and normal hexosamine sulfatase activity, tentatively designated Mucopolysaccharidosis IV B (Table III), was recently reported by Arbisser *et al.* (1977). The patient was a 14-year-old female with dysostosis multiplex, odontoid hypoplasia, short stature, and cloudy corneas, but without detectable CNS abnormalities. A conjuctivae biopsy revealed intracytoplasmic vacuoles typical of lysosomal storage disease. This disorder is further evidence for the absence of keratan sulfate in nervous tissue.

N-Acetyl-β-glucosaminidase deficiency: Patients with an apparent deficiency of *N*-acetyl-β-hexosaminidase activity (Sandhoff variant of G_{M2}-gangliosidosis) (Table III) have been studied in considerable detail (see Section 3.3), and there is no biochemical evidence for glycosaminoglycan accumulation. From a knowledge of glycosaminoglycan structure, one would anticipate the storage of chondroitin sulfates and dermatan sulfate, and this was reported in cultured fibroblasts (Cantz and Kresse, 1974). Sandhoff fibroblasts are also unable to degrade a heptasaccharide derived from chondroitin sulfate (Thompson *et al.*, 1973). However, the endoglycosidases (such as hyaluronidase) in liver, brain, and other tissues are apparently able to circumvent such a block and catabolize glycosaminoglycans to hexasaccharides or tetrasaccharides, which have escaped detection thus far.

β-Xylosidase deficiency: A deficiency of this enzyme (Table III) might be expected to result in the accumulation of the glycosaminoglycan linkage region polypeptides of xylosylerine (Table III) (Rodén, 1970), but this has not been reported. There have been some clains that β-xylosidase and β-glucosidase are the same enzyme (Öckerman, 1968), and a more detailed study of Gaucher's disease may therefore be warranted.

5.12. Generalized Mycopolysaccharidoses (the "Mucolipidoses")

The Mucolipidoses Types I, II, III, and IV are a confusing group of storage disorders with phenotypic expression resembling that of the mucopolysaccharidoses (Spranger and Wiedemann, 1970; Gilbert *et al.*, 1973; Dawson and Lenn, 1976). The original *Mucolipidosis I* classification included patients with a combination of mild Hurler-like symptoms, moderate mental retardation, normal mucopolysacchariduria, and peculiar fibroblast "inclusion bodies" similar to those seen in "I-cell disease" or Mucolipidosis II (Spranger and Wiedemann, 1970; Dawson and Lenn, 1976). Several of these patients showed a partial deficiency of β-galactosidase activity (Table III) but were subsequently found to have a specific deficiency of α-mannosidase (mannosidosis) (see Section 3.4) or neuraminidase (Cantz *et al.*, 1977; Durand *et al.*, 1977) (see Section 3.1).

Mucolipidosis II ("I-cell disease") is characterized by a clinical picture strikingly similar to that of Hurler's disease, and by the absence of mucopolysacchariduria. Some of these patients secrete sialooligosaccharides (Strecker *et al.*, 1977*a,b*) (as discussed in Section 3.1), and cultured fibroblasts show a marked deficiency (1–10% of normal) of all lysosomal hydrolases with the exception of acid phosphatase and GlcCer-β-glucosidase (Leroy *et al.*, 1971; Gilbert *et al.*, 1973). The culture media surrounding such cells as well as the patients' serum generally contains elevated levels of lysosomal hydrolases (Wiesmann *et al.*, 1971; Den Tandt *et al.*, 1974), leading to the hypothesis that the disease results from a defective high-affinity lysosomal hydrolase uptake mechanism (Hickman *et al.*, 1974). Fibroblasts cultured from some "I-cell" patients accumulate abnormally high levels of glycosaminoglycans as well as glycosphingolipids and other lipids (Matalon *et al.*, 1968; Dawson *et al.*, 1972), so that the disorder must still be regarded as a mucopolysaccharidosis that affects the nervous system, albeit an atypical one.

The term *Mucolipidosis III* (*pseudo-Hurler polydystrophy*) has been used to designate patients with moderate, but degenerating mental retardation (I.Q. 64–85), mild facial coarseness, short stature, and other skeletal abnormalities (Spranger and Wiedemann, 1970; Taylor *et al.*, 1973). A fundamental defect based on partial deficiencies of β-galactosidase or neuraminidase has been proposed but not verified as yet (Michalski *et al.*, 1978).

Mucolipidosis IV resembles the mucopolysaccharidoses, but is heterogeneous and not yet defined biochemically.

6. Effect of Lysosomal Glycosaminoglycan Storage on CNS Function and Prospects for Therapy

Light-microscopic examination of CNS tissue from patients with mucopolysaccharidoses, especially where this involves heparan sulfate accumulation, typically reveals ballooning of neurons, swollen dendrites, vacuolized cytoplasm, and excentric displaced nuclei. Histochemical stains reveal granular material that may be positive for carbohydrate (periodic acid–Schiff-positive) in cortical neurons and part of the neuroglia, but in general, negative with lipophilic stains and nonmetachromatic. Electron microscopy generally shows the presence of lamellar bodies, granulo-membrano-vacuolar bodies, and storage of amorphous material within lysosomes, a much more complex pattern than that observed in the neuronal gangliosidoses such as Tay–Sachs disease in which the multilamellar cytosomes are almost pathognomonic. The cause of neuronal dysfunction may be the sheer bulk of the lysosomal storage bodies or the formation of abnormal processes called meganeurites (Purpura and Suzuki, 1976).

There have been several attempts to treat patients with mucopolysaccharidoses and other lysosomal storage disease by some form of enzyme replacement. The infusion of human serum or leukocytes into patients with the Hurler and Sanfilippo syndromes was initially thought to result in some clinical improvement (DiFerrante *et al.*, 1971; Knudson *et al.*, 1971), but this has not been substantiated (Dekaban *et al.*, 1972; Erickson *et al.*, 1972). Similarly, the infusion of purified enzymes such as *N*-acetyl-α-glucosaminidase into patients with Sanfilippo B syndrome has not produced any remission of clinical symptoms. Although some positive biochemical changes have been reported to result from infusing purified hydrolases into patients with non-neurological disorders such as Gaucher's disease (Brady *et al.*, 1974) and from injecting normal skin fibroblasts into patients with mucopolysaccharide storage disorders, the immediate outlook is not encouraging.

References

Akasaki, M., Sugahara, K., Funakoshi, I., Aula, P., and Yamashina, I., 1976, Characterization of a mannose-containing glycoasparagine isolated from urine of a patient with aspartylglycosaminuria, *FEBS Lett.* **69:**91.

Alhadeff, J. A., Miller, A. L., Wenaas, H., Vedvick, T., and O'Brien, J. S., 1975, Human liver α-L-fucosidase: Purification, characterization and immunochemical studies, *J. Biol. Chem.* **250:**7106.

Arbisser, A. I., Donnelly, K. A., Scott, C. I., Jr., DiFerrante, N., Singh, J., Stevenson, R. E.,

Aylesworth, A. S., and Howell, R. R., 1978, Morquio-like syndrome with β-galactosidase deficiency and normal hexosamine sulfatase activity: Mucopolysaccharidosis IV B, *Am. J. Med. Genet.* **1**:195.

Austin, J. H., 1973, Studies in metachromatic leukodystrophy. XII. Multiple sulfatase deficiency, *Arch. Neurol.* **28**:258.

Autio, S., 1972, Aspartylglucosaminuria: Analysis of thirty-four patients, *J. Ment. Defic. Res. Monogr. Ser.* **1**:1.

Bach, G., Friedman, R., Weismann, B., and Neufeld, E. F., 1972, The defect in the Hurler and Scheie syndromes: Deficiency of α-L-iduronidase, *Proc. Natl. Acad. Sci. U.S.A.* **69**:2048.

Bach, G., Eisenberg, F., Cantz, M., and Neufeld, E. F., 1973, The defect in Hunter's disease: Deficiency of sulfoiduronate sulfatase, *Proc. Natl. Acad. Sci. U.S.A.* **70**:2134.

Barton, R. W., and Neufeld, E. F., 1971, The Hurler factor, *J. Biol. Chem.* **243**:7773.

Borrone, C., Gatti, R., Trias, X., and Durand, P., 1974, Fucosidosis: Clinical, biochemical, immunologic and genetic studies in two new cases, *J. Pediatr.* **84**:727.

Brady, R. O., Pentchev, P. G., Gal, A. E., Hibbert, S. R., and Dekaban, A. S., 1974, Replacement therapy for inherited enzyme deficiency: Use of purified glucocerebrosidase in Gaucher's disease, *N. Engl. J. Med.* **291**:989.

Brunngraber, E. G., Hof, H., Susz, J., Brown, B. D., Aro, A., and Chang, I., 1973, Glycopeptides from rat brain glycoproteins, *Biochim. Biophys. Acta* **304**:781.

Brunngraber, E. G., Davis, L. G., Javaid, J. I., and Berra, B., 1976, Glycoprotein catabolism in brain tissue in the lysosomal enzyme deficiency diseases, in: *Current Trends in Sphingolipidoses and Allied Disorders* (B. W. Volk and L. Schneck, eds.), pp. 31–48, Plenum Press, New York.

Cantz, M., and Kresse, H., 1974, Sandhoff disease: Defective glycosaminoglycan catabolism in cultured fibroblasts and its correction by β-*N*-acetylhexosaminidase, *Eur. J. Biochem.* **47**:581.

Cantz, M., Gehler, J., and Spranger, J., 1977, Mucolipidoses I: Increased sialic acid content and deficiency of an α-*N*-acetylneuraminidase in cultured fibroblasts, *Biochem. Biophys. Res. Commun.* **74**:732.

Carroll, M., Dance, N., Masson, P. K., Robinson, D., and Winchester, B. G., 1972, Human mannosidosis—the enzymic defect, *Biochem. Biophys. Res. Commun.* **49**:579.

Champion, M. J., and Shows, T. B., 1977, Mannosidosis: Assignment of the lysosomal α-mannosidase B gene to chromosome 19 in man, *Proc. Natl. Acad. Sci. U.S.A.* **74**:2968.

Chien, S.-F., and Dawson, G., 1978, Unpublished data.

Chien, S.-F., Weinburg, R., Li, S.-C., and Li, Y. T., 1977, Endo-β-*N*-acetylglucosaminidase from fig latex, *Biochem. Biophys. Res. Commun.* **76**:317.

Church, R. C., Tanzer, M. C., and Pfeiffer, S. E., 1973, Procollagen and collagen synthesis by a clonal line of rat Schwann cells, *Proc. Natl. Acad. Sci. U.S.A.* **70**:1943.

Coppa, G. C., Singh, J., Nichols, B. L., and DiFerrante, N., 1973, Hunter's disease: Sulfoiduronate sulfatase deficiency, *Anal. Lett.* **6**:225.

Couchot, J., Pluot, M., Schmauch, M.-A., Pennaforte, M., and Fandre, M., 1974, La mucosulfatidose, *Arch. Fr. Pediatr.* **31**:775.

Dawson, G., and Lenn, N. J., 1976, Polysaccharide metabolism disorders, in: *Handbook of Clinical Neurology* (P. J. Vinken, G. W. Bruyn, and H. L. Klawans, eds.), Vol. 27, Part I, pp. 143–168, American Elsevier, New York.

Dawson, G., Matalon, R., and Dorfman, A., 1972, Glycosphingolipids in cultured human skin fibroblasts from patients with inborn errors of glycosphingolipid and mucopolysaccharide metabolism, *J. Biol. Chem.* **247**:5951.

Dekaban, A. S., and Patton, V. M., 1971, Hurler's and Sanfilippo's variants of mucopolysaccharidosis: Cerebral pathology and lipid chemistry, *Arch. Pathol.* **91**:434.

Dekaban, A. S., Holden, K. R., and Constantopoulos, E., 1972, Effects of fresh plasma or whole blood transfusions on patients with various types of mucopolysaccharidosis, *Pediatrics* **50**:688.

Den Tandt, W. R., Lassila, E. L., and Philippart, M., 1974, Lysosomal hydrolase abnormalities in "I-cell" disease (mucolipidosis II), *J. Lab. Clin. Med.* **83:**403.

DiFerrante, N., Nichols, B. L., Donnelly, P. Y., Neri, G., Hrgovcic, R., and Berglund, R. K., 1971, Induced degradation of glycosaminoglycans in Hurler's and Hunter's syndromes by plasma infusion, *Proc. Natl. Acad. Sci. U.S.A.* **68:**303.

DiFerrante, N., Ginsberg, L. C., Donnelly, P. V., DiFerrante, D. T., and Caskey, C. T., 1978, Deficiencies of glucosamine-6-sulfate or galactosamine-6-sulfate sulfatases are responsible for different mucopolysaccharidoses, *Science* **199:**79.

Dorfman, A., and Matalon, R., 1972, The mucopolysaccharidoses, in: *The Metabolic Basis of Inherited Disease* (J. B. Stanbury, J. B. Wyngaarden, and D. S. Fredrickson, eds.), pp. 1218–1272, McGraw-Hill, New York.

Dorfman, A., Matalon, R., Cifonelli, J. A., Thompson, J., and Dawson, G., 1972, The degradation of acid mucopolysaccharides and the mucopolysaccharidoses, in: *Sphingolipidoses and Allied Disorders* (B. W. Volk and S. M. Aronson, eds.), pp. 195–210, Plenum Press, New York.

Durand, P., Borrone, C., and Della Cella, G., 1969, Fucosidosis, *J. Pediatr.* **75:**665.

Durand, P., Gatti, R., Cavalieri, S., Borrone, C., Tondeur, M., Michalski, J.-C., and Strecker, G., 1977, Sialidosis (mucolipidosis I), *Helv. Paediatr. Acta* **32:**391.

Erickson, R. P., Sandman, R., Van, W., Robertson, B., and Epstein, C. J., 1972, Inefficacy of fresh frozen plasma therapy of mucopolysaccharidosis II, *Pediatrics* **50:**693.

Feldges, A., Muller, H. J., Buhler, E., and Stalder, G., 1973, G_{M1}-gangliosidosis I: Clinical aspects and biochemistry, *Helv. Paediatr. Acta* **28:**511.

Finne, J., Krusius, T., Rauvala, H., and Hemminki, K., 1977, The disialosyl group of glycoproteins: Occurrence in different tissues and cellular membranes, *Eur. J. Biochem.* **77:**319.

Fratantoni, J. C., Hall, C. W., and Neufeld, E. F., 1969, The defect in the Hurler and Hunter syndromes. II. Deficiency of specific factors involved in mucopolysaccharide degradation, *Proc. Natl. Acad. Sci. U.S.A.* **64:**630.

Galjaard, H., Hoogeveen, A., Keijzer, W., de Wit-Verbeek, H. A., Reuser, A. J. J., Ho, M. W., and Robinson, D., 1975, Genetic heterogeneity in G_{M1}-gangliosidosis, *Nature (London)* **257:**60.

Gielen, W., 1974, The specificity of neuraminidases of higher organisms, *Behring Inst. Mitt.* **55:**85.

Gilbert, E. G., Dawson, G., ZuRhein, G. H., Opitz, J. M., and Spranger, J. W., 1973, "I-cell" disease: Mucolipidosis II, *Z. Kinderheilkd.* **114:**259.

Gilles, F. H., and Deuel, R. K., 1971, Neuronal cytoplasmic globules in the brain in Morquio's syndrome, *Arch. Neurol.* **25:**393.

Hall, C. W., Cantz, M., and Neufeld, E. F., 1973, A β-glucuronidase deficiency mucopolysaccharidosis: Studies in cultured fibroblasts, *Arch. Biochem. Biophys.* **155:**32.

Hall, C. W., Liebaers, I., DiNatale, P., and Neufeld, E. F., 1978, Enzymatic diagnosis of the genetic mucopolysaccharide storage diseases, in: *Methods in Enzymology* (V. Ginsberg, ed.), Vol. 50, pp. 439–455, Academic Press, New York.

Hickman, S., Shapiro, L. J., and Neufeld, E. F., 1974, A recognition marker required for uptake of a lysosomal enzyme by cultured fibroblasts, *Biochem. Biophys. Res. Commun.* **57:**55.

Hieber, V., Distler, J., Myerowitz, R., Schmickel, R. D., and Jourdian, G. W., 1976, The role of glycosidically bound mannose in the assimilation of β-galactosidase by generalized gangliosidosis fibroblasts, *Biochem. Biophys. Res. Commun.* **73:**710.

Hirschhorn, K., Beratis, N. G., and Turner, B. M., 1976, Alpha-L-fucosidase in normal and deficient individuals, in: *Current Trends in Sphingolipidoses and Allied Disorders* (B. W. Volk and L. Schneck, eds.), pp. 205–223, Plenum Press, New York.

Ho, M. W., O'Brien, J. S., Radin, N. S., and Erickson, J. S., 1973, Glucocerebrosidase: Reconstitution from macromolecular components, *Biochem. J.* **131:**173.

Horwitz, A. L., and Dorfman, A., 1978, The enzyme defect in Morquio's disease: The specificities of *N*-acetylhexosamine sulfatases, *Biochem. Biophys. Res. Commun.* **80:**819.

Jenner, F. A., and Pollitt, R. J., 1967, Large quantities of 2-acetamido-1-(β-L-aspartamido)-1,2-dideoxyglucose in the urine of mentally retarded siblings, *Biochem. J.* **103**:48.

Jolly, R. D., Thompson, K. G., Murphy, C. E., Manktelow, B. W., Bruere, A. N., and Winchester, B. G., 1976, Enzyme replacement therapy—an experiment of nature in a chimeric mannosidosis calf, *Pediatr. Res.* **10**:219.

Kaplan, A., Fischer, D., Achord, D., and Sly, W., 1977, Phosphohexosyl recognition is a general characteristic of pinocytosis of lysosomal glycosidases by human fibroblasts, *J. Clin. Invest.* **60**:1088.

Kelly, T. E., and Graetz, G., 1977, Isolated acid neuraminidase deficiency: A distinct lysosomal storage disease, *Am. J. Med. Genet.* **1**:31.

Kieras, F. J., 1974, The linkage region of cartilage keratan sulfate to protein, *J. Biol. Chem.* **249**:7506.

Kint, J. A., Dacrement, G., Carton, D., Orye, E., and Hooft, C., 1973, Mucopolysaccharidosis: Secondarily induced abnormal distribution of lysosomal isoenzymes, *Science* **181**:352.

Knudson, A. G., DiFerrante, N., and Curtis, J. E., 1971, Effect of leukocyte transfusion in a child with Type II mucopolysaccharidosis, *Proc. Natl. Acad. Sci. U.S.A.* **68**:1738.

Koide, N., and Muramatsu, T., 1974, Endo-β-*N*-acetylglucosaminidase acting on carbohydrate moieties of glycoproteins: Purification and properties of the enzyme from *Diplococcus pneumoniae*, *J. Biol. Chem.* **249**:4897.

Kornfeld, R., and Kornfeld, S., 1976, Comparative aspects of glycoprotein structure, *Annu. Rev. Biochem.* **45**:217.

Koster, J. F., Niermeijer, M. F., Loonen, M. C. B. and Galjaard, H., 1976, β-Galactosidase deficiency in an adult: A biochemical and somatic cell genetic study on a variant of G_{M1}-gangliosidosis, *Clin. Genet.* **9**:427.

Koto, A., Horwitz, A. L., Suzuki, K., Tiffany, C. W., and Suzuki, K., 1978, The Morquio syndrome: Neuropathology and biochemistry, *Ann. Neurol.* **4**:26.

Kraemer, P. M., 1971, Heparan sulfates of cultured cells. I. Membrane associated and cell-sap species in Chinese hamster cells, *Biochemistry* **10**:1437.

Kresse, H., 1975, New variant of Sanfilippo's disease: Sanfilippo C?, an alpha glucosaminidase deficiency, Proceedings of the Third International Symposium on Glycoconjugates, Brighton, England.

Kresse, H., Weismann, U., Cantz, M., Hall, C. W., and Neufeld, E. F., 1971, Biochemical heterogeneity of the Sanfilippo syndrome, *Biochem. Biophys. Res. Commun.* **42**:892.

Lalley, P. A., Rattazzi, M. C., and Shows, T. B., 1974, Human β-D-*N*-acetylhexosaminidases A and B: Expression and linkage relationships in somatic cell hybrids, *Proc. Natl. Acad. Sci. U.S.A.* **71**:1569.

Ledeen, R., Salsman, K., Gonatas, J., and Taghavy, A., 1965, Structure comparisons of the major monosialogangliosides from brains of normal humans, gargoylism and late infantile systemic lipidosis, *J. Neuropathol. Exp. Neurol., Pt. 1* **24**:341.

Leroy, J. G., Spranger, J. W., Feingold, M., Opitz, J. M., and Crocker, A. C., 1971, "I-cell" disease: A clinical picture, *J. Pediatr.* **79**:360.

Leroy, J. G., Ho, M. W., MacBrinn, M. C., Zielke, K., Jacob, J., and O'Brien, J. S., 1972, I-cell disease: Biochemical studies, *Pediatr. Res.* **6**:752.

Li, Y. T., Mazzotta, M. Y., Wan, G.-G., Orth, R., and Li, S. C., 1973, Hydrolysis of Tay–Sachs ganglioside by β-hexosaminidase A of human liver and urine, *J. Biol. Chem.* **248**:7512.

Lowden, J. A., Callahan, J. W., Norman, M. G., Thain, M., and Prichard, J. S., 1974, Juvenile G_{M1}-gangliosidosis, *Arch. Neurol.* **31**:200.

Margolis, R. K., and Margolis, R. U., 1970, Sulfated glycopeptides from rat brain glycoproteins, *Biochemistry* **9**:4389.

Margolis, R. U., and Margolis, R. K., 1974, Distribution and metabolism of mucopolysaccharides and glycoproteins in neuronal perikarya, astrocytes and oligodendroglia, *Biochemistry* **13**:2849.

Martin, J. J., Leroy, J. G., Farriaux, J. P., Fontaine, G., Desnick, R. J., and Cabello, A., 1975, I-cell disease (mucolipidosis II), *Acta Neuropathol.* **33**:285.

Matalon, R., and Deanching, M., 1977, The enzymic basis for the phenotypic variation of Hurler and Scheie syndromes, *Pediat. Res.* **11**:519.

Matalon, R., and Dorfman, A., 1972, Hurler's syndrome: An α-L-iduronidase deficiency, *Biochem. Biophys. Res. Commun.* **47**:959.

Matalon, R., and Dorfman, A., 1974, Sanfilippo A syndrome: Sulfamidase deficiency in cultured skin fibroblasts and liver, *J. Clin. Invest.* **54**:907.

Matalon, R., and Dorfman, A., 1976, The mucopolysaccharidoses (a review), *Proc. Natl. Acad. Sci. U.S.A.* **73**:630.

Matalon, R., Cifonelli, J. A., Zellweger, H., and Dorfman, A., 1968, Lipid abnormalities in a variant of the Hurler syndrome, *Proc. Natl. Acad. Sci. U.S.A.* **59**:1097.

Matalon, R., Arbogast, B., Justice, P., Brandt, I. K., and Dorfman, A., 1974*a*, Morquio's syndrome: Deficiency of a chondroitin sulfate *N*-acetyl-hexosamine sulfate sulfatase, *Biochem. Biophys. Res. Commun.* **61**:759.

Matalon, R., Arbogast, B., and Dorfman, A., 1974*b*, Deficiency of chondroitin sulfate *N*-acetylgalactosamine-4-sulfate sulfatase in Maroteaux–Lamy syndrome, *Biochem. Biophys. Res. Commun.* **61**:1450.

McKusick, V. A., 1972, *Heritable Disorders of Connective Tissue,* 4th ed. C. V. Mosby, St. Louis, Missouri.

McKusick, V. A., and Martin, G. R., 1975, Molecular defects in collagen, *Ann. Intern. Med.* **82**:585.

Merat, A., and Dickerson, J. W. T., 1973, The effect of development on the gangliosides of rat and pig brain, *J. Neurochem.* **20**:873.

Michalski, J. C., Strecker, G., Farriaux, J. P., Durand, P., and Maroteaux, J. P., 1978, Total or partial deficit in α-neuraminidase associated with mucolipidosis II and III and two new types of mucolipidosis, *Biochimie* (in press).

Moser, H., 1972, Sulfatide lipidosis: Metachromatic leukodystrophy, in: *The Metabolic Basis of Inherited Disease* (J. B. Stanbury, J. B. Wyngaarden, and D. S. Fredrickson, eds.), pp. 688–729, McGraw-Hill, New York.

Murphy, J. V., Wolfe, H. J., Balázs, E. A., and Moser, H. W., 1971, A patient with deficiency of arylsulfatases A, B, C and steroid sulfatase associated with storage of sulfatide, cholesterol sulfate, and glycosaminoglycans, in: *Lipid Storage Diseases: Enzymatic Defects and Clinical Implications* (J. Bernsohn and H. J. Grossman, eds.), pp. 67–110, Academic Press, New York.

Ng Ying Kin, N. M. K., and Wolfe, L. S., 1974, Oligosaccharides accumulating in the liver from a patient with G_{M2}-gangliosidosis variant O (Sandhoff-Jatzkewitz disease), *Biochem. Biophys. Res. Commun.* **59**:837.

Nordén, N. E., Lundblad, A., Svensson, S., Öckerman, P.-A., and Autio, S., 1973, A mannose-containing trisaccharide isolated from urines of three patients with mannosidosis, *J. Biol. Chem.* **248**:6210.

Nordén, N. E., Lundblad, A., Svensson, S., and Autio, S., 1974, Characterization of two mannose-containing oligosaccharides isolated from the urine of patients with mannosidosis, *Biochemistry* **13**:871.

O'Brien, J. S., 1972*a*, G_{M1}-gangliosidosis, in: *The Metabolic Basis of Inherited Disease* (J. B. Stanbury, J. B. Wyngaarden, and D. S. Fredrickson, eds.), pp. 639–662, McGraw-Hill, New York.

O'Brien, J. S., 1972*b*, Sanfilippo syndrome: Profound deficiency of alpha-acetylglucosaminidase activity in organs and skins from type B patients, *Proc. Natl. Acad. Sci. U.S.A.* **69**:1720.

O'Brien, J. S., 1977, Neuraminidase deficiency in the cherry red spot–myoclonus syndrome, *Biochem. Biophys. Res. Commun.* **79**:1136.

O'Brien, J. S., Gugler, E., Giedion, A., Wiessmann, U., Herschkowitz, N. N., Heier, C., and

Leroy, J., 1976, Spondyloepiphyseal dysplasia, corneal clouding, normal intelligence and acid β-galactosidase deficiency, *Clin. Genet.* **9:**495.

Öckerman, P. A., 1968, Indentity of β-glucosidase, β-xylosidase and one of the β-galactosidase activities in human liver when assayed with 4-methylumbelliferyl-beta-D-glycosides: Studies in cases of Gaucher's disease, *Biochim. Biophys. Acta* **165:**59.

Öckerman, P.-A., 1969, Mannosidosis: Isolation of oligosaccharide storage material from brain, *J. Pediatr.* **75:**360.

Okada, S., and O'Brien, J. S., 1968, Generalized gangliosidosis (β-galactosidase deficiency), *Science* **160:**1002.

Okada, S., and O'Brien, J. S., 1969, Tay–Sachs disease: Generalized absence of a β-*N*-acetylhexosaminidase component, *Science* **165:**698.

Orii, T., Minami, R., Sukegawa, K., Sato, S., Tsugawa, S., Horino, K., Miura, R., and Nakao, T., 1972, A new type of mucolipidosis with β-galactosidase deficiency and glycopeptiduria, *Tokohu J. Exp. Med.* **107:**303.

Palo, J., Riekkinen, P., Arstila, A. Y., Autio, S., and Kivimaki, T., 1972, Aspartylglucosaminuria II: Biochemical studies on brain, liver, kidney and spleen, *Acta Neuropathol. (Berlin)* **20:**217.

Palo, J., Pollitt, R. J., Pretty, K. M., and Savolainen, H., 1973, Glycoasparagine metabolites in patients with aspartylglucosaminuria: Comparison between English and Finnish patients with special reference to storage materials, *Clin. Chim. Acta* **47:**69.

Pinsky, L., Miller, J., Shanfield, B., Watters, G., and Wolfe, L. S., 1974, G_{M1}-gangliosidosis in skin fibroblast culture: Enzymatic differences between types 1 and 2, and observations on a third variant, *Am. J. Hum. Genet.* **26:**563.

Purpura, D., and Suzuki, K., 1976, Distortion of neuronal geometry and formation of aberrant synapses in neuronal storage disease, *Brain Res.* **116:**1.

Richman, D. P., Watts, H. G., Parsons, D., Schmid, K., and Glimcher, M. J., 1977, Familial moyamoya associated with biochemical abnormalities of connective tissue, *Neurology (Minneapolis)* **27:**382.

Rodén, L., 1970, The mucopolysaccharides, in: *Metabolic Conjugation and Metabolic Hydrolysis* (W. H. Fishman, ed.), pp. 345–442, Academic Press, New York.

Rushton, A. R., and Dawson, G., 1977, The effect of glycosaminoglycans on the *in vitro* activity of human skin fibroblast glycosphingolipid β-galactosidases and neuraminidases, *Clin. Chim. Acta* **80:**133.

Sandhoff, K., Harzer, K., Wässle, W., and Jatzkewitz, H., 1971, Enzyme alterations and lipid storage in three variants of Tay–Sachs disease, *J. Neurochem.* **18:**2469.

Schengrund, C.-L., and Rosenberg, A., 1970, Localization of sialidase in the neuronal synaptosomal membrane of steer brains, *J. Biol. Chem.* **245:**6196.

Shapiro, L. J., Cousins, L., Fluharty, A. L., Stevens, R. L., and Kihara, H., 1977, Steroid sulfatase deficiency, *Pediatr. Res.* **11:**894.

Simpson, D. L., Thorne, D. R., and Loh, H. H., 1976, Sulfated glycoproteins, glycolipids and glycosaminoglycans from synaptic plasma and myelin membranes: Isolation and characterization of sulfated glycopeptides, *Biochemistry* **15:**5449.

Sjöberg, I., Fransson, L.-A., Matalon, R., and Dorfman, A., 1973, Hunter's syndrome: A deficiency of L-iduronosulfate sulfatase, *Biochem. Biophys. Res. Commun.* **54:**1125.

Sly, W. S., Quinton, B. A., McAlister, W. H., and Rimoin, D. L., 1973, β-Glucuronidase deficiency: Report of clinical, radiological and biochemical features of a new mucopolysaccharidosis, *J. Pediatr.* **82:**249.

Snyder, P. D., Krivit, W., and Sweeley, C. C., 1972, Generalized accumulation of neutral glycosphingolipids with G_{M2}-ganglioside accumulation in brain, *J. Lipid Res.* **13:**128.

Spiro, R. G., 1969, Glycoproteins: Their biochemistry, biology, and role in human disease, *N. Engl. J. Med.* **281:**991.

Spranger, J. W., and Wiedemann, H. R., 1970, The genetic mucolipidoses, *Humangenetik* **9:**113.

Strecker, G., Michalski, J. C., Montreuil, J., and Farriaux, J. P., 1976, Deficit in neuraminidase associated with mucolipidosis II (I-cell disease), *Biomedicine* **25:**238.

Strecker, G., Peers, M.-C., Michalski, J.-C., Hondi-Assah, T., Fournet, B., Spik, G., Montreuil, J., Farriaux, J.-P., Maroteaux, P., and Durand, P., 1977*a*, Structure of nine sialyl-oligosaccharides accumulated in urine of eleven patients with three different types of sialidosis, *Eur. J. Biochem.* **75:**391.

Strecker, G., Michalski, J. C., Herlant-Peers, M. C., Fournet, B., and Montreuil, J., 1977*b*, Structure of 40 oligosaccharides and glycopeptides accumulting in the urine from patients with catabolism defect of glycoconjugates (sialidosis, fucosidosis, mannosidosis, and Sandhoff's disease), Proceedings of the 4th International Symposium on Glycoconjugates, Woods Hole, Massachusetts.

Stumpf, D. A., Austin, J. M., Crocker, A. C., and La France, M., 1973, Mucopolysaccharidosis Type VI (Maroteaux–Lamy syndrome). I. Sulfatase B deficiency in tissues, *Am. J. Dis. Child.* **126:**747.

Sugahara, K., Funakoshi, S., Funakoshi, I., Aula, P., and Yamashina, I., 1975, Characterization of one neutral and two acidic glycoasparagines isolated from the urine of patients with aspartylglucosaminuria (AGU), *J. Biochem. (Tokyo)* **80:**195.

Suzuki, Y., Nakamura, N., Shimada, Y., Yotsumota, H., Endo, H., and Nagashima, K., 1977, Macular cherry-red spots and β-galactosidase deficiency in an adult, *Arch. Neurol.* **34:**157.

Svennerholm, L., 1964, The gangliosidoses, *J. Lipid Res.* **5:**145.

Taylor, H. A., Thomas, G. H., Miller, C. S., Kelly, T. E., and Siggers, D., 1973, Mucolipidosis III (pseudo-Hurler polydystrophy): Cytological and ultrastructural observations of cultured fibroblast cells, *Clin. Genet.* **4:**866.

Thomas, G. H., Tiller, G. E., Reynolds, L. W., Miller, C. S., and Bace, J. W., 1976, Increased levels of sialic acid associated with a sialidase deficiency in I-cell disease (mucolipidosis II) fibroblasts, *Biochem. Biophys. Res. Commun.* **71:**188.

Thompson, J. N., Stoolmiller, A. C., Matalon, R., and Dorfman, A., 1973, *N*-acetyl-β-hexosaminidase: Role in the degradation of glycosaminoglycans, *Science* **181:**866.

Toole, B. P., 1976, Morphogenetic role of glycosaminoglycans (acid mucopolysaccharides) in brain and other tissues, in: *Neuronal Recognition* (S. H. Barondes, ed.), pp. 275–329, Plenum Press, New York.

Trimble, R. B., and Maley, F., 1977, The use of endo-β-acetylglucosaminidase H in characterizing the structure and function of glycoproteins, *Biochem. Biophys. Res. Commun.* **78:**935.

Tsay, G. C., and Dawson, G., 1976, Oligosaccharide storage in brains from patients with fucosidosis, G_{M1}-gangliosidosis, and G_{M2}-gangliosidosis (Sandhoff's disease), *J. Neurochem.* **27:**733.

Tsay, G. C., Dawson, G., and Li, Y.-T., 1975, Structure of the glycopeptide storage material in G_{M1}-gangliosidosis: Sequence determinination with specific endo- and exoglycosidases, *Biochem. Biophys. Acta* **385:**305.

Tsay, G. C., Dawson, G., and Sung, J. S.-S., 1976, Structure of the accumulating oligosaccharide in fucosidosis, *J. Biol. Chem.* **251:**19.

Von Figura, K., and Kresse, H., 1972, Quantitative aspects of pinocytosis and the intracellular fate of *N*-acetyl-β-D-glucosaminidase in Sanfilippo B fibroblasts, *J. Clin. Invest.* **53:**85.

Wenger, D. A., Goodman, S. I., and Myers, G. G., 1974, Beta-galactosidase deficiency in young adults, *Lancet* 1974 (Nov. 30):1319.

Weismann, U. N., Lightbody, J., Vassella, P., and Herschkowitz, N. N., 1971, Multiple lysosomal enzyme deficiency due to enzyme leakage?, *N. Engl. J. Med.* **284:**109.

Wolfe, L. S., Senior, R. G., and Ng Ying Kin, N. M. K., 1974, The structure of oligosaccharides accumulating in the liver of G_{M1}-gangliosidosis type 1, *J. Biol. Chem.* **249:**1828.

17

Perspectives and Functional Implications

Richard U. Margolis and Renée K. Margolis

Although the preceding chapters have summarized much of our current knowledge regarding nervous tissue complex carbohydrates, it is evident that most hypotheses concerning their functional roles are still highly speculative, and that firm conclusions of a functional nature must await a better understanding of their structure, localization, and metabolism. These concluding remarks will therefore attempt to set out a few of the major questions related to the biosynthesis and metabolism of complex carbohydrates in brain, their relationship to synaptic events and structure, and their possible involvement in cell recognition phenomena and cell–cell interactions.

1. Biosynthesis and Metabolism

Recent studies on the role of polyisoprenol-linked mono- and oligosaccharides as activated glycosyl carriers in the biosynthesis of glycoproteins in brain and other tissues have significantly advanced our understanding of this process. It would appear that the lipid-linked intermediates are involved primarily, if not exclusively, in the biosynthesis of the *N*-acetylglucosamine and mannose core structure of asparagine-linked oligosaccharides, to which the more peripheral sugars (galactose, fucose, and sialic acid) are later added via their nucleotide monosaccharide derivatives. The exact interrelationships between these two pathways, and the role of glucosyl residues as signals for the further "processing" of partially completed oligosaccharides and their incorporation into specific membrane glycoproteins (Chapter 4), remain among

Richard U. Margolis • Department of Pharmacology, New York University School of Medicine, New York, New York. **Renée K. Margolis** • Department of Pharmacology, State University of New York, Downstate Medical Center, Brooklyn, New York.

the many outstanding questions in this rapidly moving research area. The latter aspect is of particular interest in view of the fact that while glucose has been shown to be incorporated into a protein-bound form by enzymes present in brain and other tissues, it is not a usual constituent of animal glycoproteins.

The biosynthesis of chondroitin sulfate occurs in brain in the usual manner (Chapter 5), but relatively little is known concerning the biosynthesis of hyaluronic acid and heparan sulfate in any animal tissue. Although in brain this process apparently takes place, as expected, in the rough and smooth endoplasmic reticulum (Kiang *et al.*, 1978), the enzymology of chain initiation and elongation for hyaluronic acid and heparan sulfate is still unclear.

From the turnover half-times for the polysaccharide backbone and for sulfate in the glycosaminoglycans, and for various components of the brain glycoproteins (see Appendix II), it would appear that two major metabolic pools are present. One has a rapid turnover, with half-times in the range of 1 day to a week, while a pool of lower metabolic activity turns over with a half-time of 2–6 weeks. However, chondroitin sulfate is distinguished by being present in a single metabolic pool. These data also indicate that the sulfate groups on the glycosaminoglycans turn over independently of and more rapidly than the polysaccharide chain. The conclusion that sulfation of brain glycosaminoglycans is not necessarily coupled with chain polymerization is not based exclusively on the observed differences in turnover half-times, but is also supported by double-label studies in which the ratio of sulfate to hexosamine radioactivity in chondroitin sulfate disaccharides was found to decrease with time after an initial labeling with both precursors. This independent turnover of sulfate on an intact polysaccharide chain may have functional significance by providing a mechanism for the addition and removal of anionic groups at specialized locations in the cell.

Several studies have indicated that in embryonic chick brain, the highest activities of nucleotide sugar glycosyltransferases involved in the biosynthesis of glycoproteins and chondroitin sulfate are present in a crude mitochondrial fraction (Den *et al.*, 1975; Brandt *et al.*, 1975). These enzymes were further enriched in "synaptosomal" and smooth membrane fractions prepared from the initial crude mitochondrial fraction, leading the authors to suggest that they are components of the synaptic plasma membrane, where complex carbohydrates and their respective glycosyltransferases may be involved in synaptic transmission, and/or in the maintenance of synaptic junctions by specific intercellular adhesion.

It appears unlikely that these results can be attributed to gross contamination with microsomal membranes, since this latter fraction had both lower specific and total activity of the glycosyltransferases in question. However, since there are known to be developmental changes in the subcellular distribution of glycosyltransferases (Chapter 5), it appears most likely that these findings reflect a situation peculiar to *embryonic* brain, whereas at later ages glycoprotein and glycosaminoglycan biosynthesis occur at more conventional locations in the neuron. It should also be noted in this connection that the glycosyltransferase activities localized in synaptosomes in embryonic brain are present in only low or barely detectable levels in adult brain homogenates. Even in embryonic brain, these glycosyltransferases may actually be present in the smooth endoplasmic reticulum and Golgi complex, but undergo a redistribution and binding to synaptosomal membranes during the process of cell disruption and subsequent subcellular fractionation. In this case, cytochemical investigations may provide more accurate information concerning the localization of these enzymes *in situ.*

In adult mouse or rat brain, a much greater proportion of the total glycosaminoglycan or biosynthetic enzyme activity is found in the microsomal fraction, and studies indicating that a significant portion of the brain glycosaminoglycans are synthesized and localized at nerve endings in adult animals were generally performed before the importance of thorough washing of the crude mitochondrial fraction to remove microsomal contamination was well recognized (see Chapter 3). These results can therefore more confidently be ascribed to contamination of the synaptosomes with microsomal and Golgi membranes than can those obtained from studies of embryonic brain.

2. Axonal Transport

The very low concentration of sulfated glycosaminoglycans in purified nerve endings does, however, raise an interesting question in view of their well-demonstrated axonal transport in a number of systems (Chapter 12). It is known that the concentration of glycosaminoglycans in isolated axons (stripped of axolemma) is almost as great as that found in the nerve cell body. Approximately two thirds of this axonal glycosaminoglycan is chondroitin sulfate, with most of the remainder being hyaluronic acid (Chapter 3). Since these glycosaminoglycans are known to be present in the cytoplasm of neurons, it appears likely that at least part of the axonal glycosaminoglycan represents an axoplasmic extension of that found in the neuronal cell body.

It has also been shown that catecholamine storage granules in the adrenal medulla contain significant quantities of glycosaminoglycans (mostly chondroitin sulfate) that are released together with the other soluble granule contents during exocytosis (Chapter 15), and there are indications that synaptic vesicles from brain (and *Torpedo* electroplax) contain glycosaminoglycans that could account for the small amounts found in purified nerve endings. Since studies on thc axonal transport of glycosaminoglycans have all been performed using radioactive precursors, the resulting data are consistent with the actual transport of possibly quite small amounts of highly labeled material present in synaptic vesicles. The transport of more substantial quantities of glycosaminoglycans would require their rapid degradation on reaching the nerve ending, given their low concentration at this site. In view of these questions, the form of the axonally transported glycosaminoglycans (and glycoproteins) and their fate at the nerve ending are clearly problems requiring further study.

It is likely that axonal transport and "membrane flow" are capable of providing sufficient glycoprotein to the nerve ending without the necessity of postulating an additional mechanism involving local synthesis (Chapters 12 and 13). However, *in vivo* studies of the time-course of labeling of complex carbohydrates with [^{3}H]glucosamine indicate a rapid turnover of sialic acid (independent of hexosamine) in the synaptosomal *soluble* glycoproteins and gangliosides of brain, and studies on the turnover of the protein moiety of brain glycoproteins employing labeled threonine also indicate that some brain glycoproteins may be synthesized or modified at sites outside the endoplasmic reticulum and Golgi complex (for references, see Margolis and Margolis, 1977). Although such a process may be relatively minor in quantitative terms, it would support the hypothesis proposed by Barondes (1970), in which the local modification of nerve-ending glycoproteins might provide a mechanism for the rapid alteration of synaptic function. In this manner, the substitution, mediated by the activation of specific glycosidases or glycosyltransferases, of a neutral sugar (e.g., fucose or galactose) for a negatively charged sialic acid, or the exposure of an uncharged penultimate sugar such as galactose by removal of sialic acid, could result in a change in the charge or permeability of synaptic membranes. Addition or removal of sulfate groups, which are also known to be present in synaptic membrane and other brain glycoproteins, could have a similar effect.

3. Synaptic Membranes

There has been considerable interest in the composition and localization of glycoproteins in synaptic membranes, and much of this work is reviewed in Chapter 8. The synaptic plasma membrane is of obvious importance from several standpoints, including its involvement in a specialized type of cell–cell interaction, and because of the possible role of postsynaptic membrane glycoproteins as components of receptors for neurotransmitters, drugs, and hormones. With respect to the latter aspect, there is now considerable evidence that the acetylcholine receptor of brain and *Torpedo* electroplax, the L-glutamate receptor in brain, and [Na^+–K^+]ATPase are all glycoproteins, as is the serotonin receptor in smooth muscle and the insulin receptor in isolated adipocytes and liver plasma membranes (Cuatrecasas and Tell, 1973; Czech *et al.*, 1974; Chang *et al.*, 1975; Seto *et al.*, 1977; Salvaterra *et al.*, 1977) [see also references in Margolis and Margolis (1977) and Chapters 7 and 8].

In early studies on the composition and properties of synaptic membrane glycoproteins, it was thought that characteristic and possibly unique glycoprotein molecules might be present at this site. However, this expectation has not been fulfilled, and on the basis of the available data, it would appear that if such differences do in fact exist they are too subtle to be detected by the methods available at present. On the other hand, the often rather substantial differences in carbohydrate composition in synaptic membranes or junctional complexes prepared by different methods are probably largely attributable to the fact that by isolating such membranes in a high degree of purity, one is often studying a particular subpopulation obtained in relatively low yield. Variations in composition can also be due to the use of different analytical methods, including those for extraction of glycolipids and sugar determination, and to whether the whole membranes are analyzed, or glycopeptides purified by various methods with or without removal of any accompanying glycosaminoglycans.

It is not known to what degree the composition of synaptic membranes differs from that of plasma membranes derived from nonsynaptic areas of the neuronal surface, nor even what are the similarities and differences between plasma membranes of neurons and glia. This lack of information is largely due to our present inability to prepare purified plasma membranes from mature brain in the absence of an identifying morphological marker such as the synaptic junction (although methods for preparing purified axolemma have recently been described).

However, it is known that there are considerable similarities in the polypeptide composition of microsomal and synaptic membranes (Wannamaker and Kornguth, 1973; Gurd *et al.*, 1974; Jones *et al.*, 1975), as well as in their concentration and composition of glycoproteins and gangliosides (Krusius *et al.*, 1978) (see also Chapter 1). The presence of many common components in similar concentration and proportions in microsomal, synaptic, and possibly other types of plasma membranes is consistent with the membrane flow or endomembrane hypothesis, according to which membrane biogenesis involves the physical transfer of membranes from one subcellular compartment to another (Morré *et al.*, 1974). Thus, if synaptic membranes are derived from endoplasmic reticulum that is in turn transferred to the Golgi apparatus for transformation into plasma membrane, it would be expected that the fraction of endoplasmic reticulum destined for conversion to plasma membranes would resemble these membranes in glycoprotein composition. Another probable reason for the similarities in composition between the microsomal and synaptic membrane fractions is that the microsomal fraction is itself heterogeneous, consisting both of endoplasmic reticulum and of plasma membranes derived mostly from nonsynaptic areas of the neuronal (and glial) cell surface. Therefore, it seems unlikely that characteristic peculiarities will be found in the glycoprotein composition of synaptic membranes in the absence of methods capable of performing a finer biochemical and immunological dissection of the synapse than is currently possible.

4. Cell–Cell Interactions and Neuronal Recognition

Since it is known that complex carbohydrates are constituents of plasma membranes of many types of cells, there has been considerable speculation on the possible role of these substances in intercellular recognition phenomena and in various types of cell–cell interactions (see, for example, Chapters 8 and 11). Although most of the experimental data supporting such a role have been derived from nonnervous tissue, there has recently been a marked increase in studies employing brain, retinal cells, and sympathetic ganglia, as well as cultured neuronal and glial tumor cell lines. Certain aspects of these investigations are covered in Chapters 7, 9, and 14. Biological and biochemical studies on embryonic cell–cell recognition were also concisely reviewed by Moscona and Hausman (1977).

The general problem of investigating cell-surface complex carbohydrates in nervous tissue has been approached in a number of different ways by various laboratories, each approach having its own particular advantages and limitations. Biochemical and cytochemical studies have employed plant lectins for the isolation and fractionation of synaptic and other plasma membrane glycoproteins, or for their identification in tissue sections or dissociated cells. Although lectins are able to detect the presence of cell-surface glycoconjugates and have a considerable degree of specificity for particular sugar residues, they provide little information concerning the actual structures of the oligosaccharides to which they bind, or in many cases whether they are labeling glycoproteins, glycolipids, or both types of compounds. The latter question is assuming increasing importance in view of recent evidence that gangliosides are present on the cell surface (Gregson *et al.*, 1977) (see also Chapters 1, 2, and 9).

Since the sugar residues (mannose, *N*-acetylglucosamine, sialic acid) identified by those lectins that are most widely used and have the highest affinities are also common to a large number of different oligosaccharides, it is obvious that the type of structural information obtained from studies using lectins alone is rather limited. They therefore appear to be most useful for detecting differences (e.g., developmental changes) in the state of complex carbohydrates in a defined system, which can then be pursued in more detail using other techniques, or, in biochemical studies, for the fractionation of glycopeptides or glycoproteins preliminary to studying their detailed oligosaccharide structures by standard methods of carbohydrate chemistry. Examples of these applications can be found in Chapters 3, 7, and 9.

Studies utilizing neurotumor cell lines have provided important information concerning the chemical composition and metabolism of complex carbohydrates in relatively homogeneous cell populations (see Chapter 14). Although it is often difficult or impossible to obtain data of this type by other methods, it must be kept in mind that some of the resulting findings may be peculiar to tumor cells and not representative of normal nervous tissue. Caution in the interpretation of such studies is especially necessary in view of the now well-established changes known to occur in complex carbohydrates of malignant or transformed cells.

The recent availability of methods for the bulk isolation of neurons and glia in a high degree of purity, and often with fairly extensive processes attached, offers an alternative approach to the study of certain questions that have up to now been investigated chiefly using tumor cell lines. Although in several respects bulk-isolated cells are not as versatile as cultured cell lines, *in vivo* labeling techniques, used in conjunction with endo- and exoglycosidases and proteases, offer the promise of obtaining complementary information on cell-surface complex carbohydrates from normal nervous tissue.

A number of recent studies have also concerned endogenous "lectins" isolated from brain, muscle, and other tissues (and usually measured in terms of their hemagglutination activity). Early studies on cell-surface carbohydrate-binding proteins in relatively simple systems such as cellular slime molds, considered as a model for cell recognition in nervous tissue, were reviewed by Barondes and Rosen (1976). This approach was more recently extended to the investigation of developmentally regulated lectin activities in brain, muscle, and liver (Simpson *et al.*, 1977; Kobiler and Barondes, 1977; Den and Malinzak, 1977), and the general subject of endogenous lectins from vertebrate tissues was reviewed by Simpson *et al.* (1978). Although these lectins and other cell-aggregating factors do not merely produce a random type of clumping, but may display a considerable degree of tissue, regional, cell, and sugar specificity, it has not been demonstrated that they actually play a physiological role in cell–cell interactions in animal tissues. The evidence is more persuasive for those substances that are isolated from (presumably) the surface of dispersed embryonic cells than for lectins obtained by homogenization of tissues using methods that produce cell breakage. In the latter case, it is possible that normally intracellular constituents that possess lectin activity are isolated during the cell disruption and extraction procedure, even though they do not actually function as lectins *in situ*. A possible example of such materials is the developmentally regulated lectin isolated from embryonic chick muscle, which has been shown by immunofluorescent techniques to be present both intracellularly and on the surface of chick pectoral muscle cells (Nowak *et al.*, 1977). Since this lectin was recently found to be indistinguishable by a number of chemical and immunological criteria from similar lectins present in embryonic chick brain and liver extracts, as well as in the liver of 7-day-old chicks (Kobiler *et al.*, 1978), there is at present no basis from which to infer that such hemagglutinins (lectins) impart qualitative uniqueness to these tissues during differentiation.

Although the glycosaminoglycans and proteoglycans have not been extensively studied in nervous tissue from the standpoint of their possible roles in cell–cell interactions, it is now becoming apparent that the chondroitin sulfate proteoglycan of brain is primarily if not exclusively a cytoplasmic constituent of neurons and glia, rather than forming part of an extracellular ground substance as in cartilage. However, approximately half of the heparan sulfate is probably present in the plasma membranes of brain and other tissues, while hyaluronic acid, which is mostly membrane-associated in adult brain, may occur as a soluble extracellular matrix material during early brain development. [For a discussion of glycosaminoglycans and proteoglycans in developmental processes, see Toole (1976), Margolis and Margolis (1977), and Chapter 3.] These questions of localization may be pertinent in evaluating the functional significance of a developmentally regulated soluble agglutinin from embryonic chick brain, which has been preliminarily characterized as a proteoglycan (Barondes and Rosen, 1976).

The possibility must also be considered that the occurrence of carbohydrate in plasma membrane glycoproteins and endogenous lectins may be purely fortuitous, and unrelated to cell–cell interactions or the ability of these substances to produce their characteristic biological effects. This is known to be the case with a number of glycoprotein enzymes and hormones, and the carbohydrate moiety does not appear to be necessary for the activity of the retinal cell-aggregating factor studied by Moscona and Hausman (1977). Moreover, several macromolecules that promote cell aggregation or similar processes do not contain carbohydrate.

It is thought that the carbohydrate portion of certain enzymes, hormones, and other glycoproteins is important in "recognition" phenomena as a signal for secretion or catabolism, even if not for their intrinsic biological activity, and it is possible that the oligosaccharide moiety may perform analogous functions in nervous tissue glycoproteins (e.g., to direct insertion of the protein into a particular membrane). In this case, many glycoproteins, including those exposed at the cell surface, may be significant from the standpoint of membrane structure but unrelated to such processes as neural histogenesis and intercellular recognition. The functions at the cell surface of glycolipids and glycosaminoglycans, as compared to the more extensively studied glycoproteins, will also require further evaluation. However, since the investigation of complex carbohydrates as possible mediators of cell–cell interactions and in regulation of the nerve cell microenvironment is an area of neurobiology that is currently in exponential growth phase, we can soon expect to have more satisfactory answers to several of these questions.

References

Barondes, S. H., 1970, Brain glycomacromolecules and interneuronal recognition, in: *The Neurosciences: Second Study Program* (F. O. Schmitt, ed.), pp. 747–760, Rockefeller University Press, New York.

Barondes, S. H., and Rosen, S. D., 1976, Cell surface carbohydrate-binding proteins: Role in cell recognition, in: *Neuronal Recognition* (S. H. Barondes, ed.), pp. 331–356, Plenum Press, New York.

Brandt, A. E., Distler, J. J., and Jourdian, G. W., 1975, Biosynthesis of chondroitin sulfate proteoglycan: Subcellular distribution of glycosyltransferases in embryonic chick brain, *J. Biol. Chem.* **250**:3996.

Chang, K.-J., Huang, D., and Cuatrecasas, P., 1975, The defect in insulin receptors in obese-hyperglycemic mice: A probable accompaniment of more generalized alterations in membrane glycoproteins, *Biochem. Biophys. Res. Commun.* **64**:566.

Cuatrecasas, P., and Tell, G. P. E., 1973, Insulin-like activity of concanavalin A and wheat germ agglutinin—Direct interactions with insulin receptors, *Proc. Natl. Acad. Sci. U.S.A.* **70**:485.

Czech, M. P., Lawrence, J. C., and Lynn, W. S., 1974, Activation of hexose transport by concanavalin A in isolated brown fat cells, *J. Biol. Chem.* **249**:7499.

Den, H., and Malinzak, D. A., 1977, Isolation and properties of β-D-galactoside-specific lectin from chick embryo thigh muscle, *J. Biol. Chem.* **252**:5444.

Den, H., Kaufman, B., McGuire, E. J., and Roseman, S., 1975, The sialic acids. XVIII. Subcellular distribution of seven glycosyltransferases in embryonic chicken brain, *J. Biol. Chem.* **250**:739.

Gregson, N. A., Kennedy, M., and Leibowitz, S., 1977, Gangliosides as surface antigens on cells isolated from the rat cerebellar cortex, *Nature (London)* **266**:461.

Gurd, J. W., Jones, L. R., Mahler, H. R., and Moore, W. J., 1974, Isolation and partial characterization of rat brain synaptic plasma membranes, *J. Neurochem.* **22**:281.

Jones, L. R., Mahler, H. R., and Moore, W. J., 1975, Synthesis of membrane protein in slices of rat cerebral cortex—Source of proteins of the synaptic plasma membranes, *J. Biol. Chem.* **250**:973.

Kiang, W.-L., Crockett, C. P., Margolis, R. K., and Margolis, R. U., 1978, Glycosaminoglycans and glycoproteins associated with microsomal subfractions of brain and liver, *Biochemistry* **17**:3841.

Kobiler, D., and Barondes, S. H., 1977, Lectin activity from embryonic chick brain, heart and liver: Changes with development, *Dev. Biol.* **60**:326.

Kobiler, D., Beyer, E. C., and Barondes, S. H., 1978, Developmentally regulated lectins from chick muscle, brain, and liver have similar chemical and immunological properties, *Dev. Biol.* **64**:265.

Krusius, T., Finne, J., Margolis, R. U., and Margolis, R. K., 1978, Structural features of microsomal, synaptosomal, mitochondrial, and soluble glycoproteins of brain, *Biochemistry* **17**:3849.

Margolis, R. U., and Margolis, R. K., 1977, Metabolism and function of glycoproteins and glycosaminoglycans in nervous tissue, *Int. J. Biochem.* **8**:85.

Morré, D. J., Keenan, T. W., and Huang, C. M., 1974, Membrane flow and differentiation: Origin of Golgi apparatus membranes from endoplasmic reticulum, in: *Advances in Cytopharmacology* (B. Ceccarelli, F. Clementi, and J. Meldolesi, eds.), pp. 107–125, Raven Press, New York.

Moscona, A. A., and Hausman, R. E., 1977, Biological and biochemical studies on embryonic cell–cell recognition, in: *Cell and Tissue Interactions* (J. W. Lash and M. M. Burger, eds.), pp. 173–185, Raven Press, New York.

Nowak, T. P., Kobiler, D., Roel, L. E., and Barondes, S. H., 1977, Developmentally regulated lectin from embryonic chick pectoral muscle: Purification by affinity chromatography, *J. Biol. Chem.* **252**:6026.

Salvaterra, P. M., Gurd, J. M., and Mahler, H. R., 1977, Interactions of the nicotinic acetylcholine receptor from rat brain with lectins, *J. Neurochem.* **29**:345.

Seto, A., Arimatsu, Y., and Amano, T., 1977, A glycoprotein resembling a peripheral nicotinic acetylcholine receptor that binds $[^{125}I]\alpha$-bungarotoxin in mouse brain, *Neurosci. Lett.* **4**:115.

Simpson, D. L., Thorne, D. R., and Loh, H. H., 1977, Developmentally regulated lectin in neonatal rat brain, *Nature (London)* **266:**367.

Simpson, D. L., Thorne, D. R., and Loh, H. H., 1978, Lectins: Endogenous carbohydrate-binding proteins from vertebrate tissues: Functional role in recognition processes?, *Life Sci.* **22:**727.

Toole, B. P., 1976, Morphogenetic role of glycosaminoglycans in brain and other tissues, in: *Neuronal Recognition* (S. Barondes, ed.), pp. 275–329, Plenum Press, New York.

Wannamaker, B. B., and Kornguth, S. E., 1973, Electrophoretic patterns of proteins from isolated synapses of human and swine brain, *Biochim. Biophys. Acta* **303:**333.

Appendix I: Developmental Changes in Complex Carbohydrates of Brain

Glycoproteins

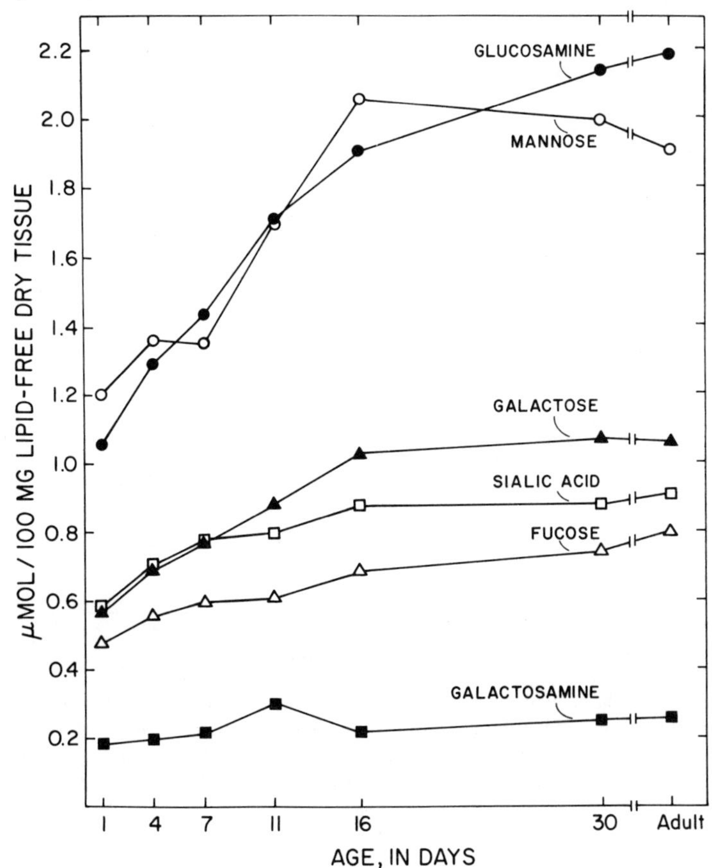

Concentration of glycoprotein carbohydrate in rat brain as a function of increasing age from neonatal to adult. From Margolis *et al.* (1976).

Glycosaminoglycans

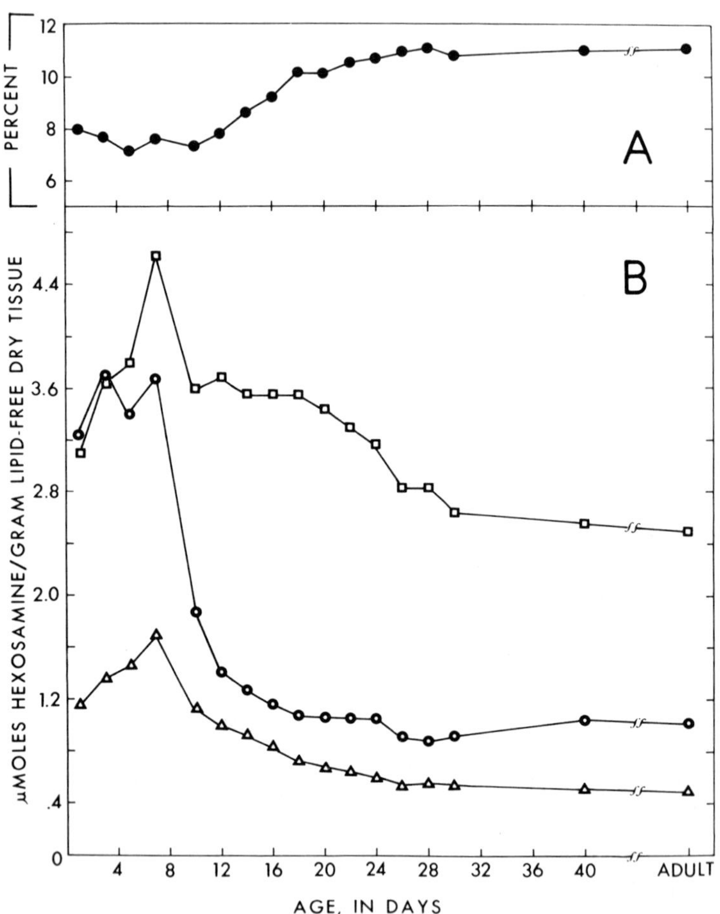

(A) Lipid-free weight as percentage of fresh weight of rat brain during postnatal development. (B) Concentration of glycosaminoglycans, expressed as micromoles of the constituent hexosamine per gram of lipid-free dry weight, as a function of increasing postnatal age. (□) Chondroitin sulfate; (○) hyaluronic acid; (△) heparan sulfate. From Margolis *et al.* (1975).

Gangliosides

From Vanier *et al.* (1971).

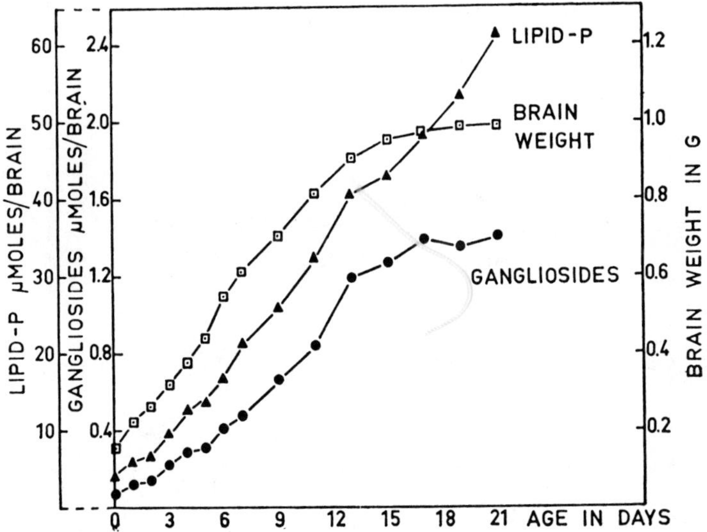

Brain weight, and content of gangliosides and phospholipids, during postnatal development of rat cerebrum.

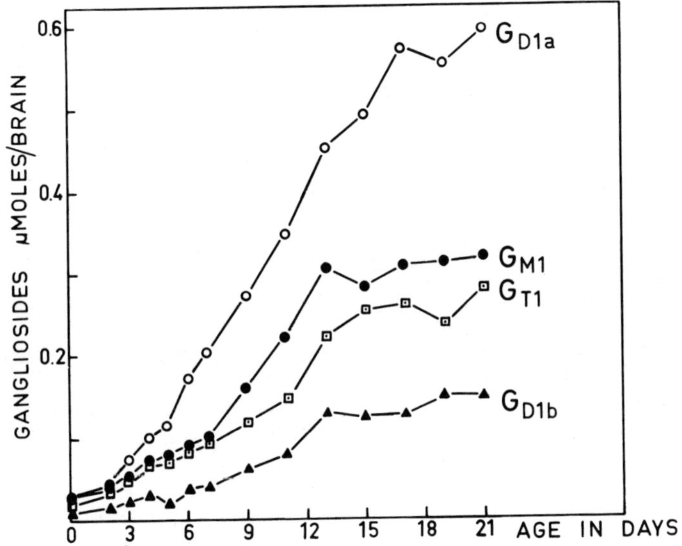

Concentration of major gangliosides during postnatal development of rat cerebrum.

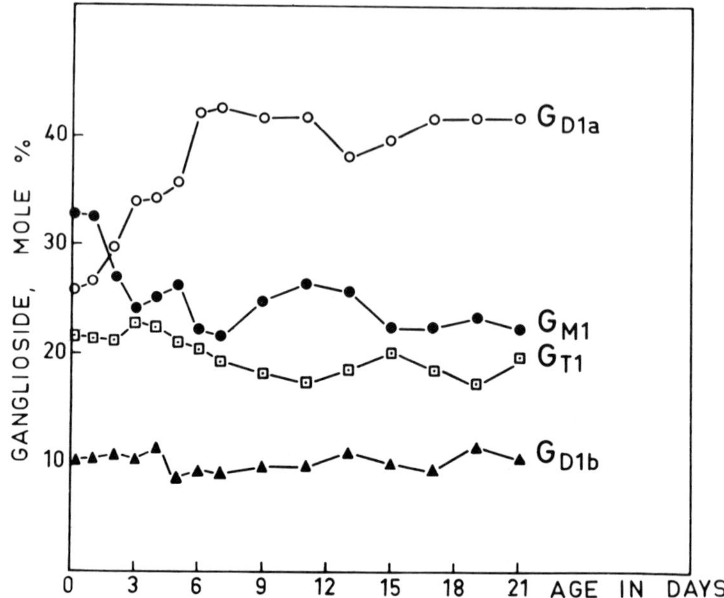

Relative proportions of major gangliosides during postnatal development of rat cerebrum.

Bibliography

Dreyfus, H., Urban, P. F., Edel-Harth, S., and Mandel, P., 1975, Developmental patterns of gangliosides and of phospholipids in chick retina and brain, *J. Neurochem.* **25**:245.

Mansson, J.-E., Vanier, M.-T., and Svennerholm, L., 1978, Changes in the fatty acid and sphingosine composition of the major gangliosides of human brain with age, *J. Neurochem.* **30**:273.

Margolis, R. K., Preti, C., Lai, D., and Margolis, R. U., 1976, Developmental changes in brain glycoproteins, *Brain Res.* **112**:363.

Margolis, R. U., Margolis, R. K., Chang, L. B., and Preti, C., 1975, Glycosaminoglycans of brain during development, *Biochemistry* **14**:85.

Merat, A., and Dickerson, J. W. T., 1973, The effect of development on the gangliosides of rat and pig brain, *J. Neurochem.* **20**:873.

Rosenberg, A., and Stern, N., 1966, Changes in sphingosine and fatty acid components of the gangliosides in developing rat and human brain, *J. Lipid Res.* **7**:122.

Suzuki, K., 1965, The pattern of mammalian brain gangliosides. III. Regional and developmental differences, *J. Neurochem.* **12**:969.

Suzuki, K., Poduslo, S. E., and Norton, W. T., 1967, Gangliosides in the myelin fraction of developing rats, *Biochim. Biophys. Acta* **144**:375.

Vanier, M. T., Holm, M., Öhman, R., and Svennerholm, L., 1971, Developmental profiles of gangliosides in human and rat brain, *J. Neurochem.* **18**:581.

Yu, R. K., and Yen, S. I., 1975, Gangliosides in developing mouse brain myelin, *J. Neurochem.* **25**:229.

Appendix II: Turnover of Complex Carbohydrates in Rat Brain

Glycoproteins

Glycoprotein constituent	$t_{1/2}$ (days)
Hexosamine	4.6, 15
Sialic acid	6, 30
Fucose	1, 30
Sulfate	2.5, 14
Threonine[a]	13, 38

For references, see Margolis *et al.* (1975).
[a] Representative of the protein moiety.

Glycosaminoglycans

	Sulfate	Hexosamine
Glycosaminoglycan	$t_{1/2}$ (days)	
Hyaluronic acid	—	9, 45
Chondroitin sulfate	7	21
Heparan sulfate	3	3.7, 10

From Margolis and Margolis (1972, 1973).

Gangliosides*

A turnover half-time of 24 days was calculated for rat brain gangliosides labeled with glucosamine, while studies using galactose and glucose as precursors gave values of 10 and 20 days, respectively (Burton *et al.*, 1964; Suzuki, 1967). The shorter half-times are thought to reflect a lower reutilization of labeled galactose and glucose, which have relatively large intracellular pools, and are thus considered more reliable. There are also age-related changes in the turnover of brain gangliosides (Suzuki, 1967), and metabolic differences between specific compartments that are not revealed by turnover studies using whole brain. For example, the turnover rate of G_{M1} ganglioside is much slower in myelin than in whole rat brain, and approaches that of other myelin lipids (Suzuki, 1970).

Compartmentation of ganglioside metabolism is reflected in the finding that G_{M1} produced by degradation of multisialogangliosides does not mix with the entire pool of G_{M1} in rat brain (Suzuki, 1967). In a biosynthetic study employing labeled glucose as precursor, it was also found that the total pool of any one of the major gangliosides of rat brain was not a precursor for any of the others (Suzuki and Korey, 1964). The same conclusion was reached by Maccioni *et al.*, (1971) using [^{3}H]glucosamine as precursor, although another study using labeled glucosamine indicated that 6-day-old rats contain a small pool of monosialoganglioside that is converted to di- and trisialogangliosides (Holm and Svennerholm, 1972).

Glucosamine is known to be an effective precursor for both the hexosamine and sialic acid residues of gangliosides, while *N*-acetylmannosamine is highly specific in labeling sialic acid. The ceramide unit has been labeled with radioactive serine and acetate. It has also been shown that there is no difference in the specific activity or turnover rate of neuraminidase-labile and neuraminidase-resistant sialic acid residues in gangliosides labeled by administration of [^{14}C]glucosamine (Holm and Svennerholm, 1972). This result suggests that there is no reutilization of neuraminidase-labile sialic acid residues. The same study showed that the specific activities of sphingosine and stearic acid, after labeling with [^{3}H]acetate, both decreased at the same rate in all four major gangliosides. Other more recent results also indicate that the entire ganglioside molecule turns over as a single unit both in whole brain (Caputto *et al.*, 1974) and in the specific pool of gangliosides that reaches the nerve endings by axonal transport (Ledeen *et al.*, 1976).

Bibliography

Burton, R. M., Balfour, Y. M., and Gibbons, J. M., 1964, Gangliosides and cerebrosides: Turnover rates in rat brain, *Fed. Proc. Fed. Am. Soc. Exp. Biol.* **23:**230.

*Contributed by R. W. Ledeen.

Caputto, R., Maccioni, H. J., and Arce, A., 1974, Biosynthesis of brain gangliosides, *Mol. Cell. Biochem.* **4**:97.

Holm, M., and Svennerholm, L., 1972, Biosynthesis and biodegradation of rat brain gangliosides studied *in vivo*, *J. Neurochem.* **19**:609.

Ledeen, R. W., Skrivanek, J. A., Tirri, L. J., Margolis, R. K., and Margolis, R. U., 1976, Gangliosides of the neuron: Localization and origin, in: *Ganglioside Function: Biochemical and Pharmacological Implications* (G. Porcellati, B. Ceccarelli, and G. Tettamanti, eds.), pp. 83–103, Plenum Press, New York.

Maccioni, H. J., Arce, A., and Caputto, R., 1971, The biosynthesis of gangliosides, *Biochem. J.* **125**:1131.

Margolis, R. K., and Margolis, R. U., 1973, The turnover of hexosamine and sialic acid in glycoproteins and mucopolysaccharides of brain, *Biochim. Biophys. Acta* **304**:413.

Margolis, R. K., Preti, C., Chang, L., and Margolis, R. U., 1975, Metabolism of the protein moiety of brain glycoproteins, *J. Neurochem.* **25**:707.

Margolis, R. U., and Margolis, R. K., 1972, Sulfate turnover in mucopolysaccharides and glycoproteins of brain, *Biochim. Biophys. Acta* **264**:426.

Suzuki, K., 1967, Formation and turnover of the major brain gangliosides during development, *J. Neurochem.* **14**:917.

Suzuki, K., 1970, Formation and turnover of myelin ganglioside, *J. Neurochem.* **17**:209.

Suzuki, K., and Korey, S. R., 1964, Study on ganglioside metabolism. I. Incorporation of D-U^{14}C-glucose into individual gangliosides, *J. Neurochem.* **11**:647.

Appendix III: The Human Sphingolipidoses

Compiled by Glyn Dawson

Disease name	Material that accumulates	Enzyme deficiency
G_{M1}-gangliosidosis Type I (pseudo-Hurler's or generalized gangliosidosis)[a]	Gal-GalNAc-[NeuAc]GalGlcCer Gal-oligosaccharide	G_{M1}- and 4-MeUmb-β-galactosidase
G_{M1}-gangliosidosis Type II	G_{M1}	G_{M1}- and 4-MeUmb-β-galactosidase
G_{M1}-gangliosidosis Types III and IV	Not characterized, but some patients secrete sialooligosaccharides or galactoglycopeptides	Partial β-galactosidase; some have glycoprotein neuraminidase deficiency
G_{M2}-gangliosidosis (Type I: Sandhoff's)[a]	GalNAc[NeuAc]GalGlcCer (G_{M2}) GalNAc-Gal-Gal-GlcCer-(globoside) GlcNAc-oligosaccharides	*N*-acetyl-β-D-hexosaminidase A and B
G_{M2}-gangliosidosis(Type II: Tay–Sachs)	G_{M2}	*N*-acetyl-β-D-hexosaminidase A
G_{M2}-gangliosidosis (Types III (juvenile) and IV (adult))	G_{M2}	Partial *N*-acetyl-β-D-hexosaminidase A
Fucosidosis[a]	H-antigen glycolipid fucosaccharides	α-L-Fucosidase
G_{M3}-gangliosidosis (not conclusively identified)	G_{M3} accumulation noted in a number of lysosomal storage diseases	Predict glycolipid neuraminidase (?)
Fabry's disease	GalGalGlcCer GalGalCer	α-D-Galactosidase

(continued)

Disease name	Material that accumulates	Enzyme deficiency
Lactosylceramidosis	GalGlcCer GlcCer G_{M3}	Enzyme deficiency unclear (partial sphingomyelinase or β-galactosidase)
Gaucher's disease[a]	GlcCer	β-Glucosidase
Metachromatic leukodystrophy[a]	Sulfo-GalCer (sulfatide)	Arylsulfatase A (cerebroside sulfate sulfatase)
Globoid cell leukodystrophy (Krabbe's disease)[a]	GalCer GalSphingosine (psychosine)	GalCer β-galactosidase
Farber's disease	Ceramide	Ceramidase
Niemann–Pick disease[a]	Sphingomyelin	Sphingomyelinase

[a]Considerable heterogeneity exists in all these disorders as a result of isoenzymic differences or differences in levels of residual enzyme activity. In general they may be classified as infantile (<5% of normal enzyme activity), juvenile (<10% of normal activity), adult (<25% of normal activity), and unaffected heterozygote (50% of normal activity), but in many cases the distinction is not clear because of technical problems in the *in vitro* assay. All the disorders listed in this table have a profound central neurodegenerative effect with the exception of Fabry's disease (which moderately affects the peripheral nervous system) and the adult form of Gaucher's disease.

Bibliography

Stanbury, J. B., Wyngaarden, J. B., and Fredrickson, D. S., (eds.), 1978, *The Metabolic Basis of Hereditary Disease*, 3rd ed., McGraw-Hill, New York.

Brady, R. O., 1973, Inborn errors of lipid metabolism, in: *Advances in Enzymology* (A. Meister, ed.), Vol. 38, pp. 293–315, John Wiley and Sons, New York.

Brady, R. O., 1976, Inherited metabolic diseases of the nervous system, *Science* **193**:733–739.

Dawson, G., 1978, Glycolipid catabolism, in: *The Glycoconjugates* (M. Horowitz and W. Pigman, eds.), Academic Press, New York.

Index